Precision Dairy Farming 2016

AF770620

Precision Dairy Farming 2016

edited by:
Claudia Kamphuis
and Wilma Steeneveld

Wageningen Academic
Publishers

EAN: 9789086862832
e-EAN: 9789086868292
ISBN: 978-90-8686-283-2
e-ISBN: 978-90-8686-829-2
DOI: 10.3920/978-90-8686-829-2

Photo cover: Dreamstime.com
© Ilya131266

First published, 2016

© Wageningen Academic Publishers
The Netherlands, 2016

This work is subject to copyright. All rights are reserved, whether the whole or part of the material is concerned. Nothing from this publication may be translated, reproduced, stored in a computerised system or published in any form or in any manner, including electronic, mechanical, reprographic or photographic, without prior written permission from the publisher:
Wageningen Academic Publishers
P.O. Box 220
6700 AE Wageningen
The Netherlands
www.WageningenAcademic.com
copyright@WageningenAcademic.com
The individual contributions in this publication and any liabilities arising from them remain the responsibility of the authors.
The publisher is not responsible for possible damages, which could be a result of content derived from this publication.

Table of contents

Preface

The dairy industry is rapidly changing, where globally the number of dairy farms is decreasing but where herd sizes are increasing. This trend poses challenges for the entire dairy industry, and particularly for farmers who have to manage this increasing number of cows with a limited time budget. The challenges farmers are facing include ensuring animal health and welfare, efficient use of feed and optimizing grazing management. These challenges lead to precision dairy farming innovations and technologies that aim at more labor efficient processes, or that aim at supporting farmers in their management decisions. Probably the most widely adopted precision dairy farming technologies involve the robotic milking systems and the automated heat detection systems. But there are numerous new innovations that vary from the use of technologies to predict the moment of calving to those that monitor cow behavior to estimate grass intake. With these precision dairy farming innovations and technologies, new questions arise that need to be answered. There is not much known about the actual economic effect when these precision dairy farming technologies are applied on-farm. Also social benefits are often unclear, and reasons why farmers invest in these technologies and what they actually expect from them are often unknown. Moreover, with the increasing amount of data that are continuously collected by these precision dairy farming technologies, we have to rethink our methods to store and manage these data, and how to retrieve useful information from it. Finally, with the increasing amount of data collected per individual animal on-farm, there are huge potentials to use these data along the entire chain. Breeding programs may benefit from automatically retrieved information, and data may be used to improve insurance of product quality along the entire chain of dairy products. At the same time, the implementation of precision dairy farming technologies is associated with risks too, such as a further alienation between society and dairy farming, and an increased economic power of the dairy processing industry and retail. The consequences of these risks are not clearly understood.

This book covers the proceedings of the material that is presented and discussed during the First International Conference on Precision Dairy Farming. It contains the papers of the keynotes and all oral presentations. We are confident that this proceedings will serve as a stepping stone for new and innovative ideas within the field of precision dairy farming. We're looking forward to see some of these new ideas being researched and results being shared during the second edition of the International Conference on Precision Dairy Farming.

A final note to say that we are happy and proud to see so many people showing interest in this subject. On behalf of the organizing committee, we are happy to invite you to use these proceedings to the best to further improve precision dairy farming technologies. We would like to acknowledge all those that made this conference and the proceedings possible including the authors, the sponsors, the scientific and organizing committee, and Wageningen Academic Publishers.

Claudia Kamphuis & Wilma Steeneveld

Organizing committee

Prof. Henk Hogeveen – Chairman
Wageningen University / Utrecht University, the Netherlands
Dr. Claudia Kamphuis – Secretary, Treasurer, Scientific affairs
Wageningen University, the Netherlands
Prof. Kees Lokhorst – Program Coordinator Farmers Day, Scientific affairs
Wageningen UR Livestock Research / Van Hall Larenstein University of Applied Science, the Netherlands
Niels Rutten MSc – Public relations, Sponsoring
Utrecht University, the Netherlands
Dr Wilma Steeneveld - Program Coordinator, Scientific Affairs
Utrecht University, the Netherlands
Dr. Tine van Werven – Program Coordinator Veterinary Day
Utrecht University, the Netherlands
Bianca Bouwman – Event Organizer
CibanaRegelt Evenement Management

Scientific program committee

Prof. Albert De Vries, University of Florida, USA
Prof. Anders Kristensen, University of Copenhagen, Denmark
Dr. Annelies Van Nuffel, Institute for Agricultural and Fisheries Research (ILVO), Belgium
Dr. Bernadette Earley, Teagasc, Ireland
Dr. Callum Eastwood, DairyNZ, New Zealand
Dr. Claudia Bahr, Agrifirm, the Netherlands
Dr. Claudia Kamphuis, Wageningen University, the Netherlands
Prof. Daniel Berckmans, KU Leuven, Belgium
Prof. Douglas J Reinemann, University of Wisconsin, USA
Dr. Hiemke Knijn, CRV, the Netherlands
Dr. Ilan Halachmi, Agricultural Research Organization – The Volcani Centre, Israel
Dr. Isabelle Veissier, INRA, France
Prof. Jeffrey Bewley, University of Kentucky, USA
Dr. Jenny Jago, DairyNZ, New Zealand
Assoc. Prof. Karin Orsel, University of Calgary, Canada
Kees De Koning, MSc, Dairy Campus, the Netherlands
Prof. Kees Lokhorst, Wageningen UR Livestock Research / Van Hall Larenstein University of Applied Science, the Netherlands
Assoc. Prof. Kendra Kerrisk, University of Sydney, Australia
Prof. Lene Munksgaard, Aarhus University, Denmark
Prof. Marcia Endres, Univerisity of Minnesota, USA
Dr. Matti Pastell, Natural Resources Institute Finland (LUKE), Finland
Prof. Nathalie Bareille, ONIRIS, France
Dr. Rudi De Mol, Wageningen Livestock Research, the Netherlands
Prof. Toby Mottram, Royal Agricultural University, United Kingdom
Prof. Trevor DeVries, University of Guelph, Canada
Dr. Vivi Thorup, Thorup consult, Denmark
Prof. Volker Krömker, Hochschule Hannover – University of Applied Sciences and Arts, Germany
Dr. Wilma Steeneveld, Utrecht University, the Netherlands

Keynotes

Animal health challenges on dairy farms increasing in size – are protocols and precision farming compatible?

Y.H. Schukken[1,2,3*], G. Van Schaik[1,3] and T.J.G.M. Lam[1,3]
[1]*GD Animal Health, Deventer, the Netherlands;* [2]*Wageningen University, Wageningen, the Netherlands;* [3]*Utrecht University, Utrecht, the Netherlands; y.schukken@gdanimalhealth.com*

Abstract

When dairy farms are increasing in size, external labor is used to manage the cows and the herd. Multiple studies have shown that such expanding farms with external labor may perform well when protocols are defined and individuals working on the farm are trained to perform the activities defined in the protocols. Alternatively, farm owners may decide to increase the use of cow-level data and develop cow-specific management strategies: a development towards precision dairy farming. Current data would indicate that precision farming does not easily result in superior performance, milk quality or animal health. Neither are there strong arguments for better economic performance in the case of precision farming when compared to herd-based protocols. The one area where precision farming may show a clear advantage is the reduced use of antibiotic treatments and hormone interventions. Such reduced use of antibiotics and hormones may be an important component of societal acceptance of growing and large dairy farms. The challenge for precision dairy farming will be to understand the use of cow-specific data on farms with superior health and performance. When the decision processes at these currently small farms can be translated to decision systems that may be applied on large and growing dairy farms, precision dairy farming will become a realistic outlook in the dairy industry.

Keywords: precision farming, animal health, milk quality

1. Introduction

As indicated on the website of precision dairy farming, the primary goals of precision dairy farming are to maximize animal performance, detect diseases in individual cows early, detect herd level health and production problems early and minimize the use of medication through preventive health measures. The use of precision dairy farming technologies enables the fulfillment of these goals without additional labor input. Examples of precision dairy farming technologies include milk yield recording systems, milk component monitors, activity monitors, lying and rumination behavior monitors, milk conductivity indicators, and heat detection monitors. The individual animal information these technologies collect supplements observations of people. By proper processing of these data in combination with decision support systems, the application of these sensors improves animal monitoring and decision making. The focus here is clearly on using data of individual animals to make the best possible decisions on these individual animals.
In apparent contrast to precision dairy farming with the emphasis on individual animal data is the increasing emphasis on the development of protocols and procedures on larger dairy farms where the emphasis is on ensuring that employees treat all cows according to a carefully drafted protocol. The importance of training employees and ensuring that employees follow protocols has shown to be an important component of success on dairy farms (Schewe *et al.*, 2015). For example, farmers that stated they ensure strict compliance of protocols were more likely to have lower bulk tank somatic cell count (BTSCC). This is in agreement with other authors who suggested that adequate workforce management is the most common challenge to producers following expansion and is a predictor of overall success of the expansion (Stup *et al.*, 2006).

The apparent contradiction between these two approaches, precision dairy farming with an emphasis on individual animal data and protocol based farming with an emphasis on treating all cows identical may be especially important in farms that are larger or that are growing in size. In Figure 1, bulk milk SCC is depicted and plotted against the kilograms delivered in the same bulk tank where SCC was measured on a monthly basis. Average bulk milk SCC was approximately 250,000 cells/ml and varied between approximately 40,000 and 2,000,000 cells/ml. In smaller herds, where milk delivered was less than 100,000 kg of milk (less than 100 cows), somatic cell counts showed an enormous amount of variation, from very low to very high. When herd size increased, variation in SCC decreased but also moved more towards the overall population mean. It is reasonable to assume that in small herds there is more individual animal based decision making while in large herds more farm level protocols are used for decision making. The data in Figure 1 would indicate that in many herds with high SCC (currently in area I of Figure 1) the use of protocols would improve milk quality and udder health of cows and move towards area II of Figure 1. It may also be argued though that the best milk quality and udder health of cows may be achieved in herds where the best individual animal based decisions are used (area III in Figure 1).

Still, the apparent contradiction remains. In larger herds, protocols and the ability to train farm workers to use protocols correctly and consistently are of increasing importance while the data of small herds indicate that individual animal based decisions may lead to enormous variation in performance with regard to milk quality and animal health. The question is then, should farmers on growing dairies choose for precision management or focus on high quality execution of generic herd based protocols?

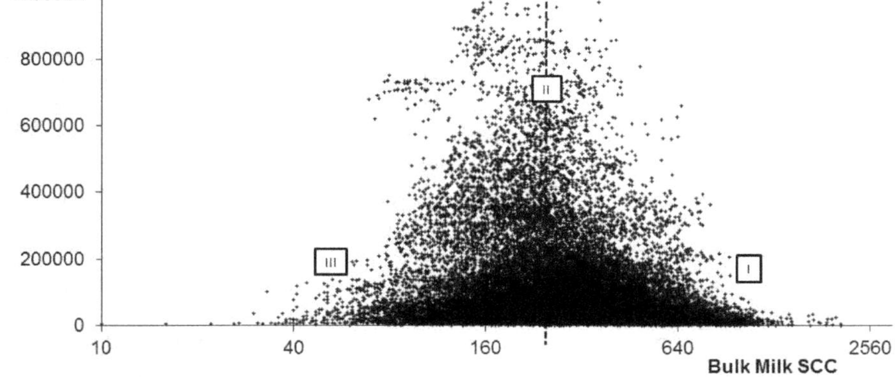

Figure 1. Bulk milk SCC in cells per ml by the volume of milk in kilograms delivered in the same bulk milk on a monthly basis. The dotted line represents the weighted mean of the data (249,000 SCC/ml). Adapted from Nightingale *et al.* (2008).

2. Protocols or precision farming

Precision dairy farming has potential in a number of areas (Rutten *et al.*, 2013) including fertility, udder health, metabolic disease and milk production and lameness. In a number of recent studies, a comparison between using individual animal data and standardized herd protocols has been reported on. Here a number of these studies are presented and evaluated for the indicated areas of health management.

2.1 Udder health

As indicated earlier, several studies show that using protocols in milk quality management are associated with improved outcome such as bulk milk somatic cell count and clinical mastitis. One area of particular interest in European countries is the selection of individual cows for either use or no use of dry cow antibiotics. Multiple studies at herd level have shown that a blanket dry cow therapy protocol is associated with a better outcome compared to using individual cow data to identify cows for treatment in a selective dry cow therapy program. A recent study by Scherpenzeel *et al.* (2014) where cows with low individual cell counts were randomized to either antibiotic treatment or no antibiotic treatment at dry-off. The results are shown in Figure 2 and these results indicate a better performance of cows treated with antibiotics both in terms of increase SCC after calving and clinical mastitis after calving.

These data would indicate that selection of individual cows based on precise data is resulting in a lower health outcome compared to a blanket herd based dry cow treatment procedure.

2.2 Metabolic diseases and milk production

Feeding high producing dairy cows may be done using a total mixed ration where all cows are fed the same diet or using a cow-specific diet where each individual cow receives an amount of concentrates that is specifically calculated based on her own daily milk production. In a recent study, these two methods of feeding cows were compared (Little *et al.*, 2016). Results indicated that mean concentrate intakes over the study period were similar with each of the 2 allocation strategies (11.5 and 11.7 kg of dry matter (DM)/cow per day for group and individual cow, respectively), although the pattern of intake with each treatment differed over time. Concentrate allocation strategy had no significant effect on either milk yield (39.3 and 38.0 kg/d for group and individual cow, respectively), milk

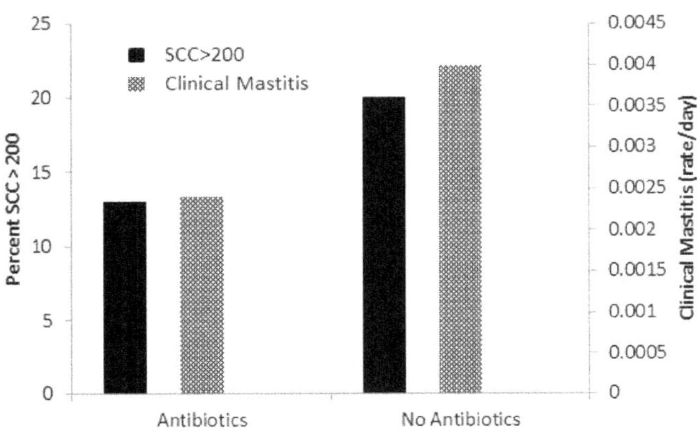

Figure 2. Clinical mastitis rate and percent SCC >200 after calving in low SCC cows randomized to either dry cow treatment with antibiotics or remain untreated (adopted from Scherpenzeel *et al.*, 2014).

composition, or milk constituent yield. However, cows on the individual cow treatment showed a greater range of concentrate intakes and milk yields than those on the group allocation strategy. Also, the relationship between mean daily milk yield and mean daily energy balance showed a more negative slope in cows that were fed a total mixed ration for all cows in the herd (Figure 3). These data would indicate that there is some beneficial impact of precision data based feeding of cows, but that no difference in overall cow or herd performance was present when comparing precision data based feeding and a total mixed ration for the entire dairy herd.

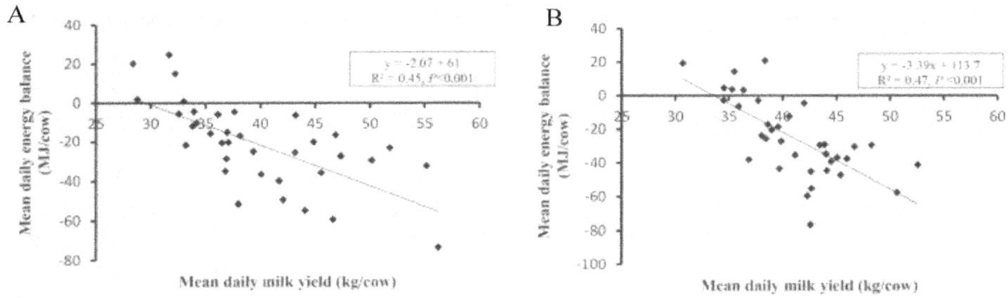

Figure 3. Energy balance versus daily milk yield in cows fed an individually balanced diet (A) and cows fed a total mixed ration for all cows (B). From Little *et al.* (2016).

2.3 Fertility

An important reason for precision dairy farming is to identify cows that are in heat and should be inseminated. Several systems are currently used, but most common are systems based on increased activity of cows as an indicator for heat. In contrast, herd based protocols for timed artificial insemination are becoming popular in larger dairy farms. Initial studies on herd based timed insemination protocols showed that these herd based protocols outperformed fertility management based on identifying individual cows (Pursley *et al.*, 1997). However, precision data collection tools were not very advanced at that point in time. In a more recent study, these two systems (precision data collection versus herd based protocols) were compared in a three arm randomized study. Cows (n=339) in treatment 1 were inseminated based on increased activity, cows (n=340) in treatment 2 were inseminated based on activity after the second PGF2α injection of a Presynch-Ovsynch protocol and cows (n=346) in treatment 3 received timed artificial insemination after completing the Presynch-Ovsynch protocol. Results are presented in Figure 4.

These data would indicate that in terms of overall fertility there was no difference in an individual cow data based system when compared to an overall herd based protocol (Figure 4B). Even more, a herd based Presynch-Ovsynch protocol resulted in the shortest time to first artificial insemination (Figure 4A).

3. Discussion and conclusions

When dairy farms are increasing in size, external labor is used to manage the cows and the herd. Multiple studies have shown that such expanding farms with external labor may perform well when protocols are defined and individuals working on the farm are trained to perform the activities defined in the protocols (Dufour *et al.*, 2011; Schewe *et al.*, 2015; Stup *et al.*, 2006). The objective of these protocols is that all cows are treated according to a standard operating procedure and that no deviations occur. Alternatively, farm owners may decide to increase the use of cow-level data and develop cow-specific management strategies: a development towards precision dairy farming.

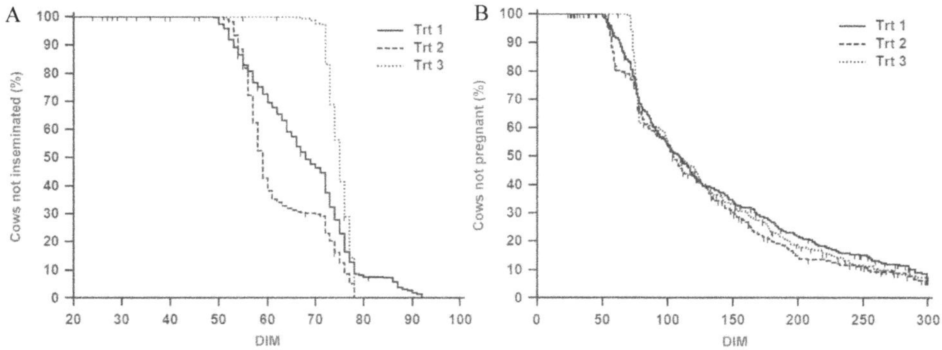

Figure 4. Kaplan-Meier survival analysis illustrating the effect of treatment on time to first AI and time to pregnancy. Treatment affected (P<0.001) median days to first AI (68, 59, and 75 days for treatment 1, 2, 3, respectively). Treatment did not affect (P=0.25) median days to pregnancy (103, 102, 102 for treatment 1, 2, 3 respectively) (Fricke *et al.*, 2014).

When comparing these two strategies, herd based protocols versus precision farming, performance data did not differ much between these two strategies. In some case animal health data were actually worse in the case of cow-specific treatment decisions relative to blanket treatment strategies for all cows in the herd.

It is of course important to define the objective of precision dairy farming. As indicated above, herd performance may not be the most important outcome. Another important outcome may be the number of antibiotic treatments and hormone interventions. Farmers may want to reduce the number of treatments and interventions to save costs. Another reason may be external societal pressure to reduce antibiotic treatments and hormone interventions. In the study on selective antibiotic treatment of low cell count cows at dry-off, the overall number of antibiotic treatments was lower in the group of cows that were not treated at dry-off, even though the number of treatments for clinical mastitis was had approximately doubled in this group of cows. Obviously, activity based heat detection rather than a standard intervention with fertility protocols such as Presynch-Ovsynch will lead to a strongly reduced use of hormones. The use of individual cow data in the case of precision dairy farming as opposed to herd level protocols is than an important tool for societal acceptance of dairy farming on larger farms.

Dairy farming is also a business and the economics of data collection, protocols, treatments and intervention will need to be attractive for farmers. A limited number of studies have been done on economic comparison of precision dairy based system compared to herd-based protocols (Fricke *et al.*, 2014; Rutten *et al.*, 2013; Steeneveld *et al.*, 2011). In these studies, no large differences in Net Present Value of either precision farming based systems or herd-based protocols were identified.

To answer the question whether there is an apparent contradiction between the use of herd based protocols and precision farming, multiple dimensions of the question need to be taken into account. In growing and larger dairies, the use of external labor often demands the development of protocols and the continuous training of farm workers in the use of these protocols. Using protocols, large dairy farms may show average or good performance and health data. As shown in Figure 1, larger dairies are able much more likely to be in area II (average to good milk quality) rather than area I or III. Current data would indicate that Precision farming does not easily result in superior performance, milk quality or animal health. Neither are there strong arguments for better economic performance in the case of precision farming when compared to herd-based protocols. The one area where precision farming may show a clear advantage is the reduced use of antibiotic treatments and hormone interventions. Such reduced use of antibiotics and hormones may be an important component of societal acceptance of growing and large dairy farms.

As shown in Figure 1, small herds that likely use cow-based decisions and probably have a very great attention to detail are able to show superior milk quality (area III). Knowing the use of individual animal data and the areas where attention to detail are crucial that are practiced by these currently small farms may be essential for larger farms to move to superior performance and health. The challenge for precision dairy farming will be to understand the use of cow-specific data on farms with superior health and performance. When the decision processes at these currently small farms can be translated to decision systems that may be applied on large and growing dairy farms, precision dairy farming will become a realistic outlook in the dairy industry.

References

Dufour, S., Fréchette, A., Barkema, H.W., Mussell, A. and Scholl, D.T. (2011). Invited review: effect of udder health management practices on herd somatic cell count. Journal of Dairy Science 94: 563-579.

Fricke, P.M., Giordano, J.O., Valenza, A., Lopes Jr., G., Amundson, M.C. and Carvalho, P.D. (2014). Reproductive performance of lactating dairy cows managed for first service using timed artificial insemination with or without detection of estrus using an activity-monitoring system. Journal of Dairy Science 97: 2771-2781.

Little, M.W., O'Connell, N.E. and Ferris, C.P. (2016). A comparison of individual cow versus group concentrate allocation strategies on dry matter intake, milk production, tissue changes, and fertility of Holstein-Friesian cows offered a grass silage diet. Journal of Dairy Science 98: in press.

Nightingale, C., Dhuyvetter, K., Mitchell, R. and Schukken, Y. (2008). Influence of variable milk quality premiums on observed milk quality. Journal Dairy Science 91: 1236-1244.

Pursley, J.R., Kosorok, M.R. and Wiltbank, M.C. (1997). Reproductive management of lactating dairy cows using synchronization of ovulation. Journal of Dairy Science 80: 301-306.

Rutten, C.J., Velthuis, A.G., Steeneveld, W. and Hogeveen, H. (2013). Invited review: sensors to support health management on dairy farms. Journal of Dairy Science 96: 1928-1952.

Scherpenzeel, C.G., Den Uijl, I.E., Van Schaik, G., Olde Riekerink, R.G., Keurentjes, J.M., and Lam, T.J. (2014). Evaluation of the use of dry cow antibiotics in low somatic cell count cows. Journal of Dairy Science 97: 3606-3614.

Schewe, R.L., Kayitsinga, J., Contreras, G.A., Odom, C., Coats, W.A., Durst, P., Hovingh, E.P., Martinez, R.O., Mobley, R., Moore, S. and Erskine, R.J. (2015). Herd management and social variables associated with bulk tank somatic cell count in dairy herds in the eastern United States. Journal of Dairy Science 98: 7650-7665.

Steeneveld, W., Van Werven, T., Barkema, H.W. and Hogeveen, H. (2011). Cow-specific treatment of clinical mastitis: an economic approach. Journal of Dairy Science 94: 174-188.

Stup, R.E., Hyde, J. and Holden, L.A. (2006). Relationships between selected human resource management practices and dairy farm performance. Journal of Dairy Science 89: 1116-1120.

Precision feeding and grazing management for temperate pasture-based dairy systems

J.L. Hills[1*], R.P. Rawnsley[1], M.T. Harrison[1], G.J. Bishop-Hurley[2], D.A. Henry[3], P. Raedts[1], M. Freeman[1] and J.R. Roche[4]
[1]Tasmanian Institute of Agriculture, University of Tasmania, Burnie, TAS 7320, Tasmania; [2]CSIRO Agriculture, St Lucia, QLD 4067; [3]CSIRO Agriculture, Werribee, VIC 3030, Australia; [1]Down to Earth Advice Ltd, Hamilton 3240, New Zealand; james.hills@utas.edu.au

Abstract

With the rapid development of precision technologies, there are increasing opportunities to measure the inherent variability in pastures and animals and use this information to optimise production and economic outcomes. Business success in pasture-based systems is driven by the production and consumption of home-grown pastures and the efficient utilisation of associated inputs. This review summarises the key factors involved in managing a successful pasture-based dairy system and highlights the precision technologies that may assist with measuring and managing variability for enhanced profitability.

Keywords: variability, dry matter intake, supplementary feed, precision technology

1. Introduction

Precision farming involves exploiting the variability that exists within components of a system, with the aim of optimising economic, social and environmental outcomes (Eastwood *et al.*, 2004). In temperate pasture-based dairy systems, significant variability exists in the quantity and nutritive value of pasture offered, both spatially and temporally (Hills *et al.*, 2015a), dry matter intake (DMI) from grazed pasture (Allison, 1985; Garcia and Holmes, 2005), feed conversion efficiency or residual feed intake, and milk production. Identifying and quantifying between animal variation is important to optimise; grazing management, supplementary feeding, feed conversion efficiency and the early identification of animal health and welfare problems.
Dairy farmers need to make accurate, timely and efficient management decisions about their farm business through access to real-time information about their farms' resources (i.e. soil, water, feed, animals and the quality and quantity of the outputs). This review summarises the key factors involved in measuring and managing this variability for pasture-based dairy production systems and highlights various precision technologies that could be used by farmers to assist with decision making. Implications around data collation and management, data visualisation, and their importance for informed decision making are also discussed.

2. Managing the pasture base

2.1. Grazing management

Effective grazing management is one of the primary drivers of farm profitability (Dillon *et al.*, 2005). Optimal pasture grazing management helps increase pasture growth rates (PGR) and nutritive values and pasture utilisation. Short-term grazing management recommendations for pastures containing perennial ryegrass are based on leaf stage, ground cover and post-grazing residuals (Chapman *et al.*, 2012; Fulkerson and Donaghy, 2001). Varying grazing rotation length, so that plants are grazed soon after canopy closure (between the second or third leaf growth stages or 2,800-3,000 kg dry matter (DM)/ha), helps maintain tiller density and optimal light interception for maximum growth.

Typical guidelines for post-grazing residuals, stipulate maintaining plants at heights of 40-60 mm or 1,500-1,750 kg DM/ha (Chapman *et al.*, 2012; Lee *et al.*, 2008). In the absence of accurate and reliable information on the quantity and quality of the pasture, the ability to optimally manage this resource is limited.

There are a number of technologies for quantifying the amount of pasture available at the paddock scale. Vehicle mounted systems that use either light beams or sonar to measure pasture height and hand-held rising plate meters have been used for many years. More recently, optical sensors such as the Greenseeker™ or Crop Circle™, have become available and provide estimates of feed quantity and quality. If combined with global navigation satellite system (GNSS) technology and near infrared spectroscopy, these technologies would allow the measurement of real-time, within paddock variation in pasture yields and quality (French *et al.*, 2014). However, all these devices require calibration, with accuracy significantly improved by using local and seasonal calibrations compared with more generic calibrations (King *et al.*, 2010; Rennie *et al.*, 2009).

Uptake of hand-held devices has been limited because of the labour requirements to capture sufficient data points across paddocks. However, satellite remote sensing systems with cloud-based data storage are available, providing pasture growth rate information directly to producers without the need for extensive on-going labour inputs. The *Pastures from Space* (Plus) product, for example, provides weekly pasture growth rate information on a regional (free) and paddock-scale (subscription basis) (http://www.pasturesfromspace.csiro.au/); these data are derived from both satellite remote sensing and climate data. New technologies, such as LiDAR, unmanned aerial vehicles and new data management and analysis techniques, will improve pasture availability estimates if they can be integrated and provide data in a readily implementable way.

2.2 Pasture growth rate forecasting

An implicit requirement of any pasture-based livestock farming system is a knowledge of seasonal weather patterns, since these regulate near-term feed availability and PGR in the absence of irrigation. Monthly forecasts of PGR based on historical climate data are produced by biophysical models such as DairyMod (Johnson *et al.*, 2008) or GrassGro (Moore *et al.*, 1997). These provide statistical information relating to pasture growth over various time intervals (e.g. daily, monthly and annual) and the accompanying inter and intra-annual variation in feed supply. This information is highly relevant for long-term strategic planning of stocking rate, calving date and supplementary feed requirements. Such approaches have limitations, however, because of variability in climate and weather (e.g. the repeatability of rainfall within fortnight across years in the Waikato region of New Zealand is only 3%; Roche *et al.*, 2009) and spatial variation between or within paddocks and/or farms. Recently, research has examined the use of sub-setting historical climate data using historical analogues of current conditions, such as the prevailing phase of the Southern Oscillation Index (e.g. neutral, strongly negative, etc.) (Rawnsley *et al.*, 2015). However, such approaches are largely statistical by nature, since they rely on analogues of past weather systems being reproduced in a similar manner to current weather systems.

New website platforms providing forecasts are available to support farmer decision making. Such platforms combine dynamic PGR models with real-time biophysical and climate data, monitored in paddocks as well as in general circulation model-based forecasts, and may also include past records, such as historically measured climate, satellite imagery and seasonal growth rates. An example of these is the use of weather predictions from global circulation models such as POAMA (the Predictive Ocean Atmospheric Model for Australia; see http://poama.bom.gov.au/). Rawnsley *et al.* (2015) demonstrated, that in general, dynamic forecasts from POAMA improve forecasts of 30-day PGR, compared with methods only based on historically measured climate data. In addition, the important climate indices of temperature, light and soil moisture (Denmead and Shaw, 1962) are also quantified and forecasted utilising this platform (http://dashboard.sense-t.org.au/).

3. Managing the animal to optimise pasture utilisation and the use of supplementary feeds

Automated identification of individual animals using radio frequency identification devices (RFID) has enabled routine collection of data relating to an individual animal in a herd. Individual animal metrics that are being utilised by producers include daily milk yields, milk components (e.g. fat, protein), milk conductivity, live weight and measures of activity using pedometers and inertial measurement units (IMU), particularly for identification of oestrus (Bewley, 2012). Other more recent technology innovations, include detailed measurements of eating and grazing behaviours (Smith *et al.*, 2015) and automated body condition score systems (Halachmi *et al.*, 2013). However, in pasture-based systems the ability to measure the intake of individual animals is the most important missing element. As with milk production, there is a large genetic and phenotypic variation in DMI by individuals within each dairy herd (Berry *et al.*, 2007).

Currently there is no accurate automated means of measuring pasture intake at the individual cow level. Accelerometers have been used to provide an estimate of grazing time (Delagarde and Lamberton, 2015; Ueda *et al.*, 2011), however, grazing time alone is not a good indicator of pasture DMI, because intake is determined by a combination of bite rate, bite mass and grazing time. A measure of bite rate in combination with grazing times may be able to improve estimates of a cow's pasture intake (Oudshoorn *et al.*, 2013) or at least enable ranking of individuals in the herd, from efficient through to inefficient grazers. Successful technologies for measuring bite rates include the IGER behaviour recorder (Rutter, 2000), sound recording (Ungar and Rutter, 2006), and accelerometers (Umemura, 2013).

Two technologies that could be used to better manage the DMI of individuals, or cohorts of animals within the herd, are virtual fencing and automated precision feeders.

3.1 Virtual fencing for managing pasture access

Autonomous animal control (AAC) using virtual fences, is a transformational technology with the potential to provide producers the ability to match feed supply and animal demand at any spatial and temporal scale (French *et al.*, 2014). The allocation of adequate pasture, of correct quality, that meets the animal's needs would not be restricted by physical infrastructure. The benefits would include improved pasture and animal production and higher pasture utilisation with reduced incidence of overgrazing. In theory, feed supply and demand could be matched at the individual animal or sub-group level, rather than at a herd level, allowing the individual to maximise genetic potential. Lower producing cows, as a result of restricted daily pasture DMI, could be individually managed within the herd to encourage higher daily intakes, leading to enhanced production. With the development of GNSS chips that require less power, batteries that are capable of storing more energy, improvements in solar panel efficiency and improved animal control algorithms (Ruiz-Mirazo *et al.*, 2011), autonomous control of animals is closer to being a reality. Other benefits of AAC include: reduced need for internal fencing and a reduction in fencing repairs and maintenance, including those damaged by fire and flood. Labour requirements for moving and mustering animals would be significantly reduced and there would be the ability to manage more intensive grazing systems without a corresponding increase in labour requirements. Environmentally sensitive areas, including riparian areas, could be protected from overgrazing, whilst allowing grazing at key times to control biomass accumulation.

3.2 Individualised feeding of supplements to optimise energy intake

The development of automated precision feeder technology has made it possible to easily feed individual cows different amounts and types of supplements. The basic premise underlying the concept of individualised feeding of dairy cows, is to exploit a natural 'advantage' that improves the cost of production, by offering cows with a greater predisposition to eat, access to more higher quality feeds and reducing the amount of expensive feeds provided to cows with lower DMI. Although logical in theory, the concept of individualised feeding on the basis of a cow's milk production (i.e. a proxy for dry matter intake), has not translated into improved herd production in pasture-based systems (Hills *et al.*, 2015b). A detailed understanding of intake regulation (see Roche *et al.*, 2008) provides reasons for the lack of measurable effect.

In a grazing system where a restricted pasture allocation is often provided, supplementary feeds such as grain-based concentrates, co-products feeds (i.e. waste from the human food industry), or dried or conserved forages (i.e. hay and silage) are used to balance and increase a herds daily feed intake. However, the consumption of these feeds affects the peripheral and neural signals of a cow's metabolic state and results in a reduction in grazing time (Bargo *et al.*, 2003; Sheahan *et al.*, 2011), grazing intensity (J.R. Roche, unpublished), and, ultimately pasture intake (Stockdale, 2000). The phenomenon where cows reduce the consumption of available pasture following the consumption of a supplement is known as substitution. As a result of substitution, total DMI does not increase by the amount of supplement provided. However, with cow variation in DMI (Berry *et al.*, 2007), it should be possible to use individualised feeding strategies to exploit the natural variation in substitution rates (Stockdale, 2000).

Although there is individual cow variation in substitution rates (Stockdale, 2000), the provision of supplements to 'exploit' this variation has not increased total herd milk production. A possible explanation, is that temperate pasture is very digestible, providing as much metabolisable energy/kg dry matter as a grain-based concentrate (Roche *et al.*, 2010). Although higher DMI cows are likely to have lower substitution rates than low DMI cows, when all cows are offered the same amount of supplement, the competition for pasture declines and the high DMI cows are given the opportunity to exploit their natural tendency to eat more (i.e. lower substitution rate). In comparison, if the high DMI cows are offered more supplement, at the expense of lower DMI cows, they consume the additional supplement, leading to higher substitution rates and the low DMI cows consume the pasture available, and there is little change in total herd milk production when compared with the situation that all cows were offered the same amount of supplement (Hills *et al.*, 2015b; Leaver, 1988). However, where cows with low DMI, due to factors such as social hierarchy or access time to pastures following milking, are offered more supplements to enable them to achieve their genetic potential for milk production with little to no effect on substitution, the marginal response to the supplement should be high. In this case offering more supplement to these cows is not likely to change total pasture consumed, but total milk production should increase due to an increase in total DMI.

Further research needs to be conducted to better understand the relationship among key factors, such as grazing and other activity based behaviours affecting pasture intake and substitution rates.

Although current feeding studies results, and what we know about the physiological regulation of intake, do not support the use of individualised feeding of grazing dairy cows based on the individual's milk production, individualised feeding of cows may still be useful. For example, under very high stocking rates (cows/ha) or stocking densities (cows/ha per day or m^2/cow per day) and, during situations when pasture growth is near zero (e.g. drought), or where individualised feeding is used and high yielding cows are consuming a high proportion of their diet as supplement the factor most limiting production may be a lack of a specific amino acid (e.g. lysine) or an over-supply of a particular feed ingredient (e.g. non-fibre carbohydrate; Auldist *et al.*, 2013). In these situations, individualised feeding of cows could balance an individual cow's diet improving overall health and productivity.

4. Managing data flows

While there is a rapidly increasing toolbox of precision technologies providing data on both the pasture base and animal that could be used to optimise the economic, social and environmental outcomes for dairy farms, unless a farmer is able to readily use this information for enhanced decision making, then investment in the technology is of no value and may actually negatively impact upon farm profitability. This phenomenon has been termed 'the bleeding edge' where money and time invested in technology has been 'bled' for no outcome (Hills *et al.*, 2015a).

Although there are currently a number of decision support tools available for pasture-based dairy systems, these are often based on static user-input information (e.g. cow numbers, feeding level, pasture availability, input and output prices, etc.). There is a paucity of decision support tools for processing real-time information and integrating data that captures the whole of farm system. The integration of data from both static and real-time sources would allow users to contrast between alternative management/feeding/financial scenarios allowing dairy farmers to make better decisions, and potentially improve both time and cost efficiencies in their operation. An example of the development and demonstration of a digital dashboard for farm management that combines access to static and real-time data in a customisable way for cattle production systems in Northern Australia is the digital homestead project (http://www.digitalhomestead.org).

In addition, there has been limited application of data mining and machine learning techniques in pasture-based livestock systems. Data mining and machine learning techniques provide an automated avenue for collating and processing real-time data flows that are too voluminous for humans to process manually and therefore would be expected to make dairy farm management more timely and cost-efficient as the technology becomes less expensive. Technology for processing multiple data flows have been used to mine cattle disease databases to identify herd disease risk (Ortiz-Pelaez and Pfeiffer, 2008) and to help animal breeders make decisions based on the likelihood of conception from previous insemination data, stage of reproductive cycle, bull semen quality and herd disease history (Grzesiak and Zaborski, 2012).

5. Conclusions

Precision dairy technologies provide significant opportunities for dairy farmers to better understand and manage variability that exists in both pasture and animal components of their dairy system. Accurate and reliable precision technologies provide information for timely, effective and efficient decision making which leads to improved economic, social and environmental outcomes for farming businesses. Of particular value for pasture-based dairy systems, is the development of technologies providing farmers with real-time autonomous data on pasture biomass and predicted pasture growth rates. In addition, accurate and reliable measures of individual cow intake, coupled with productivity and behavioural data, should allow farmers to make timely proactive decisions leading to improved animal performance, farm productivity and farm profitability. However, a significant challenge is integration of the various data sources into decision support systems that make it easier for farmers to make appropriate tactical and strategic decisions.

References

Allison, C.D. (1985). Factors affecting forage intake by range ruminants – a review. Journal of Range Management 38: 305-311.

Auldist, M.J., Marett, L.C., Greenwood, J.S., Hannah, M., Jacobs, J.L. and Wales, W.J. (2013). Effects of different strategies for feeding supplements on milk production responses in cows grazing a restricted pasture allowance. Journal of Dairy Science 96: 1218-1231.

Bargo, F., Muller, L.D., Kolver, E.S. and Delahoy, J.E. (2003). Invited review: production and digestion of supplemented dairy cows on pasture. Journal of Dairy Science 86: 1-42.

Berry, D.P., Horan, B., O'Donovan, M., Buckley, F., Kennedy, E., McEvoy, M. and Dillon, P. (2007). Genetics of grass dry matter intake, energy balance, and digestibility in grazing Irish dairy cows. Journal of Dairy Science 90: 4835-4845.

Bewley, J. (2012). How precision dairy technologies can change your world. Pennsylvania State Dairy Cattle Nutrition Workshop, Grantville, PA, USA.

Chapman, D.F., Tharmaraj, J., Agnusdei, M. and Hill, J. (2012). Regrowth dynamics and grazing decision rules: further analysis for dairy production systems based on perennial ryegrass (*Lolium perenne* L.) pastures. Grass and Forage Science 67: 77-95.

Delagarde, R. and Lamberton, P. (2015). Daily grazing time of dairy cows is recorded accurately using the Lifecorder Plus device. Applied Animal Behaviour Science 165: 25-32.

Denmead, O.T. and Shaw, R.H. (1962). Availability of soil water to plants as affected by soil moisture content and meteorological conditions. Agronomy Journal 54: 385-390.

Dillon, P.R., Roche, J.R., Shalloo, L. and Horan, B. (2005). Optimising the financial return from grazing in temperate pastures. In: Murphy, J.J. (ed.) Proceedings of a satellite workshop of the 20th International Grassland Congress, Cork, Ireland, pp. 131-147.

Eastwood, C., Chapman, D. and Paine, M. (2004). Precision dairy farming – taking the microscope to dairy farm management.

French, P., Obrien, B. and Shalloo, L. (2014). Precision technology for pasture management. Proceedings of the 5th Australasian Dairy Science Symposium, pp. 326-331.

Fulkerson, W.J. and Donaghy, D.J. (2001). Plant-soluble carbohydrate reserves and senescence – key criteria for developing an effective grazing management system for ryegrass-based pastures: a review. Australian Journal of Experimental Agriculture 41: 261-275.

Grzesiak, W. and Zaborski, D. (2012). Examples of the use of data mining methods in animal breeding. INTECH Open Access Publisher, Rijeka, Croatia.

Garcia, S.C. and Holmes, C.W. (2005). Seasonality of calving in pasture-based dairy systems: its effects on herbage production, utilisation and dry matter intake. Australian Journal of Experimental Agriculture 45: 1-9.

Halachmi, I., Klopcic, M., Polak, P., Roberts, D.J. and Bewley, J.M. (2013). Automatic assessment of dairy cattle body condition score using thermal imaging. Computers and electronics in Agriculture 99: 35-40.

Hills, J.L., Garcia, S.C., Dela Rue, B. and Clarke, C.E.F. (2015a). Limitations and potential for individualised feeding of concentrate supplements to grazing dairy cows. Animal Production Science 55: 922-930.

Hills, J.L., Wales, W.J., Dunshea, F.R., Garcia, S.C. and Roche, J.R. (2015b). Invited review: an evaluation of the likely effects of individualized feeding of concentrate supplements to pasture-based dairy cows. Journal of Dairy Science 98: 1363-1401.

Johnson, I.R., Chapman, D.F., Snow, V.O., Eckard, R.J., Parsons, A.J., Lambert, M.G. and Cullen, B.R. (2008). DairyMod and EcoMod: biophysical pasture-simulation models for Australia and New Zealand. Australian Journal of Experimental Agriculture 48: 621-631.

King, W.M., Rennie, G.M., Dalley, D.E., Dynes, R.A. and Upsdell, M.P. (2010). Pasture mass estimation by the c-dax pasture meter: regional calibrations for New Zealand. Proceedings of the 4th Australasian Dairy Science Symposium.

Leaver, J.D. (1988). Level and pattern of concentrate allocation to dairy cows. In: Garnsworthy, P.C. (ed.) Nutrition and lactation in the dairy cow. University of Nottingham, Nottingham, UK, 315-326.

Lee, J.M., Donaghy, D.J. and Roche, J.R. (2008). Effect of defoliation severity on regrowth and nutritive value of perennial ryegrass dominant swards. Agronomy Journal 100: 308-314.

Moore, A.D., Donnelly, J.R. and Freer, M. (1997). GRAZPLAN: decision support systems for Australian grazing enterprises, 3. Pasture growth and soil moisture submodels, and the GrassGro DSS. Agricultural Systems 55: 535-582.

Ortiz-Pelaez, Á. and Pfeiffer, D.U. (2008). Use of data mining techniques to investigate disease risk classification as a proxy for compromised biosecurity of cattle herds in Wales. BMC Veterinary Research 4: 24.

Oudshoorn, F.W., Cornou, C., Hellwing, A.L.F., Hansen, H.H., Munksgaard, L., Lund, P. and Kristensen, T. (2013). Estimation of grass intake on pasture for dairy cows using tightly and loosely mounted di-and tri-axial accelerometers combined with bite count. Computers and Electronics in Agriculture 99: 227-235.

Rawnsley, R.P., Harrison, M.T., Phelan, D.C., Corkrey, R. and Henry, D.A. (2015). Near-term pasture growth rate forecasts: which method works best? Proceedings of the 17th Australian Agronomy Conference, September 20-24, Hobart, Australia.

Rennie, G.M., King, W.M., Puha, M.R., Dalley, D.E., Dynes, R.A. and Upsdell, M.P. (2009). Calibration of the C-DAX rapid pasture meter and the rising plate meter for kikuyu-based Northland dairy pastures. Proceedings of the New Zealand Grassland Association 71: 49-55.

Roche, J.R., Blache, D., Kay, J.K., Miller, D.R., Sheahan, A.J. and Miller, D.W. (2008). Neuroendocrine and physiological regulation of intake with particular reference to domesticated ruminant animals. Nutrition Research Reviews 21: 207-234.

Roche, J.R., Turner, L.R., Lee, J.M., Edmeades, D.C., Donaghy, D.J., Macdonald, K.A., Penno, J.W. and Berry, D.P. (2009). Weather, herbage quality and milk production in pastoral systems, 3. Inter-relationships and associations between weather variables and herbage growth rate, quality and mineral concentration. Animal Production Science 49: 211-221.

Roche, J.R., Kay, J.K., Phyn, S., Meier, C.V.C., Lee, J.M. and Burke, C.R. (2010). Dietary structural to nonfiber carbohydrate concentration during the transition period in grazing dairy cows. Journal of Dairy Science 93: 3671-3683.

Ruiz-Mirazo, J., Bishop-Hurley, G.J. and Swain, D.L. (2011). Automated animal control: can discontinuous monitoring and aversive stimulation modify cattle grazing behaviour? Rangeland Ecology and Management 64: 240-248.

Rutter, S.M. (2000). Graze: a program to analyse recordings of the jaw movements of ruminants. Behavior Research Methods Instruments and Computers 32: 86-92.

Sheahan, A.J., Kolver, E.S. and Roche, J.R. (2011). Genetic strain and diet effects on grazing behavior, pasture intake, and milk production. Journal of Dairy Science 94: 3583-3591.

Smith, D., Dutta, R., Hellicar, A., Bishop-Hurley, G., Rawnsley, R., Henry, D., Hills, J. and Timms, G. (2015). Bag of class posteriors, a new multivariate time series classifier applied to animal behaviour identification. Expert Systems with Applications 42: 3774-3784.

Stockdale, C.R. (2000). Levels of pasture substitution when concentrates are fed to grazing dairy cows in Northern Victoria. Australian Journal of Experimental Agriculture 40: 913-921.

Umemura, K. (2013). Technical note: monitoring grazing bites and walking activity with pedometers. Journal of Dairy Science 96: 1090-1093.

Ueda, Y., Akiyama, F., Asakuma, S. and Watanabe, N. (2011). Technical note: the use of a physical activity monitor to estimate the eating time of cows in pasture. Journal of Dairy Science 94: 3498-3503.

Ungar, E.D. and Rutter, S.M. (2006). Classifying cattle jaw movements: comparing IGER behaviour and acoustic techniques. Applied Animal Behaviour Science 98: 11-27.

Novel precision dairy farming technologies

T. Norton and D. Berckmans*
Department of Biosystems, Division M3-BIORES: Measure, Model and Manage Bioresponses, KU Leuven, Kasteelpark Arenberg 30, 3001 Heverlee, Belgium; daniel.berckmans@biw.kuleuven.be

Abstract

Precision Livestock Farming (PLF) systems aim to offer a real-time monitoring and managing tool for the farmer. This is fundamentally different from all approaches that aim to offer a monitoring tool without improving the life of the animal under consideration. In dairy production the animals (cows) stay for longer in production when compared to any other farmed animals. Moreover, on an each animal unit is of a high economical value to the farmer. Therefore, for the dairy cow and consequently for the farmer it is important to detect problems as soon as possible and to take action on an individual animal level. This is idea of PLF, as it is provides real-time warning to the farmer when something goes wrong for the farmer to immediate upon. It requires real-time algorithms that are able to detect problems while the process is ongoing. To successfully develop such algorithms in an efficient way, some basic principles must be respected in the methodology being applied. This paper gives a systematic approach in using the interaction between taking field data, applying a gold standard and using labelling techniques to develop real-time algorithms that allow real time monitoring and management of individual cows.

Keywords: principles of PLF, gold standard, labelling, feature variables

1. Objectives

PLF is a multidisciplinary science which requires collaboration between animal scientists, physiologists, veterinarians, ethologists, engineers, ICT experts, etc. Since with PLF systems development we are aiming to create an early warning system, it is wise to focus on the first signs that can be monitored in a non-invasive and contactless way in an animal. When the animal experiences less than ideal conditions it will exhibit an initial response in terms of behavioural changes and these first signs should be picked up by the PLF sensing technology, such as a sensor or real-time image and/or image analysis. Therefore, the use of field data to develop real-time algorithms involves some steps that should be properly understood to make successful algorithms.

The first objective of this paper is to describe some basic principles to realise a more efficient development PLF of algorithms for monitoring and improving dairy production.

The second objective is to achieve some agreement in the terminology used by these different disciplines in order to improve communication between specialists from different disciplines.

2. Method

2.1 A living organism is a CITD system

As stated in the first EC PLF conference in 2003 (Berlin), a living organism is a so called '*CITD system*': this stands for Complex, Individually Different, Time-Varying and Dynamic (Berckmans and Aerts, 2006; Quanten *et al.*, 2006). It is obvious that a living organism is much *more complex* than any mechanical, electronic or ICT system. The complexity of information transmission in a single cell of a living organism is for example much higher than in most man-made systems (e.g. today's most powerful microchip). In biological research and the management of biological process (e.g. medical world, livestock world) in industry and society, the general trend is still to compare groups of living organisms by looking for statistical differences in experiments. Statistical methods have been developed primarily to find significant differences between the averages of groups. However, there is not a single living organism that lives or acts as the purely theoretical average of a group since all living organisms are individually different in their responses. This raises serious questions about the way a lot of research is carried out on animals and humans. Moreover the time varying character of a living organism means that a living organism's response to a (environmental) stimulus or stressor might be different each time it happens. A living organism is constantly looking for a good energy balance and as a consequence is continuously changing its physical condition and mental status. Of course living organisms are dynamic systems.

The CITD nature of living organisms has an important impact on the type of algorithms we need to develop. It implies that algorithms to monitor these time-varying individuals must continuously adapt to the individual and/or use principles that can be used in real time in the field application to adapt to the time variation of that individual.

The method of this paper is to give some principles by taking an example. In this case the example is the real-time monitoring of lameness of dairy cows by using real-time image analysis.

2.2 Field data or bio-signals

The basic methods used in PLF involve continuously measuring responses directly on the animal rather than in the environment surrounding the living organism. Since animal responses can be fast, it is useless to carry out a survey just once a year, once a month or week, or even twice a day depending on the specific objective. We need a continuous monitoring/management tool. The word 'continuous' must be interpreted in relation to the variable that is followed like in this case lameness of cows.

In this example of monitoring lameness we use image analysis for collecting real-time field data, known as 'bio-signals'. One of the advantages of image analysis is the fact that there is no need for physical contact, and so there is no risk of influencing the animal response while making the measurement, no need to recover sensors from living animals. Furthermore, costs are reduced since one camera can monitor a very large group of animals since each of them will pass under the camera every day.

2.3 Linking field data, target variable, gold standard, feature variable and labelling

The field data consist of a lot of numbers originating from the sensors (e.g. 240 samples per second for an accelerometer) or images (e.g. 25 images per second in a large image matrix). In this case to develop a monitor for cow lameness we used a series of sequential images (e.g. videos) of a cow walking away from the milk robot immediately after milking. To find the appropriate variables 3 cameras were initially used: a side view camera, a top view camera and a lower side-view camera focusing on details of the hoofs. During the experiment it became clear that a single top-view camera was sufficient (Figure 1).

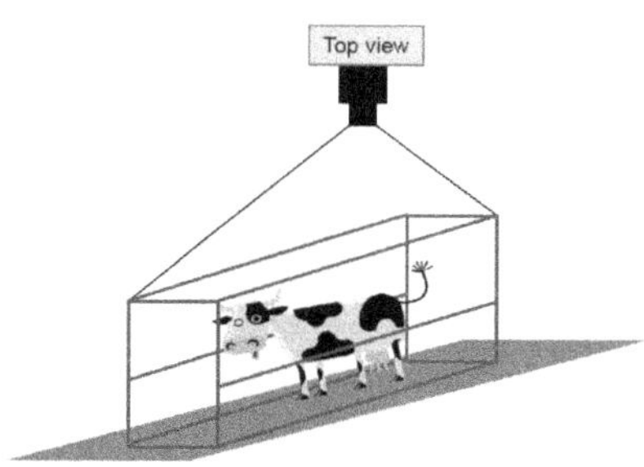

Figure 1. Measurement installation to collect field data from each cow with a top view camera.

2.4 Target variable

In each of the processes involved in developing a real-time PLF algorithm we have to take some steps in which we combine different types of variables to produce a generic method for developing algorithms. In the example case, we aim to develop a real-time lameness monitoring system for cows by using real-time image analysis of the camera-produced images of the animals. This means that in this example the target variable is the lameness status of cows. The target variable directly relates to the final objective of the algorithm. In the example of a lameness monitoring system, this means that the variable is a quantity indicating the degree of lameness of the cow. Lameness can be defined as a deviation in gait resulting from pain or discomfort from hoof or leg injuries or disease (Flower *et al.*, 2009). The prevalence of lameness can be estimated as 34% of the cows in Denmark, 12% in Poland, 16% in Germany and 31% in Austria. It can be stated that lameness is the number one problem regarding animal welfare in milking cow and in literature over 200 possible causes have been named.

2.5 Gold standard

The first question is whether we have a reliable gold standard to quantify the target variable, in this case the real health status of a cow regarding lameness. Do we have a generally accepted way of measuring the lameness status of a cow? A gold standard or reference point can be defined as a state-of-the-art scientific measurement or method which enables us to draw a conclusion relating to the final objective of the algorithm or the status of the target variable, in this case the degree of lameness of a cow. A gold standard might be an expensive and complex method but the most important point is that this gold standard should be accepted by scientists as the state-of-the-art measurement or method which will quantify the target variable in a reliable way. In the case of lameness of cows a gold standard might be the score a human experts to quantify the degree of lameness of the cow (Figure 2).

In most cases the gold standard cannot easily be applied to real time measurements and consequently has a much lower sampling frequency than the real-time solution that we are looking for. In the case of a score by a human expert, it is technically unrealistic to consider continuous scoring every day on enough cows. This can be done once a week, for example, which is already an ideal case, for a large farm with many animals. Having worked on the development of this PLF approach and methodology since 1991 (Aerts, 1991) we can say that, in most cases, establishing an accurate gold standard is one of the most difficult elements of developing PLF algorithms. In case that we do not know a clear gold standard for the target variable we aim for, we will not proceed for this target variable.

In collaboration with other teams in the world we had built up several datasets to collect field data for cow lameness where at the same time a gold standard (see further) was applied. We collected videos from walking cows at 3 different farms: 155 recordings on 70 cows with weekly gait scoring for all recordings; 4,200 recordings on 58 cows per dataset with gait scoring every 2 weeks and finally over 9,000 recordings from 965 cows over a 15 weeks period with weekly gait scoring.

Figure 2. A gold standard for lameness is the human score to quantify the degree of lameness.

This shows that the result is a huge amount of data and the transmission of so much data takes time, energy and money. Sending data wirelessly in real time involves energy and costs; we should therefore avoid transmitting too much data and develop real-time algorithms which calculate information from the data at the lowest possible level, enabling us to transmit information rather than data.

2.6 Feature variable

The idea of the real time monitoring system assumes that we can find another variable that can give an early warning of lameness in a cow; this variable is called the feature variable. In the example of lameness monitoring, the idea was to use different variables calculated from the image of the animal to find out which is the best feature variable that would indicate the lameness in an early stage. The hypothesis was that the gait analysis of an image of a cow gives the baseline for that individual cow and that a significant deviation from that baseline in time would be shown by the dynamic variation of the appropriate feature variable calculated from the image.

The feature variable is the variable that is calculated from the field measurements on the animal which are captured by in this case the image information. The idea is that the feature variable can be measured or calculated at a high sampling frequency or continuously in relation to the dynamics of the process, in this case lameness of the cow. The PLF algorithm aims for real-time calculation of the target variable so that there is no need to store any raw field data, or in this case all the image signals measured at a rate of 25 Hz, in the monitoring system.

Since it was not clear that human scoring was based on a well-defined single target variable we have tried 13 different feature variables (Table 1).

Figure 3 shows that the first objective of the algorithm must be to estimate or calculate the value of the feature variable (1 of the 13 feature variables calculated from the image) from the measured field data (the measured images the cow shed). The second objective of the algorithm is to link the value of the feature variable (1 or more of the 13 variables calculated from the image) to the target variable (the lameness status of the cow as scored by the human expert).

Automatic measured feature variables
Step Overlap: Position of hind leg – position of front leg
Swing time: Amount of time the hoof is released from the floor
Stance time: Amount of time the hoof has contact with the floor
Stride time: Amount of time the cow takes for a full stride cycle (= 1 step with each hoof)
Step length: Distance between two subsequent placements of same hoof
Back arch: Arch of the back measured with 3 points
Touch angle: Angle of the lower leg (from fetlock joint to tarsus or carpus joint) when the hoof touches the floor (from vertical line)
Release angle: angle of the lower leg (from fetlock joint to tarsus or carpus joint) when the hoof releases from the floor (from vertical line)
Active appearance model • Deformable model, fit on contour in the image • Robust • Global (holistic)
Asymmetry step overlap
Swing time/Stance time
Asymmetry step length
Fetlock range of motion

Table 1. List of the tested 13 feature variables to monitor lameness in cows in real time.

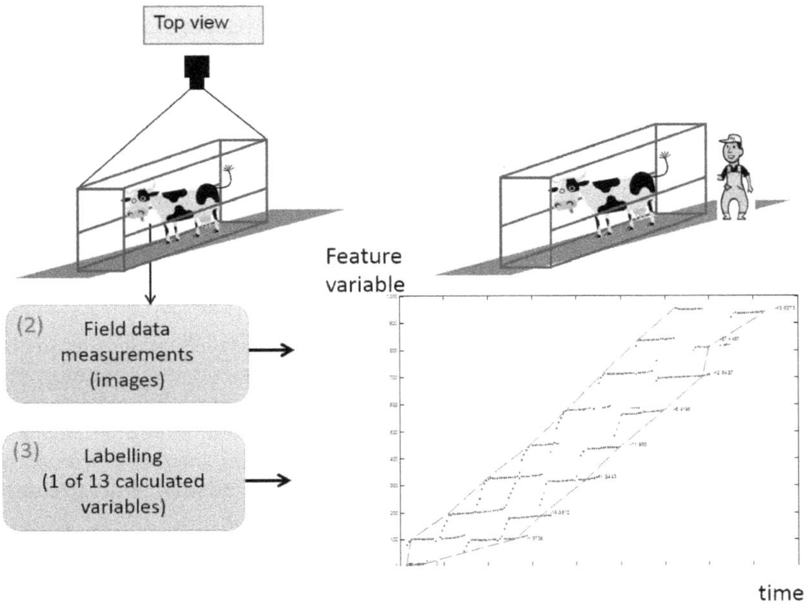

Figure 3. Example of feature variables: images collected and up to 13 feature variables calculated as a function of time.

2.7 Labelling

In this example, when collecting the bio-signals or field measurements on the animal we obtain the image data produced by the animal (Figure 2). In order to develop an algorithm that can detect the deviation from a healthy gait behaviour automatically, we again need a reference point which indicates the point in time when the field data shows these deviations from normal gait. To develop an algorithm for lameness detection it is not sufficient to know at what moment the way of walking is different; we need to know when exactly a step starts and when it ends. This information must be obtained by careful audio-visual analysis of the field data using some kind of methodology: for example a human observer on the scene who carries out audio-visual scoring of lameness or visual analyses on the scene in reality. Another method is for a human to carry out off-line visual marking of video images: image per image to mark deviations from 'normal' walking gait. This might be less expensive and easier but is it as accurate as on-the-spot observations (in the livestock house)? A main objective is to get insight in the feature variable, understand the dynamic behaviour of the feature variable and understand the biological meaning of the feature variable. This activity is called labelling: 'detailed manual audio-visual analyses of the feature variable from the measured field data to be used as a reference point for algorithm development' to calculate the feature variable (Figure 3). This accurate labelling of field data is very labour-intensive: manual labelling of a 48-hour video, involving marking start and stop points (e.g. in an image) for only 7 different activities can easily take a few man-months! Each single image or data sample has to be analysed to identify the beginning or end of one of the activities. To label, for example, the beginning and end of each deviation in normal gait, all the low quality images (sunlight, shadows, dirty lenses, moving background, etc.) in image data that were captured at a sample frequency of 25 Hz requires a serious investment of man hours. Research teams specialising in labelling have developed tools which enable this time-consuming hard work to be carried out more efficiently. It has been demonstrated that visual labelling of cow lameness on the scene in the livestock house is not very straightforward (Ismayilova *et al.*, 2013). It is clear that the accuracy of labelling will be of crucial importance in developing accurate algorithms. If the accuracy of labelling is unknown, a new problem arises with regard to how to develop an accurate algorithm. An algorithm can never be more accurate than the used gold standard or the used labelling technique.

Gold standard and labelling activities are two different things which should not be mixed up; this is evidenced by the fact that the objectives relating to the target variable and feature variable are different, and that all these variables vary over time, creating a need for real time calculation of the feature variable.

3. Algorithm development

When data are collected we print them as a function of time and apply labelling in order to obtain reference data for the feature variable. The labeller has now marked exactly which feature variable he can see in the image and when a variation of the value starts and ends in all the field data. We can now develop/run the first part of the algorithm which calculates the values of the feature variable for each image. The next step is to develop the second part of the algorithm, namely to compare the dynamic behaviour of the feature variable with the results of the gold standard in order to complete the algorithm for automatic detection of the lameness (Figure 4).

The algorithm must enable calculation of the feature variable in real time from the field data to detect the individual variation in feature variables.

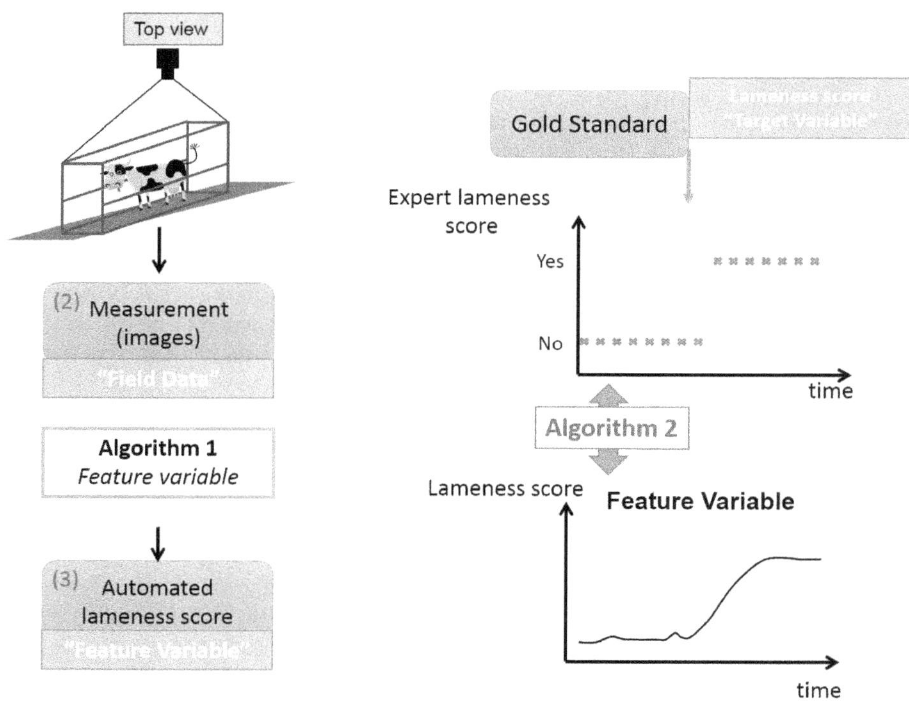

Figure 4. Measurements, labelling and gold standard to develop algorithms.

4. Algorithm testing in field situations

Whatever type of physical or mathematical simulation we use in the lab, experience shows that the hard work starts when the algorithm is implemented in a real livestock houses. The main reason for this is that animal-related processes in a commercial livestock house are much more complex than anything we can simulate in the laboratory.
In the case of real-time lameness monitoring, validation of these algorithms in the field, conducted in a test farms in practice, demonstrated that the algorithms developed were able to classify the lameness correctly in 86% of cases (Van Hertem *et al.*, 2014). The final results in terms of performance of these algorithms can be expressed quantitatively using the criteria: sensitivity, specific and overall accuracy (Exadaktylos *et al.*, 2008).

5. Conclusions

Making algorithms to work in real-life conditions is a hard job that all developers will experience before successfully developing a reliable and accurate real-time monitoring system. Producing journal publications and patents is clearly far easier than creating cheap, reliable and accurate PLF tools. As described the application of a good method is an important tool to create algorithms.
Technology nowadays offers exciting opportunities to develop automatic monitoring and management products to help farmers remain competitive in the face of the many requirements and skills that society imposes on them. Technology, however, is just a tool which supports many others. Development of suitable systems needs much more intensive collaboration between people from different disciplines, which appears to be difficult because each discipline, each team and many individuals are just hunting for more research money instead of focusing on making more progress in their field of research or the sector where their knowledge should be applied.
We hope that this paper and our efforts to define some terms will help to facilitate communication between scientists from the different disciplines that are needed to create useful PLF tools.

References

Aerts, J.M. (1991). Ontwerp van een experimentele opstelling voor het toepassen van on-line modelleringstechnieken op levende organismen. PhD thesis, KU Leuven, Leuven, Belgium, 140 pp.

Berckmans, D. and Aerts, J.M. (2006). Integration of biological responses in the management of bioprocesses. Master course in the Masters of BioSystems and of Human Health Engineering at the KU Leuven, 168 pp.

Exadaktylos, V., Silva, M., Ferrari, S., Guarino, M., Taylor, C.J., Aerts, J.-M. and Berckmans, D. (2008). Time-series analysis for online recognition and localization of sick cow (sus scrofa) cough images. Journal of the Acoustical Society of America 124(6): 3803-3809.

Flower, F.C. and Weary, D.M. (2009). Gait assessment in dairy cattle. Animal 3(1): 87-95.

Genazzani, A.D. and Rodbard, D. (1991). Use of the receiver operating characteristic curve to evaluate sensitivity, specificity, and accuracy of methods for detection of peaks in hormone time-series. Acta Endocrinologica 124(3): 295-306.

Ismayilova, G., Oczak, M., Costa, A., Sonoda, L.T., Viazzi, S., Fels, M., Vranken, E., Hartung, J., Bahr, C., Berckmans, D. and Guarino, M. (2013). How do pigs behave before starting an aggressive interaction? Identification of typical body positions in the early stage of aggression using video labelling techniques. Berliner Und Munchener Tierarztliche Wochenschrift 126(3-4): 113-120.

Quanten, S., De Valck, E., Cluydts, R., Aerts, J.M. and Berckmans, D. (2006). Individualized and time-variant model for the functional link between thermoregulation and sleep onset. Journal Of Sleep Research 15(2): 183-198.

Van Hertem, T., Viazzi, S., Steensels, M., Maltz, E., Antler, A., Alchanatis, V., Schlageter-Tello, A., Lokhorst, K., Romanini, E.C.B., Bahr, C., Berckmans, D. and Halachmi, I. (2014). Automatic lameness detection based on consecutive 3D-video recordings. Biosystems Engineering 119: 108-116.

Update on use of sensors on dairy farms

J.M. Bewley
Department of Animal and Food Sciences, University of Kentucky, Lexington, KY 40546, USA; jbewley@uky.edu

1. Setting the stage for precision dairy monitoring

During the last 40 years, the number of dairy farms in the United States has decreased considerably from over 1,000,000 dairy operations to around 75,000. Cornell University Agricultural Economists predicted that the number of dairies in the United States would decrease from 100,000 in 2000 to 16,000 by 2020. During this period, the number of large farms (>500 cows) was projected to increase from 2,700 to 3,400 while the number of small farms (<100 cows) would decrease from 84,000 to 7,000, with large farms producing over 80% of the nation's milk (LaDue *et al.*, 2003). Dairy operations today are characterized by narrower profit margins than in the past, largely because of reduced governmental involvement in regulating prices of agricultural commodities. Consequently, small changes in production or efficiency can have a major impact on profitability (Huirne, 1990). The resulting growth in competition has intensified the drive for efficiency resulting in increased emphasis on business and financial management.

Furthermore, the decision making landscape for a dairy manager has changed dramatically with increased emphasis on consumer protection, continuous quality assurance, natural foods, pathogenic free food, zoonotic disease transmission, reduction of the use of medical treatments, and the emergence of a new social ethic regarding the care of animals (Pietersma *et al.*, 1998; Rollin, 2004; Schulze *et al.*, 2007; Spilke and Fahr, 2003; Tothill, 2001). These changing demographics reflect a continuing change in the way in which dairy operations are managed. In large part, many of these changes can be attributed to tremendous technological progress in all facets of dairy farming, including genetics, nutrition, reproduction, disease control, and management. W. Nelson Philpot (2003) captured this change effectively in describing modern dairy farms as 'technological marvels'. Conceivably, the next technological marvel in the dairy industry will be in the use of technologies for individual animal status monitoring and management.

2. Characterizing dairy technologies

The use of technology for individual animal status monitoring and management has been described by many different terms in the literature. Perhaps the term that captures this concept best (and the one that will be used through the rest of this dissertation) is 'Precision Dairy Farming' (PDF). Eastwood *et al.* (2004) defined PDF as 'the use of information technologies for assessment of fine-scale animal and physical resource variability aimed at improved management strategies for optimizing economic, social, and environmental farm performance'. Spilke and Fahr (2003) stated that PDF, with specific emphasis on technologies for individual animal monitoring, 'aims for an ecologically and economically sustainable production of milk with secured quality, as well as a high degree of consumer and animal protection'. With PDF, the trend toward group management may be reversed with focus returning to individual cows through the use of technologies (Schulze *et al.*, 2007). Technologies included within PDF range in complexity from daily milk yield recording to measurement of specific attributes (e.g. fat content or progesterone) within milk at each milking. The main objectives of PDF are maximizing individual animal potential, early detection of disease, and minimizing the use of medication through preventive health measures (Schulze *et al.*, 2007). Precision Dairy Farming is inherently an interdisciplinary field incorporating concepts of informatics, biostatistics, ethology, economics, animal breeding, animal husbandry, animal nutrition, and process engineering (Spilke and Fahr, 2003).

Boehlje *et al.* (1999) described these technologies as 'intervention technologies' which allow dairy producers to intervene with an appropriate adjustment when actual performance differs from potential performance. The intervention may occur automatically or may involve a human decision as to whether or not to intervene. Van Asseldonk (1999) used the term 'information technology' defined as the 'diverse class of applications which include all the hardware and software used by farmers to capture data, to transform data into information, to help in decision making, to establish communication of data/information, internally as well as externally, and to perform production tasks'. Devir *et al.* (1993) characterized these systems as 'dairy control and management systems' which strive toward achieving maximum production with minimum resources. Finally, a sub-selection of these technologies has been referred to as biosensors. Tothill (2001) defined biosensors as 'analytical devices incorporating a biological material, a biologically-derived material or biomimic as the recognition molecules, utilized in conjunction with or integrated within a physiochemical transducer or transducing microsystems'. An electrical signal, proportional to the concentration of the analyte typically delivers the desired information. Many biosensors use naturally available antibodies, enzymes, cell receptors, or nucleic acids as sensing receptors (Tothill, 2001).

3. Potential opportunities

Perceived benefits of PDF technologies in production agriculture include increased efficiency, reduced costs, improved quality, minimized environmental impacts, and more systematically produced, biologically based, specific attribute raw materials (Boehlje *et al.*, 1999). These technologies are likely to have the greatest impact in the areas of health, reproduction, and quality control (De Mol, 2000). Realized benefits from data summarization and exception reporting are anticipated to be higher for larger herds, where individual animal observation is more challenging and less likely to occur (Lazarus *et al.*, 1990). As dairy operations continue to increase in size, PDF technologies become more feasible because of increased reliance on unskilled labor and the ability to take advantage of economies of size related to technology adoption.

A PDF technology allows dairy producers to make more timely and informed decisions, resulting in better productivity and profitability (Van Asseldonk *et al.*, 1999). Real time data can be used for monitoring animals and creating exception reports to identify meaningful deviations. In many cases, dairy management and control activities can be automated (Delorenzo and Thomas, 1996). Alternatively, output from the system may provide a recommendation for the manager to interpret (Pietersma *et al.*, 1998). Some PDF technologies may be used to control a process by measuring and comparing performance to pre-established standards. Process controls can be described as curative or preventive. With curative control, when actual performance deviates from standards, corrective action is taken. In preventive control, expected future performance, based on current performance trends, is compared to future standards. If there is a difference between the two, corrective actions are taken to ensure that desired future performance is obtained (Huirne, 1990). Devir *et al.* (1993) listed four main types of process control for dairy operations: (1) nutrition, encompassing individual feeders, mixing equipment, and water supplies, (2) production control through milking parlors and scales to weigh cows, (3) animal health control, including mastitis and body temperature, (4) fertility, estrus detection and parturition control, and (5) environmental control.

Historically, dairy producers have used experience and judgment to identify outlying animals. While this skill is invaluable and can never be fully replaced with automated technologies, it is inherently flawed by limitations of human perception of a cow's condition. Often, by the time an animal exhibits clinical signs of stress or illness, it is too late to intervene. These easily observable clinical symptoms are typically preceded by physiological responses evasive to the human eye (e.g. changes in temperature or heart rate). Thus, by identifying changes in physiological parameters, a dairy manager may be able to intervene sooner. Technologies for physiological monitoring of dairy cows have great potential to supplement the observational activities of skilled herdspersons, which is especially critical as more cows are managed by fewer skilled workers (Hamrita *et al.*, 1997). Fetrow *et al.* (1987) proposed that animal monitoring is an important part of a herd health program because it identifies problem areas and focuses priorities, shifts attention away from daily distractions toward long term needs, and identifies trends that might otherwise go unnoticed. Information obtained from PDF technologies is only useful if it is interpreted and utilized effectively in decision making (Carmi, 1992; Fetrow *et al.*, 1987; Pietersma *et al.*, 1998; Spilke and Fahr, 2003; Verstegen *et al.*, 1992). Integrated, computerized information systems are essential for interpreting the mass quantities of data obtained from PDF technologies (Pietersma *et al.*, 1998). This information may be incorporated into decision support systems designed to facilitate decision making for issues that require compilation of multiple sources of data. A monitoring system that measures information that a dairy producer already collects is much more apt to be successfully implemented (Fetrow *et al.*, 1987).

4. Description of technologies

The list of PDF technologies used for animal status monitoring and management continues to grow (Spilke and Fahr, 2003; Tothill, 2001). Because of rapid development of new technologies and supporting applications, PDF technologies are increasingly more feasible (Van Asseldonk *et al.*, 1999). Many PDF technologies are already being utilized by dairy producers including daily milk yield recording, milk component monitoring (e.g. fat, protein, and SCC), pedometers, automatic temperature recording devices, milk conductivity indicators, automatic estrus detection monitors, and daily body weight measurements. Despite its seemingly simplistic nature, the power of accurate milk weights should not be discounted in monitoring cows, as it is typically the first factor that changes when a problem develops (Philpot, 2003). Other theoretical PDF technologies have been proposed to measure jaw movements, ruminal pH, reticular contractions, heart rate, animal positioning and activity, vaginal mucus electrical resistance, feeding behavior, lying behavior, odor, glucose, acoustics, progesterone, individual milk components, color (as an indicator of cleanliness), and respiration rates. Unfortunately, the development of technologies tends to be driven by availability of a technology, transferred from other industries in market expansion efforts, rather than by need (Hogeveen and Ouweltjes, 2003). A portion of PDF technologies measure variables that could be measured manually, while the remaining technologies measure variables that could not have been obtained previously (Pietersma *et al.*, 1998).

5. Adoption barriers

Despite widespread availability, adoption of these technologies in the dairy industry has been relatively sparse thus far (Eleveld *et al.*, 1992; Gelb *et al.*, 2001; Huirne *et al.*, 1997). In fact, agricultural adoption of on-farm software packages, as a whole, has been much lower than predicted (Rosskopf and Wagner, 2003). The majority of information management systems available and utilized by many dairy producers have analytical capabilities that are underutilized. In practicality, their use is often limited to creating production tables, attention lists, and working schedules (Van Asseldonk, 1999). Perceived economic returns from investing in a new technology are likely the main factor influencing PDF technology adoption. Additional factors impacting technology adoption include degree of impact on resources used in the production process, level of management needed to implement the technology, risk associated with the technology, institutional constraints, producer goals and motivations, and having an interest in a specific technology (Dijkhuizen *et al.*, 1997; Van Asseldonk, 1999). Characteristics of the primary decision maker that influence technology adoption include age, level of formal education, learning style, goals, farm size, business complexity, increased tenancy, perceptions of risk, type of production, ownership of a non-farm business, innovativeness in production, overall expenditures on information, and use of the technology by peers and other family members. Eleveld *et al.* (1992) demonstrated that technology adoption is improved when the technology fits within the normal daily work patterns of the personnel who will be using it. Discrepancies are much larger for technologies where the producer has to interpret information produced by the system (Van Asseldonk, 1999). Farm operations with more specialization of labor are more likely to successfully adopt information technology (Eleveld *et al.*, 1992). The most progressive producers will adopt those new technologies that appear to be profitable. When a proven technology is not adopted, the operation observes a lost opportunity cost that may lead to a competitive disadvantage (Galligan, 1999).

References

Boehlje, M.D., Hofing, S.L. and Schroeder, R.C. (1999). Farming in the 21st century. Department of Agricultural Economics, Purdue University, West Lafayette, IN, USA.

Carmi, S. (1992). The performance of an automated dairy management data-gathering system. Proceedings of the International Symposium on Prospects for Automatic Milking. European Association for Animal Production, Wageningen, the Netherlands, pp. 346-352.

De Mol, R.M. (2000). Automated detection of oestrus and mastitis in dairy cows. PhD thesis, Wageningen University, Wageningen, the Netherlands, 177 pp.

Delorenzo, M.A. and Thomas, C.V. (1996). Dairy records and models for economic and financial planning. Journal of Dairy Science 79(2): 337-345.

Devir, S., Renkema, J.A., Huirne, R.B.M. and Ipema, A.H. (1993). A new dairy control and management system in the automatic milking farm: basic concepts and components. Journal of Dairy Science 76(11): 3607-3616.

Dijkhuizen, A.A., Huirne, R.B.M., Harsh, S.B. and Gardner, R.W. (1997). Economics of robot application. Computers and Electronics in Agriculture 17(1): 111-121.

Eastwood, C., Chapman, D. and Paine, M. (2004). Precision dairy farming-taking the microscope to dairy farm management.

Eleveld, B., Huirne, R.B.M., Dijkhuizen, A.A. and Overbeek, G. (1992). Users in search of farm computer information technology: what do farmers want or need? Proceedings of Farm Computer Technology in Search of Users: 4th International Congress for Computer Technology in Agriculture, Paris-Versailles, France, pp. 27-32.

Fetrow, J., Henry, E.T., Harrington, B. and Anderson, K.L. (1987). Dairy herd health monitoring. Part I. Description of monitoring systems and sources of data. Compendium on Continuing Education for the Practicing Veterinarian 9: 389-398.

Galligan, D.T. (1999). The economics of optimal health and productivity in the commercial dairy. Scientific and Technical Review of the Office International des Epizooties (Paris) 18(2): 512-519.

Gelb, E., Parker, C., Wagner, P. and Rosskopf, K. (2001). Why is the ICT adoption rate by farmers still so slow? Proceedings ICAST, Vol. VI, Beijing, China, pp. 40-48.

Hamrita, T.K., Hamrita, S.K., Van Wicklen, G., Czarick, M. and Lacy, M.P. (1997). Use of biotelemetry in measurement of animal responses to environmental stressors.

Hogeveen, H. and Ouweltjes, W. (2003). Sensors and management support in high-technology milking. Journal of Animal Science 81, Suppl. 3: 1-10.

Huirne, R. (1990). Basic concepts of computerised support for farm management decisions. European Review of Agricultural Economics 17: 69-84.

Huirne, R.B.M., Harsh, S.B. and Dijkhuizen, A.A. (1997). Critical success factors and information needs on dairy farms: the farmer's opinion. Livestock Production Science 48(3): 229-238.

LaDue, E., Gloy, B. and Cuykendall, C. (2003). Future structure of the dairy industry: historical trends, projections, and issues. Cornell Program on Agricultural and Small Business Finance (R.B. 2003-01).

Lazarus, W.F., Streeter, D. and Jofre-Giraudo, E. (1990). Management information systems: impact on dairy farm profitability. North Central Journal of Agricultural Economics 12(2): 267-277.

Philpot, W.N. (2003). Role of technology in an evolving dairy industry. Proceedings of the Southeast Dairy Herd Management Conference, Macon, Georgia, pp. 6-14.
Pietersma, D., Lacroix, R. and Wade, K.M. (1998). A framework for the development of computerized management and control systems for use in dairy farming. Journal of Dairy Science 81(11): 2962-2972.
Rollin, B.E. (2004). Annual meeting keynote address: animal agriculture and emerging social ethics for animals. Journal of Animal Science 82(3):955-964.
Rosskopf, K. and Wagner, P. (2003). Requirements for agricultural software and reasons for adoption constraints-results of empirical studies. In: Proc. EFITA 2003 Conference, Debrecen, Hungary, pp. 651-657.
Schulze, C., Spilke, J. and Lehner, W. (2007). Data modeling for Precision Dairy Farming within the competitive field of operational and analytical tasks. Computers and Electronics in Agriculture 59(1-2): 39-55.
Spilke, J. and Fahr, R. (2003). Decision support under the conditions of automatic milking systems using mixed linear models as part of a precision dairy farming concept. Proceedings of the EFITA Conference, Debrecen, Hungary, pp. 780-785.
Tothill, I.E. (2001). Biosensors developments and potential applications in the agricultural diagnosis sector. Computers and Electronics in Agriculture 30(1-3): 205-218.
Van Asseldonk, M.A.P.M. (1999). Economic evaluation of information technology applications on dairy farms. PhD thesis, Wageningen University, Wageningen, the Netherlands, pp. 123.
Van Asseldonk, M.A.P.M., Jalvingh, A.W., Huirne, R.B.M. and Dijkhuizen, A.A. (1999). Potential economic benefits from changes in management via information technology applications on Dutch dairy farms: a simulation study. Livestock Production Science 60(1): 33-44.
Verstegen, J.A.A.M., Huirne, R.B.M. and Dijkhuizen, A.A. (1992). Research outline to determine the profitability of management information systems. Farm computer technology in search of users: 4th International Congress for Computer Technology in Agriculture, Paris-Versailles, France, pp. 233-237.

Sponsor papers

Nutrition, sensors and genetics – a data driven smart feeding approach by Agrifirm

C. Bahr*, E. Bruininx and A. Van den Belt
Agrifirm Innovation Center, Landgoedlaan 20, 7325 AW Apeldoorn, the Netherlands; c.bahr@agrifirm.com

Abstract

Agrifirm a Dutch cooperative with about 17,000 farmers and horticulturists and a yearly turnover of 2.4 billion Euro summarises its vision under the name Dutch Farming, whereby all products, concepts and the innovations that are introduced are focused on three themes: 'Smart feeding', 'Healthy soil, plants and animals' and 'Strength in the chain'. These themes are strongly interlinked and can hardly be taken apart. In this context the collection, analysis, interpretation and sharing of (sensor) data within and in between themes play a major role in Agrifirm's vision on providing tailored and knowledge driven solutions for its clients, including many dairy farmers. The ongoing developments on 'Big Data', 'Internet of Things', 'Sensor Technology' etc. in dairy farming are great opportunities but at the same time big challenges for Agrifirm and its customers.

1. Introduction

Koninklijke Coöperatie Agrifirm U.A. is a Dutch cooperative founded in 1892, with about 17,000 farmers and horticulturists and a yearly turnover of 2.4 billion Euro (Figure 1). The company is market leader in the Netherlands and is globally engaged in the production and supply of products and services for feeding animals and cultivating crops.
The company summarises its vision under the name Dutch Farming whereby all products, concepts and the innovations that are introduced are focused on three themes:

- Smart feeding: The tailored nutrition and protection of animals and crops produces the best possible performance in terms of yield, quality, sustainability and return.
- Healthy soil, plants and animals: The highest resilience of soil, crop and animal is the basis of a healthy and profitable business.
- Strength in the chain: Cooperation in the chain creates dynamic, demand-driven innovation and tailored solutions. Not chain power, but chain strength (Agrifirm, 2015).

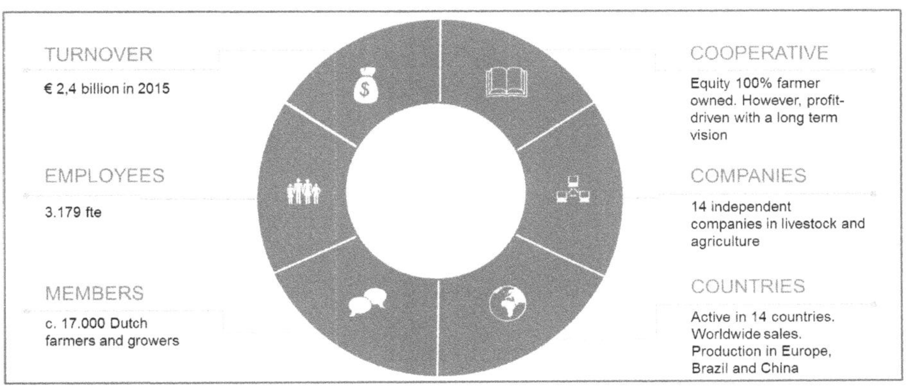

Figure 1. Overview – Agrifirm facts and figures.

Looking at the most recent developments in Europe in general and in the Netherlands in particular, the dairy cattle farming sector experiences a difficult period. The abolition of the milk quota system resulted in an increase in the dairy herd in the Netherlands. To regulate this growth, the government developed additional phosphate-related laws and regulations. Lower milk prices combined with the high investment costs and the accumulation of laws and regulations depressed the mood in this sector, considerably (Agrifirm, 2015). As a logic consequence, the pressure increases to come up with innovative products and services that allow farmers to meet these challenges, overcome economically difficult periods and be prepared for future competition. Agrifirm, producing cattle feed of about 1,218,000 tons per year and providing feed advise to dairy farmers, is convinced that the collection, analysis, interpretation and sharing of animal based (sensor) data must be incorporated in the products and services of the company. This will facilitate tailored and knowledge driven feed and feeding solutions that help farmers to stay in business.

2. Smart feeding

No question – feed is a fundamental requirement for the survival of every living organism. So it is for dairy cattle. However, it is a fact, that we don't expect a dairy cow to only maintain her body functions, we expect her to produce eight to twelve tons of milk per lactation, stay healthy and fertile and last as many lactations as possible. As a consequence we must feed and comfort her accordingly. Feeding dairy cattle is a complex matter. It does not only involve nutritive aspects and feed quality but also feeding management depending on housing, milking systems, stocking density and the farmer's own feeding philosophy and strategy. Smart feeding should be based on both high quality feed and feed management tailored to the necessities of each farm and, ideally, to the needs of each individual animal at any time in any development stage. Therefore, Agrifirm believes that besides the provision of high quality feed a comprehensive and successful smart feeding approach is also determined by incorporating the following:

1. sensor data and IT;
2. genetics;
3. animal performance at each development stage.

Agrifirm does not claim this approach to be a fully complete concept. However, the overview drafted out in Figure 2 provides a clear picture of which type of information is available and should be used to optimise feeding strategies and feed products.

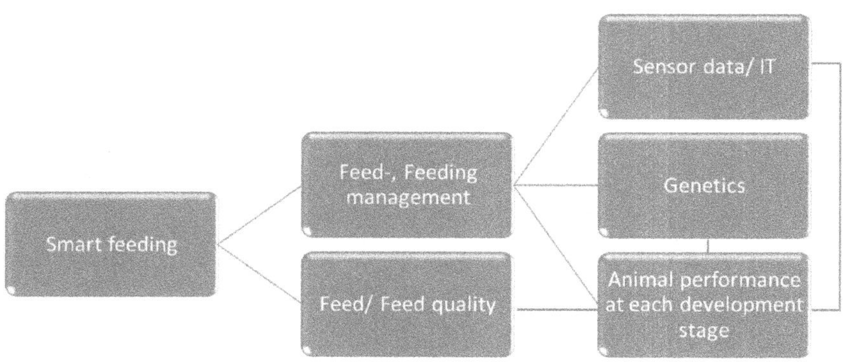

Figure 2. Components of the smart feeding approach.

3. Make use of animal based sensor data and IT

Many kinds of information from dairy cows is nowadays collected via sensor technology. The use of sensor technology and related Big Data analysis is well established in dairy industry and has reached a stage, where important processes, such as milking, are more and more automated. This allows a monitoring of dairy cows with positive effects on health and productivity. Also with respect to feeding many developments have been taken place. For example, the automation of concentrate feeding and milking enabled applications for dynamic individual concentrate feeding. Here, sensor data deliver the information to estimate the individual dynamic milk yield response to concentrate intake and milking interval. Based on certain parameters, a control algorithm calculates daily individual optimal settings, to maximize gross margin milk returns minus concentrate costs (André *et al.*, 2009). This feeding concept could be established in practise and helps farmers to optimise concentrate feeding and costs based on milk yield of individual cows. However, individual concentrate feeding is only one element in the complex world of dairy cow feeding and the full potential of all available data to optimise feeding strategies has not been used yet. That is one reason why Agrifirm engages in the Dutch project Smart Dairy Farming (SDF).

Smart Dairy Farming supports dairy farmers in the care of the individual cows, with the goal of good health, a longer life and high lifetime production. For this, Smart Dairy Farming develops sensors, decision support models, process descriptions and advisory products. With these tools dairy farmers are able to make better choices in the area of health, fertility and nutrition (http://www.smartdairyfarming.nl/nl, 2016).

The backbone of the project is the so called 'InfoBroker' a data hub or data highway that connects dairy farmers with other chain partners such as technology providers, milk processors, feed provider, etc. The InfoBroker serves as a technical component which unbundles data from different sources meaning a decoupling of the 1-to-1 relationship between the measuring apparatus / data storage and model (Figure 3). This will help in developing applications that can run real-time on a broad

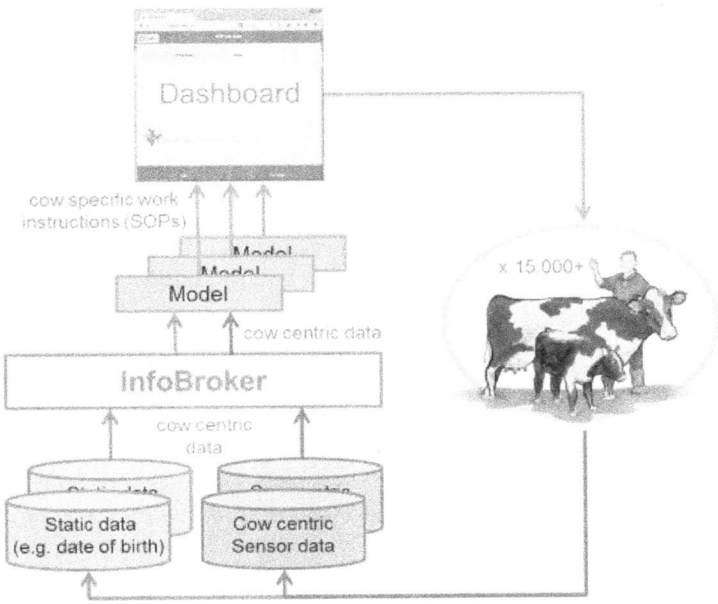

Figure 3. Schematic overview of data flow in smart dairy farming.

range of farms (connected to the InfoBroker) and it facilitates the collaboration between different chain parties in application development by making use of common data sources.

The ongoing development in sensor and information technology offer great potential to connect different types of animal related information from different production processes and fields. However, looking to daily dairy farm practise, it becomes obvious that this potential remains unused to a big extend. If we think about optimised feeding of a cow, we should feed her according to her individual needs because every animal is a complex, individual, time-varying and dynamic organism, a CITD system (Berckmans, 2004). Nowadays, the individual nutritive need of a dairy cow is mainly considered during milk production but the lifetime before and in between milk production is excluded from this individual approach. Agrifirm is convinced that smart feeding models are most beneficial, when they are individualised and optimised throughout the whole lifetime of a cow. Additional value can be gained when sensor information on animal performance (growth/ milk yield) and feed intake in different animal development stages (calf, heifer, milking cow) are connected. A crucial step towards an optimised feeding strategy covering the lifespan of a dairy cow is, for instance, to pay more attention to the calf rearing process.

4. Be aware that a calf becomes a cow (animal performance at each development stage) and consider genetics

The lifetime prior to the animal's productive life (young stock rearing) has been excluded from IT driven developments to a big extend. Although it is well known that close to 10% (Gulliksen *et al.*, 2009; Pellikan, 2014; Silva Del Río *et al.*, 2007) of calves do not survive the rearing period and that rearing young stock is crucial for the later lifetime production of dairy cows, farmers often pay less attention to the development stages of young stock. It seems that dairy farmers tend to underestimate the importance of calf mortality or morbidity, partly because they do not have good information on how the performance of calves affects their later performance as lactating cows (Rushen and De Passillé, 2013) as well as the relatively long 'return on investment' period. However, rearing young stock in good and efficient fashion is crucial. Knowing the animal's needs and monitoring their developments at any moment are important elements contributing to a healthy and highly productive life of dairy cows throughout lactations. Nowadays dairy cows in the Netherlands achieve on average 3.8 lactations (Zijlstra *et al.*, 2013), which is already higher compared to other countries such as the U.S. with 2.6 lactations (De Vries, 2013). However, the longer a cow is productive the higher the economic value of the animal and everything starts with a good raised calf.

Although, we may think that sensor information on feeding (feed intake, feeding behaviour, feed composition, etc.) and animal performance throughout the different development stages seem appropriate and sufficient for developing and/or optimising feeding models, we cannot neglect the impact of genetics on performance measures such as weight, growth and milk yield because they are moderately and highly heritable traits (Gosey, 2016; Massey and Vogt, 2016; Veerkamy and Brotherstone, 1997). A combination of genetic data and sensor based data to estimate the nutritional needs of individual cows throughout their whole lifetime seems a logic concept for smart feeding but it can only be realised with a sufficient data availability and quality. Sensors and related IT will therefore play a key role if we want to realise this ambitious vision.

References

Agrifirm (2015). Annual Report. Koninklijke Coöperatie Agrifirm UA, Apeldoorn, the Netherlands.

André, G., Bleumer, E.J.B. and Van Duinkerken, G. (2009). Evaluation of an application for dynamic feeding of dairy cows. In: Precision livestock farming 2009. Wageningen Academic Publishers, Wageningen, the Netherlands, pp. 25-32.

Berckmans, A. (2004). Automatic on-line monitoring of animals by precision livestock farming. Keynote in the ISAH conference 'Animal Production in Europe': the way forward in a changing world. October 11-13, 2004. Saint-Malo, France, pp. 27-31.
Gosey, J. (2016). Selecting for carcass marbeling and muscling benefits and pitfalls. Available at: http://tinyurl.com/zhhenhj.
Gulliksen, S.M., Lie, K.I., Løken, T. and Østerås, O. (2009). Calf mortality in Norwegian dairy herds. Journal of Dairy Science 92(6): 2782-2795.
Rushen, J. and De Passillé, A.M. (2013). The importance of improving cow longevity. In: Conference Proceedings of the cow longevity conference. August 28-29, Tumba, Sweden, pp. 5-21.
Massey, J.W. and Vogt, D.W. (2016). Heritability and its use in animal breeding. Available at: http://extension.missouri.edu/p/G2910.
Pellikan, F. (2014). Grote en groeiende bedrijven hebben niet minder levend geboren kalveren van vaarzen. Veeteelt September 2014: 10-14. Available at: http://tinyurl.com/z8gkx9m.
Silva Del Río, N., Stewart, S., Rapnicki, P., Chang, Y.M. and Fricke, P.M. (2007). An observational analysis of twin births, calf sex ratio, and calf mortality in Holstein dairy cattle. Journal of Dairy Science 90(3): 1255-1264.
Smart Dairy Farming (SDF) (2016). Doelen. Available at: http://www.smartdairyfarming.nl/nl
Veerkamy, R.F. and Brotherstone, S. (1997). Genetic correlations between linear type traits, food intake, live weight and condition score in Holstein Friesian dairy cattle. Journal of Animal Science 64: 385-392.
De Vries, A. (2013). Cow longevity economics: the cost benefit of keeping the cow in the herd. In: Conference Proceedings of the cow longevity conference. August 28-29, Tumba, Sweden, pp. 22-52.
Zijlstra, J., Boer, M., Buiting, J., Colombijn-Van der Wende, K. and Anne-Andringa, E. (2013). Routekaart levensduur, rapport 668, Wageningen UR Livestock Research, Wageningen, the Netherlands.

Recent advances in mastitis detection with automatic milking systems

J.R. Baines
Fullwood Ltd, Grange Road, Ellesmere, Shropshire, SY12 9DF, United Kingdom; jbaines@fullwood.com

Abstract

Mastitis, together with lameness and failure to reproduce, is one of the main production diseases of dairy cows. Presence of high levels of mastitis pathogens, in raw milk at least, can constitute a risk to human health and is regarded as an indication of less than adequate milking methods. Recent survey results indicate that annual incidence remains, stubbornly, at an average of 47-65 clinical cases per 100 cows per year. Dairy Hygiene Legislation requires that 'that milk from each animal is checked for organoleptic (taste, colour, feel, smell) or physico-chemical abnormalities by the milker or a method achieving similar results and that milk presenting such abnormalities is not used for human consumption'. The clear implication is that, with automatic milking systems where a human milker is not present, sensors are provided, used as a basis to divert abnormal milk and that the 'gold standard' is that which can be achieved by a human. A patent search on the words 'mastitis sensor automatic milk' reveals that there are at least 1,205 patents already applied for or granted. There are also many published scientific papers discussing a myriad of possibilities for automated sensors for mastitis. In spite of this, the range of technologies being used in practice is relatively limited. This paper sets out to discuss the application of sensor technology as a tool for compliance with legislation and to manage udder health.

1. Mastitis – the disease and symptoms

Mastitis is not one disease but can be caused by a large number of pathogens. Clinical mastitis signs vary from seeing a few clots in the milk with perhaps some swelling of the infected quarter to severe signs which include swollen quarter or whole udder. The cow may have a high temperature, fever, loss of appetite, and be dehydrated. In severe cases, the disease can result in the death of the animal. Sub clinical mastitis is when there is bacterial infection present with no clinical signs. The cow could have a raised somatic cell count (number of white blood cells/ml of milk) and there may be bacteria in milk cultures. However, legislation does not demand that such milk is rejected.

It is generally accepted in the UK that a somatic cell count of >200,000/ml of milk is an indication that a cow or quarter is infected. Care has to be taken when using this figure as a cow could have a cell count just below 200,000 and have ¾ with a cell count of less than 50,000 and ¼ with a cell count well over 200,000/ml. Where a herd risks being penalised on the basis of somatic cell counts, rejection of milk from cows known to have a high somatic cell count can be a useful milk quality management tool. The cost of that milk must be weighed against the likely penalty.

One significant trend is the change in the relative proportions of contagious and environmental mastitis over the last 40 years (Figure 1). Successful implementation of mastitis control strategies has reduced contagious infection level to less than 20% of the total.

2. Sensor technology for detecting mastitis

There are a large number of patents relating to the use of mastitis sensors on automatic milking units. In the wider context, a patent search on 'mastitis sensor' indicates that there are at least 2,695

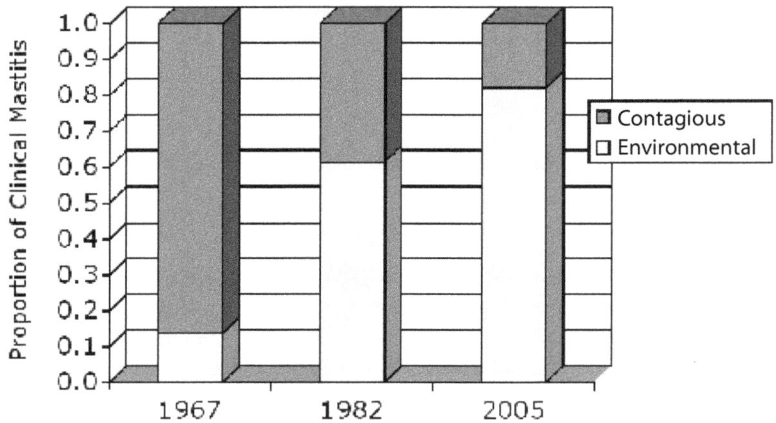

Figure 1. Contagious and environmental clinical infections 1967-2005.

patents applied for or granted. It is not practical to list all of the technologies which have been the subject of these applications. Some of the technologies proposed as possible sensors include:

- milk lactate (Davis *et al.*, 2004);
- acute phase proteins milk amyloid A (MAA) (Eckersall *et al.*, 2001);
- NAGase (Mottram *et al.*, 2000);
- electrical conductivity of milk (Hamann and Zecconi, 1998; Mein *et al.*, 2004);
- colour/blood (Reinemann and Helgren, 2004; Whyte *et al.*, 2004);
- somatic cell count (Reinemann and Helgren, 2004);
- viscosity, California mastitis test (Whyte *et al.*, 2004).

When considering the application, relevance and relative merits of sensor technology in respect of mastitis detection, it is useful to refer to the mastitis cause-effect spectrum explained by Claycomb *et al.* (2002).

Whilst very useful from the point of view of identifying the level of pathogenic infection and corresponding immune responses, those sensors aimed at the stages prior to the onset of clinical symptoms prompt the 'so what' question. In other words, what action could be taken and what are the costs and benefits, if any. It is highly arguable that the Response Options identified at the PreClinical Mastitis stage should be being implemented anyway.

At all stages, an assessment has to be made in respect of the cost and corresponding benefit, if any, of intervention. It is evident from Table 1 that those effects occurring earlier in the cause effect spectrum are much more likely to be associated with 'self-cure' where therapeutic intervention may not be justified.

In the context of automatic milking, the last thing an operator wants or needs is to be burdened by mastitis alerts where there are no visible symptoms, where intervention may not to be justified or has a negative cost benefit. However, there is a contrasting argument, from the point of view of udder health management, that early warning of impending problems can be useful. This highlights the fact that it is essential to educate users in respect of the correct use of sensor technologies.

3. Sensors in practical use for mastitis detection

In 2004, in a paper with a similar brief, Reinemann and Helgren (2004) reported on the basis of European and American studies that electrical conductivity and milk colour were the most widely used on-line milk sensing methods. He also noted that, deviation in milk yield and milking interval

Table 1. Mastitis cause effect spectrum (Claycomb *et al.*, 2002).

Cause ⟶ Effect

Reducing likelihood of self-cure ⟶

Biological cause	Bacteria enter teat	Defence system triggered	Immune response	Epithelial tissue damage	Blood/ milk barrier breached	Serious infection/ illness	Untreated acute infection
Physiological effect	Bacterial metabolites, e.g. lactate	Acute phase protein MAA	SCC rise due to white blood cells	SCC rise due to dead cells NAGase	Increased ions, conductivity increase	Swelling, clots, yield loss	Lower activity, temp rise sick cow
No. of milkings after infection (Hillerton, 2000)	Bacteria in milk *Streptococcus aureus* 1.1 *Streptococcus uberis* 2.3		10 fold rise in SCC *S. aureus* 2.3 *S. uberis* 3.1		Conductivity rise *S. aureus* 2.7 *S. uberis* 3.5	Clots in milk *S. aureus* 4.2 *S. uberis* 5.2	
Condition	Pre-clinical mastitis		Sub-clinical mastitis			Clinical mastitis	
Response options	Preventative attention to hygiene and teat preparation		Preventative to curative strip quarters to remove pathogens non-antibiotic treatment			Curative antibiotic treatment diversion of milk	

MAA = milk amyloid A; SCC = somatic cell count.

are widely used supporting diagnostic techniques. He also predicted that cow side measurements of somatic cell count were likely to be applied to automatic milking in the future.

Whilst, superficially, the situation would seem to be largely similar today, a number of significant observations can be made.

Most sensor technology implemented today has been driven by the need to detect, or at least predict, onset of clinical mastitis symptoms.

An ideal sensor should be non-invasive, cleanable, avoid damage to the milk and not require the use of reagents, particularly where those reagents may be dangerous to the environment or difficult to dispose of. This point has ensured that sensor technologies such as electrical conductivity and optical sensing methods, including colour, continue to be attractive. The use of such sensors also strikes a balance between effectiveness and low running cost.

Although the use of electrical conductivity has been discussed, supported and condemned for many years, the main point to note is that there have been significant developments in the refinement of its use. Significantly, Hamman and Zecconi (1998) concluded that 'within cow comparison between quarters seems to be the best way to use conductivity measurements' and 'measurement of electrical conductivity in milk requires a high degree of standardisation'. These conclusions indicate that the main challenge with electrical conductivity measurement is in standardisation of the measurement technique and interpretation of the physical data.

Mein *et al.* (2004), using temperature correction and more sophisticated algorithms to take account of milking interval and other factors, have shown a sensitivity of 92% and specificity of 95% where a true positive was defined as the presence of a major pathogen. A further study was carried out by the same authors on 4 herds and reported sensitivity of 80% and specificity of 93%. In that study a true positive was defined as a cow with visible clots in the foremilk of one or more quarters at a monthly herd visit, reported in the same paper reported. True negatives had no visible signs of clots in their milk and low CMT (scores of 1 or 2). The sensitivity and specificity data achieved are favourably comparable with that reported for human observation by Hillerton (2000).

The predicted use of cow-side somatic cell counting or estimating is not regarded as an indicator of abnormal milk and has not, by any means, been universally adopted (Rasmussen, 2003; Reinemann and Helgren, 2004).

The use of optical technology to measure various milk components, including blood, is already showing promise whilst still satisfying the ideal of being non-invasive and not requiring reagents. The final comment from Reinemann and Helgren (2004), that inputs from a number of sensors, including milk composition, animal behaviours and milking characteristics need to be co-ordinated via a 'smart' system, is as true now as it was then. This is particularly relevant when considering the predominance of environmental pathogen related mastitis. Often the first symptoms can be failure to attend the automatic milking machine and/or loss of milk production. For many farmers, one of the most important sensors is the list of cows not attending the machine.

4. Conclusions

An enormous array of technologies are possible for detecting mastitis related characteristics within milk. The application of such sensing technology is driven by legal requirements to detect and divert abnormal milk. Other technologies must provide a very clear benefit at a cost which is recoverable. Sensors should, ideally, not damage milk, nor require discard of milk and not require reagents. The most important factor in sensor technology for mastitis is an intelligent management system to combine the output of various sensors and permit 'management by exception'.

References

Claycomb, R., Mein, G., Whyte, D. and Wilson, A. (2002). On-line milk component sensing. In: Proceedings of the 33rd ICAR Session, Interlaken, Switzerland.

Davis, S.R., Farr, V.C., Prosser, C.G., Nicholas, G.D., Turner, S.-A., Lee, J. and Hart, A.L. (2004). Milk L-lactate concentration is increased during mastitis. Journal of Dairy Research 71: 175-181.

Eckersall, P.D., Young, F.J., McComb, C., Hogarth, C.J., Safi, S., Weber, A., McDonald, T., Nolan, A.M. and Fitzpatrick, J.L. (2001). Acute phase proteins in serum and milk from dairy cows with clinical mastitis. Veterinary Records 148: 35-41.

Hamann, J. and Zecconi, A. (1998). Evaluation if the electrical conductivity of milk as a mastitis indicator. Bulletin of the International Dairy Federation, No. 334, pp. 5.

Hillerton, J.E. (2000). Detecting mastitis cow-side. National Mastitis Council Annual Meeting 39: 48-53.

Mein, G.A., Sherlock, R.A. and Claycomb, R.W. (2004). Making sense of electrical conductivity sensing. Automatic Milking. Wageningen Pers, Wageningen, the Netherlands.

Mottram, T., Hart, J. and Pemberton, R. (2000). Biosensing techniques for detecting abnormal and contaminated milk. In: Proceedings of the International Symposium, Robotic Milking. Wageningen Press, Wageningen, the Netherlands, pp. 108-113.

Rasmussen, M.D. (2003). Definition of normal and abnormal milk at time of milking. EU Project Automatic Milking.

Reinemann, D.J. and Helgren, J.M. (2004). Sensing issues for automatic milking. In: Proceedings of the ASAE/CSAE Annual International Meeting.

Whyte, D.S., Johnstone, P.T., Claycomb, R.W. and Mein, G.A. (2004). On-line sensors for earlier, more reliable mastitis detection. In: Proceedings of the British Mastitis Conference, Stoneleigh, UK, pp. 61-67.

Whyte, D., Walmsley, M., Liew, A., Claycomb, R. and Mein, G. (2005). Chemical and rheological aspects of gel formation in the California mastitis test. Journal of Dairy Research 72(1): 115-121.

Precision livestock farming to unburden the dairy farmer

A.J. Van der Kamp* and R. Van der Tol
Lely Innovations, Cornelis van der Lelylaan 1, 3147 PB, Maassluis, the Netherlands; avanderkamp@lely.com

Abstract

A new generation of dairy equipment and subsequent standard operating procedures (SOPs) promise to unburden the dairy farmer. This can be done either physically, by pre-defined actions, or mentally, by only giving the dairy farmer information that requires an action. In addition, when it comes to repetitive farm-management actions, decisions can be made for the dairy farmer based on standardized protocols. In research many authors focus on precision livestock technology and decision support systems. The quality of these varies a lot, e.g. accuracy ranges from 10 to 90%. Although most technologies show promise, their adoption is slow. For example, the electrical conductivity of milk, available for over thirty years, is now adopted by around 70% of Dutch farmers. How can science be brought into practice? The challenge is to add value to a dairy farm with precision livestock technology. For example, electrical conductivity of milk in itself does not bring much value, but the alert that a cow has an increased risk of udder infection does. The next step is that the farmer is provided with a pre-defined protocol: how to handle conductivity alerts. Here value is added by creating a farm-management support tool. In other words, a common attention is personalized by using SOPs depending on the farmer's strategy and goals, allowing the next step in dairy farming to be reached. The vast majority of dairy farmers are not looking for precision livestock technology. Farmers search for a way to make their work easier and smarter. Ideally an automatic decision support (ADS) system is used, providing mental relief. For the successful implementation of ADS, a problem-oriented rather than a technology-driven approach is required. Looking at the needs of the dairy farmer, on-farm implementation, farm-management routines and interfacing are just as important as the technology. Precision livestock technology is not a goal in itself: it is the path towards unburdening the dairy farmer to provide a bright future in dairy farming.

Keywords: decision support, easy to use, sensor technology

1. Sensor technology

For precision livestock farming, sensor technology is an important driver. Sensors make it possible to take individual measurements per cow. Based on this data, cow-specific actions can be performed. Simple sensors like milk yield measurements have been around for a long time, but recently the sensor-technology range has been expanding and systems like in-line SCC measurements and rumination activity sensors are now available.
Steeneveld and Hogeveen (2015) found a large variation in uptake of sensor technology on dairy farms. This ranged from 93% uptake for electrical conductivity sensors to 9% uptake for rumination activity sensors on farms with an automatic milking system (AMS) (Table 1). An AMS offers an ideal platform for implementing sensor technology, as many cows during the day are milked by it. On the other hand, the sensor technology on the AMS is also more important as it replaces the senses of the dairy farmer.

2. Decision support system

Relatively little is known about decision supports systems in the dairy sector, probably due to the biological nature of farms. It is easier to automate a factory than a farm with live animals. However,

Table 1. Overview of sensor systems used on farms with an automatic milking system (AMS) (Steeneveld and Hogeveen, 2015).

Type of sensor system on the farm	% of AMS farms using the sensor system
Milk color sensor	60
SCC sensor	17
Electrical conductivity	93
Weighing platform	27
Rumination activity sensor	9
Activity meters for dairy cows	41
Fat and protein sensor	20
Milk temperature sensor	46

many sources can be used to implement a decision support system (DSS), e.g. sensor data, farmers' input, advisors' inputs, etc. The options are limitless.

However, in practice there are also multiple definitions possible for what a farmer perceives as a DSS. Some require only the information from individual sensors to draw their own conclusions, while others desire a short list of cows with a fair chance of being sick. A properly designed DSS is an interactive software-based system intended to help decision makers compile useful information from a combination of raw sensor data, documents, and personal knowledge in order to make decisions. To develop these DSS, one needs good references from cows that were assessed by the farmer and treated for disease as well as a firm knowledge of sensor structures (information sources) available to the system. The difficulty in this is: what is considered the best reference? Is it the cows that received treatments (assessed by a vet and/or a farmer), or the ones that the farmers wish to check physically and decide for themselves? Often vets use their diagnosis as a true gold standard, but farmers are mainly interested in the animals that need treatment. To build a correct model-based predictive DSS that holds in practice, these parameters need to be defined.

3. The farmer's wishes

A modern dairy farmer has three targets. Firstly, good welfare for the cows: the cows are the crucial part of the dairy farm and taking good care of them will reduce costs on health treatments, increase the cows' fertility and ultimately increase milk production. Second is a good income from the farm: this is important in order to keep developing the dairy farm and to make a living from it. And third is flexibility: it is important to have sufficient time to spend with family and friends. Depending on the dairy farmer, the focus can be more towards one of the three areas, but there is also a strong interaction between the three wishes.

With mechanization, automation and sensor technology, farmers are being supported to reach these targets. The right interpretation of the information coming from sensor technology can bring the dairy farmer further in his quest to reach his personal targets. A step forward in flexibility is gained with the introduction of an AMS, but also, for instance, with the introduction of smartphones in the barn, making it possible to see the performance of your cows at every moment of the day. With sensors like activity and rumination activity, more insight is gained into health status and fertility status which, provided that the farmer takes action on this, will improve cow welfare. In addition, milk yield sensors, for example, give the farmer insight into the profitability of the cows, but also allow changes to be made, for instance concentrate feeding to optimize the fed amount per 100 kg milk production.

4. Development of sensor information

Over the years the aim has been to give the dairy farmer optimal information. This started out with raw data: just showing the dairy farmer a value, for instance, 'Milk production is 41.3 kg/day'. The dairy farmer had to know his cows and could see that this cow was in need of attention. Over time, 'intelligence' was added to the data. By comparing the current milk production with the milk production over the last week, the dairy farmer got direct insight into changes in milk production. The higher the drop in milk production, the greater the need to give the cow attention and check her out for diseases.

Later, other milk sensors came into play, such as color, conductivity and SCC measurements. With all sensors having their own attention, redundancy becomes an issue. This is due to the desire to show all cows that needed attention, resulting in the creation of very long lists of cows for the farmer to check, as an attention on one sensor could already have placed a cow on an attention list. Although the quality of the sensors and the attentions is going up, there is still a challenge for the farmer to filter the attention list for his own efficiency, according to his personal strategy. Having multiple sensors gives opportunities to combine data and create more reliable information and attentions. This way the attention lists shorten, and as such the sensitivity, specificity and positive predictive value (PPV) of the algorithms go up (Song and Van der Tol, 2010).

In future, combining data from multiple sensors will be more and more helpful to dairy farmers. The first systems are appearing that use all of the sensor information available to be as fast, accurate and precise as possible. This is a huge step towards unburdening the dairy farmer: he will only see the cows that are in need of attention and can take appropriate actions for these cows.

5. Interfacing

5.1 Easy to use

Having good sensitivity, specificity and PPV is crucial for bringing a new sensor (system) to the market. Besides this, the visual interface plays a big part in the adoption rate. If the output is unclear and confusing for the end user then he will struggle to use it. This could end up in it being used the wrong way or not being used at all. The system usability score (SUS) gives a measure of the quality of the user interface (Brooke, 1996). It focuses on effectiveness, so that the right cows are brought

Animal Number	Group	Lactation days	Day Production (24h)	Sick Chance	Attention		
					Sensor	Value	Severeness
70	Melkkoeien	196	22,4	28	Conductivity RR	120	
					Time Away	36:03	
81	Melkkoeien	4	19,3	44	Activity(eating)	63	
					Time Away	36:12	
12	Melkkoeien	185	17,3	97	Milk Drop	-6.6	
					Conductivity LF, RR	127	
					Time Away	47:27	
					Mastitis LF		
					Activity(eating)	81	

Figure 1. Possible interfacing of combined sensor attentions.

to the attention of the dairy farmer, efficiency, so that it is easy for the dairy farmer to perform the task, and satisfaction. If these factors are sufficiently in place then the tool is ready to go to market. When putting precision livestock farming into practice, this is a method for increasing the chance of being successful.

5.2 Automation

When a cow is brought to the attention of the dairy farmer, he has to perform multiple tasks that are needed for making decisions. These can be listed in a standard operating procedure (SOP) (Table 2). A dairy farmer will try to follow the SOP to come to a solution for the case presented. A decision support system can help him in this. Two different levels of decisions can be defined. The first is the actual decision on how to treat the cow and the second is decisions to make it easier for the farmer to come to a treatment decision. The cows brought to the attention of the dairy farmer should in theory all need a physical check, because based on sensor data the attention has been generated. If there is the possibility of separating the cow into a separation area then this decision can be made automatically, making it easier for the dairy farmer to follow up on the attention. By actually making these kinds of decision for him, the system is making it easier for the dairy farmer to come to a final solution on how to treat the cow. The physical check is still needed for the final decision on the treatment.

An automatic decision support (ADS) system helps the dairy farmer further. When decisions are required repeatedly, following identical deductions, these tasks can be automated. Consequently a rules-based ADS system can be made. Administering medicines might still be a bridge too far, but milder treatments for a cow with no side effects, supporting self-curing, can be started automatically. A mild mastitis case can be cured by the cow herself; changing milk settings can help support the cow in this. An automated decision for this is to decrease the teat-cup take-off level when the electrical conductivity of the milk is increased. To do this, a sensor is needed to measure the electrical conductivity and a system is needed that can alter the take-off level, but when these are in place a very basic ADS system can be implemented. More complex ADS systems can be thought of that will unburden the farmer.

Table 2. Simple example of an standard operating procedures for a cow with a health attention.

Action	Decision/question
Check the sensor information for the cow	Is a physical check needed?
	Do I need to separate the cow?
Physically check the cow	Is a treatment needed?
Decide on the treatment	No action taken for this cow
	No action taken for the cow but separate the milk
	Apply a medical treatment for the cow
	Change milk settings

6. Conclusions

Sensor technology is becoming more embedded in dairy farming, but the majority of dairy farmers are not focused on precision livestock technology. Dairy farmers aim to make their life and work more profitable, enjoyable and sustainable. Precision livestock technology can support this by implementing (automatic) decision support systems, giving the farmer mental relief. The focus and approach is currently technology-driven, but for a successful implementation a problem-oriented approach is needed and the technology should fill a gap experienced by the farmer. This requires not only solid technology, but interfacing, farm-management routines and implementation on the farm are also equally important. Therefore precision livestock technology is not a goal in itself: it is the path towards unburdening the dairy farmer to provide a bright future in dairy farming.

References

Brooke, J. (1996). SUS: a quick and dirty usability scale. In: Jordan, P.W., Thomas, B., Weerdmeester, B.A. and McClelland, A.L. (ed.) Usability evaluation in industry. Taylor and Francis, London, UK.

Hogeveen, H., Buma, K.J. and Jorritsma, R. (2013). Use and interpretation of mastitis alerts by farmers. Proceedings of the 6th European Conference on Precision Livestock Farming, pp. 313-319.

Song, X. and Van der Tol, R. (2010). Automatic detection of clinical mastitis in astronaut A3™ milking robot. Proceedings of the 1st North American Conference on Precision Dairy Management.

Steeneveld, W. and Hogeveen, H. (2015). Characterization of Dutch dairy farms using sensor systems for cow management. Journal of Dairy Science 98(1): 709-717.

Unique multiple monitoring system using the ear

G.M. Griffioen[1*] and R.L. Nebel[2]
[1]*Agis Automatisering, Gerverscop 9, 3481 LT Harmelen, the Netherlands;* [2]*Select Sires Inc., 42 North, Plain City, OH 11740, USA; g.griffioen@agis.nl*

Abstract

Precision dairy monitoring technologies are changing the shape of the dairy industry across the globe. Human observation limits dairy producers' ability to identify sick or lame cows or cows in estrus for insemination. Precision monitoring is the use of technologies to measure physiological, behavioral, and production indicators on individual animals to improve management strategies and farm performance. The CowManager® system provides tremendous opportunities for improvements in individual animal management by providing activity, high activity, resting, ruminating, eating, and temperature measurement that are up-dated hourly. Estrus and health alerts are based on previous within animal baselines using proprietary algorithms. The CowManager system consists of three modules: Fertility, Health and Nutrition. The end user customizes the modules and the graphs purchasing only what they desire to monitor. Data is downloaded every 15 minutes to the server and locally, the on-farm PC is up-dated every hour. Three levels of alerts for both Fertility and Health are available. Health alerts are calculated using a drop in rumination and increase in resting time. Fertility or in Heat alerts are determined using a seven day baseline of activity and two thresholds for Suspicious and in Heat alerts.

Keywords: cowmanager, sensors, fertility, health, nutrition

1. Hardware

The sensors are lasting for at least 5 years based on a very low power microprocessor which measures every second of every day throughout the year. The wireless ear sensor measures multipoint temperature, eating, ruminating, inactive, active and high active for heat- and health detection and nutrition (Figure 1). The sensor is waterproof and very light and durable. The location of the sensor in the ear is important, it should be close to the middle between two cartilage ribs.
The sensors communicate with one or more routers, the routers with a coordinator and the coordinator is connected to the PC, so that the cow can be in the network continuously. Internet

Figure 1. The ear sensor.

connection of the PC is required (Figure 2). The sensors have a range of about 300 feet to the router or coordinator and send their data every 15 minutes. This means that at bigger farms with possibly more than one barn they would contain multiple routers for the best connection between sensors and coordinator. The range between routers itself and the coordinator is 3,000 feet in clear sight of line. For data collecting in the field there is a solar router available.

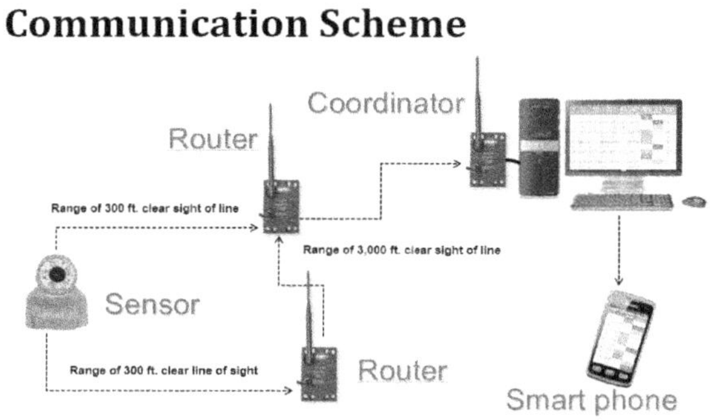

Figure 2. The network.

2. Software

By the process of collecting data through hardware, software shows available information about the cows. There are three module options available: Fertility, Health and Nutrition. The software makes the difference in functionality. It is easy, scalable, flexible on PC, smartphone or tablet.
The Fertility module provides information on the optimal moment for insemination in graphs or tables at the main dashboard (Figure 3). The arrow which is pointing up or down indicates if the cow is coming in heat and the level of high activity is increasing or, when the arrow is pointing down it indicates the cow is going out of heat and the level of high activity is decreasing. The module also shows insights into the cow's cycle and gives you all of the information you need to make a pregnancy.

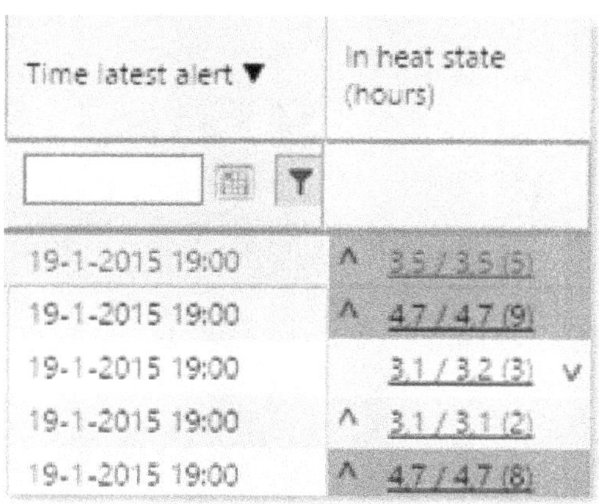

Time latest alert ▼	In heat state (hours)
19-1-2015 19:00	^ 3.5 / 3.5 (5)
19-1-2015 19:00	^ 4.7 / 4.7 (9)
19-1-2015 19:00	3.1 / 3.2 (3) v
19-1-2015 19:00	^ 3.1 / 3.1 (2)
19-1-2015 19:00	^ 4.7 / 4.7 (8)

Figure 3. Fertility alert.

The Health module provides early and reliable detection of cow ailments (Figure 4). Early detection means there are less health issues, reduced use of antibiotics, decreased production loss and lower labor costs. The CowManager® system can alert you to possible ailments including: mastitis, *Escherichia coli*, uterine infections, hardware disease, displaced abomasums, diarrhea. By providing insight into substandard cow behavior and ear temperature, the CowManager® system is also an exceptional tool to monitor the recovery of sick cows.

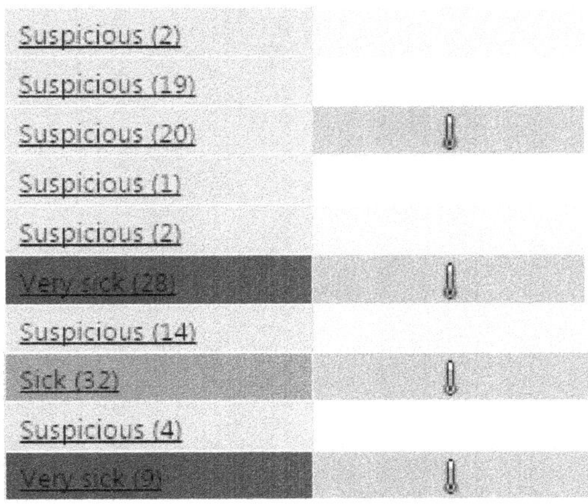

Figure 4. Health alerts.

The Nutrition module provides insight into the eating behavior of individual cows, management groups and the entire herd (Figure 5). The CowManager® system gives the perfect assessment of the right ration for your herd as well as an excellent tool to monitor cows in the transition period. The CowManager® system provides: Eating and rumination minutes, Group overview drying off, Group overview calving, and Group overviews of your management groups.
The information of these modules per cow are also available on an Android App and iOS app. By this you can act directly when an alert pops up on your mobile phone. When acting to these alerts finding a cow has never been easier by the FindMyCow app (Figure 6). Using this app, cows on a

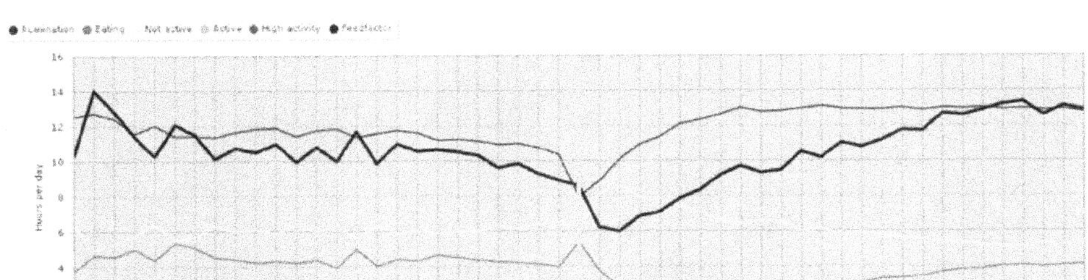

Figure 5. Nutrition graph of a group of calving cows, the FeedFactor is a combination of rumination, eating and non-active and it extrapolates the behavior of the group.

Figure 6. FindMyCow app.

module alert list are identified as you walk along the headlocks. The sensor communicates directly with the Locator connected to your smartphone, there is no need for any extra routers. You can find specific cows for: Insemination, Health checks, Pregnancy checks, Group checks.

3. Conclusions

It is important to acknowledge that a farmer or employee needs to own the system. On-site management of the CowManager® system is critical for success. The unique multiple monitoring system using the ear opens a new era of farming, based on sensor data it helps the producer as a decision support system at expert level. It even works better than before (Bikker, 2014). Today farmers need an extra eye, even when they are not around observing, CowManager does this for them. Because More Milk means More Money.

Reference

Bikker, J.P. (2012). Evaluation of an ear-attached movement sensor to record cow feeding behavior and activity. American Dairy Science Association 97(5): 2974-2979.

Real-time automatic cow monitoring – new practices

A. Arazi*, D. Rak and E. Ishay
Research Department, Afimilk Ltd., Kibbutz Afikim, Israel; alon_a@afimilk.co.il

Abstract

Calving is a crucial event in a productive cow's life cycle. It has significant influence on herd profitability and cow's welfare. Detecting calving event is a key factor to ensure successful calving with minimal harm to the calf and the dam. It is useful to determine if an intervention is needed, when to move cows to a maternity pen and obtain proper colostrum administration soon after calving. Lameness in dairy cattle is a major source of economic loss and impairs individual cow welfare. Lameness can be assessed using locomotion scoring. However, this method is time consuming and highly depends of the observer skills. The two studies presented below, review validation tests of automatic cow monitoring systems measuring cow's lying and activity behavior (AfiAct and AfiAct II, Afimilk Ltd., Kibbutz Afikim, Israel) for detection of calving onset and lame cows. Calving detection up to four hours before calving was demonstrated in more than 80% of the successful calving alerts, in all four farms included in the test. Average time from detection to calving was about two hours for tags, fitted both on front and rear leg. Lameness detection both for early recognition of new cases of locomotion score (LS)=2 and for lame cows LS≥3 was established with moderate to high sensitivity and high specificity. Using the system, the percentage of cows detected with LS≥3 was doubled compared with the routine farm observation method. The use of such monitoring systems can help improving calving management and lameness detection. Hence, allows for reduction in production losses, treatment costs, calves mortality and morbidity as well as improving cows' reproduction, longevity and welfare.

Keywords: lameness, calving detection, lying behavior

1. Calving detection

1.1 Introduction

Calving is a significant event in a productive cow's life cycle. It is a crucial time for both the dam and the calf. Difficult calving (dystocia) is prevalent in dairy cows, range from 10.7% to 29.4% in multiparous and 28.6% to 51.2% in primiparous (Lombard *et al.*, 2007; Meyer *et al.*, 2001). Dystocia has substantial effect on dairy herd productivity and is associated with increased calf mortality and morbidity around calving (Lombard *et al.*, 2007). Dystocia also increases the risk of retained placenta (Oltenacu *et al.*, 1988), uterine diseases (Sheldon *et al.*, 2009) and lower milk production (Rajala and Grohn, 1998). Additionally, dystocia adversely affects fertility and increases culling rate (Tenhagen *et al.*, 2007). Appropriate calving assistance and prevention of dystocia can reduce dystocia's damages (Palombi *et al.*, 2013).

Detection calving event on time allows intervention and assistance if needed, and to obtain a proper colostrum management soon after calving. Most cows deviate considerably from the expected calving date (Jensen *et al.*, 2013). Therefore, accurately identification of calving onset is a great challenge. External signs such as teat filling, pelvic ligament relaxation, udder distension, vulva edema, vaginal discharge and restlessness behavior are used to predict calving onset. Nevertheless such observations are time consuming and subjective.

On the day of calving, cows decrease lying time, increase the number of standing bouts and are more active in comparison with the previous days (Jensen, 2012; Maltz and Antler, 2007). In the last 2-6 hours preceding calving, the number of standing/lying transitions and restlessness episodes are significantly increasing (Gatien *et al.*, 2012; Jensen, 2012). Therefore, a cow monitoring system

recording cow's activity and lying behavior, could be a potential tool for accurately time manner detection of calving events.
The aim of the current study was to test timing accuracy (i.e. time from detection to calving) of an automatic monitoring system. Furthermore a comparison was done to evaluate the performance of the system when tags were fitted on front or rear leg.

1.2 Materials and methods

The study was conducted on four Israeli dairy herds, from 10/08/2015 to 22/10/2015. Herds ranging from 356 to 1,012 Israeli Holstein milking cows.
Cows were fitted with two tags (AfiTag II, Afimilk Ltd., Kibbutz Afikim, Israel) on front and rear leg, when moved to the close-up pen. Lying and activity data was recorded by the AfiAct II monitoring system (Afimilk Ltd.).
Calving times (date and hour) were recorded by the herds' teams. Calving detection alerts which generated by the AfiFarm herd management SW (Afimilk Ltd.) were recorded as well.

1.3 Results

In total 231 and 187 successful calving detection alerts was recorded for cows fitted with tags on rear and front leg, respectively (not all the cows were fitted with tags on front leg). Detection timing prior to calving were similar for front and rear leg. The distribution were about 35.5, 28, 26.5, 8 and 2% respectively for the last hour, one-two hours, two to four hours, four to eight hours and more than eight hours before calving, respectively (Table 1).
In all four herds 50% and more of the alerts were provided in the two hours preceding calving for both legs (range: 50-79%) and more than 80% of the alerts were in the last four hours before calving (range: 81.9-94.8%). The percentage of alerts of more than eight hours previous to calving were 2.5% or less in all herds except for 6.1% for front leg in herd A.
The average time from detection to calving was about two hours both for front and rear leg. It was ranged from 01:18 hours for front leg in herd C up to 02:38 hours for front leg in herd A. In three out of the four herds there was a very small (≤6 minutes) difference between average detection time to calving, in front and rear leg, while in herd C there was higher difference with average time to calving, longer by 19 minutes for rear leg compare with front leg (Table 1).

Table 1. Distribution of successful calving alerts timing and average time from detection to calving by herd and by leg fitted with tag.

Farm	Leg	Time from alert to calving (hours)						Average time from detection to calving (hh:mm)
		0-1	1-2	2-4	4-8	>8	Total	
A	Front	9 (27.3%)	9 (27.3%)	9 (27.3%)	4 (12.1%)	2 (6.1%)	33	02:38
	Rear	11 (27.5%)	9 (22.5%)	13 (32.5%)	6 (15.0%)	1 (2.5%)	40	02:37
B	Front	36 (37.1%)	30 (30.9%)	25 (25.8%)	5 (5.2%)	1 (1.0%)	97	01:43
	Rear	44 (43.6%)	24 (23.4%)	26 (25.7%)	5 (5.0%)	2 (2.0%)	101	01:49
C	Front	12 (63.2%)	3 (15.8%)	3 (15.8%)	1 (5.2%)	0 (0%)	19	01:18
	Rear	10 (40.0%)	8 (32.0%)	5 (20.0%)	2 (8.0%)	0 (0%)	25	01:37
D	Front	10 (26.3%)	11 (28.9%)	13 (34.2%)	4 (10.5%)	0 (0.0%)	38	02:13
	Rear	16 (24.6%)	24 (36.9%)	17 (26.2%)	7 (10.8%)	1 (1.5%)	65	02:12
Total	Front	67 (35.8%)	53 (28.3%)	50 (26.7%)	14 (7.5%)	3 (1.6%)	187	01:56
	Rear	81 (35.1%)	65 (28.1%)	61 (26.4%)	20 (8.7%)	4 (1.7%)	231	02:02

1.4 Discussion

In all four herds more than 80% of the alerts were provided up to four hours before calving. This allows a close monitoring of detected cows in the last few hours prior to calving, providing assistance in case a human intervention is needed.

High similarity was found between the system results in the four herds (particularly the percentage of calving up to four hours from detection) and for front and rear leg. These findings are indicative of the robustness of the system to detect calving under different conditions (herds) and for both legs.

2. Lameness detection

2.1 Introduction

Lameness is one of the most common and economically production diseases of dairy cattle worldwide (Bennett *et al.*, 1999). The prevalence of lameness of dairy cows is ranged from 5 to 48% (Becker *et al.*, 2014; Bichalo *et al.*, 2009; Braker *et al.*, 2010; Dippel *et al.*, 2009; Ito, 2006) with the majority of estimates within the range of 20 to 25% (Annex to the ESFA Journal, 2009). Lameness is associated with reduce in milk yield and fertility and increase in culling rate and medical costs (Bichalio *et al.*, 2007, 2008; Green *et al.*, 2002).

Commonly accepted methods to identify lame cows are based on visual observations and locomotion scoring. Nevertheless these methods are time consuming and subjective and frequently result in inconsistencies between and within observers (Blackie *et al.*, 2011; Schlangeter-Tello *et al.*, 2014). Moreover, in many cases it is underestimating the real prevalence of lame cows. Reports which didn't relay on self-reporting of lameness by farmers, found annual lameness incidences of approximately double the incidence on reports based on veterinary and farm treatment records (Annex to the ESFA Journal, 2009).

Early detection is a key factor in reducing the negative impact of lameness. Lame cows spent more time lying and had longer duration of lying bouts (Solano *et al.*, 2015; Yunta *et al.*, 2012). Therefore a cow monitoring system records cow's activity and lying behavior, could be a potential tool for early detection of lame cows.

The objective of the current study was to test the accuracy of an automatic monitoring system to detect lame cows, under field conditions. In addition the system performance was compared with the farm's routine method for detecting lame cows.

2.2 Materials and methods

The study was conducted on a commercial Spanish herd, milking 1,530 cows from 30/09/2012 to 11/11/2012. Cows were milked three times a day, fed total mixed ration (TMR) and housed in free stalls pens. Behavior tags (Pedometer Plus™, Afimilk Ltd.) were fitted to 651 milking cows.

Weekly lists of cows suspected for lameness, based on the recorded lying parameters from the AfiFarm herd management SW (Afimilk Ltd.), were created. These cows were scored on the same week for locomotion score (LS) scaled 1-5 (1 sound, 5 severe lame) by the herd's staff. In addition any lame cow or cow with hoof damage, which was observed by the herd's staff during that week, was recorded and considered as 'lame cow'. Every 'suspected' cow that was found with LS≥2 was removed from the subsequent weeks of the study. Only cows with lying data for at least one week were included.

Data analysis: every week, cows with LS ≥2 and cow reported lame by the herd's staff were classified as lameness event (LS+). All cows with LS=1 (sound) or cows with no LS records on that week were classified not lame (LS-). All cows detected by the system were classified system score positive (SS+), cows not detected by the system was classified system score negative (SS-). The accuracy of the system (sensitivity, specificity, and accuracy) for detecting lame cows with LS≥2 or LS≥3 were evaluated, both for cow by week event (CBW) and for cow by study event (CBS).

2.3 Results

Totally 73 (11.2%) cows were observed LS=2 and 81 (12.4%) were observed LS≥3. The sensitivity and specificity for detecting lame cows for the entire study (CBS) was 91.6, 86.1 and 94.2, 82.2% for cows with LS≥2 and LS≥3, respectively (data not presented). The CBW accuracy, for detecting cows with LS≥2 is summarized in Table 2. The sensitivity and specificity for LS≥2 were 74.8 and 91.7%, respectively. The sensitivity for 1st lactation cows (92.6%) was slightly higher than that of 2nd lactation 66.7% and mature cows 70.7%. No difference was found for specificity, 94.9, 97.8 and 98.8% for 1st lactation, 2nd lactation and mature cows, respectively. The sensitivity and specificity for LS≥3 were 60.0 and 94.4%, respectively. The sensitivity for 1st lactation cows 95.0% was higher than that of 2nd location 50.0% and mature cows 55.9%. No difference was found for specificity, 92.5, 94.8 and 96.1% for 1st lactation, 2nd lactation and mature cows, respectively. Comparing the prevalence of reported LS≥3 cows, during the study period, between the 651 cows monitored in the study, with 917 cows which were not monitored by the system (not fitted with Pedometer Plus tags), the detection of LS≥3 cows was almost doubled for cows monitored by the system, 12.4 vs 6.7% (data not presented).

2.4 Discussion

The accuracy obtained by the model was moderate to high sensitivity (50-95%) and high specificity (92.5-98.8%) both for LS≥2 and LS≥3 cows. These results are indicating that lameness could be detected by the automatic monitoring system. The prevalence of LS≥3 cows (12.4%) in this herd was on the lower scale compared with other studies. This could be due to the fact that only suspected cows were observed specifically, while other cows were observed as part of the herd lameness detection routine. This could have an influence on system sensitivity performance. Current results are in agreement with former studies demonstrating that records based on farmer treatment are about half of the actual incidence of lame cows (Annex to the ESFA Journal, 2009). The capability of the system to detect cows with LS=2 allows for early detection of lame cows, which could lead for early veterinary intervention and faster recovery.

Table 2. System accuracy for LS≥2 for cow by week event.[1]

Parity	TP	TN	FP	FN	N	N posit	Positive (%)	Sensitivity (%)	Specificity (%)	Accuracy (%)
1	50	1,088	58	4	1,200	54	4.5	92.6	94.9	94.8
2	48	759	17	24	848	72	8.5	66.7	97.8	95.2
≥3	65	1,141	14	27	1,247	92	7.4	70.7	98.8	96.7
All	163	2,988	89	55	3,295	218	6.6	74.8	97.1	95.6

[1] TP = true positive; TN = true negative; FP = false positive; FN = false negative; N = number of cow week's events; N posit = number of cow week's events with LS≥2.

3. Conclusions

The results demonstrating that real-time automated cow monitoring system measuring activity and lying behavior, could be a valuable tool for detecting calving and lameness events in dairy cows. The use of such a system can help improving calving management and lameness detection. Hence, allows for reduction in production losses, treatment cost, calves mortality and morbidity as well as improving cows' reproduction, longevity and welfare.

References

Barker, Z.E., Leach, K.A., Whay, H.R., Bell, N.J. and Main, D.C. (2010). Assessment of lameness prevalence and associated risk factors in dairy herds in England and Wales. Journal of Dairy Science 93: 932-941.

Becker, J., Steiner, A., Kohler, S., Koller-Bahler, A., Wuthrich, M. and Reist, M. (2014). Lameness and foot lesions in Swiss dairy cows: I. Prevalence. Schweizer Archiv für Tierheilkunde 156: 71-78.

Bennett, R.M., Christiansen, K. and Clifton-Hadley, R.S. (1999). Estimating the costs associated with endemic diseases of dairy cattle. Journal of Dairy Science 66: 455-459.

Bicalho, R.C., Vokey, F., Erb, H.N. and Guard, C.L. (2007). Visual locomotion scoring in the first seventy days in milk: impact on pregnancy and survival. Journal of Dairy Science 90: 4586-4591.

Bicalho, R.C., Warnick, L.D. and Guard, C.L. (2008). Strategies to analyze milk losses caused by diseases with potential incidence throughout the lactation: a lameness example. Journal of Dairy Science 91: 2653-2661.

Bicalho, R.C., Machado, V.S. and Caixeta, L.S. (2009). Lameness in dairy cattle: a debilitating disease or a disease of debilitated cattle? A cross-sectional study of lameness prevalence and thickness of the digital cushion. Journal of Dairy Science 92: 3175-3184.

Blackie, N., Bleach, E., Amory, J. and Scaife, J. (2011). Impact of lameness on gait characteristics and lying behaviour of zero grazed dairy cattle in early lactation. Applied Animal Behavior Science 129: 67-73.

Dippel, S., Dolezal, M., Brenninkmeyer, C., Brinkmann, J., March, S., Knierim, U. and Winckler, C. (2009b). Risk factors for lameness in freestall-housed dairy cows across two breeds, farming systems, and countries. Journal of Dairy Science 92: 5476-5486.

Gatien, J., Le Broc, M., Philipot, J. and Salvetti, P. (2012). Behavioural changes during the 12h before calving and predictors of dystocic delivery in Holstein cows. In: Proceeding of the 16th Annual Conference of the European Society for Domestic Animal Reproduction (ESDAR). August 29 – September 1, 2012. Dublin, Ireland.

Green, L.E., Hedges, V.J., Schukken, Y.H., Blowey, R.W. and Packington, A.J. (2002). The impact of clinical lameness on the milk yield of dairy cows. Journal of Dairy Science 85: 2250-2256.

Jensen, M.B. (2012). Behaviour around the time of calving in dairy cows. Applied Animal Behaviour Science 139: 195-202.

Jensen, M.B., Proudfoot, K.L. and Von Keyserlingk, M.A.G. (2013). Using behaviour to improve housing and management around the time of calving. WCDS Advances in Dairy Technology 25: 63-73.

Lombard, J.E., Garry, F.B., Tomlinson, S.M. and Garber, L.P. (2007). Impacts of dystocia on health and survival of dairy calves. Journal of Dairy Science 90: 1751-1760.

Maltz, E. and Antler, A. (2007). A practical way to detect approaching calving of the dairy cow by a behavior sensor. In: Precision livestock farming. Wageningen Academic Publishers, Wageningen, the Netherlands, pp. 141-146.

Meyer, C.L., Berger, P.J., Koehler, K.J., Thompson, J.R. and Sattler, C.G. (2001). Phenotypic trends in incidence of stillbirth for Holstein in the United States. Journal of Dairy Science 84: 515-523.

Oltenacu, P.A., Frick, A. and Lindhe, B. (1988). Use of statistical modeling and decision analysis to estimate financial losses due to dystocia and other disease in Swedish cattle. In: Proceedings of the 5th International Symposium on Veterinary Epidemiology and Economy, Copenhagen, Denmark, pp. 353-355.

Palombi, C., Paolucci, M., Stradaioli, G., Corubolo, M., Pascolo, P.B. and Monaci, M. (2013). Evaluation of remote monitoring of parturition in dairy cattle as a new tool for calving management. BMC Veterinary Research 9: 191.

Rajala, P.J. and Gröhn, Y.T. (1998). Effects of dystocia, retained placenta, and metritis on milk yield in dairy cows. Journal of Dairy Science 81: 3172-3181.

Schlageter-Tello, A., Bokkers, E.A., Groot Koerkamp, P.W., Van Hertem, T., Viazzi, S., Romanini, C.E., Halachmi, I., Bahr, C., Berckmans, D. and Lokhorst, K. (2014). Effect of merging levels of locomotion scores for dairy cows on intra- and interrater reliability and agreement. Journal of Dairy Science 97: 5533-5542.

EFSA (2009). Report on the effects of farming systems on dairy cow welfare and disease. Annex of the EFSA Journal 1143: 1-7.

Sheldon, I.M., Cronin, J., Goetze, L., Donofrio, G. and Schuberth, H.J. (2009). Defining postpartum uterine disease and the mechanisms of infection and immunity in the female reproductive tract in cattle. Biology of Reproduction 81: 1025-1032.

Solano, L., Barkema, H.W., Pajor, E.A., Mason, S., LeBlanc, S.J. Nash, C.G.R., Haley, D.B., Pellerin, D., Rushen, J., Passille, A.M., Vasseur, E. and Orsel, K. (2015). Association between lying behavior and lameness in Canadian Holstein-Friesian cows housed in freestall barns. Journal of Dairy Science 99: 2086-2101.

Tenhagen, B.A., Helmbold, A. and Heuwieser, W. (2007). Effect of various degrees of dystocia in dairy cattle on calf viability, milk production fertility and culling. Journal of Veterinary Medicine. A, Physiology, Pathology, Clinical Medicine 54: 98-102.

Yunta, C., Guasch, I. and Bach, A. (2012). Lying behavior of lactating dairy cows is influenced by lameness especially around feeding time. Journal of Dairy Science 95: 6546-6549.

Nedap precision dairy farming – tools for operational, tactical and strategic decision-makers

R.J.H. Lammers and L. Meijer*
Nedap, P.O. Box 104, 7140 AC Groenlo, the Netherlands; linda.meijer@nedap.com

Nedap Precision Dairy Farming technology makes it easier for cattle farms to focus on animal management. By offering this technology, the sector is basically given new magnifying glass. The farmer can switch faster and discover problems sooner. This is obviously very practical, but at the same time it also requires a lot from the farmer personally: spending more time on monitoring, but also having a certain flexibility and the willingness to learn new things.

1. Feeding monitoring

A good practical example is monitoring the feeding times of cows that are in the dry period. For a large number of cows, reduced feeding times can already be seen pre-partum. Even though no illness can be diagnosed in the cow at that time, it will have an increased risk of all kinds of transition-related problems after calving. By providing an insight into this at an earlier stage, these cows receive the attention they deserve. This means that the farmer can already intervene before the actual clinical picture is diagnosed. The feeding sensors help to implement preventive measures, which makes watching helplessly and acting curatively a thing of the past. As a farmer, however, you must learn to apply the information provided by the sensor data. Furthermore, it can be very difficult to move away from the mindset of acting curatively.

2. Powerful extension

The use of sensors and technology in dairy farming is regularly presented as some kind of holy grail. This sometimes creates a distorted picture: 'New technologies will make sure that the farmer can manage his entire farm from his comfy chair and will always have a handle on everything'. Among other things, this idyllic picture comes from the idea that innovative technologies will make sure that we will always know exactly what a cow needs at a certain time. It's a very nice idea and an equally good thing to pursue, but we are not there yet, not by a long shot. At the moment we are in a situation where sensors are not yet replacing the '*master's eye*', but where precision technology is a very powerful extension for the expert. If the farmer can handle the new tools, he will be able to lift his management level to a higher plane and reap the rewards of the possibilities offered by Precision Dairy Farming technology.

3. Support of cow management

The value of the data generated lies in the information and points for attention provided by the system. The way in which personnel responds to the information and applies it is also crucially important. In addition, the results, the ease with which the actions can be performed and the verification of the data are decisive. The aim is to support dairy farms in improving cow management: better results, fewer problems and working more easily.

The big advantage of using sensor technology is that it continuously pays full attention to each cow, 24 hours a day and seven days a week. The Nedap Smarttags measure the behaviour of the dairy cows. Behaviour is a key indicator for the wellbeing of a cow, but also for the events involving the cow. External factors largely determine behaviour, naturally in connection with the behavioural characteristics of the cow itself. Graphically this could be represented as in the Figure 1.

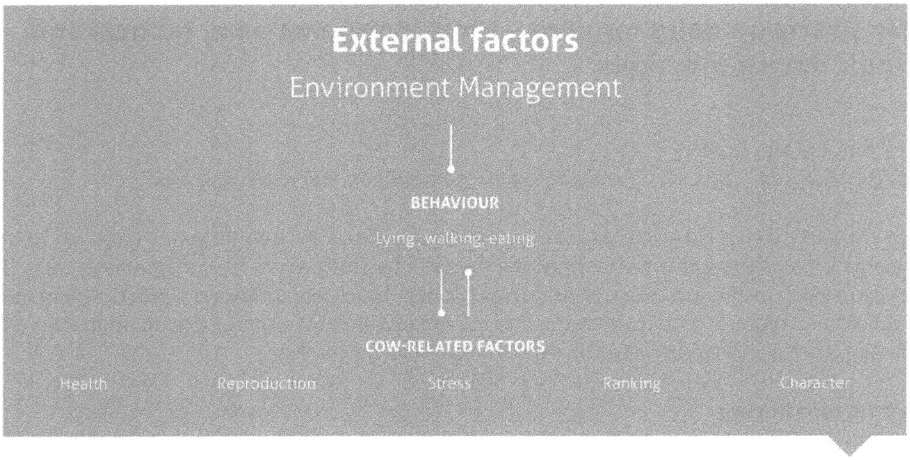

Figure 1. Dairy cow behavior and its relations.

By monitoring all of these factors, farmers receive a structured image of the events in the barn. This allows them to respond to them directly and consistently.

4. External factors

A characteristic of the external factors is that they affect the behaviour of a cow, but the behaviour of the cow does not affect these factors. Examples of these are the barn surroundings or the feeding regime at a dairy farm. By definition, these external factors are picked up by behavioural sensors. Correctly analysing and visualising the behavioural information makes it possible to gain a good insight into these external factors.

5. Cow-related factors

A characteristic of cow-related factors is that they do affect the behaviour exhibited. Two cow-related factors, reproduction and health, are the key parameters when using Precision Dairy Farming technology. These factors have a direct link to the day-to-day work of the farmer, making these factors the most valuable ones in the short term. The trick is to accurately single out these factors and to distinguish them from other behavioural effects.

6. Behaviour

In addition, cow-related factors are also affected by the external factors through a change of the animal's own behaviour. This applies, for example, to an animal's health. If unwanted behaviour is exhibited due to non-optimum barn conditions, the animal will often not be able to consume enough feed. This may result in a cow's health deteriorating, especially in the long term.
Not all factors affecting the behaviour of dairy cows are relevant to the farmer in day-to-day work. This partly depends on the nature of the work and the management level at which the work is being performed. During the day-to-day work, it is important to monitor animals more closely if they require attention with regard to their health and reproduction. In the longer term it would be valuable, for example, to see the effects of the milking times on group behaviour.

Roughly speaking, the information provided by Precision Dairy Farming technology can be classified into three management levels: strategic, tactical and operational (Figure 2).
Precision Dairy Farming takes each level of this management pyramid into account. It is important to ask the right questions at each level. Only then will the farmer get a useful answer that reflects the effect his intervention has on the business processes. In addition, it is important to measure the complexity and frequency. The data generated here should coincide with whether a job is a daily, monthly or even yearly task. That can also be found in the schematic representation in Figure 2.

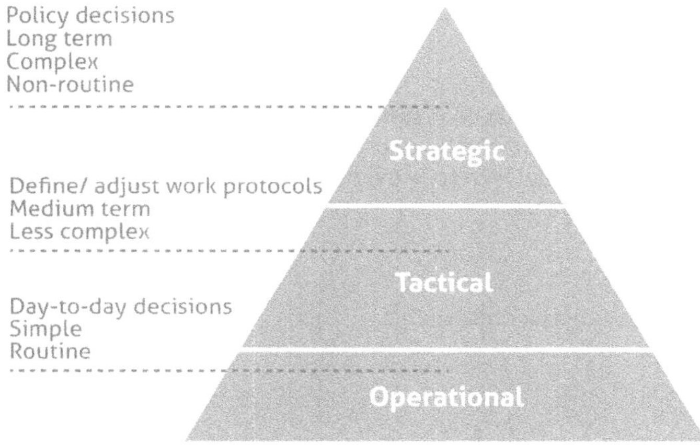

Figure 2. Management levels and information demand.

7. Management levels

An example of an strategic task could be: Using an indicator like the percentage of cows eating simultaneously for the decision to reduce overstocking at the feeding fence by adding more feeding places. An example of an tactical task could be the decision to raise the number of feed push ups per day to allow the cows to eat more frequently and distributed evenly over the course of the day. Data from Precision Dairy Farming technology can help showing the direct effects of the current and future management protocols. An example of an operational task is to find cows that did not eat when expected. These cows could be ill or another reason could prevent them from eating.
Successful use of sensor technology with regard to operations not only means that the sensor and detection algorithms must be excellent; the processes surrounding them must also be well-coordinated. A system for detecting heat can be the best in the world; if the system is not optimally incorporated in the business protocols, the farmer will still not be able to maximise the benefits. The farmer/employee must be aware of what should be done if a certain point for attention arises. Additionally, he must have all the relevant information to make a well-considered decision. He/she must also have the right tools to actually implement this decision. This is all about the overall picture. The design of the systems must be based on optimum interaction between farmer and technology (Figure 3).
Figure 4 shows how measuring individual lying behaviour can yield information in various areas. Each individual cow (as presented at the bottom of the figure) has its own behavioural pattern during the day. This individual pattern can be used to see how the cow feels or what its status is. Is she in heat? Does she have a health issue? Or is everything fine?

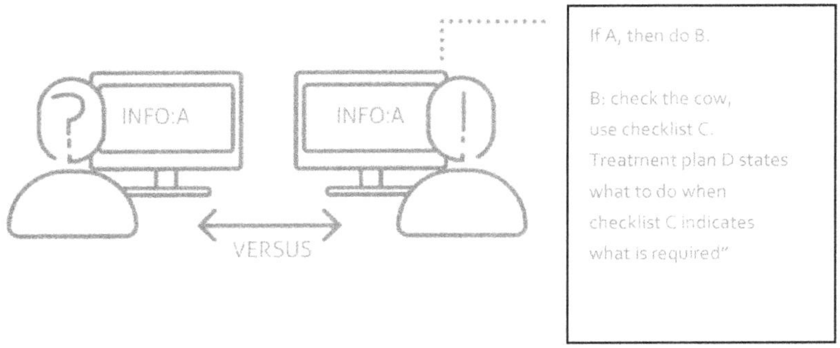

Figure 3. If, then? Operational information presented to the farmer without him knowing what to do has a very low priority and often remains unused. On the other hand, information presented to the farmer when he does know how to use it or how to respond to it is used more quickly and has a higher priority and value.

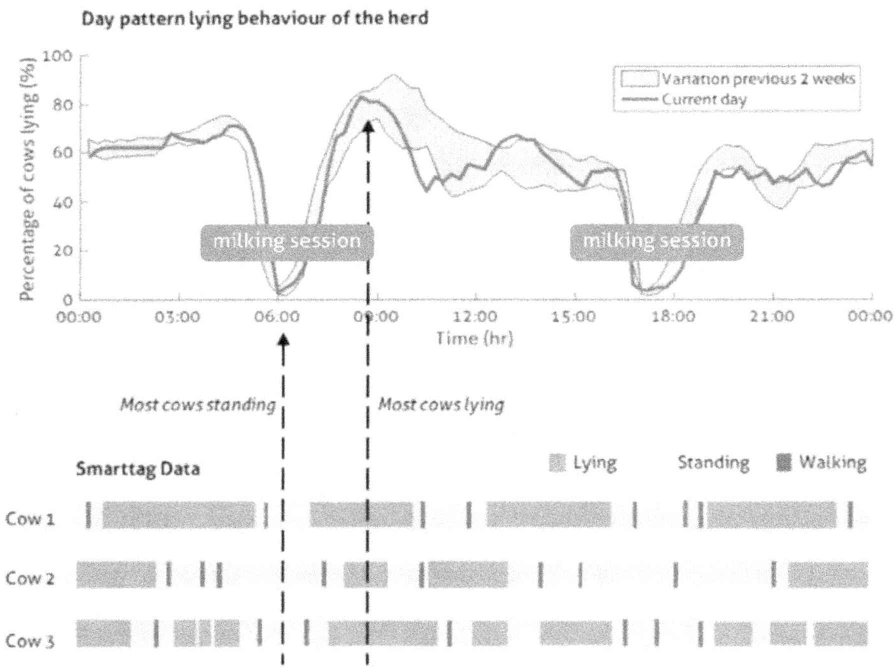

Figure 4. How to move from individual cow data to group information.

Combining these individual measurements into a 'herd' pattern provides completely different insights. One very notable aspect is that there are activities that affect all the cows. In the above example, the two milking sessions are the clearest to identify. The information derived from this is: how long cows have to stand for milking process. It also presents the maximum percentages of times when cows can lie down at the same time. This information is useful for calculating barn capacity or lying cubicle use. Business protocols can be modified to suit this or barn modifications can be made to improve the lying pattern. All of this has one final aim: improving general animal health and productivity.

Another example in which the same source data yields different information is cow positioning. This can be used to easily determine where the cows are in the barn and is therefore a solution to a very practical operational issue (Figure 5). Positioning data over a period of time provides a very different type of information. Creating a '*heat map*' of the barn conveniently allows the most intensely used locations in the barn to be seen. This data provides more information on the local climate and (possible) problem areas in the barn (Figure 6).

In the various analyses, one is often interested in the effects of external factors on the behaviour of an individual animal or a group of animals. A farmer will mainly want to purchase Precision Dairy Farming technology to simplify the daily (operational) work. This does not mean that the data provided by this sensor technology is not highly relevant to the farmer when performing the tactical and/or strategic management. It's precisely there that sensor technology can make a huge (possibly the greatest) contribution to improving general management at a dairy farm.

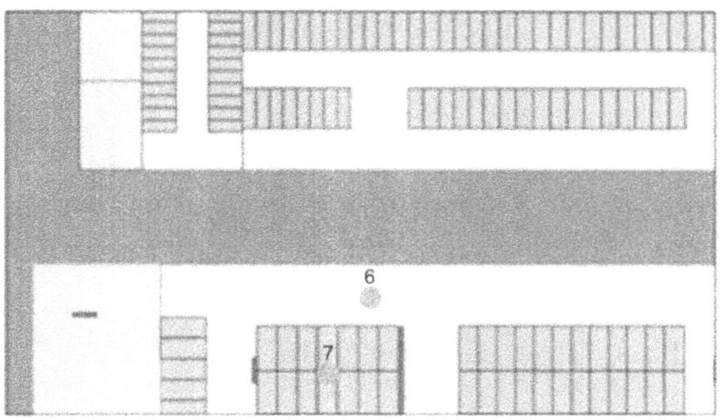

Figure 5. Finding cows.

Figure 6. Heat map of barn.

8. Company-specific interpretation

For suppliers, advisors and those on the periphery, the information is also highly valuable on a tactical and strategic level. They can support the farmer when translating the sensor information into specific actions. Suppliers and advisors only take a snapshot when they visit a farm. By using sensor technology, one can also easily look back and identify deviating patterns that are not (or no longer) visible at the time of the visit. The information from sensor technology on a tactical and/or strategic level is very closely related to specialist disciplines. In addition, information from sensors must often also be interpreted in a company-specific manner. Nedap believes in collaboration with other parties to maximise the value of this functionality for the farmer. We make sure that the various specialisms are brought together, ensuring that the farmer receives the best information and support.

9. Integration into daily routine

Nedap wants to encourage agricultural workers, suppliers and advisors to integrate Precision dairy farming solutions into their daily routine and to define possible actions arising from the information obtained. One possible action is: checking the animal and giving it the correct treatment. This is determined based on decision trees, treatment plans, standard work procedures, knowledge of cattle farming and obviously based on the individual cow's health.

10. Nedap – technology that matters

For over 35 years, we have been developing this technology and expertise for managing large groups of animals as if they were a single animal. Focusing on the individual animal is the basis for a healthy group. We help large businesses and integrators in cattle farming to grow by optimising the performance of the animals through individual animal management and by providing the right information to the right person.

Cow time budget and beyond – experience with the Smartbow® system

D. Tomic[1], M. Iwersen[2] and W. Auer[3*]
[1]Smartbow GmbH, Maria-Jakobi-Gasse 1, 1030 Wien, Austria; [2]Clinical Unit for Herd Health Management, University Clinic for Ruminants, University of Veterinary Medicine, Vienna, Austria; [3]Smartbow GmbH, Jutogasse 3, 4675 Weibern, Austria; wolfgang.auer@smartbow.at

Abstract

Smartbow® is an advanced animal localization, activity and health monitoring system currently offering estrus and rumination monitoring. The Smartbow® vision is to provide a detailed time budget information for all activities of dairy cows – the 'cow time budget' (CTB) information, based on localization and acceleration data obtained from active ear-tags. Understanding the power of data integrated from many different sources, and their role in a holistic animal monitoring, Smartbow® is offering a solution that enables a simple access to, and integration of data of different equipment, by means of service-based open interfaces using a unified data model. In addition, Smartbow® is developing an information management solution to verify the benefits of data coming from many different farms, based on a multidimensional model of the most representative cow's data. Using this model to analyze the cows' localization and activities' data from the Smartbow® system, together with the barn climate data, cows' milk yield and quality, feed rations and intake data, body condition data, etc., promises to provide valuable insights into how a dairy cow's performance, health and fertility is interrelated with, or impacted by the daily schedule of cow's activities within different functional areas, partially imposed by management practices, a barn setup and/or overstocking. This paper presents a study in which an experimental barn is instrumented with a Smartbow® system, and the Smartbow® interface technology is used to implement automatic collection of data from different sources: the herd management data come from DairyComp 305, additional milk related data come from the milk quality reports, environmental data come from climate loggers, and feed data come from feed reports. Collected data present an input for a semantic data warehouse, which is being designed based on a multidimensional data model that facilitates analysis along different dimensions. This information management approach represents a strategic extension to the Smartbow® system aiming at facilitating knowledge creation in dairy farming, in particular related to early diagnosis of health risks, especially of subclinical diseases. This paper also briefly presents a complementary strategic extension of the Smartbow® system that aims at providing a business interaction platform for different actors in the dairy sector, including farmers, consultants, veterinarians, scientists, equipment vendors and other data providers, and data integrators including providers of herd management systems and dairy farming software applications. This concept promises to simplify: (1) access to data for data integrators; (2) the way the consultants deliver a value – monitoring as a service; (3) monitoring for the farmers; and (4) justification of an ad-value for precision dairy farming equipment, increasing its adoption.

1. Introduction

Precision farming technology is becoming indispensable in modern dairy barns, as the size of herds is growing, and time that a farmer can spend directly interacting with animals is declining. Animal localization and monitoring systems are essential tools that can not only support or replace direct visual observations and manual examinations, but can also deliver insights beyond visible.

Today, there are numerous sensor solutions that monitor different activity, health and production parameters of cows such as: (1) ear, leg or neck mounted systems for cow localization and activity monitoring, e.g. standing/laying and feeding, estrus monitoring, rumination or body temperature monitoring; (2) vaginal temperature monitoring systems predicting calving events; (3) orally administered reticulorumen boluses that monitor temperature or pH fluctuations for rumen health monitoring; and a (4) a variety of in-line systems that monitor milk yield and flow, conductivity, temperature, somatic cell counts, etc.[1] Each of these systems aims at replacing the need for observations or manual actions by using the automatic sensor readings. The combination of these systems can deliver a body of data that capture different, potentially unknown causalities to be elicited by further data analysis.

Systems currently gaining particular interest are monitoring systems that holistically track cow's activities during the day – a cow time budget (CTB) – as the cow moves, stands in the alley or in the stall, rests in the stall, ruminates, eats or drinks, visits a milking robot or a parlor, or is being confined for calving or for checks or care within different functional areas in the barn. The benefits of using a normalized cow time budget of healthy, highly productive cows in well managed barns as a benchmark to detect problems and infer about the comfort, wellbeing and health, has already been analyzed and identified (Cook, 2008). Consequently, one can already work with some known values, e.g. the dairy cow requirements for resting of minimal 12 h/d, standing in the allay for socializing of an average of 2.4 h/day, and feeding of an average of 4.4 h/day. This leaves residual 5 h for milking, drinking, and care or check. A holistic CTB monitoring system would be able to capture an activity budget per individual cow, create individual benchmark times, and generate alarms when these benchmark times are violated. With a CTB system it would be easier to track the impact of overstocking for instance on the reduction of the lying time, the impact of individual laying time on productivity and fertility, as well as detect a sweet spot at which the CTB does not deviate from the benchmark despite of overstocking.

Today, there is no single sensor system that can holistically monitor all the activities of the cow and provide an accurate automatically accessed cow time budget. Most of the published CTB studies are conducted with multi-sensor systems, which have to deal with the problem of data time synchronization (Behmann *et al.*, 2014). Furthermore, the CTB systems are often specially designed to support scientific studies in experimental setups and are therefore composed of costly high-end sensors with a narrow application field, which are not suitable for commercial farms (Porto *et al.*, 2015).

2. Smartbow® for animal localization and CTB monitoring

Smartbow® is an animal identification, localization and activity monitoring system based on active eartags with radio transponders. The communication and data processing infrastructure includes a number of wireless receivers and a central server, which are locally installed in a barn. The Smartbow® eartag integrates a transponder, a three-axes acceleration sensor, and temperature sensors. Eartags periodically send data to routers, which forward them to the central server for real-time localization, activity monitoring, estrus detection and rumination monitoring. The localization is based on time-of-arrival (TOA) algorithms (Pichler *et al.*, 2014). The position of a cow in a barn is determined in real-time, with 2D accuracy of 1 meter. The position is mapped on a digital plan of the barn, more specifically, within a specific functional area, such as cubicles for resting, alleys, area around waterer, feeding alley, milking area (parlor, milking robot), calving box, etc. The Smartbow® solution currently offers partial 'cow time budget' monitoring, by combining the localization data, the spatial definition of the functional areas in the barn, and the readings of the accelerometer sensor from the eartags. The Smartbow® solution can already distinguish between activities such as standing, laying and moving (walking distance), rumination and no rumination, as well as classifying activities into low, high and medium activity. The areas of current innovation in activity monitoring include improvements in distinguishing between standing and laying, and detection of feeding and

[1] Precision Dairy Farming Technology, University of Kentucky, http://bit.ly/1Mz4UiI.

drinking. The Smartbow® vision is to provide a detailed time budget information for all activities of dairy cows, based on localization and acceleration data obtained from eartags, and is currently improving pattern recognition algorithms in different experimental setups.

3. Beyond CTB monitoring – Smartbow® for data-driven dairy production improvement

Extending the Smartbow® system in the realm of data-driven dairy production we are addressing new innovation needs at the level of (1) data integration – innovation in achieving interoperability at the interface and data level at a single farm; (2) decision support – innovation in methodology for creation of new process knowledge; and (3) security and business interactions – innovation in implementing security, data access, and service offerings to match business constraints related to collecting and using data from many farms.

Our innovation in data integration is an open interface and data integration solution project agriOpenLink that enables a simple access to, and integration of data of different equipment at a farm, by means of service-based interfaces using a unified data model (Tomic, 2014).

Our innovation in process knowledge generation is being developed in the project agriProKnow. To leverage opportunity of accessing new process parameters which become more and more tangible as new sensors become available, we are developing information management tools for modeling and testing of process hypotheses based on semantic data warehouse (DWH) depending on a multidimensional model of the most representative cow's data. However, as the process of knowledge generation requires large data samples, i.e. data from many farms and data sources, security, and business as well as regulatory constraints of all involved stakeholders also need to be taken into account.

Our innovation in security and business interactions is the 'Dairy Data Aggregation and Integration platform' developed within an innovation project DADAFI. The platform aims at supporting business interactions between different actors in the dairy sector, including farmers, consultants, veterinarians, scientists, equipment vendors and other data providers, as well as data integrators, including providers of herd management systems and dairy farming software applications. Business interactions include consulting service offerings, in form of (1) customized monitoring procedures; and (2) recommendations for precision dairy farming technology based on specific data offered by this technology, and the monitoring procedure utilizing this specific data; (3) catalogues service for technology providers; and (4) customized APIs for data integrators for controlled access to specific set of data. The core of the platform is based on FIWARE technology. The central security component comprises FIWARE Identity Manager, PEP Proxy Wilma GE and Auth PDP GE[2].

4. An information system for process knowledge creation

To strategically extend the Smartbow® offerings with a decision support system functions with the goal to significantly improve animal health and milk production efficiency, we are developing a novel methodology for process related information management for dairy farming, with the focus on modelling, monitoring, and control of health and welfare of dairy cows. This is currently developed and tested in the research project agriProKnow. The major innovation is in a procedure for process knowledge generation, which combines methods of stochastic analysis of sensor data, and semantic data-warehousing. This system shall help agri-industry companies, farmers and consultants to create new process knowledge via data analysis.

The basis for knowledge generation via semantic data-warehousing is a multidimensional model. The central concept of this model is an animal – a dairy cow. Different cow's data are systemized within different informational components, as illustrated in Figure 1. For aggregation, filtering and statistical analysis of data, several dimensions are defined. The first dimension includes all

[2] FIWARE Security GEs, http://catalogue.fiware.org/chapter/security.

the master record parameters ('master record', Figure 1): the unique identification number, the so called life number, the identification of the breed(s), unique identification of the dam and the sire, and the cow's birthday. This dimension supports finding parents, offspring and siblings of a specific animal, filtering based on the breed, or finding animals born on the same day, or within the same month, or the year. As the model aims at cross-farm data integration, we introduce the global geographic dimension ('location', Figure 1) – the country, the region, the area, the farm, and the business / organizational dimension – the business, the farm, the group ('business', Figure 1). Thirdly, the model includes the functional area dimension, which identifies distinct areas within the barn including the milking area (milking parlor, milking robot), the feeding area, the alley, the cubicles, the calving box, the lock-out area, the area around the water trough, etc. Fourthly, the model defines two time dimensions. The first time dimension describes the data that can be either collected or generated on the daily bases in which the aggregation is possible within the month, the quarter, and the year ('day', Figure 1).

This dimension is relevant for data such as the number of milking, milk yield and quality per milking, body condition scoring, health examinations, e.g. blood screening, hoof care, daily diagnosis events such as detection of mastitis, or examination of a fresh cow including body temperature, rumen score, total daily rumination time, etc. The second time dimension describes the data that are created each hour, and the data can be further aggregated within different phases of a day, a day, a month, a quarter and a year ('hour', Figure 1). To this category belong different 'cow time budget data', belonging to the information components 'Activity' describing the duration of activities such as walking, standing, lying, drinking and milking assessed in different or corresponding functional areas, 'Feeding' describing feed intake and quality. A feed type and source dimension ('type, source', Figure 1) is further defined to allow aggregation per different types of feeds such as roughage, concentrate, or total mixed rations, and different sources of data such as laboratory analyses or estimations. To provide for filtering and aggregation of data based on the lactation cycle the calving and lactation dimensions are also defined. These include: the calving date (the first day of the lactation), the lactation day (day in milk), and the lactation phase. An additional dimension is related to milkings including the number of the milking and the type of the data source such as laboratory, or a milking robot. The calving data have an additional dimension describing the calf

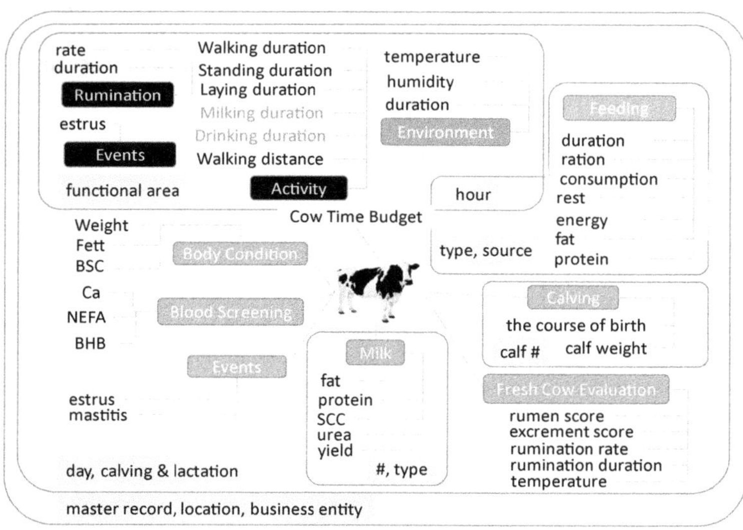

Figure 1. A knowledge creation model.

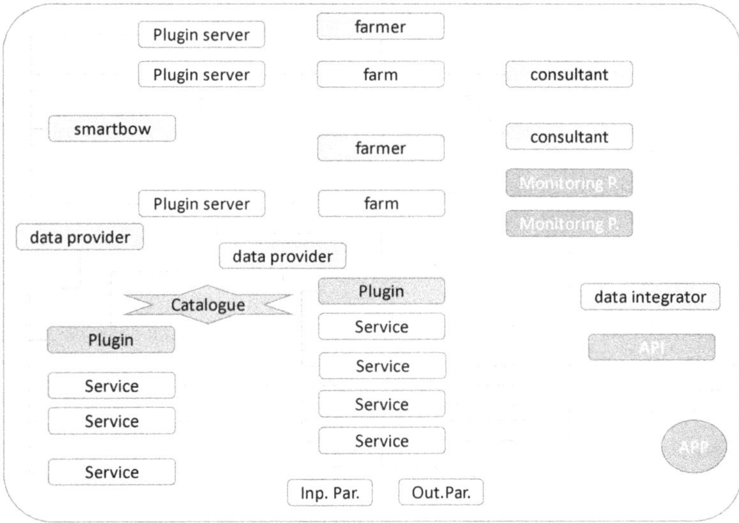

Figure 2. A business interaction model.

number to capture multiple gestations. The multidimensional model illustrated in Figure 1, is a blueprint for a cross-farm data warehouse which has to be filled-in with data from many farms. In the project agriProKnow an initial data collection setup is created at the Teaching and Research Farm Kremesberg, Vetmeduni Vienna, Austria[3]. This farm uses the Smartbow® localization and activity monitoring system, as well as the Smartbow® open interface solution for integration of data from other systems and data sources.

In the extraction, transformation and load (ETL) process the DWH is filled-in with authentic data extracted from the files created by different devices and accessed via a file sharing interface. The files include: 1) files automatically dumped by the herd-management software DairyComp 305 including the information about milk yield, the information about the events, e.g. calving, body condition, blood screening, and fresh cow evaluations. The master cow data are automatically obtained from the national register, and the milk quality data are processed from the milk quality reports (of a milk quality monitoring organization) manually uploaded in the file sharing directory. The files with the temperature and humidity data from the loggers are also manually uploaded. A customized interface is created for accessing the files from the Smartbow® system and translating this data into a CTB data. The feed related data are extracted from files of the feeding system, as well as from the manually created files describing the feed components. A file with static location and business entity data is manually created in this test system, and used to initialize the DWH.

The data that are uploaded into the DWH via the ETL process are not complete, as for special data parameters that may be generated only sporadically. The data creation process within DWH, amends the ETL process, by creating missing data, replicating the values already existing in the database. With this approach, the old value is re-used until the new value is available. In this way created historic data provide good basis for data analysis. Each repeated value is also annotated with its age and the maximum accepted age before an ETL alarm is created. The data warehousing technology provides a flexible means to support expressive queries that can have high impact. Data warehousing combines business intelligence and the possibility to search and find important information. The results of queries can be analyzed and used directly or used to filter out information for further data mining.

[3] Kremesberg, Experimental Farm, VETMED, Austria, http://bit.ly/1RtwjTG.

5. Platform for Data Aggregation and Business Interactions

DADAFI is a platform for business interactions between farmers, consultants, equipment and data providers and data integrators. The interaction model of the DADAFI platform is illustrated in Figure 2. The interaction model is created around the service-based data-integration model developed in agriOpenLink (Tomic, 2015). The data access capability of the integrated equipment is modeled by means of Plugins, Plugin services and data accessed via plugin services. The data integration capability of a particular farm is modeled in terms of Plugin servers that control different plugins. A Plugin is created by a specific equipment supplier or data provider. The relationships between plugins, plugin-servers, -services, and -providers are depicted in Figure 2, and present the basis for defining interactions among platform users.

In DADAFI data access is implemented with the FIWARE GE Keyrock that integrates OpenStack Identity Management (IdM) Keystone and an OAuth2.0[4] server. The data access model is based on the notion of a 'common project' – a farm – and so the farmer and all his consultants are equally authorized to access the data of this common project.

Interactions between consultants and their clients (farmers) are based on the model of a monitoring procedure. A monitoring procedure is a query that identifies specific data to be monitored, the monitoring interval and the frequency of query execution, as well as the monitoring set in which the results of monitoring are stored for further inspection. Both, farmers and their consultants can access these sets. A monitoring procedure ('Monitoring P', Figure 2) maps into an alarm as well as into a description of an action that needs to be followed upon the alarm. Monitoring procedures are created by consultants for their customers, based on the knowledge about the data available on the farm. The procedures are interfaces to data that only authorized consultants and a farmer can access. As monitoring procedures are on the one hand linked to specific parameters, i.e. data from specific devices within the barn, and on the other to a specific purpose of the monitoring, such as improved detection of subclinical diseases, e.g. ketosis, the impact of monitoring procedures can be verified in terms of their immediate results. Accordingly, a monitoring procedure, i.e. equipment supporting it can be recommended to a farmer who currently does not use it.

DADAFI platform also offers a catalogue service ('Catalogue' in Figure 2) for data providers who use the plugin based integration model. The plugin models described in the catalogue can be searched for and used to create monitoring procedures and related recommendations.

The unique offering of data integrators is a new knowledge provided via specialized software applications ('APP' in Figure 2). Herd management systems, feeding applications, process optimization applications, or similar, all belong to data integrators. The end beneficiaries of these applications can be farmers or consultants. To create new knowledge, these applications need access to data. DADAFI provides controlled data access by utilizing the FIWARE security framework as already mentioned. Following the OAuth2.0 protocol, the data integrators can register their applications with a customized dynamically created and deployed API ('API' in Figure 2). An API is similar to a monitoring procedure, over which the whole or only a selected part of the farm data context can be accessed.

6. Conclusions

Dairy farming today is characterized by the use of various precision technology solutions, which integrated together, should assist towards healthy herds, high quality of milk, optimized use of resources, and a sustainable environment. This paper presented innovation solutions and currently ongoing research efforts that utilize Smartbow® technology. Our innovations in cow time budget monitoring, data integration, knowledge creation and decision support as well as business

[4] The OAuth 2.0 Authorization Framework, https://tools.ietf.org/html/rfc6749.

interactions aim at reaching the vision of data-driven dairy farming based on holistic monitoring and understanding of dairy cows,

Acknowledgments

The work presented here is co-funded by the Austrian Research Funding Agency (FFG) within projects agriOpenLink (no. 840185) and agriProKnow (no. 848610) and by innovation activity DADAFI endorsed by the European FIWARE and Open Data Accelerator FINODEX. We thank our project partners for their contribution to the presented work.

References

Behmann, J., Hendriksen, K., Müller, U., Walzog, S., Büscher, W. and Plümer, L. (2014). Recognition of activity states in dairy cows with SVMs and graphical models. GIL 34: 221-224.

Cook, N.B., 2008. Time budgets for dairy cows: how does cow comfort influence health, reproduction and productivity? Available at: http://tinyurl.com/jnz4nhl.

Pichler, M., Rudic, B., Breitenberger, W. and Auer, W. (2014). Robust positioning of livestock in harsh agricultural environments. 14th Mechatronics Forum, Mechatronics.

Porto, S.M.C., Arcidiacono, C., Anguzza, U. and Cascone, G. (2015). The automatic detection of dairy cow feeding and standing behaviours in free-stall barns by a computer vision-based system. Biosystems Engineering 133: 46-55.

Tomic, S.D.K. (2014). Semantic services for adaptive processes in livestock farming. International Conference of Agricultural Engineering. Zurich, Switzerland.

Tomic, S.D.K. (2015). Experiences with creating a Precision Dairy Farming Ontology (DFO) and a knowledge graph for the data integration platform. Journal of Agricultural Informatics 6(4): 115-126.

Precision dairy technology on the way – and how it can really break through

S. Klimpel
GEA Farm Technologies GmbH, Siemensstraße 25-27, 59199 Boenen, Germany; susanne.klimpel@gea.com

Abstract

Precision dairy technologies are widely spread in almost all parts of the world. Acceptance and penetration rates differ between countries. Some technologies are already mature and considered as commodities. Nevertheless, in the century of digitalization and farming 4.0 still many factors have to be considered in the stages from technology to product respective solution. This is even more important as new technologies are leaving the heat detection and automatic milking behind and therefore interactions that are more complex are needed. This paper discusses the animal – technology – human interaction for different aspects of precision technology. Adaption rates of technology are also depending on return on invest for the users. Early intervention studies to support this are limited. Furthermore, public perception in times of growing farm sizes and reduction of antibiotic use often does not consider technology as being animal friendly and supporting welfare. Further work is needed to underline this.

Keywords: precision dairy, PLF

1. Introduction

Benjamin Disraeli, Prime Minister of the United Kingdom in the 19th century said: 'In a progressive country change is constant; change is inevitable' (Disraeli 2015).
The 18th and 19th centuries were the times of the first industrial revolution. It had a major impact on the working and social life esp. of farmers and workers. Human labor was supported or replaced by horses and machines, farming focused on increasing productivity in crop and animal production. Today his quote is even more valid than ever. We are facing the industrial revolution 4.0 comprised of automation, data exchange and changes in manufacturing (Figure 1). One of the major challenges of this latest development is also the biggest advantage: the speed of invention has increased dramatically. Until the 20th century, society had app. 100 years to adapt and adopt major technology breakthroughs. Only 50 years ago automation and robotics were introduced on a larger scale and this was called the 3rd industrial revolution.
Still in this fast changing environment, the technology is only one part of this development. To be successful with any development we must not forget the human and in our environment the animal that are involved as well. The success of any invention is strongly related with the advantages that people using this are able to generate (Flett *et al.*, 2004). Some of these advantages can be calculated and are financial, others are non-financial and must not be underestimated (Bijl *et al.*, 2007). Arnold Bennett said: 'Any change, even a change for the better, is always accompanied by drawbacks and discomforts' (Bennett, 2015). This was already described 100 years ago, so the challenge today is: What can we do to support the adoption of technology in the 21st century?

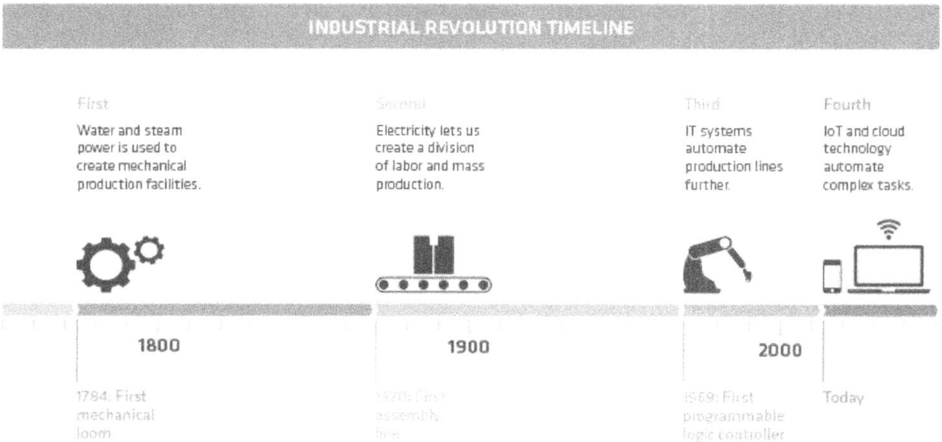

Figure 1. Industrial revolution timeline (http://tinyurl.com/jkdzjs8).

2. Technology adoption automatic milking system and heat detection

Everett Rogers (2003) described the innovation diffusion model (Figure 2). Additionally he pointed out the influencing factors of this model:

- the innovation itself;
- the adopters;
- the communication channel;
- time;
- social system.

If an innovation shall reach a critical mass, these key elements have to be considered.

Automatic milking and heat detection systems have reached this critical mass in the dairy industry: In 2014 5,180 units of milking robots were sold worldwide – this is an increase of 8% compared to 2013 (IFR Statistical Department, 2015).

Both systems were elaborated in the 1980s (Peter and Bosu, 1986; Rossing *et al.*, 1985; Williams *et al.*, 1981). Since beginning 1990s the first automatic milking as well as heat detection systems have been commercially available. Following the investigation of performance, the economic benefit could be calculated (Cooper and Parsons 1999; Van Asseldonk *et al.*, 1999, Wade *et al.*, 2004).

Looking at the influencing factors listed by Rogers the innovation as itself has proven the economic benefit, the perceived usefulness and perceived ease of use. Early adopter farmers have been using the technology now since almost 25 years and second and third generation technology is already available.

The communication channel is a very crucial point esp. in farming. Scientific studies to underline the objective evaluation of the systems have to be available. Additionally, farmers like to learn from other farmers and this model has been adapted in many intervention studies like mastitis control programs. Furthermore, opinion leaders multiply experiences to farmers. Heat detection systems are widely recognized as being a helpful tool to improve productivity in reproduction on farms without influencing the animal. The 'social system' as the last key element has also adopted the technology. Although we are seeing a wide acceptance today, we have to acknowledge that it has been more than 2 decades since the first systems were commercially available. As heat detection is a technology that is strongly related to physiological condition of the animal and the economic benefit is easy to calculate and can be generated fast, this timeline can still be considered as long.

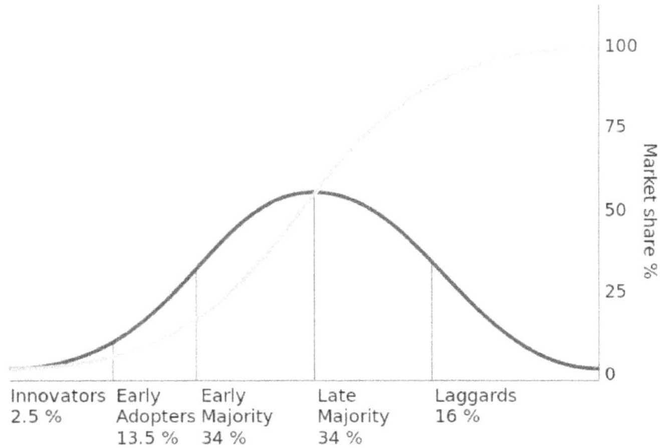

Figure 2. Technology diffusion model (https://en.wikipedia.org/wiki/Diffusion_of_innovations)

3. Future applications

Since 2000 research has started to look deeper into possibilities to identify diseases like lameness in cows using different technologies (Chapinal *et al.*, 2011, Poursaberi *et al.*, 2010, Richard *et al.*, 2015; Van der Tol *et al.*, 2002). In parallel welfare quality programs and assessments (Welfare Quality Network, 2015) have been established to support good animal husbandry. The challenging task of these new detection methods is to turn the information into decision and consequent action (Rutten *et al.*, 2013). In contrast to other technologies like e.g. automatic teat dipping these systems do not prevent but detect a condition or disease. The success of these systems is highly correlated to the actions that are being taken after the alarm has been raised. Furthermore conditions might be revealed that have not been known to the farmer to this extent (Leach *et al.*, 2010).

The costs of production diseases in dairy cattle and the distribution of direct and indirect costs are well known (Bruijnis *et al.*, 2010; Kossaibati and Esslemont, 1997). The impact of prophylactic treatment has been investigated already (Hernandez *et al.*, 2007), whereas early detection and intervention on production diseases especially with the use of sensor technology has only been examined in some papers (Geishauer *et al.*, 2001, Leach *et al.*, 2012, Milner *et al.*, 1997,). Economical evaluations of the impact of investment in detection technology and subsequent management adaptations are still missing.

An additional point Davis (1989) has included in the 'perceived ease of use' is the necessity to display the information in a user friendly and relevant way. We are facing the situation where information is available in a surplus but the user has to be able to apply this information for his needs.

The last point mentioned by Rogers, the social system, is somehow disconnected as well. Less people are living in the countryside and have a connection or background in farming (United Nations, 2014). Large animal husbandry setups are being observed critically and smaller farms with less technology are perceived as being more animal friendly.

Apart from reliable and precise technology the analysis and display of the information is the key to success for future systems. Additionally the impact of early disease detection on farm economics and animal welfare needs to be investigated deeper. If the benefit of early disease detection and intervention can be shown to the farmer and the society, the adaptation of technology in dairy farms can be accelerated. The next industrial revolution is yet to come and the timeline might be even shorter than before.

References

Benett, A. (2015). Quotations. Available at: https://en.wikipedia.org/wiki/Arnold_Bennett#Quotations.

Bewley, J. (2013). Exciting dairy breakthroughs: science fiction or precision dairy farming. In: Proceedings of the precision dairy conference and expo. Rochester, MN, USA, pp. 26-27.

Bijl, R., Kooistra, S.R. and Hogeveen, H. (2007). The profitability of automatic milking on Dutch dairy farms. Journal of Dairy Science 90: 239-248.

Bruijnis, M.R.N., Hogeveen, H. and Stassen, E.N. (2010). Assessing economic consequences of foot disorders in dairy cattle using a dynamic stochastic simulation model. Journal of Dairy Science 93(6): 2419-2432.

Chapinal, N., De Passille, A.M., Pastell, M., Hanninen, L., Munksgaard, L. and Rushen, J. (2011). Measurement of acceleration while walking as an automated method for gait assessment in dairy cattle. Journal of Dairy Science 94: 2895-2901.

Cooper, K. and Parsons, D.J. (1999). An economic analysis of automatic milking using a simulation model. Journal of Agricultural Engineering Research 73: 311-321.

Davis, F.D. (1989). Perceived usefulness, perceived ease of use, and user acceptance of information technology. MIS Quarterly 13: 319-339.

Disraeli, B. (2015). Encyclopaedia Britannica: Benjamin Disraeli, additional information. Available at: http://tinyurl.com/ztehoe9.

Flett, R., Alpass, F., Humphries, S., Massey, C., Morriss, S. and Long, N. (2004). The technology acceptance model and use of technology in New Zealand dairy farming. Agricultural Systems 80: 199-211.

Geishauser, T., Leslie, K., Kelton, D. and Duffield, T. (2001). Monitoring for subclinical ketosis in dairy herds. Compendium: S65-S71.

Hernandez, J.A., Garbarino, E.J., Shearer, J.K., Risco, C.A. and Thatcher, W.W. (2007). Evaluation of the efficacy of prophylactic hoof health examination and trimming during midlactation in reducing the incidence of lameness during late lactation in dairy cows. Journal of the American Veterinary Medical Association 230(1): 89-93.

IFR statistical department (2015). World robotics survey: service robots are conquering the world. Available at: http://tinyurl.com/jdmrrju.

Kossaibati, M.A. and Esslemont, R.J. (1997). The costs of production diseases in dairy herds in England. The Veterinary Journal 154(1): 41-51.

Leach, K.A., Whay, H.R., Maggs, C.M., Barker, Z.E., Paul, E.S., Bell, A.K. and Main, D.C.J. (2010). Working towards a reduction in cattle lameness: 1. Understanding barriers to lameness control on dairy farms. Research in Veterinary Science 89(2): 311-317.

Leach, K.A., Tisdall, D.A., Bell, N.J., Main, D.C. and Green, L.E. (2012). The effects of early treatment for hindlimb lameness in dairy cows on four commercial UK farms. The Veterinary Journal 193(3): 626-632.

Milner, P., Page, K.L. and Hillerton, J.E. (1997). The effects of early antibiotic treatment following diagnosis of mastitis detected by a change in the electrical conductivity of milk. Journal of Dairy Science 80: 859-863.

Peter, A.T. and Bosu, W.T.K. (1986). Postpartum ovarian activity in dairy cows: correlation between behavioral estrus, pedometer measurements and ovulations. Theriogenology 26: 111-115.

Poursaberi, A., Bahr, C., Pluk, A., Van Nuffel, A. and Berckmans, D. (2010). Real-time automatic lameness detection based on back posture extraction in dairy cattle: shape analysis of cow with image processing techniques. Computer Electronic Agriculture 74: 110-119.

Richard, M.M., Sloth, K.H. and Veissier, I. (2015). Real time positioning to detect early signs of welfare problems in cows. EC-PLF, Milan, Italy, 4 pp.

Rogers, E. (2003). Diffusion of innovations. Simon and Schuster, New York, NY, USA.

Rossing, W., Ipema, A.H. and Veltman, P.F. (1985). Perspectieven voor het melken in een voerbox. Instituut voor Mechanisatie, Arbeid en Gebouwen, Wageningen, the Netherlands.

Rutten, C.J., Velthuis, A.G.J., Steeneveld, W. and Hogeveen, H. (2013).

Sensors to support health management on dairy farms. Journal of Dairy Science 96(4): 1928-1952.

United Nations (2014). World's population increasingly urban with more than half living in urban areas. Available at: http://tinyurl.com/q44pzh8.
Van Asseldonk, M.A.W., Jalvingh, J. Huirne, R.B.M. and Dijkhuizen, A.A. (1999). Potential economic benefits from changes in management via information technology applications on Dutch dairy farms: a simulation study. Livestock Production Science 60: 33-44.
Van der Tol, P.P.J., Metz, J.H.M., Noordhuizen-Stassen, E.N., Back, W., Braam, C.R. and Weijs, W.A. (2002). The pressure distribution under the bovine claw during square standing on a flat substrate. Journal of Dairy Science 85: 1476-1481.
Wade, K.M., Van Asseldonk, M.A.P.M., Berentsen, P.B.M., Ouweltjes, W. and Hogeveen, H. (2004). Economic efficiency of automatic milking systems with specific emphasis on increases in milk production. In: Meijering, A., Hogeveen, H. and De Koning, C.J.A.M. (ed.) Automatic milking, a better understanding. Wageningen Academic Publishers, Wageningen, the Netherlands.
Welfare Quality Network (2015). Welfare quality assessment protocols. Available at: http://www.welfarequalitynetwork.net/network/45848/7/0/40.
Williams, W.F., Yver, D.R. and Gross, T.S. (1981). Comparison of estrus detection techniques in dairy heifers. Journal of Dairy Science 64: 1738-1741.

Part 1. Novel precision technology developments

Detection of lameness in dairy cattle using limb motion analysis with automatic image processing

K. Zhao[1], D. He[1] and J.M. Bewley[2*]
[1]*College of Mechanical and Electronic Engineering, Northwest A&F University, Yangling, China;*
[2]*Department of Animal and Food Sciences, University of Kentucky, Lexington, KY 40546, USA;*
jbewley@uky.edu

Abstract

Lameness has become a frequent and serious problem for dairy herd productivity and animal welfare. The walking pattern of cows is the most significant and direct characteristic related to lameness. Gait characteristics have been used to estimate lameness by human experts. The objective of this study was to develop an automatic system of locomotion score in cows by analyzing the leg swing using vision techniques to detect and predict the lameness with high accuracy and practicability. The focus was on quantifying the moving performance of cows and demonstrating the possibility of classifying lameness using the features extracted from movement analysis. Side-view images, with a resolution of 704×576 pixels, were acquired when cows passed through a narrow aisle to a water trough. Cows were scored visually by an expert on a scale from 1 (normal) to 3 (severely lame), totaling in 621 videos from 98 cows. The motion curve was extracted using the position of the moving limb by image processing. The rising period was extracted from the curve to generate six features referring to gait asymmetry, tenderness, etc. The box-plot of the features within 3 classes showed that the dataset was near linear, separable under the six features. The Decision Tree classifier was applied to the dataset, and 10-fold cross validation was used to verify the accuracy of the classification algorithm. 94.20% of cows in the dataset were correctly classified, and the averages of true positive rate and false positive rate were 94.54 and 3.17%, respectively. The results showed that it may be possible to classify dairy cow lameness based on the six extracted motion features in this study.
Keywords: cows, locomotion score, leg swing

1. Introduction

Lameness is a frequent and serious problem for dairy herd productivity and animal welfare (Bruijnis *et al.*, 2012). Early detection of lameness in cows is important and needed. Visual observation by the farmer based on subjective locomotion scoring is widely used to diagnose lameness in cows at an early stage (Flower and Weary, 2009). However, this method is time-consuming and expensive to perform continuously on the farm. The subtle differences in walking patterns of cow are difficult to be perceived by visual observation (Cook, 2003; Pluk *et al.*, 2012).

Computer vision technology is becoming increasingly used in the dairy industry. Poursaberi *et al.* (2010) utilized image analysis techniques for early lameness detection in dairy cattle. By detecting the arc of back posture and fitting a circle through selected points on the spine line, the average inverse radius was assigned to cow. Based on this curvature value, a score representing the status of lameness was given. Viazzi *et al.* (2013) designed a descriptor called BMP (body movement pattern) based on the back posture of cows, and Decision Tree was used to classify the BMP output into 3 lameness classes (not lame, lame, and severely lame) according to the given scores. In a data set of 223 videos from 90 cows, 76% of cows were correctly classified, with an 83% true positive rate and 22% false positive rate when using the population approach.

In the visual lameness scoring system, seven indicators, including tenderness, arched back, reduced speed, irregular gait, etc. (Van Nuffel *et al.*, 2009), were used to determine the lameness. Studies based on back posture only contained parts of the lameness indicators. Besides, as the boundary between cows and background, the back posture is difficult to be separated from images correctly, especially in the complex farming situation. However, the lameness was defined as a deviation in gait to reduce pain (Scott, 1989), and such deviation can be obviously observed on the leg swing. Because over half of the lameness indicators are related to walking patterns (Van Nuffel *et al.*, 2009), the analysis of cows' leg swing has the potential to classify locomotion and lameness.

The objective of this study was to develop an automatic and continuous system of locomotion score in cows for detecting and predicting lameness based on leg swing analysis using computer vision techniques. First, image processing for extracting the leg swing features was explored. Second, lameness classification based on the leg swing features was studied, and the classification performance was verified against the expert score.

2. Materials and methods

2.1 Animals and housing

Video data were collected at a commercial dairy farm in Yangling, China. The cows were milked 3 times a day in a milking parlor. After milking, side-view images were acquired while cows were passing through a 2-m-wide, 7-m-long aisle with a concrete floor to a water trough. To avoid any further intrusion into the farm's routine and any interference with cow traffic, the cameras were positioned on the supported beam of a feeding shed at a height of 1.8 m and 35 m away from the alley. The dataset of 621 videos from 98 lactating Holstein cows were acquired over a period of 8:00 h to 18:00 h on sunny days from 9 July to 5 August 2013.

2.2 Reference locomotion scoring method

The observer evaluates gait by watching the recorded videos, and then gives scores based on the presence of one or more of the lameness indicators, including irregular gait in time or place, reduced tracking up, reduced speed, tenderness, increased abduction and head bobbing. The cows that did not show any of the 'lameness indicators' were given a score of '1'. The cows with the one 'lameness indicator' were scored '2'. When cows have a severe 'lameness indicator' or multiple 'lameness indicators', the cows were assigned a score of '3' (Van Nuffel *et al.*, 2009).

2.3 Image processing

The image processing in shown in Figure 1. The block matching algorithm (Acharjee and Chaudhuri, 2012) was used to calculate the motion field in images to detect the target of cows. After target

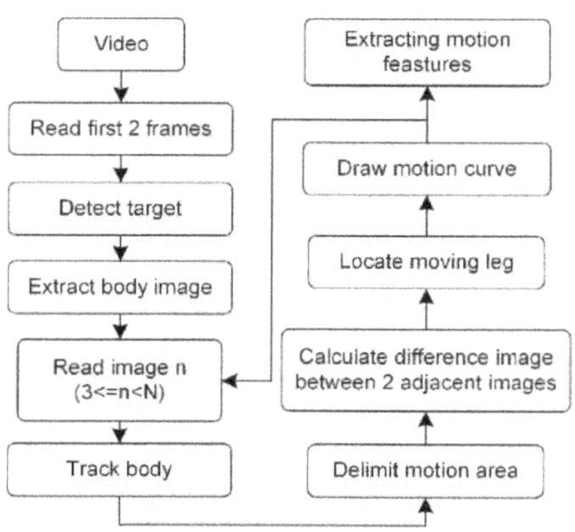

Figure 1. Flowchart of image processing.

detection, the body image from the target image was located by the shape matching method. The body area of the cows was tracked through the video.

2.4 Extraction of motion curve

When cows were walking, only one limb was swinging sharply with slight motion on the others. The position of the moving limb was detected to draw the motion curve of cows.

The difference between several consecutive frames is feasible to detect the motion in the video. To enhance the detection result of limb motion, three consecutive frames were subtracted and combined with each other to generate a motion image. This study only focused on the leg swing. Therefore, in each frame, the moving area of 4 limbs was delimited by the position of body area which was tracked in this frame using the method in the last section. The leg motion image, defined as bw_L, was extracted from the entire motion image using the moving area of limbs.

A rectangle sized $m \times n$ was moved through the bw_L to find the area with a maximum binary intensity as the position of the moving limb, where the m was set to the one-tenth of the width of bw_L, and n was the same as the height of bw_L.

Every frame from the video of the cows walking through the vision field was processed to generate the motion curve of cows. A low level binary intensity indicated that the cow was moving slightly or stopping, which was defined as stopping frame. An example of the process to extract motion curve is shown in Figure 2. Due to the size of the moving area that was calculated by the size of the body, the position of the limb is expressed relative to the size of the body. Therefore, the influence of individual diversity on movement features was prevented from the motion curve.

2.5 Extracting features

When cows are moving the legs on the same side swing successively, this is defined as a motion period indicated by the ring segment in motion curve, where local minimum is the sign of beginning of the hind leg's moving, and local maximum is that of the forelimbs stopping. The local extremums in the motion curve were calculated and matched to extract the rising segment, which indicated a moving period from the moving of the hind leg to the stopping of the forelimb. A moving period indicates the movements of a half gait cycle of the limbs on the same side. The rising segment which contains the lame limb has differences from the neighboring one. The drawing of the motion curve and rising periods of one cow are shown in Figure 3.

The swing speed of the limb within period, defined as the slope of rising segment, indicates the moving speed of cows, and is calculated as follows:

$$S_i = \frac{MC_{Mf_i} - MC_{Nf_i}}{Mf_i - Nf_i} \times fps \quad i = 1, \cdots, m \tag{1}$$

where S_i is the swing speed of limb within i-th period (% s^{-1}), m is the number of periods through a video, MC_i is the position of the moving limb in i-th frame, Mf_i is the frame number of i-th maximum, Nf_i is the frame number of i-th minimum, and fps is the frame rate of the video.

The swing speed of the limb before touching the ground is calculated as follows:

$$Fs_i = \frac{MC_{Mf_i} - MC_{Mf_i - 5}}{5} \times fps \quad i = 1, \cdots, m \tag{2}$$

where Fs_i is the swing speed of the forelimb before touching ground in the i-th period (% s^{-1}).

The proportion of the stopping frames in a period is calculated as follows:

$$Rf_i = \frac{Ts_i}{T_i} \quad i = 1, \cdots, m \tag{3}$$

where Rf_i is the proportion of stopping frames within the i-th period (%), Ts_i is the number of stopping frames in the i-th period, and T_i is the number of entire frames in the i-th period.

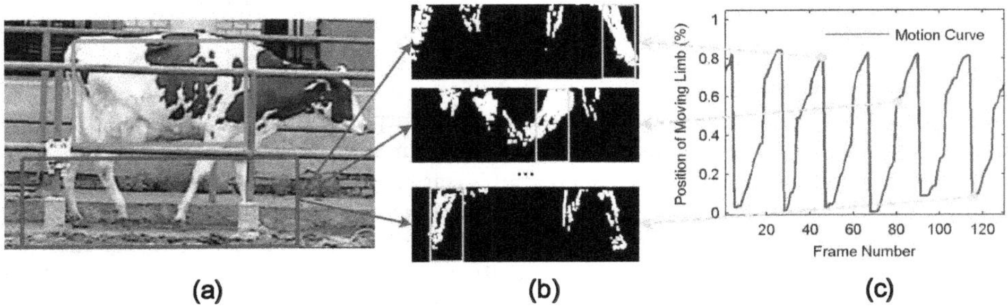

Figure 2. The process to extract motion curve: (A) moving area of 4 limbs (in lower rectangle), (B) calculating the maximum binary density (in rectangle), (C) the motion curve through a video.

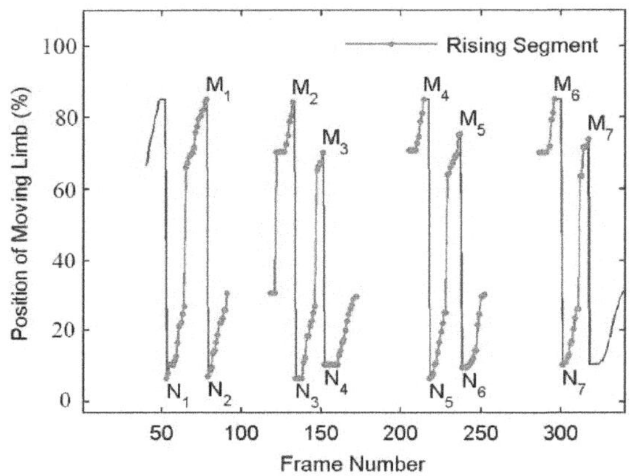

Figure 3. Drawing of motion curve, rising periods of one cow, and local extremums calculated (in which the N_i and M_i are local minimum and maximum used to emerge the rising segment, and discontinuity points or blank indicate stopping frames).

The gait asymmetry can be measured by the difference of S_i and Rf_i, which are calculated as follows:

$$\begin{cases} Ds_i = |S_{i+1} - S_i| & i = 1,\cdots,m-1 \\ Drf_i = |Rf_{i+1} - Rf_i| & i = 1,\cdots,m-1 \end{cases} \tag{4}$$

where Ds_i is the difference between S_i and S_{i+1} (% s^{-1}), and Drf_i is the difference between Rf_i and Rf_{i+1} (%).

To evaluate the lameness, six features of walking performance were designed using the parameters described above. The features and its relationship to the lameness indicator are shown in Table 1.

Table 1. The feature values of lameness.

Features	F1	F2	F3	F4	F5	F6
Description	min $(S_1, ..., S_m)$	mean $(Ds_1, ..., Ds_{m-1})$	max $(Rf_1, ..., Rf_m)$	mean $(Drf_1, ..., Drf_{m-1})$	$1/m$	mean $(Fs_1, ..., Fs_m)$
Related lameness indicator	Reduced speed	Gait asymmetry	Reduced tracking up	Stance time	Stride length	Tenderness

3. Results and discussion

3.1 Verification result of motion curve

Because of the negative effect of the protection fence between cows and camera on image processing, the correlation coefficient between the motion curve extracted by image analysis and the manual reference was 85.33%.

RV coefficient is a multivariate generalization of the squared Pearson correlation coefficient. It measures the closeness of two sets of points that may each be represented in a matrix (Josse *et al.*, 2008). The average of the RV coefficient between the point sets of the manually selected and the extremum in the motion curve from 60 random samples was 98.99%, which shows that the local extremums extracted by vision-based method were highly correlated to the real situation.

3.2 Distribution of features among classes

Features have different distributions in the three classes of lameness. Figure 4 shows box-plot of the 6 features under the grouping measurements by the visual gait scores given by observers.

The median of *F1* in score 1 and 2 are 81.79 and 78.57, respectively, which shows a slight decline on moving speed in score 2. When cows are severely lame scoring 3, the moving speed and stride length reduce further, and the median of *F1* drops from 78.57 in score 2 to 58.21 in score 3. The cows with two or more infected hoofs walk constrainedly, and the difference of moving speed between consecutive periods declines. Because of the increasing tenderness, the median of *F6* drops 2.09% compared with score 2. The condition of cows in severe lameness degenerates rapidly, leading to a long tail in the box-plot in Figure 4B, 4C, and 4D. Increasing interval is the exclusive symbol of severe lameness, which can be used to distinguish the cows in severe lameness from others.

The result showed that the cows had different lameness indicators in different stage. Therefore, multi-features were capable to express the lameness information more accurately through all the lameness stages.

3.3 Classification result and evaluation

Table 2 illustrates the confusion matrix of the Decision Tree Classifier using 10-fold cross-validation. As shown in Table 2, 171 not lame, 261 lame, and 153 severely lame cows were correctly classified. 585 out of 614 samples were correctly classified, which resulted in an accuracy of 94.20% without any error classification between score 1 and score 3.

The confusion matrix is shown in Table 3. In the score 2, the true positive rate was lower (92.55%), and false positive rate was higher compared with that for the other 2 classes, which shows that this class is the most difficult to detect.

When a binary classification (lame or not lame) was used to analyze the result instead of 3 classes, severely lame and lame cows were both regarded as lame, and the confusion matrix using binary classification is shown in Table 4. The average of the true positive rate, false positive rate, and precision of the binary classification was 97.42, 3.33, and 96.69, respectively. The overall accuracy was 97.42%.

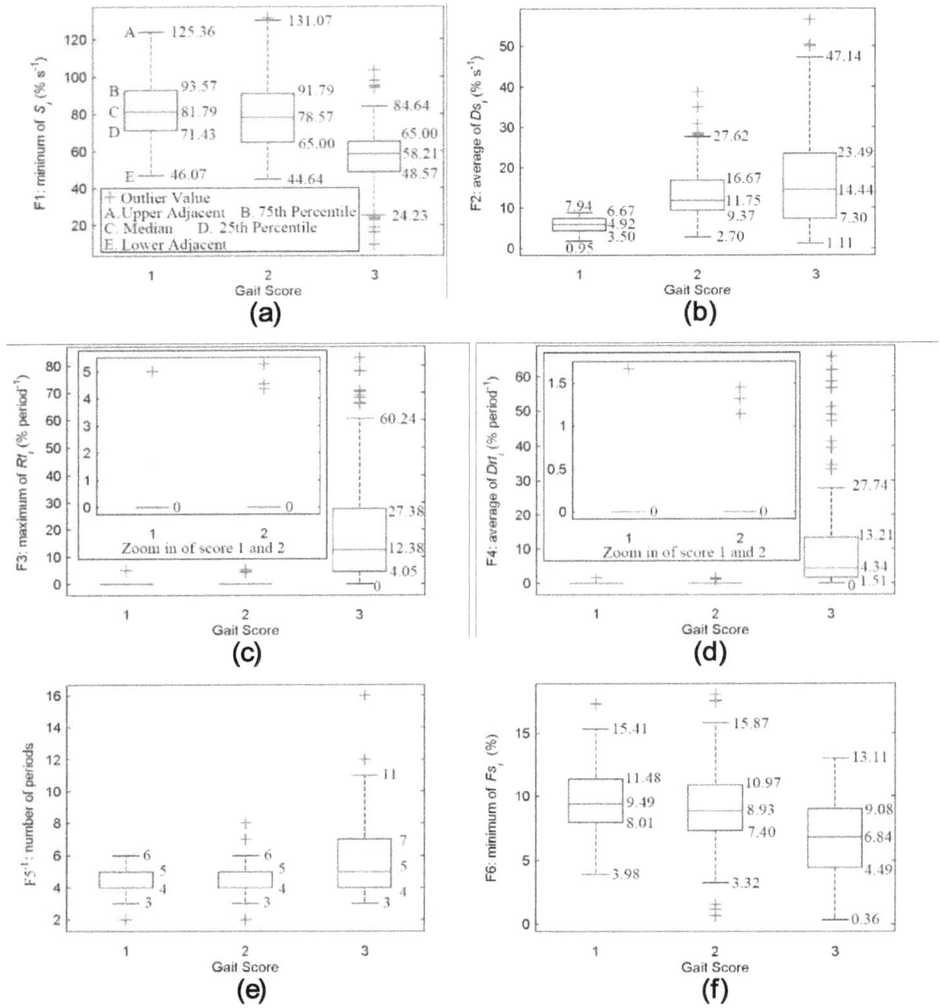

Figure 4. Relationship of features and gait score: (A) boxplot of *F1*, S_i is the swing speed of limb, (B) boxplot of *F2*, Ds_i is the difference between S_i and S_{i+1}, (C) boxplot of *F3*, Rf_i is the proportion of stopping frames, (D) boxplot of *F4*, Drf_i is the difference between Rf_i and Rf_{i+1}, (E) boxplot of *F5*, (F) boxplot of *F6*, Fs_i is the swing speed of fore limb before touching ground.

Table 2. Confusion matrix of the decision tree classifier using 10-fold cross validation.

Classified by the expert	Classified by the algorithm		
	Score 1	Score 2	Score 3
Score 1	171	6	0
Score 2	11	261	10
Score 3	0	9	153

Table 3. The evaluation of classification of lameness using algorithm.

Classified by the expert	True positive rate (%)	False positive rate (%)	Precision (%)
Score 1	96.61	1.35	93.96
Score 2	92.55	6.19	94.57
Score 3	94.44	1.96	93.87
Average	94.54	3.17	94.13

Table 4. Confusion matrix of the decision tree classifier using 10-fold cross validation and binary classification.

Classified by the expert	Classified by the algorithm	
	Not lame	Lame
Not lame	170	7
Lame(lame + severely lame)	9	435

4. Conclusions

This paper proposed a system for classifying the lameness in cows automatically in dairy using video analysis of movement analysis. The results suggested that the motion curve extracted by image processing had a great relationship with manual reference, and the motion features were feasible to quantify the lameness in cows. Multi-features were capable to express the lameness information more accurately through all the lameness stages. This study proved that movement analysis method with image processing technique has great potential to be used for analyzing and perceiving walking behavior of animals. Decision tree learning was used as a classifier, and the result of classification tests showed that the overall accuracy was 94.20% of all samples when cows were scored into 3 classes, and the recognition accuracy in binary classification was 97.10%.

References

Bruijnis, M., Beerda, B., Hogeveen, H. and Stassen, E.N. (2012). Assessing the welfare impact of foot disorders in dairy cattle by a modeling approach. Animal 6(6): 962-970.

Cook, N.B. (2003). Prevalence of lameness among dairy cattle in Wisconsin as a function of housing type and stall surface. Journal of the American Veterinary Medical Association 223(9): 1324-1328.

Flower, F.C. and Weary, D.M. (2009). Gait assessment in dairy cattle. Animal 3(1): 87-95.

Josse, J., Pages, J. and Husson, F. (2008). Testing the significance of the RV coefficient. Computational Statistics and Data Analysis 53(1): 82-91.

Pluk, A., Bahr, C., Poursaberi, A., Maertens, W., Van Nuffel, A. and Berckmans, D. (2012). Automatic measurement of touch and release angles of the fetlock joint for lameness detection in dairy cattle using vision techniques. Journal of Dairy Science 95(4): 1738-1748.

Poursaberi, A., Bahr, C., Pluk, A., Van Nuffel, A. and Berckmans, D. (2010). Real-time automatic lameness detection based on back posture extraction in dairy cattle: shape analysis of cow with image processing techniques. Computer and Electronics in Agriculture 74(1): 110-119.

Scott, G.B. (1989). Changes in limb loading with lameness for a number of Friesian cattle. The British Veterinary Journal 145(1): 28-38.
Van Nuffel, A., Sprenger, M., Tuyttens, F. and Maertens, W. (2009). Cow gait scores and kinematic gait data: can people see gait irregularities? Animal Welfare 18(4): 433-439.
Viazzi, S., Bahr, C., Schlageter-Tello, A., Van Hertem, T., Romanini, C.E.B., Pluk, A., Halachmi, I., Lokhorst, C. and Berckmans, D. (2013). Analysis of individual classification of lameness using automatic measurement of back posture in dairy cattle. Journal of Dairy Science 96(1): 257-266.
Zhu, S. and Ma, K.K. (2000). A new diamond search algorithm for fast block-matching motion estimation. IEEE Transaction on Image Processing 9(3): 287-290.

Automated footbaths to implement a standardized protocol: impact digital dermatitis dynamic

K. Orsel[1*], L.M. Solano[1], C. Pickel[1], D. Döpfer[2] and H.W. Barkema[1]
[1]*University of Calgary, Faculty of Veterinary Medicine, Department of production animal health, 3330 Hospital Drive NW, Calgary, AB, T2N 4N1, Canada;* [2]*University of Wisconsin, School of Veterinary Medicine, Department of food animal production medicine, 2015 Linden Drive, Madison, WI 53706, USA; karin.orsel@ucalgary.ca*

Abstract

Digital dermatitis (DD) is the most prevalent foot lesion in Canada and commonly found around the world. The most common on-farm prevention strategy for DD is the use of footbaths. However, there is a wide variability in footbath practices regarding frequency of use, products and concentration, refreshing of solutions and footbath dimensions. Despite a wealth of information in scientific literature, there seems to be a knowledge translation gap between research and on-farm application. A longitudinal trial was performed on 9 Alberta dairy farms for 7 months. Each farm served as its own historical control with data collection before and after the intervention. The intervention consisted of the implementation of a computerized automated footbath on all farms with a standardized protocol based on literature recommendations. DD scoring using the M-scoring system was done every 2 weeks in the milking parlor and confirmed with observations in the trimming chute during 3 sessions. Controlled concentration and programmed refreshing of footbath solutions resulted in changed dynamics of DD on farms. All farms had a significant increase of M0 scores (no lesions; increase from 31 to 40%), whereas cows diagnosed with DD lesions transferred from active M2 to more chronic stages of the infection (M3-M4). Prevalence of cows affected with M2 lesions reduced from 41 to 25% overall, and M3-M4 from 28 to 35% overall). Computerized footbaths programmed with a standardized protocol are effective under field conditions to control the dynamics of DD on farm into a steady state with limited new cases and controllable chronic lesions. Practical implications: Automated footbaths can have a significant contribution to the control of infectious foot diseases on dairy farms.

Keywords: digital dermatitis, footbath, standardized protocol, automated

1. The use of footbaths to control infectious foot disorders

Worldwide, footbaths are considered to be an important component of a preventive foot health programs. They are essential in controlling potential contagious foot disorders such as digital dermatitis (DD; aka hairy heel warts, strawberry heel warts, Mortellaro's disease). With the dairy industry moving away from tiestall housing, management in freestall housing with slatted or concrete flooring has led to a more frequent occurrence of infectious foot lesions.

Footbaths are an important component of the management of both pastured cows and cows housed indoors. The first step in installing a footbath is to determine the location of the bath; it must allow for undisturbed cow flow. Most often a position is chosen where cows exit the milking parlor; in pasture settings, outdoor locations can be identified as well. These locations are chosen where all cows need to pass on a daily basis, without the opportunity to avoid the footbath. To allow for easy use of footbaths, filling and draining ease must also be considered when identifying the location. Preferably, cows should enter the footbath with minimal manure on their feet, and exit onto a clean floor to allow the product to work on clean skin. The use of footbaths for non-lactating animals or animals newly added to the herd might be a driver to identify an alternative location that allows for heifers and dry cows to walk through the bath.

The size of the footbath is also very important. Ideally, the solution in the bath should reach the skin of the foot over the coronary band. This meets the recommendation of a footbath solution depth of approximately 10 cm (4 inches) when the last cow walks through the footbath.

University of Wisconsin researchers recommend a footbath 3.0-3.7 m long and 0.5-0.6 m wide, with a 28 cm step-in height (length: 10-12 feet, width: 20-24 inches and ~11 inch deep; Cook *et al.*, 2012). These dimensions optimize the number of foot immersions per cow pass, while limiting the footbath volume. The bath should not be perceived by the cow as an obstacle; it should allow cows to freely step in and out of the bath. The footbath should have a textured bottom or otherwise provide enough grip (e.g. footbaths with a sponge) for the cows to safely walk through without the risk of slips and falls.

Footbaths should be used for the control of infectious foot/skin diseases and potentially for hardening of the claw horn. Certain solutions like formaldehyde, may help in hardening the horn of the claws, thus making the claws more resistant to non-infectious claw disorders (Arkins *et al.*, 1986). When footbaths are used in temperatures below 10 °C, the efficacy of the product (e.g. formalin) will drastically decrease. This can be mitigated using warm(er) water and storing solution at room temperature. If the solutions sticks to the skin of the cow's foot, the interdigital space is likely back to body temperature within minutes.

If the solutions are left overnight or between milkings in colder temperatures, reduced efficacy can be anticipated. In high temperatures solutions might evaporate or precipitate, also negatively impacting the efficacy. A general recommendation is to refresh the solution after 200 cow passes. This number is somewhat variable and should be optimized for each farm, taking into account herd size, contamination with organic material, temperatures, etc.

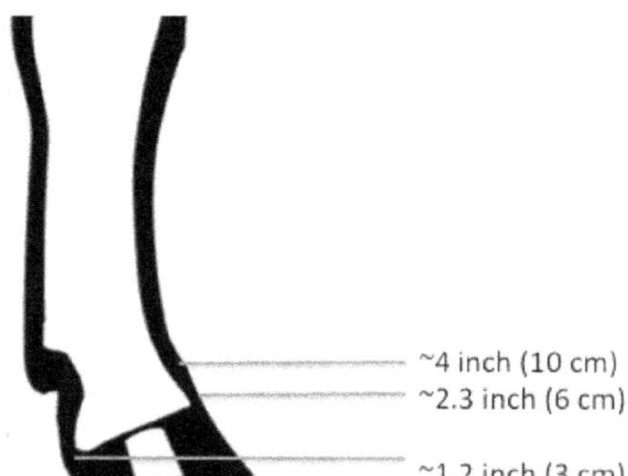

Figure 1. Preferred depth of the footbath solution.

2. Current use of footbaths in the dairy industry in Canada

In 2011-2012, a cluster of Canadian researchers investigated cow longevity and comfort. Part of this assessment focused on lameness and lameness management. In Quebec, Ontario and Alberta, a footbath was routinely used on 122 farms, which represented 87% of all farms visited. Smaller herds (below 100 cows) were less likely to not use a footbath (Solano *et al.*, 2015). One to 4 products were used per farm, the most common being copper sulphate and formaldehyde, with median concentrations of 4.5% (ranging from 0.3 to 12.5%) and 5% (ranging from 1 to 10%), respectively. On farms where footbaths are present, half of the farms visited used footbaths two days/week, whereas the other half used them more than two days. Footbaths were more frequently used on farms with lower lameness prevalence (Table 1).

Cook *et al.* (2012) studied freestall-housed dairy herds with an average herd size of 1,023 milking cows in 5 different countries (US, Spain, Japan, UK and New Zealand), and found that footbaths were used 1-4 times per day for 1-7 days per week, with between 80 and 3,000 cows passing through the bath between changes of the chemical solution. Similar to that found in Canada, the most common agents used were copper sulphate (63%) and formaldehyde (34%). Twenty-seven herds (42%) used more than one chemical. The median footbath was 2.03 m long by 0.81m wide (80×32 in), and was filled to a depth of 0.11 m (4.3 in) (Cook *et al.*, 2012).

3. Footbaths used in a standardized matter

With the large variation in frequency of use, products used, concentration and refreshing footbath solutions, we decided to perform a follow-up study measuring the impact on DD prevalence of a scientific based, standardized footbath protocol.

Ten farms were selected with at least 10% prevalence of DD within the lactation herd. Nine continued participation in the trial. All farms are in the Canadian province of Alberta, representing the average of the dairy industry (Table 2).

The prevalence of DD in the lactating herd was followed over a period of 4 months. DD scoring was done every 2 weeks in the milking parlor and confirmed with the observations in the trimming chute. For DD diagnostics we used the so-called M-score which classifies the foot in 5 categories: M0 (no

Table 1. Footbath management variables for dairy herds with low, medium and high lameness prevalence (Solano *et al.*, 2015).[1]

	Herd level lameness prevalence			
	Low (≤10%) n=24	Medium (10-30%) n=94	High (≥30%) n=23	Overall n=122
Footbath product (%)				
Copper sulphate	38	38	35	37
Formaldehyde	5	16	20	15
Copper & formaldehyde	52	40	35	41
Other	5	6	10	7
Number footbath products (%)				
1	29	41	35	38
2	57	44	50	47
≥3	14	15	15	15
Footbath frequency of use (%)				
≤2 d/wk	38	51	50	48
>2 d/wk	62	49	50	52
Footbath dimension (median)				
Length (cm)	186	186	220	208
Width (cm)	76	72	73	74
Depth (cm)	15	16	15	15

[1] The overall category excludes 19 farms that had a footbath, but rarely used it.

Table 2. Distribution of continuous (mean ± SD) and categorical (n(%)) herd-level variable of 9 freestal dairy farms and the average freestall farms in Alberta, Canada.

	Study farms	Average Alberta freestall farm
No. lactating cows	152±33	123
Daily milk yield (kg)	33.3±9.5	32
Parity	2.1±1.3	2.3
305 d milk yield (kg)	10,294±1,984	9,960

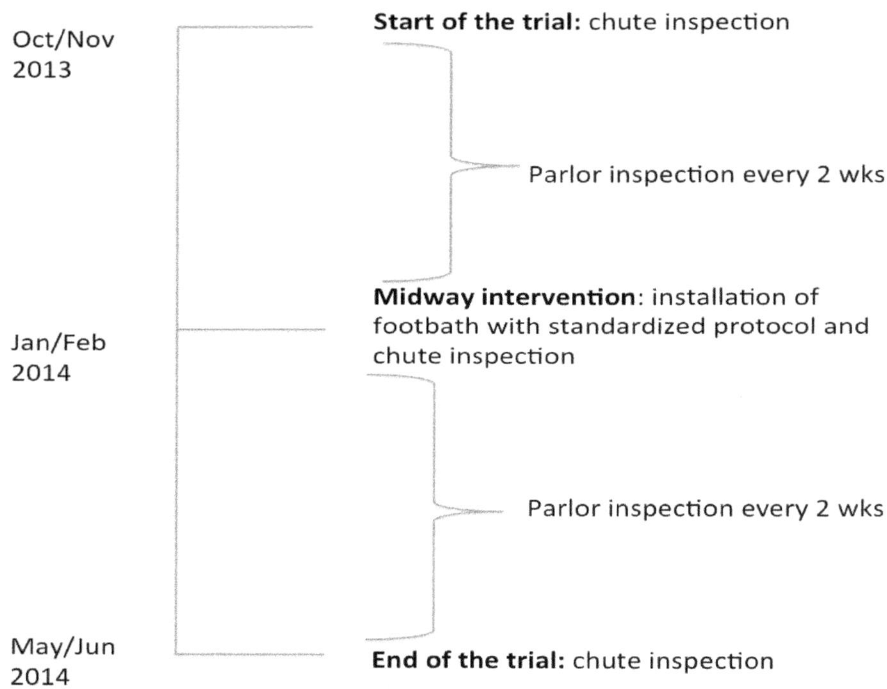

Figure 2. Timeline for DD prevalence study with implementation of standardized computerized footbath protocol.

lesions), M1 (small early lesion), M2 (larger active lesion), M3 (healing lesion), M4 (chronic lesion) and M4.1 (chronic lesion, with an active M1 lesion) (Berry *et al.*, 2012). Two months after the start of the study an automated footbath was implemented, with a computer based weekly protocol of 2 consecutive days (4 milkings) using 5% copper sulphate, refreshed after approximately 200 cow passages (Cook *et al.*, 2006, Relun *et al.*, 2012, Speijers *et al.*, 2010, 2012).

Controlled concentration and programmed refreshing of footbath solutions resulted in a decrease of active stages of DD on these dairy farms. All farms had a significant increase in M0 scores from 31 to 40%, whereas cows diagnosed with active M2 lesions transferred to more chronic stages of the infection (M3-M4). Cows affected with active M2 lesions decreased from 41 to 25%, whereas M3-M4 lesions increased from 28 to 35%.

4. Conclusions

Well-designed footbaths that are carefully programmed will result in a decrease of the prevalence of active DD lesions.

References

Arkins, S., Hannan, J. and Sherington, J. (1986). Effects of formalin footbathing on foot disease and claw quality in dairy cows. Veterinary Record 118(21): 580-583.

Berry, S.L., Read, D.H., Famula, T.R., Mongini A. and Döpfer, D. (2012). Long-term observations on the dynamics of bovine digital dermatitis lesions on a California dairy after topical treatment with lincomycin HCl. Veterinary Journal 193(3): 654-658.

Cook, N.B. (2006). Footbath alternatives. Available at: http://tinyurl.com/gktwq2x.

Cook, N.B., Rieman, J., Gomez, A. and Burgi, K. (2012). Observations on the design and use of footbaths for the control of infectious hoof disease in dairy cattle. Veterinary Journal 193(3): 669-673.

Relun, A., Lehebel, A., Bareille, N. and Guatteo, R. (2012). Effectiveness of different regimens of a collective topical treatment using a solution of copper and zinc chelates in the cure of digital dermatitis in dairy farms under field conditions. Journal of Dairy Science 95(7): 3722-3735.

Solano, L., Barkema, H.W., Pajor, E.A., Mason, S., LeBlanc, S.J., Zaffino Heyerhoff, J.C., Nash, C.G.R., Haley, D.B., Vasseur, E., Pellerin, D., Rushen, J., De Passillé, A.M. and Orsel, K. (2015). Prevalence of lameness and associated risk factors in Canadian Holstein-Friesian cows housed in freestall barns. Journal of Dairy Science 98(10): 6978-6991.

Speijers, M.H., Baird, L.G., Finney, G.A., McBride, J., Kilpatrick, D.J., Logue, D.N. and O'Connell, N.E. (2010). Effectiveness of different footbath solutions in the treatment of digital dermatitis in dairy cows. Journal of Dairy Science 93(12): 5782-5791.

Speijers, M.H., Finney, G.A., McBride, J., Watson, S., Logue, D.N. and O'Connell, N.E. (2012). Effectiveness of different footbathing frequencies using copper sulfate in the control of digital dermatitis in dairy cows. Journal of Dairy Science 95(6): 2955-2964.

Assessment of the possibilities to downscale the Gaitwise system by reducing the length of the sensor surface

T. Van de Gucht[1,2], W. Saeys[2], S. Van Weyenberg[1], L. Lauwers[3,4], K.C. Mertens[1], L. Vandaele[5], J. Vangeyte[1] and A. Van Nuffel[1*]

[1]Institute for Agricultural and Fisheries Research - ILVO, Technology and Food Sciences Unit, Burgemeester van Gansberghelaan 115, 9820 Merelbeke, Belgium; [2]KU Leuven Department of Biosystems, MeBioS, Kasteelpark Arenberg 30 Box 2456, 3001 Leuven, Belgium; [3]Institute for Agricultural and Fisheries Research - ILVO, Social Sciences Unit, Burgemeester van Gansberghelaan 115, 9820 Merelbeke, Belgium;[4]Ghent University, Faculty of Bio-Engineering, Department of Agricultural Economics, Coupure Links 653, 9000 Gent, Belgium; [5]Institute for Agricultural and Fisheries Research - ILVO, Animal Sciences Unit, Scheldeweg 68, 9090 Melle, Belgium; annelies.vannuffel@ilvo.vlaanderen.be

Abstract

Although automatic lameness detection systems seem promising, their high cost and low on-farm applicability impede their adoption in practice. The object of this study was to find out whether lameness detection sensors such as Gaitwise can be downscaled to smaller and cheaper systems by using less sensors. The reduction in sensors was achieved by reducing the measurement zone length. This was simulated by using only part of the data gathered with the original sensor. Next, gait variables were calculated using the modelled dataset and detection algorithms were built using linear discriminant analysis to discriminate between non lame, mildly lame and severely lame cows. Analysis of the minimum length needed to monitor a full gait cycle indicated that the measurement zone length could be reduced by maximum 30% compared to the original sensor system. Using these configurations, comparison of the overall accuracy showed little difference between the original sensor system and the simulated sensor with a shorter measurement zone. Using a shorter version of the Gaitwise decreases the space needed on farm and also the cost. Hence, this might support the adoption of such lameness detection systems in practice.

Keywords: lameness, detection, automatic, adoption, profitability

1. Introduction

Lameness in dairy cattle is an important health problem causing high economic losses and a serious deterioration of the cows' welfare (Booth *et al.*, 2004; Bruijnis *et al.*, 2010). Since visual lameness detection is difficult and time consuming (Brenninkmeyer *et al.*, 2007; March *et al.*, 2007), and herds tend to become larger (Bewley *et al.*, 2001), the need for other, more automated methods to score cow gait increases. Therefore, several prototypes of automatic lameness detection systems have been developed, using load cells (Neveux *et al.*, 2006; Thorup *et al.*, 2014), pressure mats (Maertens *et al.*, 2011; Van der Tol *et al.*, 2002), camera systems (Van Hertem *et al.*, 2014; Viazzi *et al.*, 2014) and accelerometers (Chapinal *et al.*, 2011; Munksgaard *et al.*, 2007; Pastell *et al.*, 2009).
Most of these prototypes deliver very detailed information for a high cost. In practice however, such systems are too expensive. Therefore, these systems should be downscaled in both economic and spatial terms in order to be affordable and practically installable on farm. One system that could potentially be downscaled is the Gaitwise pressure mat. This sensor provides detailed information of hoof imprints in two spatial dimensions, as well as the timing and the pressure exerted on the leg. Since Gaitwise needs a walkway with enough space before and after the sensor for cows to be able to walk at a normal pace, the original sensor surface length of 4.88 m renders the total needed space too long for installation on dairy farms in practice. Possibly, this large length and high cost

could be reduced by decreasing the amount of sensors. However, during the downscaling process, the system performance should be assessed as well, as the accuracy should not be reduced excessively. Therefore, in this study, was investigated whether the amount of sensors used could be reduced by decreasing the length of the measurement zone without an excessive negative effect on the system's detection performance.

2. Materials and methods

2.1 Data collection

Gait data of 45 lactating Holstein cows were collected during seven months on the ILVO experimental farm using the original Gaitwise sensor in the setup described by Maertens *et al.* (2011). The Gaitwise sensor consists of a pressure sensitive mat of 488 by 61 cm with a resolution of 1.27 cm that monitors footfalls in place (X and Y), time (60 Hz) and pressure (8 levels). Gait scoring was performed by a trained observer on a scale from 1 to 3 based on the presence of visual lameness indicators, which were: head bobbing, irregular hoof placement in place or time, abduction, stiff joints, tender hoof placement, low speed, arched back and tracking up. After removing measurements of cows standing still on the Gaitwise, a dataset of 1,240 measurements was retained.

2.2 Sensor downscaling

In order to reduce the amount of sensors in the length, the minimal length of the measurement zone should be known. In order to be able to calculate gait variables derived from the sensor data, at least one complete gait cycle needs to be monitored. The minimum length is thus defined by the first eight imprints of the cow on the sensor surface, since at least two imprints of each leg are needed to monitor a full gait cycle.

The length of a full gait cycle of a cow is influenced by several factors. A first factor is defined by the body proportions of the cow. Large cows take larger steps compared to smaller cows, and hence need a longer distance to complete a full gait cycle (e.g. comparison average body size Holstein – New Jersey). This effect is visualized in Figure 1, where measurements of different cows are represented. The length needed for eight imprints is indicated by the grey line. cow 1 had a slightly larger step length compared to cow 2, hence increasing the length needed to register eight imprints.

A second factor is the position of the legs in relation to the position of the pressure mat before entering the measurement zone. In case the previous imprints of the cow fell right before the start of the measurement zone, the first part of the sensor surface will not register any imprints. This implies that a larger length is needed to monitor the eight imprints of a full gait cycle. If in contrary the first imprints on the sensor surface are very close to the beginning of the measurement zone, the needed length is a lot smaller. This effect is presented in Figure 1, where cow 3 put her feet down for the first time on a different position on the sensor surface compared to cow 4, hence influencing the needed length for a full gait cycle. The position of the first imprints on the sensor surface are indicated by the red circle.

The needed minimum length was determined for each measurement in the dataset. From this analysis was concluded that a length of 328 cm was needed to monitor at least eight imprints in 99% of the measurements. The loss of 1% measurements was due to cows putting their feet down just before the start of the measurement surface, hence needing a larger length to complete a full gait cycle (cfr. cow 3 in Figure 1). In order to reach 100%, an additional 30 cm was needed, which contradicts the goal of this study. Therefore, the minimum length of the measurement zone was fixed to 328 cm for further analysis. Data of such a shorter sensor mat was simulated by retaining only the data monitored with the first 328 cm of the mat, hence simulating a shorter version of the measurement zone.

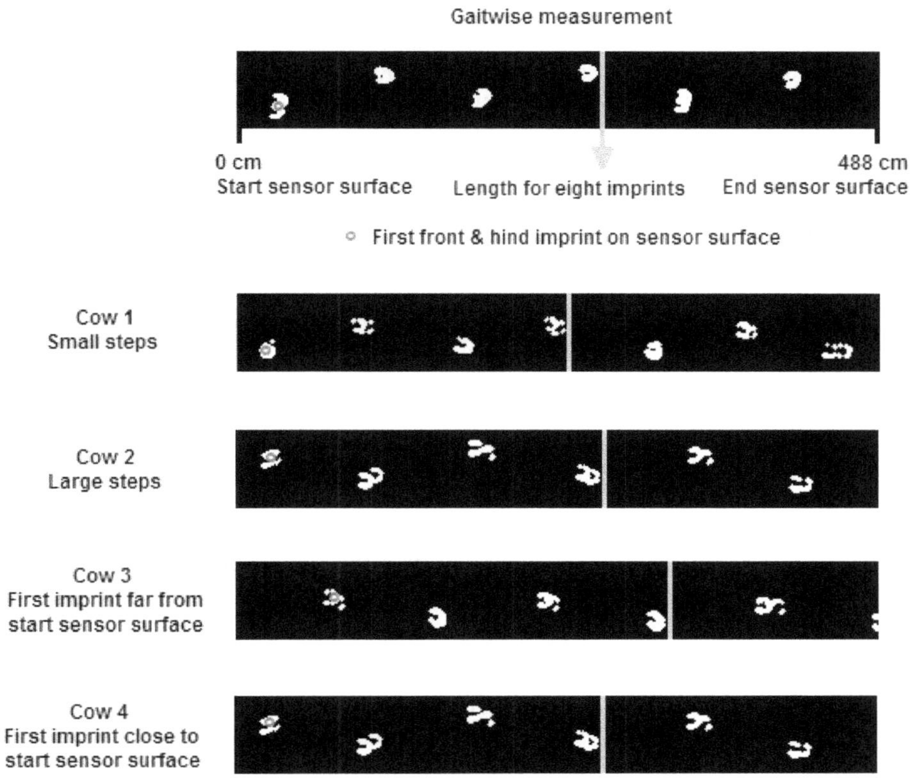

Figure 1. Visual representation of the influence of cow body proportions and the place of the first recorded imprints on the length needed to monitor one complete gait cycle (eight imprints). The length needed to monitor eight imprints is indicated by the grey line. The position of the first imprints on the sensor surface is indicated by the circle in the hoof print.

2.3 Variable calculation and lameness detection models

Gait variables were calculated based on the gait variables described by Maertens *et al.* (2011) and Van Nuffel *et al.* (2013) for both the original data and the simulated data. In each dataset, the data was split into a training set and a test set. The training set was composed of 100 randomly chosen measurements of each lameness status, thus resulting in a training set with 300 records. The remaining 940 records were used as a test set for model validation.

Linear discriminant analysis was used to build a lameness detection model for both training datasets. An ln-transformation was applied to assure normality. For both models, the test set data was used to validate the model and obtain the results for the comparison between the original sensor mat an the simulated shorter sensor mat.

3. Results

The results of the lameness detection system are presented in Table 1. For the original sensor configuration, the overall accuracy was58%. For the reduced length configuration, the accuracy

Table 1. Results of the lameness detection system based on the test data. For both the original sensor surface length and the reduced length configuration, the overall accuracy (bold), the total positive rate (TPR), the 1-False Positive Rate (FPR) and the cross table with the number of measurements in each class for the expert's gait score and for the detection system's gait score is shown.

		Original length 488 cm Gait score detection model				Minimum length 328 cm Gait score detection model			
		1	2	3	TPR (%)	1	2	3	TPR (%)
Gait	1	418	201	78	60	427	214	52	62
score	2	44	97	51	51	52	100	40	52
expert	3	3	19	29	57	6	18	27	53
	1-FPR (%)	81	71	85	58	76	69	90	59

was 59%. In this configuration, the number of cows that were falsely detected as severely lame was lower compared to the original configuration. However, the number of cows falsely detected as non-lame was higher.

4. Discussion

The minimum length of 328 cm was determined based on the herd at the ILVO farm. However, since average cow body proportions can vary between farms, this minimum length may not be suitable for all herds. Therefore, the minimum length should be determined based on data of several herds with different average body proportions in order to find a general minimum length that is applicable for all farms and cow breeds.

The results of the detection system indicated that there were some shifts in the number of false alarms between false positives and false negatives when comparing the reduced sensor system to the original. The overall detection accuracy was not decreased, indicating that it is possible to reduce the length of the measurement zone without decreasing the lameness detection performance. However, more research is needed to improve the overall detection accuracy, as the accuracy of the original pressure mat was only 58%.

5. Conclusions

The results in this study indicated that it is possible to downscale the length of the Gaitwise sensor by reducing the number of sensors in the length of the measurement surface without an extensive deterioration of the lameness detection performance. The minimum length of the sensor system was set to 328 cm, but should be confirmed on farms with different cow body sizes. Future research should focus on determining the difference in cost of the smaller system compared to the original system size, and on investigating whether the sensor resolution can be downscaled as well in order to reduce the cost further.

References

Bewley, J., Palmer, R.W. and Jackson-Smith, D.B. (2001). An overview of experiences of Wisconsin dairy farmers who modernized their operations. Journal of Dairy Science 84(3): 717-729.

Booth, C.J., Warnick, L.D., Grohn, Y.T., Maizon, D.O., Guard, C.L. and Janssen, D. (2004). Effect of lameness on culling in dairy cows. Journal of Dairy Science 87(12): 4115-4122.

Brenninkmeyer, C., Dippel, S., March, S., Brinkmann, J., Winckler, C. and Knierim, U. (2007). Reliability of a subjective lameness scoring system for dairy cows. Animal Welfare 16(2): 127-129.

Bruijnis, M.R.N., Hogeveen, H. and Stassen, E.N. (2010). Assessing economic consequences of foot disorders in dairy cattle using a dynamic stochastic simulation model. Journal of Dairy Science 93(6): 2419-2432.

Chapinal, N., De Passille, A.M., Pastell, M., Hanninen, L., Munksgaard, L. and Rushen, J. (2011). Measurement of acceleration while walking as an automated method for gait assessment in dairy cattle. Journal of Dairy Science 94(6): 2895-2901.

Maertens, W., Vangeyte, J., Baert, J., Jantuan, A., Mertens, K.C., De Campeneere, S., Pluk, A., Opsomer, G., Van Weyenberg, S. and Van Nuffel, A. (2011). Development of a real time cow gait tracking and analysing tool to assess lameness using a pressure sensitive walkway: The GAITWISE system. Biosystems Engineering 110(1): 29-39.

March, S., Brinkmann, J. and Winkler, C. (2007). Effect of training on the inter-observer reliability of lameness scoring in dairy cattle. Animal Welfare 16(2): 131-133.

Munksgaard, L., Van Reenen, C.G. and Boyce, R. (2007). Automatic monitoring of lying, standing and walking behavior in dairy cattle. In: ADSA/ASAS 2006 Annual Meeting, Minneapolis, MN, USA.

Neveux, S., Weary, D.M., Rushen, J., Von Keyserlingk, M.A.G. and De Passille, A.M. (2006). Hoof discomfort changes how dairy cattle distribute their body weight. Journal of Dairy Science 89(7): 2503-2509.

Pastell, M., Tiusanen, J., Hakojarvi, M. and Hanninen, L. (2009). A wireless accelerometer system with wavelet analysis for assessing lameness in cattle. Biosystems Engineering 104(4): 545-551.

Thorup, V.M., Do Nascimento, O.F., Skjoth, F., Voigt, M., Rasmussen, M.D., Bennedsgaard, T.W. and Ingvartsen, K.L. (2014). Short communication: changes in gait symmetry in healthy and lame dairy cows based on 3-dimensional ground reaction force curves following claw trimming. Journal of Dairy Science 97(12): 7679-7684.

Van der Tol, P.P.J., Metz, J.H.M., Noordhuizen-Stassen, E.N., Back, W., Braam, C.M. and Weijs, W.A. (2002). The pressure distribution under the bovine claw during square standing on a flat substrate. Journal of Dairy Science 85(6): 1476-1481.

Van Hertem, T., Viazzi, S., Steensels, M., Maltz, E., Antler, A., Alchanatis, V., Schlageter-Tello, A.A., Lokhorst, K., Romanini, E.C.B. and Bahr, C. (2014). Automatic lameness detection based on consecutive 3D-video recordings. Biosystems Engineering 119: 108-116.

Van Nuffel, A., Vangeyte, J., Mertens, K.C., Pluym, L., De Campeneere, S., Saeys, W., Opsomer, G. and Van Weyenberg, S. (2013). Exploration of measurement variation of gait variables for early lameness detection in cattle using the GAITWISE. Livestock Science 156: 88-95.

Viazzi, S., Bahr, C., Van Hertem, T., Schlageter-Tello, A., Romanini, C., Halachmi, I., Lokhorst, C. and Berckmans, D. (2014). Comparison of a three-dimensional and two-dimensional camera system for automated measurement of back posture in dairy cows. Computers and Electronics in Agriculture 100: 139-147.

Development of a non-invasive system for monitoring dairy cattle sleep

J.M. Klefot[1], J.L. Murphy[1], K.D. Donohue[2], F. O'Hara[3], M.E. Lhamon[4] and J.M. Bewley[1*]
[1]*Department of Animal and Food Sciences, University of Kentucky, Lexington, KY, USA;* [2]*Center for Visualization and Virtual Environments, University of Kentucky, Lexington, KY, USA;* [3]*Department of Biology, University of Kentucky, Lexington, KY, USA;* [4]*Digital Key Consulting, Lexington, KY, USA; jbewley@uky.edu*

Abstract

Limited research has been conducted to assess sleep in production livestock, primarily because of limitations with monitoring capabilities. Consequently, biological understanding of production circumstances and facility options that affect sleep was limited. The objective of this study, conducted at the University of Kentucky Coldstream Dairy in September 2013, was to assess the proof of concept usefulness of a non-invasive system using a 3-axis accelerometer device to correlate between sleep-like behaviors and wake behaviors by using signals from the device. Four Holstein dairy cows were visually observed for two 24-hour periods where two people each observed two cows, recording the times of each behavior change. Head and body movements were then compared to signals from the device. The accelerometer device was attached to a harness positioned on the right side of each cow's neck. Times of classified behaviors of wake or sleep-like behaviors were recorded continuously. The radial signal was extracted from three different axes of the accelerometer to obtain a motion signal independent of direction of movement. Radial signal features were examined for maximizing the performance of detecting sleep-like behaviors using a Fisher's linear discriminant analysis (LDA) classifier. The study included 652 min of high activity wake behaviors and 107 min of sleep-like behavior among 4 cows. Results from a bootstrapping analysis showed an agreement between human observation and the LDA classifier with accuracy of 93.7±0.7% for wake behavior and 92.2±0.8% for sleep-like behavior, with a 95% confidence interval. This first prototype shows promise in assessing sleep-like behaviors. Improvements to both hardware and software should allow more accurate determinations of subtle head movements and respiratory movements that will further improve the assessment of these sleep-like behaviors, including estimates of deep sleep, light sleep, and REM amounts. These future studies will require simultaneous EEG/EMG measures and perhaps additional measures of arousal thresholds, in order to fully validate this non-invasive system.

Keywords: behavior, accelerometer, circadian rhythm

1. Introduction

Monitoring dairy cattle physiology and behavior has extensive implications within both research and on-farm settings. Dairy consumers and producers often express concern over animal well-being. Sleep is essential to the welfare of the animal and is vital to brain function and maintenance of homeostatic control (Hobson, 2005). According to Siegel (2005), sleep can be defined as a state of immobility with greatly reduced responsiveness. Both non-rapid eye movement (NREM) and rapid eye movement (REM) sleep are homeostatically regulated in separate ways suggesting that each has important functions for optimal health. Increasing evidence suggests a clear link with the immune system and disease resistance. Sleep patterns may also reflect the adaptation of a cow to her environment (Hokkanen *et al.*, 2011). Sleep research in human medicine has advanced considerably in recent years largely using activity based monitoring systems (Chen *et al.*, 2013.). Although the basic physiological functions of sleep are not well understood, a breakthrough in this area has recently been made by Maiken Nedergaard's lab at Rochester University. It now appears that a primary function of sleep is to flush the brain of toxic waste products that accumulate during wake (Xie *et al.*, 2013). This new feature of sleep appears to be a fundamental and core function and may be critical to other sleep functions including improved learning, memory, optimal immune function, endocrine function, and many aspects of general health.

Research specifically on dairy cattle sleep is limited and has consisted primarily of brain electrophysiology electroencephalograms (EEG) (Hänninen *et al.*, 2008). Very few of these studies exist due to cost and physical invasiveness of these procedures. Dairy cattle sleep research has been limited due to the inability to accurately and easily measure sleep versus wake states. It could also be argued that most researchers have not thought to measure sleep. Most current research looks at lying time, which focuses on quantity of resting behaviors rather than quality of rest behaviors. Understanding and monitoring dairy cattle sleep under typical dairy farm conditions may improve health, increase milk production, and reduce involuntary culls resulting from disease (Gamaldo *et al.*, 2012). Research from Ruckebusch in 1972 used EEG and electrocorticographic (ECoG) patterns to classify slow-wave sleep (characterized with low muscle tone) and paradoxical sleep (absence of muscle tone) along with drowsy states. The results showed that cows spend most of their time awake in a drowsy state when accustomed to their environment and are not sleeping or eating.

The work of this project seeks to develop and test a proof of concept device to efficiently track sleep-like behaviors and design experiments to assess sleep behavior or its lack thereof on dairy cow immune function, health, and productivity. The long-term objective is to continue developing this device to characterize sleep patterns in dairy cattle to assess sleep quality in research and on farm.

2. Materials and methods

The study was conducted at the University of Kentucky Coldstream Dairy using lactating Holstein (n=4) cows in September 2013. The sample size sufficient to allow for development of the classifier with one set, and the other set could test the classifier. The cows were housed in a tiestall barn for the duration of the study. The tiestall barn consisted of tiestalls measuring 125 cm wide × 170 cm long with Dual Chamber Cow Waterbeds (Advanced Comfort Technology Inc. Reedsburg, WI) as the freestall base. Kiln-dried sawdust (5 cm) was added for bedding twice daily. Cows were provided *ad libitum* water and total mixed rations. Fans were on during most observation periods for extra ventilation. Natural light was used to observe cows during the day and at night, two barn lights created a dim light that allowed observers to see the cows, with the least disturbance. Observers sat approximately two meters directly behind the four cows, and were urged to remain quiet and move forward only to confirm whether their eyes were opened or closed.

Cows were observed continuously and times of wake behaviors, drowsy behaviors, and sleep-like behaviors were recorded. The observation periods were a total of 24 hours, with two observers per two cows for a shift of four hours. Each observer was trained and given specific instructions on how to record each individual behavior. Wake behaviors were classified as the cow being alert (eating, drinking, grooming, ruminating, head movement, eyes open) while standing or lying. Drowsy behaviors were classified as the cow lying still, with their head down, ruminating, with eyes closed or partially closed. Sleep-like behaviors were classified with the movements of lying with head relaxed either resting on the ground or flank with eyes closed.

The three-axis accelerometer device, developed by the co-authors, sampled at 32 Hz. The system consisted of a low-cost battery-powered accelerometer that fit on a halter attached to the cow's neck. Data were stored on a removable SD card. Two AA batteries located both internally and externally powered the device. The direct outputs of the 3-axis accelerometer are voltages proportional to accelerations associated with motion in the x, y, and z orthogonal directions. Because the orientation of the device on the side of the cow's neck was not exactly the same for all cows, a motion signal independent of direction was derived. This captured the energy and force excreted by the cow and quasi-periodic motions from rumination and breathing. classifier (Donohue *et al.*, 2008; McLachlan, 2004). The signal features were examined in a Fishers Linear Discriminant Analysis (LDA) classifier to discriminate sleep-like behaviors from all other wake behaviors.

The ability for the features to discriminate these behaviors was dependent on the window length *N*, used to compute the features. States of relative stillness occurred when the cow was alert and were similar to the stillness observed in sleep-like behavior. Figure 1 illustrates this, where s_N was computed over 4 second windows (n=128). Three types of behavior (from different times) were overlaid on the same plot for comparison. Note that the state of sleep-like behavior remains at a low level throughout the 350-second time period. However, during the observed wake interval there were up to 30 second intervals where the stillness was similar to sleep-like behavior. The drowsing behaviors were characterized by heavier breathing; subtle movements of rumination were at a consistent higher level with transient dips toward the level of the sleep-like behavior. Although it was initially thought that drowsing would be a greater challenge for distinguishing it from sleep, this result (shown at a higher time resolution that in Figure 3 shows that for some periods of standing the accelerometer picks up stillness similar to sleep-like behavior.

The movements recorded by the accelerometer in this wake state are less frequent than in other states, they are very strong. To enhance the ability of the feature to distinguish this state from a sleep-like state, the window size can be increased to increase the likelihood that a movement during this wake state will occur. Figure 2 compares the same time periods with a 64 second interval used for computing $s_{N.}$ Note in this case there is a clear separation between wake and sleep behaviors. This suggests a trade-off between resolution and accuracy of the classifier. Therefore, the classifier was tested with multiple windows sizes (32, 64, and 96 seconds) to better understand this trade-off. A bootstrapping method was used to assess performance. Subsets of the feature vectors in each class were selected for training and testing multiple times to estimate the repeatability of the estimate. Sets of feature vectors for each testing and training phase are exclusive (i.e. vectors used to train classifier and not used in the testing to ensure the classifier generalizes and that it was not simply

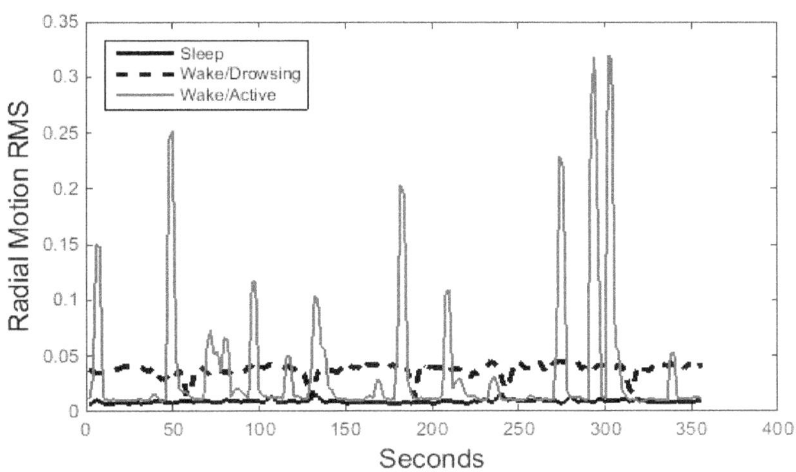

Figure 1. Radial acceleration Standard Deviation (RMS power) from an accelerometer-based sleep device computed over 4 s sliding windows.[1]

[1]The peaks represent sudden movements such as a jerk of the head while the cow was alert. The wake/drowsing line show consistent dips indicating movements while ruminating, and the sleep line shows a state in which the cow was still.

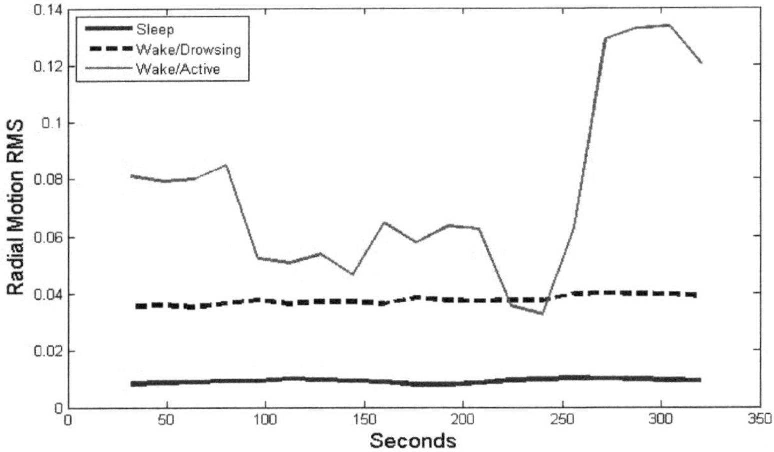

Figure 2. Radial acceleration RMS values from an accelerometer-based sleep device computed over 64 s sliding windows.[1]

[1]This data shows motivation for using a larger segment to compute RMS values. This allows recognition of the different levels of motion and separation between wake-active and sleep data.

learning the data set). In this analysis the training and testing phase was repeated 100 times. The mean correct classification rate for each class and the 95% confidence limits were computed for 3 different window sizes. Features were applied individually and in combination with each other.

3. Results and discussion

Results from the bootstrapping analysis between human observation and the LDA are presented in Table 1. As expected, the best overall performance came for the large window size with a 96.2±0.7% agreement when using s_N by itself; however p_N and the combined set (feature vector) were within each other's 95% confidence limits. The smallest window, 32 s, achieved a classification agreement of 89.0±1.3% for the combined features. All cases showed no statistical difference between using the feature individually or combined in a vector. All were within their 95% confidence ranges. In addition, there were no strong imbalances between classification errors in the sleep-like or wake state. The greatest difference between sleep-like behaviors and wake accuracy came from the 32 s window showing a 3 to 5% difference favoring more accuracy for the sleep-like state.

The impact of a larger window size improved accuracy, it lowered the resolution of making a sleep-wake decision. The shorter 4-second window had a higher resolution and showed specific peaks in each line where the larger window was accurate in showing the difference in the radial signal between the sleep-like behaviors and the wake behaviors (Figure 1). Sleep-like bouts of about 20 s would likely not be detected with the 64-second window (Figure 2), but would with the 32-second window, as it was able to pick up shorter bouts of sleep-like behaviors. Improving the resolution additional features that contribute independent information is needed. Other features from the radial signal were initially examined, such as entropy, and spectral peak statistics, but these showed little to no correlation with the observed sleep-like and wake states. Most of the performance was driven by the movements detected in the neck area (Hänninen *et al.*, 2008). Both features used were highly correlated, and that was why their combination in a feature vector did not result in a significant improvement. Other features independent or complementary features would be needed for exploring

Table 1. The mean correct classification rate for each class and the computed 95% confidence limits for features extracted from different window lengths with ±95% confidence intervals from an accelerometer-based sleep device.

Feature(s)	Window length	% agreement sleep-like behaviors	% agreement wake behaviors	% agreement overall
p_N	32 s	91.1±1.1	86.4±1.2	88.8±1.2
σ_N	32 s	90.5±1.2	87.4±1.3	88.9±1.3
σ_N, p_N	32 s	90.6±1.2	85.6±1.2	89.0±1.3
p_N	64 s	92.2±1.0	92.5±1.0	92.4±1.0
σ_N	64 s	92.3±1.1	93.8±0.9	93.1±1.0
σ_N, p_N	64 s	91.5±1.0	91.3±1.0	92.7±1.0
p_N	96 s	97.0±0.6	95.5±0.9	96.2±0.8
σ_N	96 s	96.9±0.7	95.5±0.8	96.2±0.7
σ_N, p_N	96 s	96.4±0.7	94.8±0.9	95.6±0.8

improvements in accuracy. The current device was limited to detecting motion independent of direction and orientation. More effective mounting procedures, multiple accelerometers, and other biological estimates from the animal could potentially improve performance and allow multiple behaviors to be detected and tracked, but this would potentially increase the complexity of the device and mounting procedures.

Sleep in animals is often estimated by behavioral observations, or recorded on restrained animals with invasive electroencephalogram (EEG) techniques, which might affect sleep patterns (Ternman *et al.*, 2012). In the current study, the accelerometer was placed on a halter, providing minimal irritation to the animals, thus minimally affecting sleep patterns. The change in surroundings from a freestall barn, where these cows were normally housed, to a tiestall barn could have altered sleeping times. Human presence also may have reduced normal amounts of sleep. Previous studies using observed behaviors validated with EEG recordings provided subjective classification methods of behaviors as sleep versus wake behaviors in calves but not cows (Hänninen *et al.*, 2007; Ternman *et al.*, 2012). These researchers demonstrated that the behaviors were not as useful for cows as for calves.

Limitations of the experiment included battery life and recording consistency of the monitors. Development of future versions of the device will include consideration of alternative battery locations and a protective cover. Data analysis was limited due to complications with the monitors and the inability of them to get wet; however, enough data was provided to validate sleep behaviors and sleep time.

4. Conclusions

This newly developed proof of concept-tested device was successful in classifying sleep-like behaviors compared to human observations. However, the device needs to be validated with an EEG to actually know if the cow is sleeping or not. It could be valuable information in knowing that the cow is getting 'quality rest' based on those behaviors even if we do not know for sure if she is actually in a sleep state. The accuracy of the device suggests that this type of device could be used to track animal health and well-being in research and on farm in the future but at this point could not be used in large settings.

References

Chen, Z. (2013). Unobtrusive sleep monitoring using smartphones. In: Pervasive computing technologies for healthcare. 7th International Conference, pp. 145-152.

Donohue, K.D., Medonza, D.C., Crane, E.R. and O'Hara, B.F. (2008). Assessment of a non-invasive high-throughput classifier for behaviours associated with sleep and wake in mice. Biomedical Engineering Online 7: 14.

Fisher, A.D., Verkerk, G.A., Morrow, C.J. and Matthews, L.R. (2002). The effects of feed restriction and lying deprivation on pituitary-adrenal axis regulation in lactating cows. Livestock Production Science 73(2-3): 255-263.

Gamaldo, C.E., Shaikh, A.K., McArthur, J.C. (2012). The sleep-immunity relationship. Neurologic Clinics 30(4): 1313-1343.

Hänninen, L. (2007). Sleep and rest in calves-relationship to welfare, housing and hormonal activity. Academic Dissertation, University of Helsinki, Helsinki, Finland.

Hänninen, L., Mäkelä, J.P., Rushen, J., De Passillé, A.M. and Saloniemi, H. (2008). Assessing sleep state in calves through electrophysiological and behavioural recordings: a preliminary study. Applied Animal Behaviour Science 111: 235-250.

Hobson, J.A. (2005). Sleep was of the brain, by the brain and for the brain. Nature 437: 1254-1256.

Hokkanen, A.-H., Hänninen, L., Tiusanen, J. and Pastell, M. (2011). Predicting sleep and lying time of calves with a support vector machine classifier using accelerometer data. Applied Animal Behaviour Science 134: 10-15.

McLachlan, G.J. (2004). Discriminant analysis and statistical pattern recognition. Wiley Interscience, New York, NY, USA.

Ruckebusch, Y. (1972). The relevance of drowsiness in the circadian cycle of farm animals. Animal Behaviour 20: 637-643.

Siegel, J.M. (2005). Clues to the functions of mammalian sleep. Nature 437: 1264-1271.

Ternman, E., Hänninen, L., Pastell, M., Agenäs, S. and Nielsen, P.P. (2012). Sleep in dairy cows recorded with a non-invasive EEG technique. Applied Animal Behaviour Science 140: 25-32.

Xie, L., Kang, H., Xu, Q., Chen, M.J., Liao, Y., Thiyagarajan, M., O'Donnell, J., Christensen, D.J., Nicholson, C. and Iliff, J.J. (2013). Sleep drives metabolite clearance from the adult brain. Science 342: 373-377.

Pressure-indicating film to measure the pressure between a collapsing liner and an artificial teat

S. Demba[1*], S. Rose-Meierhöfer[2], K. Schröter[1], S. Elsholz[1] and R. Brunsch[1]
[1]*Leibniz Institute for Agricultural Engineering Potsdam-Bornim, Department of Engineering for Livestock Management, Max-Eyth-Allee 100, 14469 Potsdam, Germany;* [2]*Hochschule Neubrandenburg, University of Applied Sciences, Department of Agricultural Machinery, Brodaer Str. 2, 17033 Neubrandenburg, Germany; sdemba@atb-potsdam.de*

Abstract

The pressure caused by a liner during milking is currently mainly detected by estimation methods. There seem to be no direct method to measure this pressure. The aim of this investigation was to test the usability of a pressure-indicating film by Fujifilm for measuring the pressure between a liner and an artificial teat during liner collapsing. The Ultra Super Low (Film 1) and the Extreme Low (Film 2) pressure-indicating films were tested on an artificial teat made of silicone. The experiment was performed by using a conventional milking cluster equipped with round, rubber liners. For each film type, 30 repetitions were done and the liner collapsed thirty times per repetition. Each repetition was done with a new piece of film. The pressure was analyzed for the whole teat as well as for the teat tip. To determine the most suitable film type for measuring the load between a teat and the collapsing liner, the areas at which the pressure indication color was developed were calculated for both film types. Generally, both film types measured the pressure between the collapsing liner and the artificial teat. The pressure ranged between 0.20 and 0.64 MPa for Film 1. For Film 2, the measurements were between 0.08 MPa and 0.23 MPa. The comparison of the film types shows, that Film 2 seems to be more suitable than Film 1, because the colored area for Film 1 and Film 2 was 22-24% and 75-88%, respectively. Nevertheless, a pressure-indicating film that includes the measurement ranges of both film types would be optimal for the current investigation. In further studies the usability of the pressure-indicating film to measure the pressure between a collapsing liner and an elastic teat that is more similar to a natural teat should be tested. This would open up the opportunity to get more information about the teat load caused by a liner during milking.

Keywords: teat load, liner collapsing, pressure-indicating film

1. Introduction

Gentle udder milking is the evacuation of the udder by a milking technique that causes the lowest possible stress to the udder to preserve udder health. Gentle milking is influenced by several factors. One factor is the milking technique because the teat cup liner is the interface between the teats of dairy cows and the milking technique during milking. For the proportion of udder health disorders caused by milking technique, a current accurate estimate is not available, but a milking system that does not work properly can cause udder infections and damage the teats (Rose-Meierhoefer *et al.*, 2009). Therefore, the teat load caused by the liner during milking has to be determined. There are several methods to estimate and calculate the teat load caused by a liner during milking. The detection of the Touched Point (TP) is one of these methods and was developed for round liners. It is the needed pressure difference to collapse the liner till the opposite walls first touch each other (Mein and Reinemann, 2009). The TP is frequently used in the literature (Alejandro *et al.*, 2014; Mein and Reinemann, 2009; Rosca *et al.*, 2012; Spencer and Jones, 2000; Spencer *et al.*, 2007; Zucali *et al.*, 2008). The TP, however, cannot be used for triangular and square liners and it is measured without a teat in the liner. Another method is the calculation of the residual vacuum available for massage.

It is calculated by subtracting the vacuum needed for liner collapsing from the mean claw vacuum (Mein and Reinemann, 2009) and it is used in several investigations as well (Bade *et al.*, 2009; Mein *et al.*, 2003). Alejandro *et al.* (2014), Mein and Reinemann (2009), and Zucali *et al.* (2008) used the Liner Compression to detect the teat load caused by liner collapsing in their investigations. It is defined as the average compressive pressure which is applied to the tissue of the teat end by the liner during the d-phase of pulsation cycle (Mein and Reinemann, 2009). Another used method is the detection of over-pressure which is the average compressive pressure applied to the teat tip to stop milk flow from the teat (Mein and Reinemann, 2009; Mein *et al.*, 2003; Neijenhuis *et al.*, 2005). The assessment of the ratio between the time of milk flow and the time of stopping milk flow (true milk : rest ratio) is also a used method to estimate the teat load during milking. Thereby, within each pulsation cycle, the proportion of time that milk flows from the teat, relative to the proportion of time that milk flow is stopped by liner collapsing is estimated (Mein and Reinemann, 2009; Mein *et al.*, 2003). Muthukumarappan *et al.* (1993) tested several film sensors regarding the detection of the pressure applied to a teat during milking, but they evaluated them as inacceptable for pressure measurements on live teats. Tol *et al.* (2010) used a pressure-sensitive sensor to investigate the pressure and the pressure distribution at the teat liner and teat-calf interface with the help of a silicone teat. They found it usable to measure the pressure between an artificial teat and a liner.
The Prescale film by Fujifilm (KAGER Industrieprodukte GmbH, Dietzenbach, Germany) is a mono- or two-sheet pressure-indicating film which is used to measure pressure, pressure distribution and pressure balance. It contains color developing material which indicate the applied pressure. As a result of an applied pressure to the pressure-indicating film, the microcapsules are broken, the color-forming material reacts with the color-developing material and red areas appear on the film. The different pressure levels were pointed out with the intensity of the red color. According to Cuffaro (2013) the Prescale film have a pressure range between 0.05 and 300 MPa and it does not support shear stress (Rodriguez-Martinez *et al.*, 2012). After measuring the films could be analyzed with the associated software which recorded six parameters: the effective rate (ER) in percent, the pressed area (PA) in square millimeters, the average pressure (AP) in megapascals, the maximum pressure (MP) in megapascals, the load (L) in newton, and the measured area (MA) in square millimeters. The aim of this investigation was to test the usability of the Prescale pressure-indicating film for direct measurement of the pressure between a liner and an artificial teat during liner collapsing.

2. Material and methods

2.1 Study design and data collection

During this investigation the Ultra Super Low (Film 1) and the Extreme Low (Film 2) pressure-indicating films (KAGER Industrieprodukte GmbH, Dietzenbach, Germany) were tested. The films are two-sheet types with an A-film and a C-film. The components of the A-film are a polyester base and a microencapsulated color-forming layer. In contrast, the C-film composed of a polyester base and a color-developing layer. The pressure ranges measurable by the films are 0.2-0.6 MPa (Film 1) and 0.05-0.2 MPa (Film 2), respectively.
Both film types were tested on an artificial teat made of silicone rubber. The length and diameter of the teat were 56 mm and 21 mm, respectively. According to the producer, the silicone rubber had a shore hardness of 25.
The experiments were performed with a conventional milking cluster in the laboratory milking parlor of the Leibniz Institute for Agricultural Engineering in Potsdam-Bornim. The milking cluster was equipped with round rubber liners and all measurements were performed in the same teat cup. The machine vacuum was adjusted at 40 kPa. The pressure-indicating film was cut into pieces (35×45 mm) and all pieces were attached in the identical position on the teat. The teat with the pressure-indicating film was inserted into the teat cup, so that it was aligned in the teat cup with the collapsed liner and the pressure-indicating film pressed together. For both film types 30

repetitions were done and within each repetition the liner collapsed thirty times. Each repetition was done with a new film piece.
After measurements, the C-films were analyzed using the FDP-8010E software. To detect the usability of the Prescale pressure-indicating film the values of AP and MP as well as the colored area (CA) were used. AP, MP and CA were calculated for the whole area of the film peace and for the area of the teat tip. The area of teat tip was defined as the lower third of the teat barrel. CA was calculated as follows:

$$CA=PA/MA$$

with CA is the colored area; PA is the pressed area in square millimeters; MA is the measured area in square millimeters.

2.2 Statistical analysis
All calculations were done with the SAS software package 9.4 (SAS Institute Inc., Cary, NC, USA). The UNIVARIATE procedure was used to detect the applied pressure on the teat. The descriptive statistic for AP and MP was calculated. The GLIMMIX procedure was used for a one-way ANOVA to examine the percentage of CA for each film type. As the observations were percentage values, binomial distribution with a logit link function was used.

3. Results

Film 1 as well as Film 2 measured the pressure and the pressure distribution between a collapsing liner and a teat made of silicone. The results for the descriptive statistic for AP and MP for both tested areas and film types are given in Figure 1 and Figure 2, respectively.

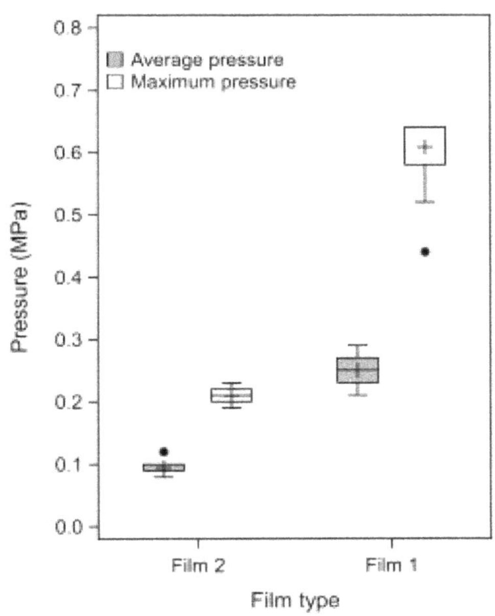

Figure 1. Average pressure and maximum pressure for the whole area and both film types.

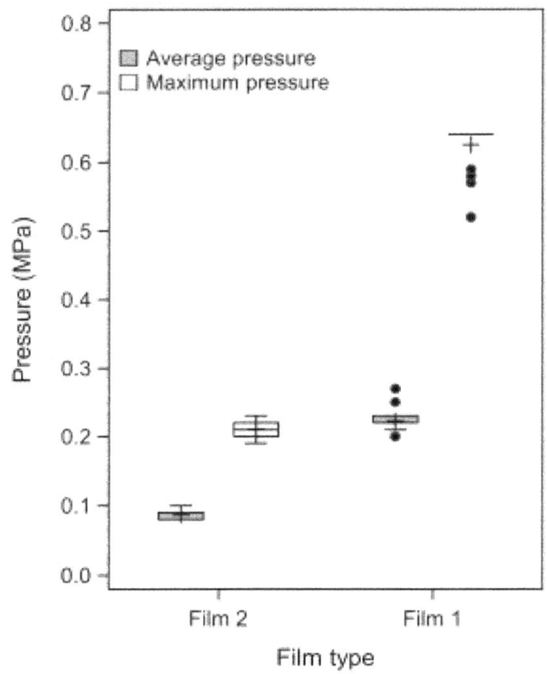

Figure 2. Average pressure and maximum pressure for the teat tip area and both film types.

For the whole area the pressure on the teat ranged between 0.20 and 0.64 MPa for Film 1. For Film 2 the applied pressure varied between 0.08 and 0.23 MPa. The mean pressure measured by Film 1 and Film 2 was 0.22 MPa and 0.09 MPa, respectively.

The applied pressure on the teat tip for Film 1 and Film 2 was 0.21-0.64 MPa and 0.08-0.23 MPa, respectively. On the teat tip an average pressure of 0.25 MPa was detected with Film 1. Film 2 measured a mean pressure of 0.09 MPa on the teat tip area. The mean, minimum and maximum values of AP and MP for both tested areas and film types are given in Table 1 in detail.

The color developing area was for the whole area 22% for Film 1 and 75% for Film 2. For the teat tip area color developed on 24% for Film 1 and on 88% for Film 2. Thus, more color has developed on Film 2 and in the teat tip area. This means that more pressure was applied on the teat tip.

4. Discussion

Depending on film type and measuring area the applied pressure due to a collapsing liner ranged between 0.08 and 0.64 MPa in our investigation. More color has developed, so more pressure was applied in the teat tip area, compared with the whole teat barrel. Our results agree with these of Tol *et al.* (2010). They found that the maximum pressure was always exerted to the teat end. They detected pressure values between 99 and 180 kPa for the teat end as well. The pressure measured in our investigation was higher than this of Tol *et al.*(2010). The artificial teat used in this investigation could be the reason, because it was not hollow. Thus, it was very stiff, although it was made of silicone. In further investigations a hollow silicone teat will be used to detect the pressure caused by a collapsing liner. The study of Muthukumarappan *et al.* (1993) showed that the maximum pressure

Table 1. Mean, minimum and maximum values of average pressure (AP in MPa) as well as maximum pressure (MP in MPa) for both tested areas and film types.

Area	Film	Variable	Mean±SD[1]	Minimum	Maximum
Whole	Film 1	AP	0.22±0.01	0.20	0.27
		MP	0.63±0.03	0.52	0.64
	Film 2	AP	0.09±0.01	0.08	0.10
		MP	0.21±0.01	0.19	0.23
Tip	Film 1	AP	0.25±0.02	0.21	0.29
		MP	0.61±0.05	0.44	0.64
	Film 2	AP	0.09±0.01	0.08	0.12
		MP	0.21±0.01	0.19	0.23

[1] SD = standard deviation.

was applied within 1 or 2 mm of the teat tip. They detected a progressive decrease of the applied pressure over the upper 3 or 4 mm of the teat end. Their results confirm to our results as well. Film 2 seems to be more suitable to measure the pressure between a collapsing liner and an artificial teat than Film1 because more color has been developed on Film 2. The specific pressure range of Film 2 could be an explanation for this. Nevertheless, Film 2 showed white pixels in the area of the teat tip as well. This could be also explained by the pressure range of this film type. The pressure seems to be higher at these points than Film 2 can measure. A pressure-indicating film that included the measurement ranges of both film types would be optimal for the current investigation.

5. Conclusions

In general, it can be concluded that Film 1 as well as Film 2 measured the pressure and the pressure distribution between a silicone teat and a collapsing liner. Based on our results, a pressure-indicating film that included the measuring ranges of both film types would be optimal to detect the load caused by liner collapsing. Further studies are needed to evaluate the usability of the pressure-indicating film for measuring the pressure between a collapsing liner and a teat which is more similar to the natural teat of a cow. The influence of milking settings and different liner types on the teat load will be examined as well.

References

Alejandro, M., Roca, A., Romero, G. and Diaz, J.R. (2014). Effects of overmilking and liner type and characteristics on teat tissue in small ruminants. Journal of Dairy Research 81: 215-222.

Bade, R.D., Reinemann, D.J., Zucali, M., Ruegg, P.L. and Thompson, P.D. (2009). Interactions of vacuum, b-phase duration, and liner compression on milk flow rates in dairy cows. Journal of Dairy Science 92: 913-921.

Cuffaro, V. (2013). Prediction method for the surface damage in splined couplings. PhD thesis, Politecnico di Torino, Turin, Italy.

Mein, G.A. and Reinemann, D.J. (2009). Biomechanics of milking: teat-liner interaction. ASABE, Reno, Nevada, USA.

Mein, G.A., Williams, D.M.D. and Reinemann, D.J. (2003). Effects of milking on teat-end hyperkeratosis: 1. Mechanical forces applied by the teatcup liner and responses of the teat, National Mas titis Council Regional Meetings Proceedings. National Mastitis Council, Forth Worth Texas, USA, pp. 114-123.

Muthukumarappan, K., Reinemann, D.J. and Mein, G.A. (1993). Compressive load applied by the teatcup liner to the bovine teat. American Society of Agricultural Engineers, Meeting, Chicago, IL, USA.

Neijenhuis, F., Klungel, G.H., Hogeveen, H. and Noordhuizen, J.P.T.M. (2005). Machine milking risk factors for teat end callosity in dairy cows on herd level. In: Mastitis in dairy production: current knowledge and future solutions. Wageningen Academic Publishers, Wageningen, the Netherlands, pp. 376-382.

Rodriguez-Martinez, R., Urriolagoitia-Sosa, G., Torres-San Miguel, C.R., and Carbajal-Romero, M.F. (2012). Development of an experimental apparatus for testing a total knee prostheses focused on Mexican phenotype. International Journal of Physical Science 7: 5779-5786.

Rosca, R., Carlescu, P., Tenu, I. and Chirila, C. (2012). Evaluation of a data acquisition system for measuring the milking machine process parameters. Lucrari Stiintifice – Universitatea de Stiinte Agricole si Medicina Veterinara, Seria Zootehnie 58: 285-290.

Rose-Meierhoefer, S., Brunsch, R. and Jakob, M. (2009). Reduction of forces on the teats by single tube guiding in conventional milking parlours. South African Journal of Animal Science 39: 161-164.

Spencer, S.B. and Jones, L.R. (2000). liner wall movement and vacuum measured by data acquisition. Journal of Dairy Science 83: 1110-1114.

Spencer, S.B., Shin, J.W., Rogers, G.W. and Cooper, J.B. (2007). Short communication: effect of vacuum and ratio on the performance of a monoblock silicone milking liner. Journal of Dairy Science 90: 1725-1728.

Van der Tol, P.P.J., Schrader, W. and Aernouts, B. (2010). Pressure distribution at the teat-liner and teat-calf interfaces. Journal of Dairy Science 93: 45-52.

Zucali, M., Reinemann, J.D., Tamburini, A. and Bade, R.D. (2008). Effects of liner compression on teat-end hyperkeratosis. ASABE, RI, USA.

Maximizing the biogas production of manure-based biogas plants with a dynamic input advice system

M. Timmerman* and J.W. Van Riel
Wageningen UR Livestock Research, P.O. Box 338, 6700 AH Wageningen, the Netherlands; maikel.timmerman@wur.nl

Abstract

A lot of commercial manure-based biogas plants have suboptimal biogas productions, experience frequent process disturbances and achieve disappointing financial returns. One of the causes is suboptimal operational management of the input of manure and coproducts to the digesters. The daily optimization of the input is a hard task for the farmer due to the complexity of the anaerobic digestion process, varying composition of coproducts and the change in composition of manure over time, which have to be taken into account with the optimization. When the daily input to the biogas plant is being optimized, direct (financial) improvement is feasible. To assist the farmer in the daily optimization Wageningen UR Livestock Research has developed a dynamic input advice system based on an adaptive model with an Bayesian foundation which uses plant specific process data. The advice system determines daily the actual response of the input on the biogas production. This response curve determines the optimal input which gives the highest achievable biogas production under the current operational conditions. Through a control algorithm the current input is adjusted into the direction of the optimal input. The prototype of the advice system has been tested on two dairy farms with a full-scale biogas plant. The results show that the theoretical approach of operational modelling of the relationship between input of feedstock and output of biogas production is feasible. Adjusting the input resulted in the expected increase in biogas production during all advice periods. The estimated influence of a feedstock on the biogas production varied over time. This endorses the necessity for a model with enough flexibility to deal with varying quality of feedstock. The Dynamic Input Advice System has been proven to work under practical circumstances. By developing the advice system into a commercial software application gives farmers a management tool to improve the results of their biogas plant.

Keywords: adaptive model, anaerobic digestion, coproduct, manure, optimization

1. Problem statement

The input costs of coproducts are around half of the operational costs of agricultural biogas plants. Due changing market and growing circumstances the composition, volumes and prices of coproducts vary. Together with the complexity of the biological anaerobic digestion process makes the daily optimization of the input not an easy task. Therefore a lot of commercial manure-based biogas plants have suboptimal biogas productions, experience frequent process disturbances and achieve disappointing financial returns (Peene *et al.*, 2011; Van den Boom and Van der Elst, 2013; Velghe and Wierinck, 2013). One of the causes is suboptimal operational management of the input of manure and coproducts to the digesters. The daily optimization of the input is a hard task for the farmer due to the complexity of the anaerobic digestion process, varying composition of coproducts and the change in manure composition over time, which have to be taken into account with the optimization. When the daily input to the biogas plant is being optimized, direct (financial) improvement is feasible. To assist the farmer in the daily optimization Wageningen UR Livestock Research has developed a Dynamic Input Advice System based on an adaptive model with an Bayesian foundation which uses plant specific process data for the daily adjustment of the input to the biogas plant. The advice system determines daily the actual response of the input on the biogas production. This response curve determines the optimal input which gives the highest achievable biogas production under the current operational conditions. Through a control algorithm the current input is adjusted into the direction of the optimal input, in this research the highest achievable biogas production. In principal the control algorithm can also be set to determine the highest feed balance (energy returns minus feed costs), but was not tested in this research. The research objective was to determine the proof of principle of Dynamic Input Advice System under practical circumstances.

2. Methods

The prototype of the advice system has been tested on the full-scale biogas plant at the experimental dairy farm Dairy Campus and on the full-scale biogas of a commercial dairy farm. The biogas plant at Dairy Campus consisted of several digesters, for this research one main digester (1,100 m^3) was used of the which the biogas production could be directly led into a cogeneration unit (345 kWe) by shutting valves in the biogas lines in order to get a direct relationship between the input of the digester and biogas production. The biogas plant at the commercial dairy consisted of two digesters (900 m^3) in series and one cogeneration unit (345 kWe).

The Dynamic Input Advice System is based on a dynamic linear model, developed by West and Harrison (1997), for on line estimation of parameters that describe the efficiency of the biogas production from the input of manure and coproducts. This method is a Bayesian approach of adaptive modeling with sequential learning techniques to investigate the time-specific relationship between amount of input and output. The general univariate dynamic linear model is written as:

Observation equation $$G_t = F_t^{`}\beta_t + v_t \qquad v_t \sim (0, V_t) \qquad (1)$$

System equation $$\beta_t = M_t\beta_{t-1} + \omega_t \qquad \omega_t \sim (0, W_t) \; (\qquad (2)$$

The biogas production G_t on day t is dependent on the coproduct input, the regression vector X_t. This relationship is being given in the observation equation. In this notation, the Ft is the matrix with regression vectors X (number of columns is dependent of the number of groups of coproducts), and including the unit vector for modelling the intercept. The observation error of the biogas production at day t is normally distributed v_t~$(0,V_t)$. V_t is used in the preceding since the variance of the observation error can change in time. β_t is the vector with parameter estimates at time t. M_t is a matrix of known coefficients that defines the systematic evolution of the vector β_t and w_t is the system variance matrix which can change in time.

For the conversion of coproducts into biogas the law of diminishing returns is applicable. At high levels of input the retention time decreases due the fixed volume of the digester. Therefore the added value of extra input will decreases when the amount of input increases. This can represented by a quadratic response of the biogas production on the input. The response of the biogas production G_t on the input of a single coproduct X (or a group of coproducts X) can be represented by the following equation:

$$G_t = \beta_{0,t} + \beta_{1,t}X_t + \beta_{2,t}X_t^2 \qquad (3)$$

Wherein:

- G_t=biogas production at day t
- X_t=input coproduct X (or group of coproducts X) at day t
- $\beta_{0,t}$=intercept of the base level (m^3/day), i.e. manure without coproduct
- $\beta_{1,t}$=linear effect of the input of coproduct X (m^3/kg/day)
- $\beta_{2,t}$=quadratic effect of the input of coproduct X (m^3/kg^2/day)

The parameter estimates were obtained by a Bayesian analysis of time series (West and Harrison, 1997) consisting of the daily measurements of the input and biogas production. Typical for this method is the repeated cycle of the data processing at each new measurement, in this case the daily input and the daily biogas production. Disturbances of the process, such as outliers, are automatically detected.

The control algorithm calculates the optimal setting of the input of the coproducts. As optimization criteria the 'actual maximum biogas production' was chosen. The optimization criteria of coproduct X is calculated from the parameter estimates for the linear effect

$\beta_{1,t}$ and the quadratic effect $\beta_{2,t}$ by:

$$G_{max_t}: X_{opt,t} = \frac{-\beta_{1,t}}{2\beta_{2,t}} \quad (4)$$

Wherein:

- $Gmax_t$=maximum achievable biogas production at day t
- $X_{opt,\ t}$=optimal input of coproduct X (or group of coproducts X) at day t

For the biogas plant at Dairy Campus an adaptive model was chosen which could independently of each other estimate the response of the total amount of liquid coproducts (L) and the total amount of dry coproducts (D) on the biogas production (G). The adaptive model consisted of the following dynamic linear model:

$$G_t = \beta_{0,t} + \beta_{1,t}L_t + \beta_{2,t}L_t^2 + \beta_{3,t}D_t + \beta_{4,t}D_t^2 + v_t \qquad v_t \sim N(0, V_t) \ (5) \quad (5)$$

$$\beta_{.,t} = M_t\, \beta_{.,t-1} + \omega_t \qquad \omega_t \sim N(0, U_t) \quad (6)$$

Wherein:

- L_t=the total input of liquid coproducts (3-day average) at day t
- D_t=the total input of dry products (3-day average) at day t

The control algorithm calculated the optimal setting of the total input of liquid coproducts and the total input of dry coproducts. In this study, there has been weekly advised on the basis of the optimization criteria the 'actual maximum biogas production'. The optimization criteria was:

$$G_{max_t}(Liquid): L_{opt,t} = \frac{-\beta_{1,t}}{2\beta_{2,t}} \quad (7)$$

$$G_{max_t}(Dry): D_{opt,t} = \frac{-\beta_{3,t}}{2\beta_{4,t}} \quad (8)$$

The calculated optimal settings were not blindly followed, since otherwise too large changes could take place which could upset the digestion process. Changes in the input were gradually made. The used step size (increase or decrease) per day was usually 50 kg per coproduct-group (dry or liquid). In the research at the commercial dairy farm a prototype adaptive model was chosen which could estimate the response of the total amount of the liquid coproduct Ecofrit (L) on the electricity production (E). The adaptive model consisted of the following dynamic linear model:

$$E_t = \beta_{0,t} + \beta_{1,t}L_t + \beta_{2,t}L_t^2 + v_t \qquad v_t \sim N(0, V_t) \quad (9)$$

$$\beta_{.,t} = \beta_{.,t-1} + \omega_t \qquad \omega_t \sim N(0, U_t) \quad (10)$$

Wherein:

- L_t=the total input of Ecofrit (9-day average) at day t.

The control algorithm calculated the optimal setting of the total input of Ecofrit. In this study, there has been weekly advised on the basis of the optimization criteria the 'actual maximum biogas production'. The optimization criteria was:

$$E_{max_t}(Ecofrit): L_{opt,t} = \frac{-\beta_{1,t}}{2\beta_{2,t}} \qquad (11)$$

Wherein:

- $Emax_t$=maximum achievable electricity production at day t
- $L_{opt,\,t}$=optimal input of Ecofrit at day t

The calculated optimal settings were not blindly followed, since otherwise too large changes could take place which could upset the digestion process. Changes in the input were gradually made. The used step size (increase or decrease) per day was usually 100 kg.

3. Results

In the first advice period at Dairy Campus which lasted from June 11 until September 2009, a mixture of manure (dairy slurry and solid dairy manure), liquid coproducts (Ecofrit and glycerin) and dry coproducts (maize, grass silage, onions, potato pressing fibers, grain waste mix and cookie mix) was used to feed the biogas plant. Figure 1 gives the real fed and calculated optimal amounts of the liquid and dry coproducts during the first advice period.

Figure 1 shows that the input of dry coproducts during the first half of the advice period was too high. In the second half of the research period the input of dry coproducts comes in line with the

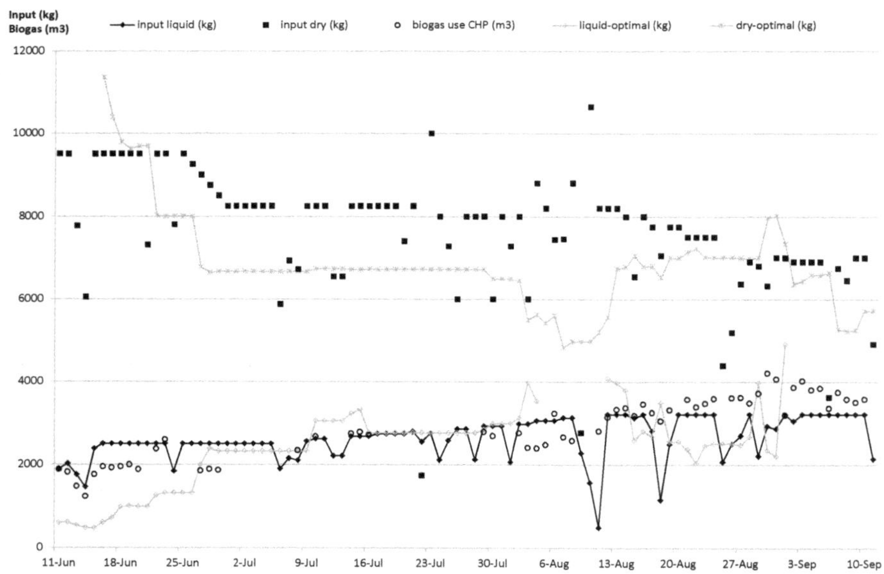

Figure 1. Amounts of fed liquid and dry coproducts, the optimal input of the liquid and dry coproducts with the expected maximum biogas production and the realized biogas production of the biogas plant at Dairy Campus during the first advice period from June 11 to September 11, 2009.

optimal input which would produce the maximum achievable biogas production. The outliers in the amounts of dry coproducts on July 23 and August 10 were caused by extra feeding since the day before a lot less dry coproducts were fed. The input of liquid coproducts was at the beginning too high, but while the optimal input increased in June the real fed input of liquid coproducts came in line with the optimal input. In order to operate on full load the combined heat and power plant (CHP) needed around 4,000 m^3 biogas per day which was de production at the end of the first research period. At the end of the advice period it turned out that some fine-tuning of the settings of the CHP were necessary to do in order to make optimal use of the produced biogas. The analysis of the digestate during the research did not show any sign of process disturbances and were within standard limits, only the dry matter was a little above the limit of 11%. From the technical and economical results it was shown that at the start-up of the advice period the biogas production was 30% lower than expected which resulted in high feed costs of €0.125 per kWh and a feed balance of €140 per day. Halfway the advice period, the biogas production had increased with 50% and was almost in line with the expected biogas production. The feed balance had increased to €434 per day. At the end of the advice period, the biogas production was 10% higher than expected, which resulted in a feed cost of €0.079 per kWh and a feed balance of €646 per day. So at the end of the advice period, the biogas production had increased with 120% and the feed balance with 361% compared with the start of the advice period. During the second advice period from March 1 to June 5, 2010 (data not shown) the biogas production had increased with 31% and the feed balance with 86% compared to the start of the advice period.

In the advice period at the commercial dairy farm which lasted from April 14 until September 1, 2012, a mixture of manure (dairy slurry), Ecofrit, Lecithine and potato steam peelings were fed. Figure 2 gives the real fed and calculated optimal amounts of Ecofrit during the advice period.

At the start the increase in Ecofrit was too large, because the analysis results of the digestate on May 9 (data not shown) showed that the digestion process was disturbed. This led to a sharp decrease in

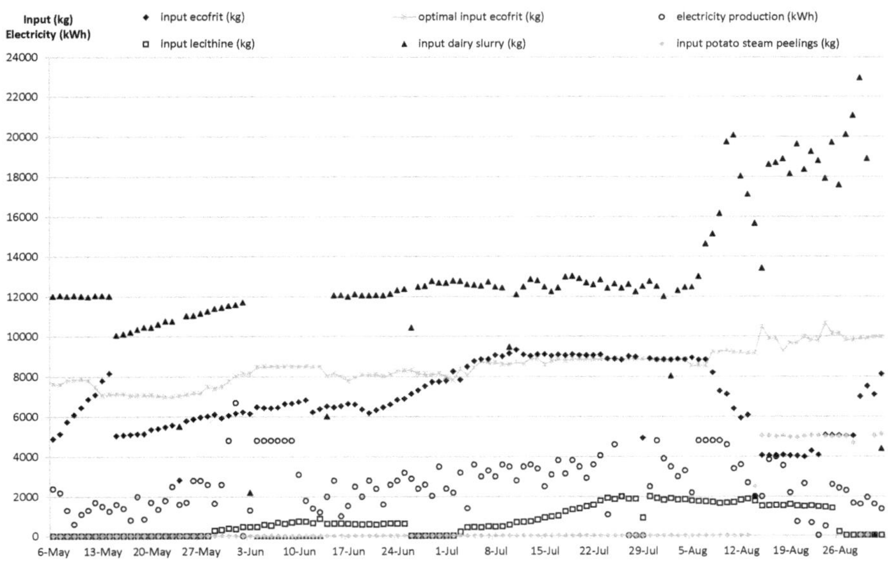

Figure 2. Amounts of fed manure and coproducts, the optimal input of Ecofrit with the expected maximum electricity production and the realized electricity production to the biogas plant of the commercial farm during the advice period from May 6 to September 1, 2012.

the input of manure and Ecofrit and was slowly increased during the month of May which resulted in a slow increase of the electricity production. At the end of May the operator began slowly feeding Lecithine. As a consequence of a technical malfunction of the manure pomp in the beginning of June, during a week no dairy manure could be fed which probably was the main cause for the drop in electricity production in the first half of June. In the course of June the input of Ecofrit was further increased until the input was at its optimum. During this period the electricity production also increased, but gave also a distorted view since the Lecithine was used up at the end of June and wasn't fed for a week. At the beginning of July Lecithine was once again being fed and slowly increased. During August the supply of Ecofrit was limited due shortage in the production of Ecofrit, causing the input of Ecofrit to drop. The operator wanted to compensate for the decreased input by a sharp increase in the input of manure and by starting with feeding of a substantial amount of potato steam peelings. These changes in feed ration were too big and resulted in a decreased methane content of the biogas and a decreased biogas production and also a decrease in the feed balance. From the technical and economical results it was shown that at the start-up of the advice period the feed cost was €0.100 per kWh and increased to €0.108 per kWh just before the limited supply of Ecofrit, but during this time the electricity production increased by 149% and the feed balance increased from €82 to €173 per day.

The results of the research show that the theoretical approach of operational modelling of the relationship between input of feedstock and output of biogas production is feasible. Adjusting the input resulted in the expected increase in biogas production during all advice periods. The results were also showing that the estimated influence of a coproduct on the biogas production varied over time. This endorses the necessity for a model with enough flexibility to deal with varying quality of coproducts.

Full details of the research at both biogas plants are given in Timmerman and Riel (2015).

4. Practical implications

The Dynamic Input Advice System has been proven to work under practical circumstances. By developing the advice system into a commercial software application gives farmers a management tool to improve the results of their biogas plant.

References

Peene, A., Velghe, F. and Wierinck, I. (2011). Evaluatie van de vergisters in Nederland – fase I. Organic Waste Systems NV, Belgium, 185 pp.

Timmerman, M. and Van Riel, J.W. (2015). Prototype van een dynamisch input advies systeem voor biogasinstallaties. Wageningen UR Livestock Research, Wageningen, the Netherlands, 58 pp.

Van den Boom, H. and Van der Elst, C. (2013). Toekomst biogas: van laagwaardige input naar hoogwaardige output. Rabobank Cijfers en Trends 37: 1-12.

Velghe, F. and Wierinck, I. (2013). Evaluatie van de vergisters in Nederland – fase II. Organic Waste Systems NV, Belgium, 136 pp.

West, M. and Harrison, J. (1997). Bayesian forecasting and dynamic models. Springer, New York, NY, USA, 682 pp.

Basics for the energy distribution in a dairy barn with an energy management system

M. Höld[1,2*], A. Gräff[3], J. Stumpenhausen[1] and H. Bernhardt[3]
[1]Hochschule Weihenstephan-Triesdorf, University of Applied Sciences, Agriculture and Food Economy, Am Hofgarten 1, 85354 Freising, Germany; [2]Technical University of Munich, Munich School of Engineering, Boltzmannstaße 17, 85748 Garching, Germany; [3]Technical University of Munich, Agriculture Systems Engineering, Am Staudengarten 2, 85354 Freising, Germany; manfred.hoeld@hswt.de

Abstract

Dairy Farmers often not only produce milk but also electrical energy by using sun and wind energy or by burning biogas from the slurry of the dairy cows. For the farmers it is cheaper to use self-produced energy than to buy electrical energy from energy suppliers. Therefore, an energy management system (EMS) must be integrated in the dairy barn for distributing the energy within the dairy barn at the right time to the right place. However, the welfare of the cows must be guaranteed constantly in a high automated dairy barn. A model dairy barn in Bavaria has to provide space for 125 dairy cows. Daily load profiles of all electrically operated machines have to be created by measurements on different dairy barns in Bavaria with as much electrically operated machines as possible. For these measurements, a power meter with time intervals of one minute is necessary during recording, because some machines, such as automatic feeding systems, only work for a few minutes and a 15 min interval is not descriptive. By overlapping the load profiles of the individual machines, a daily load profile can be generated for the whole dairy barn. The biogas can be stored and burned when it is necessary but the sun and wind power depends on the weather. The amount of energy and the load profiles of these plants are simulated with a program. The EMS must work based on algorithms. First of all, the dairy barn must be supplied with energy and then the residual energy can be delivered to the energy supplier or stored in batteries. On a dairy barn with an asymmetric shed roof a photovoltaic system with an installed power of 180 kW_{peak} can be mounted. The electrical energy produced from one installed kW_{peak} per year in Bavaria is about 1,100 kWh. The slurry of each cow provides about 1,500 kWh of energy. In a high automated dairy barn approximately 3,100 kWh per dairy cow can be produced while about 500 to 700 kWh per cow are consumed. Together with neighbor dairy farms a dairy farm officiates as a virtual power plant for the energy supplier when the connection to the public grid can be realized in an intelligent way. Demand side management (DSM) is the result.

Keywords: integrated dairy farming, stall 4.0, demand side management, smart grid, energy storage

1. Introduction

Since the year 1990, when the Kyoto-protocol was ratified, the production of renewable energy is increasing in Germany. Renewables are now the most important energy source for the generation of electricity. The government set the following statement: renewable energy sources continue to grow until they reach the targets: 40 to 45% of gross electricity consumption by the year 2025 and 55 to 60% by the year 2035 (BMWi, 2015).

Farmers are often the owners of land and therefore can use sun or wind energy to produce electricity from photovoltaic systems or wind power plants. If there are animals on the farm, biogas can be produced from the slurry. This biogas can be burned and electricity can be produced with an electric generator. Therefore, the grid is no longer only a distribution network for electricity but a network where electricity goes in either direction: from the supplier to the consumer and from the renewable

energy producer into the grid and then to consumers. The main problem is that the electricity production from photovoltaic systems and wind power plants depends on the weather. Hence, it is increasingly difficult for the transmission system operators to guarantee stability of the grid.
Furthermore, since the year 2012, it is cheaper for the farmers in Germany to use their self-produced energy from photovoltaic systems than to buy electrical energy from the energy suppliers (Pfeifroth *et al.*, 2012). Therefore, the main target must be to use as much self-produced energy from photovoltaic systems as possible in the dairy barn. To realize this requirement, the amount of electricity produced from the photovoltaic system on the roof of the dairy barn, the small wind power plants and the biogas plant must be known. Likewise, the load profiles of all machines within the dairy barn driven by electricity must be known.
To realize an energy distribution within the dairy barn, an Energy Management System (EMS) must be installed. However, no bus system for technique within the dairy barn exists. That means, the EMS cannot communicate with the electricity production unit and with the electric driven machines. This problem leads to the following project: 'Integrated Dairy Farming – Stall 4.0'.
At the end of this project, a model for distribution energy should exist which considers the facts of energy production, energy supply, storage of energy, stability of the grid, economics and especially the welfare of the dairy cows.
The project should give answers to the following questions:

- Which combination of electricity production plants can provide the demand of the electricity in the dairy barn?
- Which load profiles exist of each electric driven machine and how does the total load profile of the dairy barn look like?
- Which algorithms are necessary for an economical operation of the EMS with an agreeable welfare for the cows?

The model dairy barn should be provided with a maximum dimensioned photovoltaic system on the roof. The biogas plant should be dimensioned according to the amount of slurry. The load profile of the model dairy barn has to be preferably constant with less or only small peaks. Furthermore, the electricity supply of the dairy barn should be flexible, because in this case the farm can offer positive and negative minute reserves and the farm can thus help the transmission system operator to stabilize the public grid. For this, the biogas produced from the biogas plant can be stored in times when less electricity is needed and can be burned in times when much electricity is needed.

2. Material and methods

In cause of the incompatibility of the electrically operating machines in the dairy barn the EMS has to distribute the electrical energy produced by the photovoltaic system, a small wind power plant or burning the biogas. The main unit of the EMS is the 'central control unit' which has to monitor and manage the energy consumption and production. The energy consumption is given to the system by product data sheets containing the following aspects:

- priority with which each consumer has to be supplied with electrical energy;
- flexibility in time of each consumer;
- maximum power of the electrical engines;
- load profile of each consumer and producer.

The 'central control unit' has to be in contact with all electrical consumers and all plants and systems for electrical energy production. In addition, an intelligent connection to the public grid must exist because demand side management can help to realize the stability of the grid. The opportunity for the farmer to perform changes in the system because of declining welfare of the dairy cows must be given at all times (Figure 1). Finally, the central control unit must be in contact with the weather forecast due to the photovoltaic system.

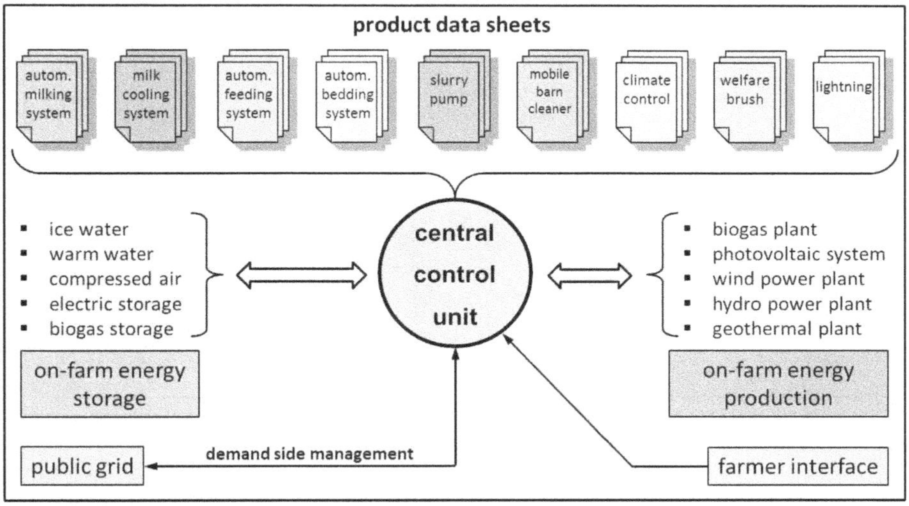

Figure 1. Links of the central control unit.

First of all, the dairy barn is to be supplied with energy and then the residual energy can be delivered to the energy supplier. If an energy storage system should be integrated, algorithms must be determined how the charging process should be done and which battery capacity is needed. These algorithms must guarantee the welfare of the dairy cows.

Within this project a model dairy barn has to be planned for 125 dairy cows, a typical size for Bavarian dairy farms operated as family farms. The model dairy barn should allow installing different kinds of electrical machines. As a result, feed stalls are necessary to allow the use of a cleaning robot or dung scraper for cleaning the floors without disturbing the dairy cows when feeding (Benz *et al.*, 2014). The feeding of the cows should be realized by an automatic feeding system or a mobile feeding system. Figure 2 shows a ground plan for the model dairy barn.

The construction of the model dairy barn has an assymetric shed roof with a southbound main roof area. An area for a photovoltaik system of 1,363 m^2 (length 67 m, width 20.35 m) is available where a photovoltaic system can be mounted with an installed power of 180 kW_{peak}. The amount of slurry from 125 dairy cows can be calculated with 2,500 m^3.

For the model dairy barn daily load profiles of all electrically operated machines are created by measurements on different dairy barns in Bavaria. For the measurements of the load profiles power meter 'econ sense+' from the company 'econ solutions' are used. Time intervals of one minute are necessary during recording, because some machines in the dairy barn work for a few minutes only. That means, during short intervals there are high load peaks which cannot be seen during intervals of 15 min. By overlapping the load profiles of the individual machines, a daily load profile can be generated for the whole dairy barn.

The photovoltaic systems on the roof of the dairy barn only provide to the electrically consumers within dairy barn when the sun shines. On the other side, the biogas plant can supply electrical energy when it is needed in the dairy barn because biogas can be stored and burned as necessary. The amount of energy and the load profiles of the biogas plants are calculated with theoretical values. For the photovoltaic system the load profiles for an average day and for each month are simulated by a computer program. If there is a small wind power plant on the farm, an average load profile can be realized by using the Bavarian Wind Atlas (Bavarian Ministry of Economic Affairs and Media, Energy and Technology, 2014). Storage capacities such as batteries, volume of ice water for milk

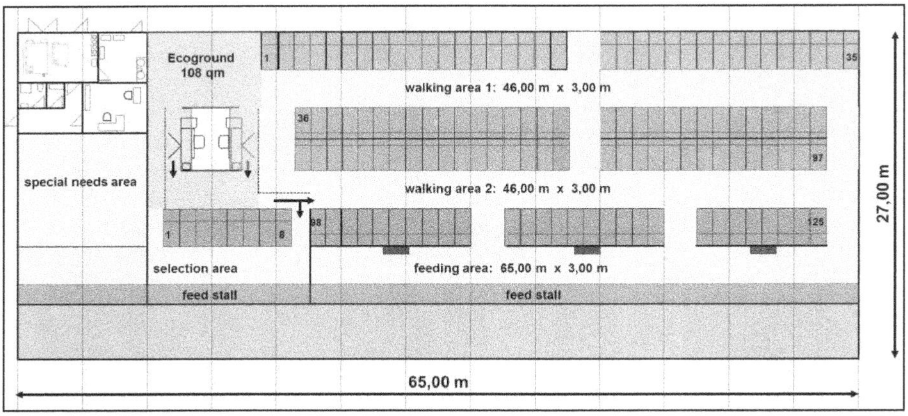

Figure 2. Ground plan for a model dairy barn for 125 dairy cows.

cooling, volume of hot water for cleaning the milking system, volume of compressed air and volume of biogas to be stored can be calculated when the demand of electricity is known.

3. Results and discussion

A first calculation with a computer program shows that the photovoltaic system on the roof of the dairy barn produces more energy than the machines in the dairy barn use assumed that the consumption of electrical energy is about 500 to 700 kWh per dairy cow and year. The consumption depends on the level of automation in the dairy barn. On a dairy barn with a southbound asymmetric shed roof the photovoltaic system produces about 1,100 kWh per year with one installed kW_{peak} provided that the dairy barn is built in Freising, a town in the northeast of Munich. For each cow an amount of electrical energy of about 1,600 kWh can be produced. Figure 3 shows the average load profile of the photovoltaic system and the average load profile of the dairy barn in January. The month of January was chosen because the power of the photovoltaic system is much less in the winter months than in the summer. It shows that the energy demands of the dairy farm can only be

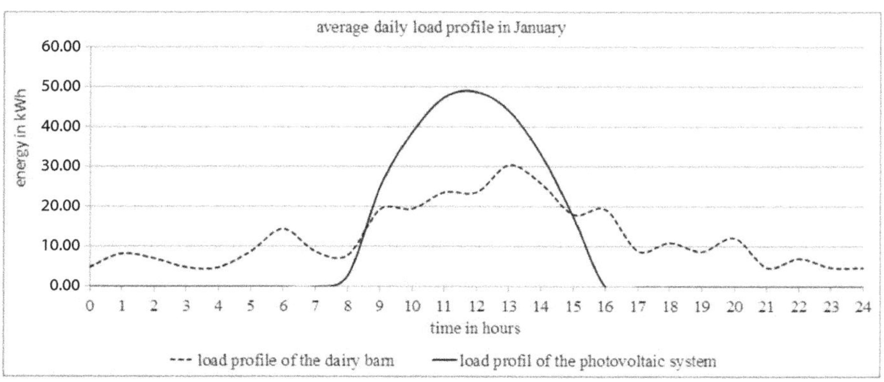

Figure 3. Average daily load profile of the dairy barn and the photovoltaic system.

met for eight hours even though ice water for milk cooling and hot water for cleaning the milking system is produced with electrical energy from the photovoltaic system.
In addition, the energy from the biogas plant can be used in the dairy barn. In case of a permanent operation of the biogas engine the amount of biogas produced from the slurry of the dairy cows together with the photovoltaic system is sufficient to supply the dairy barn.
Figure 4 shows the load profile of the total yield and the load profile of the dairy barn. The course of the total yield curve resembles the load profile of the photovoltaic system but the value is about 22 kWh higher. The upward displacement is generated by the electricity supply from the biogas plant. If the biogas plant does not produce permanent electricity, a storage volume for biogas is.
All throughout the year, the biogas plant supplies the dairy barn with an amount of energy of about 1,500 kWh per cow. The combination of a biogas plant and a photovoltaic system can produce about 3,100 kWh per cow and year. The demand of about 500 to 700 kWh per cow and year can be subtracted from the total yield of about 3,100 kWh. Thus, the difference of the electrical energy can be sold or can be used to help the transmission grid operators to stabilize the grid. For this the farm needs an intelligent grid connection and an economically dimensioned capacity for storing the energy in ice water, hot water or simply in batteries.

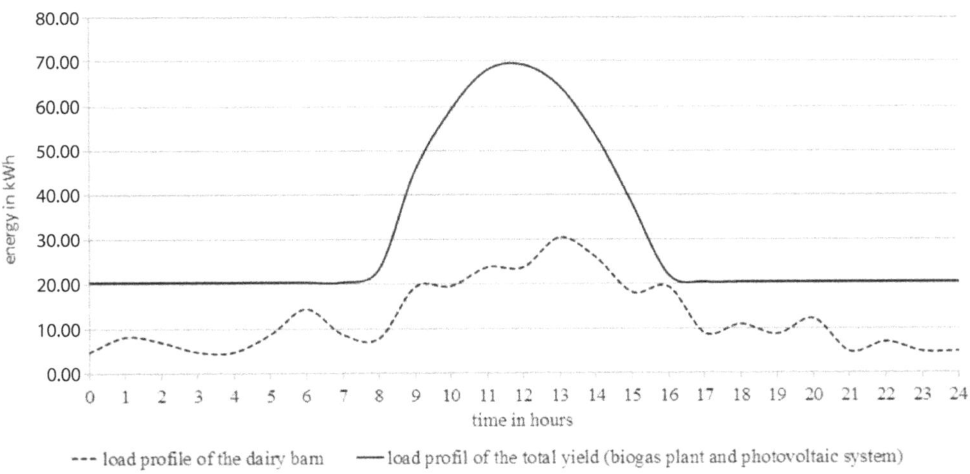

Figure 4. Average daily load profile of the dairy barn and the total yield of the biogas plant and the photovoltaic system.

4. Conclusions

After calculating the amount of energy which can be produced with a biogas plant and the simulation of the average daily load profile of a photovoltaic system the next step is to define the algorithms for the distribution of the energy to the consumers. When the 'central control unit' works within the dairy barn, the EMS should be connected with the public grid and can thus contribute to stabilizing the electricity grid. The same method will be done with thermal energy.

References

Benz, B., Ehrmann, S., Richter, T. (2014). The influence of elevated feed stalls on feeding behaviour of lactating dairy cows. Landtechnik 69(5): 232-238.

Bavarian Ministry of Economic Affairs and Media, Energy and Technology (StmWi) (2014). Bavarian wind atlas. Asterion Germany GmbH, Viernheim, Germany.
Federal Ministry for Economic Affairs and Energy (2015). Renewable energy sources in figures. Silber Druck oHG, Niestetal, Germany, 6 pp.
Pfeifroth, P., Schmid, T., Samweber, F. (2012). Grid parity and internal consumption at rising electricity prices. Available at: http://tinyurl.com/gnevbge.

Evaluation of an ear-based real-time location system

K. Orsel[1*], B. Wolfger[1,2], B.A. Wadsworth[3] and J.M. Bewley[3]
[1]University of Calgary, Faculty of Veterinary Medicine, Department of production animal health, 3330 Hospital Drive NW, Calgary, AB, T2N 4N1, Canada; [2]Current employment with veterinarian Dr H. Wolfger, 12 Februar strasse 7, St Michael, Obersteinmark, Austria; [3]University of Kentucky, Department of Animal and Food Sciences, 407 WP Garrigus Building, Lexington, KY, USA; karin.orsel@ucalgary.ca

Abstract

Position tracking of cows within the barn environment allows for determining behavioral patterns and activities. Such data might be used for detection of estrus and disease and for real-time cow microenvironment modification. Methods: A newly marketed location monitoring system (Smartbow®, MKW Electronics GmbH, Weibern, Austria) was tested. Cow location was continuously monitored with the Smartbow® tags mounted to the cows ear. This tag sends low-frequency signals to receivers further transmitting the information to a server. Through incoming data, the server triangulates the location of the cow within the barn environment in real-time. The validation of the system was carried out in four steps. The first two steps served as static testing steps (tags and one cow positioned at 30 reference points) and step 3 and 4 were dynamic steps with cows moving in the barn environment. For 48 hours, locations of 15 cows were confirmed hourly by laser measurements performed by a team (step 3) or one observer (step 4). Results: Inter-observer variability was 0.83 m (range: 0.05 to 2.87 m), and intra-observer variability had a range of 0.02 to 0.31m. In the four validation steps, the mean distance between observer laser measurements and Smartbow® was between 1.22 and 1.80 m. Step 4 with 334 observations resulted in a mean distance difference of 1.22 m (SE 1.32 m). Practical Implications: Data can be used for development of algorithms to detect sick cows with changed behavioral patterns. Data may also be used to monitor cow responses to physical environment, potentially improving facility design. Time budgets in proximity of important barn features (i.e. feed bunk and water trough), and distances traveled can be calculated and used to identify cows in need of caretaker's attention and identify the cows exact location in the barn.

Keywords: position tracking, dairy facility, Smartbow

1. Introduction

In a recent review by Barkema *et al.* (2015), an overview is provided on the adoption of technology on dairy farms. The authors reflect on the profound implications for how cattle are housed and managed on dairy farms (Barkema *et al.*, 2015). Many technologies now have become broadly available. These technologies can have a great impact on improved animal welfare and well-being. Where technology initially was used in pasture settings (Turner *et al.*, 2000), in our study we focused on a new technology that allows position tracking in dairy facilities.

2. Position tracking in dairy facilities

A position tracking system can locate an animal in real-time in a facility. The system needs to be robust factoring in the impact of metals used in the construction. Proper location of the animal can be beneficial to a producer, when trying to locate an individual. Another potential use of position tracking is the detection of abnormal behavioural patterns as indicators for health disturbances or estrus.

The Miner Agricultural Research Institute has reviewed many research trials and has recommended a daily time budget (Table 1) for a typical lactating cow that meets their requirements for feeding, resting and ruminating.
Every activity is most likely occurring in a specific part of the dairy facility, forcing the animal to move around. This will result in a normal movement pattern of the cow in her environment. It also allows generating alerts for any abnormal behavior. Abnormal behaviors can be a sign of health disturbance and allows for identification of sick cows and the health of the herd over time.

Table 1. Daily time budget for a lactating dairy cow (http://tinyurl.com/jesj7rb).

Activity	Time spent/day
Eating	5.5 hours (9 to 14 meals/day)
Resting	12-14 hours with 6 hours of rumination
Standing or walking in alleys (includes grooming, rumination, other)	2 to 3 hours
Drinking	30 minutes
Total time needed	21 to 23 hours

3. Smartbow technology validation

3.1 The system

A recently marketed location monitoring system (Smartbow®, MKW electronics GmbH, Weibern, Austria) enabled individual animal behavior monitoring by tracking cow movements within the barn environment over time. Tags mounted to the ear send low-frequency signals to receivers, which transmit the information to a server. Through incoming data, the server triangulates the location of the cow within the barn environment in real-time. Time budgets in proximity of important barn features (i.e. stall area, feed bunk and water trough), and distances traveled can be calculated and used in algorithms to identify cows in need of caretaker's attention. The real-time location additionally enables farm personnel fast localization of cows.

3.2 The validation

The importance of the accuracy of a location monitoring device depends on the application of the system. While lower accuracy might be sufficient to locate cows, determining behavior and interactions between cows might require higher accuracy.
The real-time positioning device was validated against a laser distance measurer as reference tool. The laser measure device was attached to a stick with also a level attached, assuring a parallel position to the floor.
All validation steps were performed in the research dairy barn at the University of Kentucky, Lexington, USA. The facility was a freestall barn with solid concrete flooring. A continuous monitoring of all cows equipped with the position tracking ear-tag device was performed.

Step 1: static with activated tags
Using the map of the barn, 30 static points were identified. The X and Y coordinates of these points were determined with the reference tool. Secondly two tags were strapped onto the stake at different

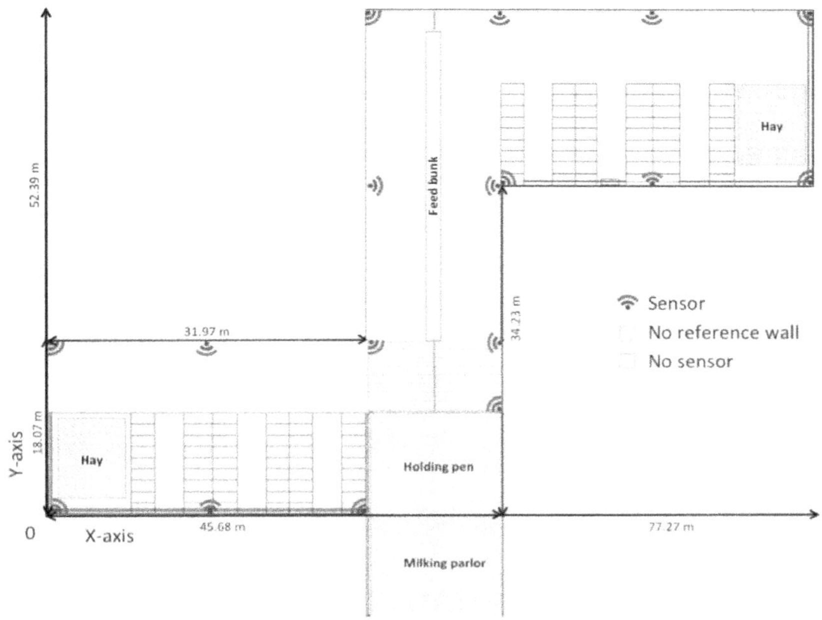

Figure 1. Barn lay-out and position of sensors for tracking system.

heights to mimic a standing and lying cows head position. The stake was placed for three minutes at all 30 locations to allow for undisturbed real-time position tracking.

Step 2: static with one cow

One cow was equipped with the position tracking device and moved through the facility lead by a halter. At all 30 reference points, the cows remained for at least 2 minutes to allow for the real time position tracking.

Step 3: 48-hour observation of 15 cows in barn environment by several researchers

For the final step, 15 cows where equipped with the position tracking device and housed in the same facility. Every hour all 15 cows were identified and their head position measured with the reference tool. The exact time was noted on a sheet before the laser measurements were performed. The cows head was slowly approached allowing for the distance to the wall (x-ax) and floor (y-ax) to be determined with the laser measure device and noted on the sheet. This was continued every hour for 48 hours consecutive.

Step 4 was a repetition of step 3 with only one observer to exclude the inter-observer differences.

4. Results

In all validation steps, the mean distance between observer laser measurements and Smartbow® was between 1.22 and 1.80 m. Step 4 with 334 observations resulted in a mean distance difference of 1.22 m (SE 1.32 m).

5. Conclusions and future directions

The system accurately identified the cows' positions in a dairy facility. This position tracking allows for research and commercial applications as well as development of algorithms to detect sick cows with changed behavioral patterns. Further investigation should also include exploration of an integrated alarm system to detect estrus as well as disease.

Acknowledgements

The authors thank the graduate and undergraduate students at the University of Kentucky for their assistance in collecting data. Additionally, the authors thank Smartbow for their financial and technical assistance in working with the system.

References

Barkema, H.W., Von Keyserlingk, M.A.G., Kastelic, J.P., Lam, T.J.G.M., Luby, C., Roy, J.P., LeBlanc, S.J., Keefe, G.P. and Kelton, D.F. (2015). Invited review; changes in the dairy industry affecting dairy cattle health and welfare. Journal of Dairy Science 98: 7426-7445.

Turner, L.W., Udal, M.C., Larson, B.T. and Shearer, S.A. (2000). Monitoring cattle behavior and pasture use with GPS and GIS. Canadian Journal of Animal Science 80: 405-413.

A multi-Kinect-system for monitoring and measuring functional traits in dairy cows

J. Salau*, J.H. Haas, W. Junge and G. Thaller
Institute of Animal Breeding and Husbandry, Kiel University, Olshausenstraße 40, 24098 Kiel, Germany; jsalau@tierzucht.uni-kiel.de

Abstract

Camera based systems in dairy cattle account for a high portion within the field precision livestock farming. Overcoming the limited application range of single camera systems, a system equipped with six Microsoft Kinect cameras is presented. It provides 3D data on a high percentage of the cow's surface, thus various kinds of functional traits could be measured. As a start, this study deals with automated analysis of the udder. Heights of udder bases above ground were determined automatically with precision between 0.5 and 1.8 cm.

Keywords: computer vision, 3d camera, database, object recognition, linear traits

1. Introduction

Objectively analyzing dairy cows' movement and body traits gains more and more importance. The field of precision dairy farming has thus become a substantial aspect of agricultural science, dealing with topics ranging from lameness detection to body condition determination. With regard to computer vision application in this field, various camera types were introduced during the last years. E.g. Viazzi *et al.* (2013) used digital 2D cameras in side view position. In Azzaro *et al.* (2011) and Halachmi *et al.* (2013) 2D data or thermal images were recorded from top view. Also different devices to collect 3D information on dairy cows have been studied. Salau *et al.* (2014) and Weber *et al.* (2014) used Time-Of-Flight cameras, and a Microsoft Kinect (Microsoft, 2010) was presented in Viazzi *et al.* (2014) and Kuzuhara *et al.* (2015). In all mentioned studies, the setup was designed for one specific aspect, i.e. the monitoring of either lameness or body condition. As there is the necessity for holistic monitoring, this study presents concept and first results of a 3D cow scanner based on multiple Microsoft Kinect cameras. The cameras' fields of view were combined to collect 3D information on a high percentage of the animal's surface, making various applications possible. The objective evaluation of the animals' body measures is one specific goal of the cow scanner's development. Precise conformation recording is the basis for breeding values in dairy cattle and necessary for healthy and productive animals for breeding. The linear description according to the International Committee for Animal Recording (ICAR, 2015) is meant to 'score traits individually and to measure degree rather than desirability'. As the traits are lengths and angles along the cow's body, measuring them is well suited to be carried out with methods of computer vision, and the influence of the classifying person would be removed. Except from analyzing singled out body parts (i.e. Zwertvaegher *et al.*, 2011), no approach to automate conformation recording has been made so far. To gain orientation on the cow's surface, this study focuses on the automatic recognition of body parts starting with the udder and a calculation of the udder depth.

2. Materials and methods

The Microsoft Kinect is a combined RGB camera and depth sensor (Figure 1A). It gathers depth information using 'Structured Light' (Andersen *et al.*, 2012). A framework with pass line height and width of approximately 2 m was built (Figure 1B). Each side of this prototype was equipped with three Kinect cameras: One is placed 0.6 m above the ground in horizontal side view, and two were fixated back to back and crossed to each other (Figure 1B) in the diagonal sections of the framework and were facing the ground in an angle of 45°.

2.1 Data collection and preparation

The prototype was firmly installed in a room connected to the cow barn at the research farm Karkendamm of Kiel University. The room's dimensions were 12.5 m and 4.9 m. A runaround to guide the cows through the framework and preventing them from getting too close to the cameras was built. Between November 2015 and February 2016 nineteen randomly chosen cows were recorded for several runs each to test this fixed installation. All Kinect cameras recorded simultaneously while cows walked through the framework freely. As additional information the unique life number, dates of birth and last calving, and current lactation were noted for all cows. Four cows, one in first, second, fourth, and fifth lactation, respectively, were chosen from the dataset. Ten runs each were available. The recordings were sorted and segmented. Only images that showed a cow were used for the analysis. In one side view depth map of an additional cow (second lactation, 147 cm height) the udder was isolated (Figure 2A) to be used as an udder model later. The images chosen for analysis were manually sorted regarding whether the corresponding cow was in the same posture as in the depth map where the model was taken from.

2.2 Application of computer vision methods

The udder model was taken as a starting point for a database of body parts (paragraph Creating a database of body parts) and compared to the recordings of the cows chosen for analysis via the object recognition pipeline described in paragraph 3D Object recognition. The results of the object recognition were checked visually. On the set of images showing the posture in Figure 2, automatically recognized and not recognized udders were counted true positive (TP) and false negative (FN), respectively. Outside this set, they were regarded false positive (FP), respectively true negative (TN). Accuracy ACC=(TP+TN)/number_of_images, true positive rate TPR=TP/(TP+FN), and true negative rate TNR=TN/(TN+FP) were calculated. Using a RANSAC (Fischler and Bolles, 1981 and PCL tutorial, 2015) based approximation of the scenery's floor, the height of udder base above ground in centimeters was calculated from the images with correctly recognized udders.

2.3 Software

Software to operate six Kinect cameras and store the data was implemented in C++. Details on hardware and further information on recording software, developed data types (KDM streams),

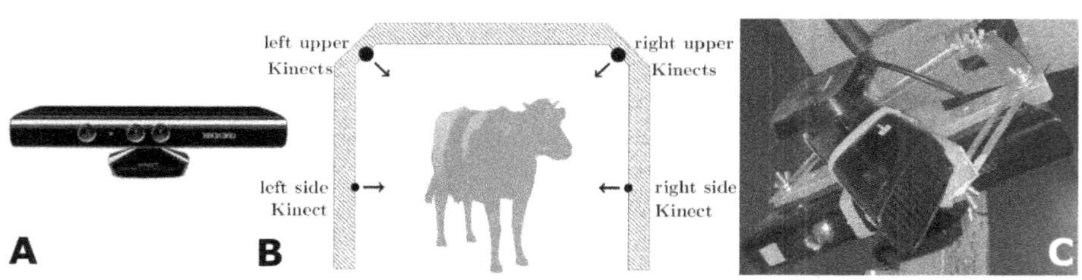

Figure 1. (A) The Microsoft Kinect. (B) Schematic representation of the prototype. A wooden framework was equipped with six Kinect cameras. On each side was one camera mounted in 0.6 m height and two in ≈2 m height fixated crossed to each other in diagonal top view position; the black arrows indicate their views. (C) Close up of a pair of crossed Kinect cameras fixated in the framework.

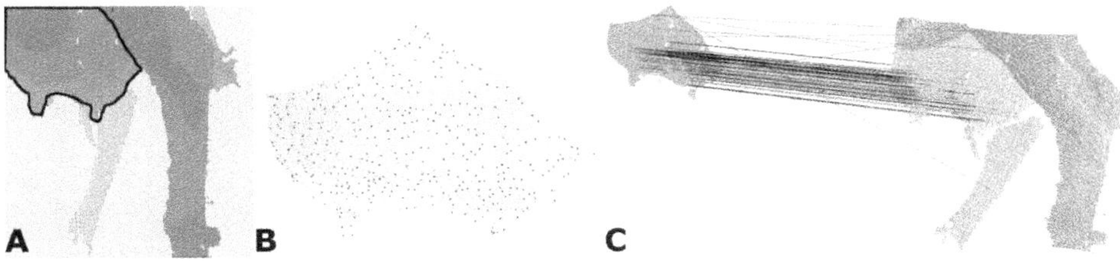

Figure 2. (A) Side view depth map with outlined udder. (B) Point cloud of the isolated udder with key points. (C) Scene and model (shifted outwards). Point correspondences were marked. Gray and black lines indicate wrong and correct correspondences, respectively.

synchronizing, as well as sorting the images and defining the cow as image foreground (segmentation) can be found in Salau *et al.* (2015, 2016). Preprocessing was implemented in MATLAB.

2.4 3D Object recognition

The Point Cloud Library (PCL, Rusu and Cousins, 2011) is a large collection of open source C++ libraries holding state-of-the-art algorithms and tools to process 3D data. A 3D object recognition pipeline was implemented (Aldoma *et al.*, 2012a). To find the object of interest in a given 3D point cloud (scene), a 3D model point cloud of the object of interest and a way to unambiguously compare point clouds are needed. The pipeline started by extracting keypoints from scene and model cloud (Figure 2B). Keypoints should be repeatably detectable and highly descriptive with respect to the surface. A main reason for this, was to reduce the point cloud to its most distinctive and recognizable points to save computational time in further processing. Vectors holding surface characteristics in the vicinity of keypoints (i.e. curvature, gradients in different directions) were computed afterwards. These descriptors represented reduced information on scene respective model point cloud that could easily be compared numerically. Whether underlying points in scene and model were corresponding to each other (Figure 2C), was decided based on the differences between descriptors. Using Hough voting (Tombari *et al.*, 2010a), the found correspondences were afterwards clustered to proposals, where the sought for object was located in the scene. As a last step, these proposals were verified (Aldoma *et al.*, 2012b). From a wide variety available in literature, the authors implemented three keypoint-descriptor pairs: uniform downsampling and SHOT descriptors (Tombari *et al.*, 2010b), NARF keypoints and descriptors (Steder *et al.*, 2011), ISS keypoints and FPFH descriptors (Rusu *et al.*, 2009; Zhong, 2009).

2.5 Creating a database of body parts

To store the models, a database (DB) server was set up and software to address and manage the DB from within the object recognition pipeline was implemented. For every body part of interest (BPI), i.e. udder or hind leg, a new DB table was created, in which the information of each model generated for a BPI was stored in a new row. For each table a primary key had to be defined to make every row uniquely identifiable. The primary key was drawn from the image, from which the model was generated. It consisted of the image's timestamp in milliseconds (ms), the camera's location (wall or middle), and its type (role in the group of three). The procedure of adding body part information to the DB is illustrated in Figure 3 (DIN, 1966). While recording, the depth data was stored in KDM streams, from which the preprocessing software extracted and segmented only the images showing a cow (Figure 3, dashdotted box). On the one hand, primary key information for every image was read from the filename and was then used to add a new row to the table corresponding to BPI. On the other hand, the images were displayed on the screen, to enable the user to manually draw a closed polygon course (Figure 2, left) around the BPI. All pixels not belonging to BPI were set to

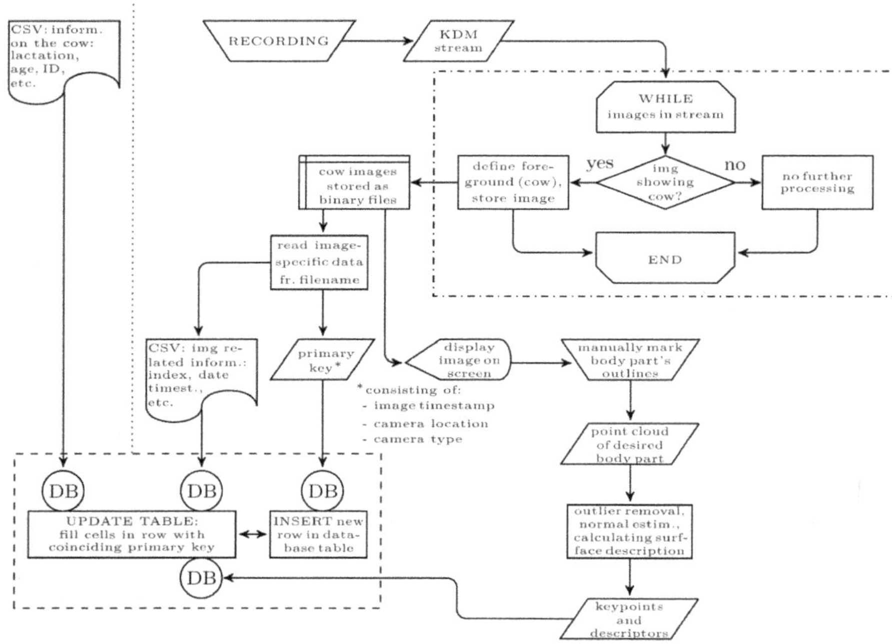

Figure 3. Procedure's flow chart of storing body part information in database (DB). Left of the vertical line: external cow informations were directly written to DB. Parts carried out manually (trapezoidal symbols) were recording and cutting out body parts, before handing the data to PCL processing. The dashdotted and the dashed box hold the procedure's parts implemented in MATLAB and the part dealing with writing to DB, respectively.

zero, and BPI depth information was handed to the PCL software for the calculation of keypoints and descriptors. This information was afterwards stored in the DB (Figure 3, dashed box) and loaded during the object recognition pipeline.

3. Results

The averaged values for the floor height showed standard errors below 0.03 cm. Table 1 presents the results of the 3D object recognition and the automated calculation of the udder base's height above ground for the four chosen cows.

4. Discussion

This study presented how a database of cow's body part models could be set up and used to achieve orientation on the cow's surface given in 3D data. A pipeline for 3D object recognition was established as well. As object recognition and database access worked stably, the software infrastructure has been proven ready to be used on a larger number of cows under analysis and a wider range of models from different cows and postures. Object recognition resulted in medium TPR (Table 1). This was to be expected and can be considered satisfactorily, because the natural variability in udders is unlikely to be represented by the single udder model used in this test. Although high TPRs are desirable, to calculate reliable udder traits, it is not necessary to detect the udder in every image,

Table 1. Mean udder bases' height (UBH, standard error in brackets) above ground calculated from true positive (TP) images as well as accuracy (ACC), true positive and negative rates (TPR, TNR) of the object recognition, and the total number of images per cow (N).

Cows	Lactation	Height (cm)	UBH (cm)	TP	ACC	TPR	TNR	N
Cow 1	5	144	42.3 (1.7)	13	0.72	0.48	0.86	71
Cow 2	4	147	43.0 (1.8)	9	0.75	0.32	0.92	100
Cow 3	2	148	44.3 (0.5)	18	0.84	0.56	0.93	138
Cow 4	1	144	58.9 (0.5)	42	0.66	0.56	0.70	242

but few reliably detected udders are needed. The discovered TNR were high. This is favorable, because falsely detected udders produce errors in further measurements of udder traits. Since the floor height could be considered a very reliable reference for the calculation of mean UBHs, the UBH standard errors given in Table 1 were thus caused solely by udder recognition imprecision and animal movement. Precision in mean UBH was high for cows 3 and 4, but also sufficiently high for cows 1 and 2. A connection between the successful application of a model and similarity of age is suggested by the fact that cow 3 showed highest ACC, TPR, TNR, and precision in the calculation of mean UBH while being the most similar to the model cow of all analyzed cows. The same TPR and UBH precision were observed for the only one year younger cow 4, but TNR and ACC were smaller. The older cows 1 and 2 showed lower TPR and lower UBH precision, although TNR and ACC were superior to the values for cow 4. The combination of object recognition pipeline and model database could on the one hand provide orientation on the cow's surface and the possibility to calculate body traits. On the other hand, quality of recognition as a measure of similarity might allow classification of cows' traits according to information on models held in the database.

Acknowledgements

Gratitude is expressed to Stiftung Schleswig-Holsteinische Landschaft and to Rinderzucht Schleswig-Holstein eG for advice and financial support.

References

Aldoma, A., Marton, Z.C., Tombari, F., Wohlkinger, W., Potthast, C., Zeisl, B., Rusu, R.B., Gedikli, S. and Vincze, M. (2012). Point cloud library – three-dimensional object recognition and 6 dof pose estimation. IEEE Robotics and Automation Magazine 19: 80-91.

Aldoma, A., Tombari, F., Di Stefano, L. and Vincze, M. (2012). A global hypotheses verification method for 3d object recognition. Lecture Notes in Computer Science 7574: 511-524.

Andersen, M.R., Jensen, T., Lisouski, P., Mortensen, A.K., Hansen, M.K., Gregersen, T. and Ahrent, P. (2012). Kinect depth sensor evaluation for computer vision applications. Technical Report Technical Report ECE-TR6, Department of Engineering, Aarhus University, Denmark.

Azzaro, G., Caccamo, M., Ferguson, J.D., Battiato, S., Farinella, G.M., Guarnera, G.C., Puglisi, G., Petriglieri, R. and Licitra, G. (2011). Objective estimation of body condition score by modeling cow body shape from digital images. Journal of Dairy Science 94: 2126-2137.

Deutsche Industrienorm (DIN) (1966). Deutsche Industrienorm 66001. Available at: http://tinyurl.com/jc7lmpj.

Fischler, M.A. and Bolles, R.C. (1981). Random sample consensus: a paradigm for model fitting with application to image analysis and automated cartography. Communications of the ACM 24: 381-395.

Halachmi, I., Klopcic, M., Polak, P., Roberts, D.J. and Bewley, J.M. (2013). Automatic assessment of dairy cattle body condition score using thermal imaging. Computers and Electronics in Agriculture 99: 35-40.

International Committee for Animal Recording (ICAR) (2015). Conformation recording. Available at: http://tinyurl.com/zdk35x2.

Kuzuhara, Y., Kawamura, K., Yoshitoshi, R., Tamaki, T., Sugai, S., Ikegami, M., Kurokawa, Y., Obitsu, T., Okita, M., Sugino, T. and Yasuda, T. (2015). A preliminarily study for predicting body weight and milk properties in lactating Holstein cows using a threedimensional camera system. Computers and Electronics in Agriculture 111: 186-193.

Microsoft (2010). PrimeSense Supplies 3-D-Sensing Technology to 'Project Natal' for Xbox 360.

PCL tutorial (2015). How to use random sample consensus model. Available at: http://tinyurl.com/zxue9f7.

Rusu, R.B., Blodow, N. and Beetz, M. (2009). Fast point feature histograms (fpfh) for 3d registration. In: IEEE International Conference on Robotics and Automation. May, 2009. Kobe, Japan, pp. 3212-3217.

Rusu, R.B. and Cousins, S. (2011). 3D is here: Point Cloud Library (PCL). IEEE International Conference on Robotics and Automation. May, 2011. Shanghai, China, pp. 1-4.

Salau, J., Haas, J.H., Junge, W., Bauer, U., Harms, J. and Bieletzki, S. (2014). Feasibility of automated body trait determination using the SR4K time-of-flight camera in cow barns. Springer Plus 3(225): 1-16.

Salau, J., Haas, J.H., Junge, W., Leisen, M. and Thaller, G. (2015). Precision livestock farming applications – making sense of sensors to support farm management. Development of a multi-Kinect-system for gait analysis and measuring body characteristics in dairy cows. Wageningen Academic Publishers, Wageningen, the Netherlands, pp. 55-64.

Salau, J., Haas, J.H., Thaller, G., Leisen, M. and Junge, W. (2016). Developing a multi-kinect-system for monitoring in dairy cows: object recognition and surface analysis using wavelets. Animal 3: 1-12.

Steder, B., Rusu, R.B., Konolige, K. and Burgard, W. (2011). Point feature extraction on 3d range scans taking into account object boundaries. In: IEEE International Conference on Robotics and Automation. May, 2011. Shanghai, China, pp. 2601-2608.

Tombari, F. and Di Stefano, L. (2010). Object recognition in 3d scenes with occlusions and clutter by hough voting. In: Fourth Pacific-Rim Symposium on Image and Video Technology. November, 2010. Singapore, pp. 349-355.

Tombari, F., Salti, S. and Di Stefano, L. (2010). Unique signatures of histograms for local surface description. Lecture Notes In Computer Science 6313: 356-369.

Viazzi, S., Bahr, C., Schlageter-Tello, A., Van Hertem, T., Romanini, C.E.B., Pluk, A., Halachmi, I, Lokhorst, C. and Berckmans, D. (2013). Analysis of individual classification of lameness using automatic measurement of back posture in dairy cattle. Journal of Dairy Science 96: 257-266.

Viazzi, S., Bahr, C., Van Hertem, T., Schlageter-Tello, A., Romanini, C.E.B., Halachmi, I., Lokhorst, C. and Berckmans, D. (2014). Comparison of a three-dimensional and two-dimensional camera system for automated measurement of back posture in dairy cows. Computers and Electronics in Agriculture 100: 139-147.

Weber, A., Salau, J., Haas, J.H, Junge, W., Bauer, U., Harms, J., Suhr, O., Schönrock, K., Rothfuβ, H., Bieletzki, S. and Thaller, G. (2014). Estimation of backfat thickness using extracted traits from an automatic 3d optical system in lactating Holstein-Friesian cows. Livestock Science 165: 129-137.

Zhong, Y. (2009). Intrinsic shape signatures: a shape descriptor for 3d object recognition. In: IEEE 12th International Conference on Computer VisionWorkshops. September, 2009. Kyoto, Japan, pp. 689-696.

Zwertvaegher, I., Baert, J., Vangeyte, J, Genbrugge, A. and Van Weyenberg, S. (2011). Objective measuring technique for teat dimensions of dairy cows. Biosystems Engineering 110: 206-212.

Internet of cows: a novel sensor bolus for oral application

R. Jorritsma*, M. Nielen, J. Heesterbeek and H. Aardema
Department of Farm Animal Health, Utrecht University, Yalelaan 7, 3584 CL Utrecht, the Netherlands; r.jorritsma@uu.nl

Abstract

Due to current farm characteristics, there is an increasing demand from farmers for cheap and reliable devices to measure the behaviour and the health of cows for the benefit of optimal animal husbandry. There is also an increasing demand from the society to demonstrate the health and welfare of milk producing cows. Currently available sensor devices are usually not able to provide this information, due to technical limitations and some constraints in the use of the sensor devices. The new LoRa technology developed for the Internet of Things, enables the development of a novel sensor bolus for oral application in cows, that addresses these issues. In collaboration with our commercial partner Invenit/Clickey, we have confirmed the performance of this LoRa technology for transmitting data from the reticulum of a cow to the LoRaWAN network. We are currently validating the sensor bolus for measurements of several parameters that are generally used to assess animal health, such as body temperature, respiration rate, pulse frequency, and rumen contractions. We have planned the evaluation of the sensor bolus for the quantification of behavioural parameters, such as social interaction and collective and individual feeding and drinking behaviour, as well as for the detection of oestrous behaviour and calving.

Keywords: cow, sensor, behaviour, health, monitoring

1. Background

1.1 Farm characteristics

In many European countries, the average milk production per cow is still increasing and this coincides with a decline in the number of herds and a growing size of the remaining herds. This may be illustrated by some preliminary Dutch statistic data on dairy herd size up to 2015. From this data, it appears that there was an increase in the number of herds with more than 300 cows from 69 in 2010 to 182 in 2015 (CBS, 2016). Reason for these developments is to reduce the fixed costs per animal and this explains that there is also a trend towards a higher number of cows per farmworker. Today's farmworkers are therefore responsible for the health and welfare of a larger number of cows with a higher milk production. This urges the need for efficient work processes and creates a demand for cheap and reliable devices for the observation of cows.

1.2 Measuring and quantifying behaviour

The decreasing number of people involved in dairy farming possibly drives the increasing curiosity of society on how the milk production is realized. While consumption of milk is generally considered and frequently tested to be safe, there is increased attention and concern for the health and welfare of the cows that produce the milk. In the absence of direct and personal observations by consumers on the behaviour and welfare of cows, there is an increasing demand for objective and reliable information on these issues. Transparent and automated digital monitoring of animal health and welfare could in the future meet this demand, but is currently only in the early stages of development and is fundamentally more complicated than the monitoring of milk for product safety purposes. Of course monitoring to safeguard the absence of e.g. specific infections is possible, but absence of disease does not necessarily prove that animal welfare is warranted. While there have been several initiatives and evaluation attempts for the objective welfare classification at herd level using observations on animal behaviour (Bartussek *et al.* 2000, De Vries *et al.* 2013), others have questioned these scoring systems as they use subjective value judgements of observers (Fraser 1995).

1.3 Current situation

Given the constraints of value judgements and the time required for the collection of such data by independent and trained observers, objective and automated measurements of animal behaviour would largely contribute to an accurate animal welfare classification of cow herds. Existing devices in the dairy industry with sensors that collect data on animal behaviour, such as lying time or activity, could be used for this classification. Unfortunately, these devices are developed and validated for other purposes such as detection of heat, activity or lying time and not for classification of animal welfare. Notwithstanding the importance of these purposes for the farmer, the devices provide no information on other behaviour associated with animal welfare, such as social interaction and the collective and individual feeding and drinking behaviour of cows. The existing sensor devices also require an expensive on-farm infrastructure, including a reading equipment to collect the information, which hampers the widespread implementation of sensor devices in general (Kamphuis *et al*, 2015, Rutten *et al*, 2014). Finally, these reading devices collect data two or three times a day data at the entrance or exit of the milking parlour. This frequency and timing of data transmission might be enough for some observations, but continuous information might be valuable for the detection of social activities during feeding, grazing, and resting.

1.4 Constraints

Thus, while some aspects of the current sensor devices are very useful for farmers, there are important constraints with regard to the associated costs, its use outside the barn and the limited information on animal behaviour. Of additional concern is the limited performance of many monitoring techniques to detect clinical diseases. This is in general partly due to their lack of specificity. Even when relatively common diseases such as mastitis are considered, specificity constraints will generate many false positives results, given the high volume of measurements (Hogeveen *et al.* 2010). It could therefore be more appropriate to use such measurements for the detection of healthy instead of diseased animals.

1.5 Internet and LoRaWAN

To improve the infrastructure part of this situation, transmission of data via the 4G-internet may look attractive, but connections are not suited for the use of data transmission on animal behaviour as subscriptions per animal are too expensive. Other important complications of the use of 4G-internet are the combination of a relatively high energy requirement compared to the energy source of most devices, and the suboptimal connections when cows are housed inside a barn. Recently developed LoRa technology seems an attractive alternative for the currently used technology. The LoRaWAN network is developed for a bi-directional and wireless communication and will be used for the Internet of Things (IoT). The IoT is especially dedicated to the cheap and efficient transmission of data and nicely fits with the demands of sensor technology in cows. The sensor devices are able to transmit short data messages to the network and require only limited energy. Moreover, the transmission of data messages is hampered less by physical barriers, such as walls and buildings.

2. Project

To address the limitations of the current sensor devices, we are collaborating with a commercial company, Invenit/Clickey, specialized in IoT-technology and data capture. The collaboration aims at the development of an oral sensor bolus for the monitoring of animal health and welfare that is connected to the LoRaWAN network. The fact that transmission of data messages is not hampered by physical barriers and has very low energy requirements, created possibilities for the development of a sensor bolus for oral application. To achieve a high retention in the cow, we will use the existing

knowledge on the oral application of other orally applied devices. This knowledge shows that retention close to 99% is realistic and that a bolus is not harmful for the animal. The particular weight and size of the device determines that it will arrive and stay in the reticulum of cows.
The objective is to equip the oral sensor bolus with different sensors for temperature, electric conductivity, sound, position, and movements. The sensors are currently validated in fistulated cows, for the measurements of several parameters that are generally used to assess animal health, such as body temperature, respiration rate, pulse frequency, and rumen contractions. Moreover, the combined measurements of different sensors will likely provide additional information on behavioural parameters, such as social interaction and collective and individual feeding and drinking behaviour and we also expect that the combined sensors of the sensor bolus will be able to indicate when a cow is in oestrous or when a cow is calving.

References

Bartussek, H., Leeb, C.H.M. and Held, S. (2000). Animal needs index for cattle: ANI35L/2000 cattle. Federal Research Institute for Agriculture in Alpine Regions, BAL Gumpenstein, Irdning, Austria.

CBS, Centraal Bureau voor de Statistiek (2016). CBS, Heerlen, The Netherlands.

De Vries, M., Bokkers, E.A., Van Schaik, G., Botreau, R., Engel, B., Dijkstra, T. and De Boer, I.J. (2013). Evaluating results of the Welfare Quality multi-criteria evaluation model for classification of dairy cattle welfare at the herd level. Journal of Dairy Science 96: 6264-6273.

Fraser, D. (1995). Science, values and animal welfare: exploring the "inextricable connection". Animal Welfare 4: 103-117.

Hogeveen, H., Kamphuis, C., Steeneveld, W. and Mollenhorst H. (2010). Sensors and clinical mastitis--the quest for the perfect alert. Sensors (Basel) 10: 7991-8009.

Kamphuis, C., Steeneveld, W. and Hogeveen, H. (2015). Economic modelling to evaluate the benefits of precision livestock farming technologies. In: Ilan Halachmi (ed.) Precision livestock farming applications. Wageningen Academic Publishers, Wageningen, the Netherlands.

Rutten, C.J., Steeneveld, W., Inchaisri, C. and Hogeveen H. (2014). An ex ante analysis on the use of activity meters for automated estrus detection: to invest or not to invest? Journal of Dairy Science 97: 6869-6887.

Part 2. Automatic milking

Effects of a free-stall barn layout on the efficiency of Dutch dairy farms with an automatic milking system

D. Heurkens[1,2], C. Kamphuis[1] and A.J. Van der Kamp[2*]
[1]*Business Economics Group, Wageningen University, Hollandseweg 1, 6706 KN, Wageningen, the Netherlands;* [2]*Lely International NV, Farm Management Support, Cornelis van der Lelylaan 1, 3147 PB, Maassluis, the Netherlands; avanderkamp@lely.com*

Abstract

When using an automatic milking system (AMS), labor efficiency is an important factor as it is one of the main reasons for automatic milking. For a well-functioning AMS and high labor efficiency, it is important that cows display good visit behavior towards the AMS. Factors influencing this visit behavior include general farm management, feeding management, feed quality and barn layout. The objective of this research was to find out which free-stall barn layout indicators influence performance indicators, and thus AMS and labor efficiency. Barn layouts of 127 Dutch dairy farms with a total of 240 AMS units were quantified in 30 barn layout variables. Performance data were collected from October 2014 to March 2015 and included 'milk yield per minute cow spent in the AMS' as a measure of AMS efficiency, and 'number of cows to collect manually' as a measure of labor efficiency on an AMS farm. Performance data also included indicators that affect AMS efficiency and that were used as random factors, e.g. 'kg concentrates per 100 kg milk'. Simple and multiple regression analyses were performed using a mixed model approach. Preliminary results indicated that having a split-entry waiting area positively affected milk yield per minute cow spent in the AMS. Furthermore, situating the AMS units at a right angle to the feed fence and having multiple AMS units per group of cows had a positive effect on milk yield per minute cow spent in the AMS. The split-entry gate also had a positive influence on milking frequency, whereas the positioning of the crossovers interacts with the number of refusals, which indirectly indicates a positive effect on labor efficiency. Results demonstrated that the effects of general farm management are more relevant to AMS and labor efficiency than barn layout, but barn layout indicators did affect AMS and labor efficiency to some extent. Therefore, it is important to keep barn indicators in mind when designing a new barn to ensure AMS and labor efficiency.

1. Introduction

In 1992, the first automatic milking system (AMS) was sold commercially on the Dutch market (De Koning, 2010). Despite this early adoption, large-scale acceptance of this technology only started at the beginning of the 21st century. Over 3,400 dairy farms (19%) in the Netherlands were milking with an AMS in 2014 and this number is still increasing (SKOM, 2015). Using an AMS instead of a milking parlor requires a different approach in getting the cows to the milking machine. When using a milking parlor, the farmer brings the cows to the parlor two or three times a day. With an AMS, the cows have to go for milking by themselves (Scott *et al.*, 2014). Each of these two types of milking imposes different requirements on barn layout. The optimal utilization of barn layout indicators for an AMS, such as feed bunk space, cubicles and stocking density, is yet to be found (Halachmi *et al.*, 2002).
Optimal usage of the AMS unit can be achieved by optimizing performance indicators, such as maximizing milk yield per minute spent in the AMS unit (Castro *et al.*, 2012), improving milking frequency, and minimizing the free milk time of an AMS unit (Dooren *et al.*, 2004). Milking frequency improves the usage of the AMS unit because cows achieve a higher milk production per hour (Hogeveen *et al.*, 2001). The free milk time is the time that the AMS unit is available for

milking. If this time is used for milking additional cows, or milking the same number of cows more frequently, AMS efficiency can be improved. In addition, the number of refusals can be an indication of the willingness of the cows to come voluntarily to the AMS system. A relatively high number of refusals may be an indication that cows are visiting the AMS often. Cows coming to the AMS often may indicate that a farmer has to fetch and take fewer cows to the AMS. This saves the farmer time, thus making the AMS unit more efficient.
Considerable research has been done on the layout of barns with a conventional milking parlor. For instance, milk yield will increase when competition at the feed bunk is reduced. This reduced competition can be achieved by increasing feed bunk space (DeVries and Keyserlingk, 2006). Naess *et al.* (2011) found that two or more dead-end alleys will reduce milk yield. Specific studies on AMS barn layout are mostly done on cow traffic. The distinction can be made between free, semi-forced and forced cow traffic (Bach *et al.*, 2009; Hermans *et al.*, 2003). Semi-forced cow traffic tends to be better for feeding-area usage compared to forced cow traffic (Hermans *et al.*, 2003). With forced cow traffic, eating behavior worsened but cows increased their AMS visits compared to free cow traffic (Bach *et al.*, 2009), whereby cows enter the AMS voluntarily. Jacobs *et al.* (2012) described how gates and alleys positioned around the AMS may influence cow traffic and cow behavior, which potentially affects the time that the AMS is available for milking. No quantitative research has been conducted on the effect of barn layout on AMS and labor efficiency. The objective of this research is to define important barn layout factors that influence labor and AMS efficiency.

2. Materials and methods

Barn layout drawings were collected for 127 farms located across the Netherlands. From these drawings 30 variables were derived, containing information on barn set-up, AMS position, walkways and cubicle rows. These data were combined with farm information (e.g. AMS type, year of installation) and performance data. The performance data consisted of six performance indicators: 'milk yield per AMS', 'milk yield per minute spent in the AMS unit', 'number of refusals', 'milking frequency', 'percentage of cows with a milking interval of over 14 hours' and 'free milk time of the AMS unit'. For this research, data were used from two different periods: October 2014 until March 2015, and January 2016 until March 2016. Data were available per day and were averaged over per month.
From the aforementioned performance indicators, two were selected to include in the model as dependent variables. 'Milk yield per minute spent in the AMS unit' had normal distribution of the residuals, so this was analyzed with a Linear Mixed Model (LMM). For these variables, data analyses were performed with a PROC MIXED procedure in SAS (version 9.3, SAS Institute Inc., Cary, NC, USA). Repeated measures per month, for each AMS unit nested within a farm, were included in the model. 'Percentage of cows with a milking interval of over 14 hours' also had a normal distribution of the residuals. For this variable the analysis was performed using Linear Mixed-Effects Models in R (R 2.15.0, R Core Team (2014), Austria). The independent variables consisted of barn layout variables. From the simple regression analysis, variables were selected to be included in the multiple regression model on the basis of their significance level ($P<0.10$). The final model was constructed using a backward elimination procedure. The simple regression analysis for 'Milk yield per minute box time' and 'Percentage of cows with a milking interval over 14 hours' are not presented in this article.

3. Results

For the results, the focus was on milk yield per minute and percentage of cows with a milking interval of over 14 hours, as these are main indicators for work load for the dairy farmer and the capacity potential of the AMS.

3.1 Milk yield per minute box time

The multiple regression analysis (Table 1) demonstrated that the AMS type had a significant effect on the milk yield per minute, with the A3 and A4 models having lower results compared to the A3 Next. On the other hand, the A3 and A4 models show a positive interaction effect with kg concentrates per cow, which counteracts the differences between the models.

Table 1. Effect of barn layout variables on milk yield per minute (multiple regression procedure).

Effect	Class	Estimate	SE	DF	t value	*P*-value
Intercept		1.6649	0.1395	252.0	11.93	<0.0001
Kg concentrates/cow	Continuous	0.0031	0.0350	350.0	0.09	0.9293
AMS type	A3	-0.6187	0.2648	379.0	-2.34	0.0200
	A4	-0.2211	0.0919	197.0	-2.41	0.0170
	A3N	0.0000				
Concentrates stall occupation	Continuous	0.0000	0.0013	337.0	0.01	0.9904
Kg concentrates/cow × AMS type	A3	0.2126	0.0902	326.0	2.36	0.0190
	A4	0.0437	0.0235	262.0	1.86	0.0636
	A3N	0.0000				
Kg concentrates/cow × concentrates stall occupation	Continuous	-0.0005	0.0004	371.0	-1.19	0.2362
Random effects						
Month		0.9322	0.0116			<0.0001
Random error		0.0178	0.0028			<0.0001

3.2 Percentage of cows with a milking interval of over 14 hours

Table 2 shows the results of the multiple regression analysis for the percentage of cows with a milking interval of over 14 hours. The number of cubicles per AMS has a small decreasing effect on this parameter. The AMS length position in the middle of the barn also has a decreasing influence compared to its positioning at the front. A distance of more than 15 cubicles between the AMS and the crossover position, and having 3 AMS units per group, were shown to have a decreasing effect compared to having 1 robot per group. Having no straw area or no split-entry waiting area slightly increased the percentage of cows with a milking interval of over 14 hours. The alignment of the AMS parallel to the feed bunk with the front not at the feed bunk showed an increase of 9.24 percentage points compared to a right-angled alignment to the feed bunk. One walkway dead end increased the percentage of cows with a milking interval of over 14 hours by 2.62 percentage points.

4. Discussion

This research is the first to mention a split-entry waiting area as a barn layout variable. The presence of a split-entry waiting area had a significant positive effect on milk yield per minute. This can be explained by low-dominance cows being less disturbed when it comes to entering the AMS. Less disturbance can cause a reduction in stress, which will increase milk yield per minute (Szentléleki *et al.*, 2015). Having a split-entry gate also resulted in a lower percentage of cows with an interval of more than 14 hours without being milked, which could be because it is easier to milk a cow that is overdue for milking.

Table 2. Effect of barn layout variables on the percentage of cows with a milking interval of over 14 hours (multiple regression procedure).[1]

Effect	Class	Estimate	SE	t value	*P*-value
Intercept		17.7600	1.8760	9.47	<2e-16
Cubicle number per AMS	Continuous	-0.0532	0.0215	-2.48	0.0135
Distance AMS to cubicles row	Continuous	0.0001	0.0000	2.68	0.0076
Straw area	No	1.0410	0.4409	2.36	0.0185
	Yes	0.0000			
Split-entry waiting area	No	0.6592	0.4475	1.47	0.1413
	Yes	0.0000			
AMS type	A4	-3.6550	1.1280	-3.24	0.0013
	A3	-3.7460	1.2380	-3.02	0.0026
	A3N	0.0000			
AMS length position	Middle	-4.9690	1.2830	-3.87	0.0001
	Front	0.0000			
AMS alignment	Parallel to feed bunk, front at feed bunk	2.7410	1.5770	1.74	0.0828
	Parallel to feed bunk, front not at feed bunk	9.2430	1.2210	7.57	0.0000
	Right-angled to feed bunk	0.0000			
AMS per group	2	0.1629	0.4910	0.33	0.7401
	3	-3.2490	1.0030	-3.24	0.0013
	1	0.0000			
Crossover position	>15 cubicles	-1.0750	0.4376	-2.46	0.0143
	≤15 cubicles	0.0000			
Walkway dead ends	1	2.6280	0.7028	3.74	0.0002
	>1	0.7503	1.1270	0.67	0.5059
	0	0.0000			

[1] AMS = automatic miling system; SE = standard error.

The presence of a straw area for separation and the surface of the straw area (in m^2) had a positive effect on milk yield per minute. These results were in line with Naess *et al.* (2011), who stated that the milk yield per cow will increase if a separation pen of at least 10 m^2 is available. Farms with a straw area had fewer cows with a milking interval of over 14 hours. Cows with a long milking interval are usually freshly calved or sick cows. These cows are kept in the straw area from where they can easily access the AMS, thus reducing the percentage of cows with a milking interval >14 hours. Aligning the AMS at a right angle to the feed bunk induced a higher milk yield per minute. This may be related to the fact that the cow is able to look at her herd mates while she is in the AMS unit. Being able to see other cows may reduce stress levels. Reduced stress levels have been associated with increased milk yield per minute (Szentléleki *et al.*, 2015). The percentage of cows with a milking interval of over 14 hours was also lower when the alignment was right-angled. When the AMS causes less stress for the cow, she may also be more likely to revisit the AMS. From these results it can be seen that barn layout indicators in the area around the AMS (straw area and split-entry waiting area) have a significantly positive effect on performance, indicating that the area around the AMS is important in the design of a new barn.

In this research the multiple regression analysis was performed to gain an insight into the predictability of the dependent variables. Unfortunately, it appears that barn layout variables did not explain much of the variation in the dependent variables. On the other hand, the random herd effect and the month effect explained a large part of the variation in the dependent variables. This may be because feed and management factors may have a bigger effect on performance indicators than barn layout variables. For example, feeding concentrates has an effect on the visit behavior of the cows (Halachmi *et al.*, 2002). Moreover, good management ability on the part of the farmer and good overall farm conditions have a positive influence on herd productivity. In addition, Jacobs *et al.* (2012) described how management plays an important role in the performance of an AMS, because previous studies described contradicting results regarding the same aspects of an AMS. The number of cows in a group can have an influence on the performance indicators. For example, more cows in a group may have a positive influence on the milk yield per AMS unit and this may reduce the free milk time of the AMS unit and increase the percentage of cows with a milking interval of over 14 hours.

5. Conclusions

Barn layout variables contribute to the performance indicators on a farm, although the variation is mainly explained by other variables, such as feed management. Barn layout variables associated with the AMS area, such as the alignment and a straw area, had the strongest effects on performance indicators.

References

Bach, A., Devant, M., Igleasias, C. and Ferr, A. (2009). Forced traffic in automatic milking systems effectively reduces the need to get cows, but alters eating behavior and does not improve milk yield of dairy cattle. Journal of Dairy Science 92(3): 1272-1280.

Castro, A., Pereira, J., Amiama, C. and Bueno, J. (2012). Estimating efficiency in automatic milking systems. Journal of Dairy Science 95(2): 929-936.

De Koning, C. (2010). Automated milking – common practice on dairy farms. Wageningen UR Livestock Research, Lelystad, the Netherlands.

DeVries, T. and Keyserlingk, M. (2006). Feed stalls affect the social and feeding behavior of lactating dairy cows. Journal of Dairy Science 89(9): 3522-3531.

Dooren, H., Heutinck, L., Biewenga, G. and Zonderland, J. (2004). The influence of three grazing systems on AMS performance. In: Meijering, A., Hogeveen, H. and De Koning, C. (ed.) A better understanding: automatic milking. Wageningen Academic Publishers, Wageningen, pp. 292-297.

Halachmi, I., Metz, J.H., Land, A., Halachmi, S. and Kleijnen, J.P. (2002). Case study: optimal facility allocation in a robotic milking barn. American Society of Agricultural Engineers 45(5): 1-8.

Hermans, G., Ipema, A., Stefanowska, J. and Metz, J. (2003). The effect of two traffic situations on the behavior and performance of cows in an automatic milking system. Journal of Dairy Science 86(6): 1997-2004.

Hogeveen, H., Ouweltjes, W., Koning, C. and Stelwagen, K. (2001). Milking interval, milk production and milk flow-rate in an automatic milking system. Livestock Production Science 72(1-2): 157-167.

Jacobs, J.A., Ananyeva, K. and Siegford, J.M. (2012). Dairy cow behavior affects the availability of an automatic milking system. Journal of Dairy Science 95(4): 2186-2194.

Naess, G., Boe, K. and Osteras, O. (2011). Layouts for small freestall dairy barns: effect on milk yield for cows in different parities. Journal of Dairy Science 94(3): 1256-1264.

Scott, V., Thomson, P., Kerrisk, K. and Garcia, S. (2014). Influence of provision of concentrate at milking on voluntary cow traffic in a pasture-based automatic milking system. Journal of Dairy Science 97(3): 1481-1490.

Stichting Kwaliteitszorg Onderhoud Melkinstallaties (SKOM) (2015). Statistiek; overzicht aantallen veehouders en rapporten. Available at: http://tinyurl.com/he82e24.

Szentléleki, A., Nagy, K., Széplaki, K., Kékesi, K. and Tözsér, J. (2015). Behavioural responses of primiparous and multiparous dairy cows to the milking process over an entire lactation. Annals of Animal Science 15(1): 185-195.

The impact of a partially shaded laneway on voluntary traffic between pasture and an automatic milking system

A. Wildridge[1*], S.G. Garcia[1], P. Thomson[2], C.E.F. Clark[1], E. Jongman[3] and K.L. Kerrisk[1]
[1]*Dairy Science Group, Faculty of Veterinary Science, School of Life and Environmental Sciences, University of Sydney, Camden, NSW, 2570, Australia;* [2]*ReproGen Animal BioScience Group, Faculty of Veterinary Science, University of Sydney, Camden, NSW, 2570, Australia;* [3]*Animal Welfare Science Centre, Faculty of Veterinary and Agricultural Sciences, University of Melbourne, Parkville, Victoria, 3010, Australia; ashleigh.wildridge@sydney.edu.au*

Abstract

Hot weather reduces dairy cow feed intake and locomotion, with increased water consumption and time spent using shade. These behaviors often lead to declines in milk production and overall cow performance. The impact may be exacerbated in automatic milking systems (AMS), due to the reliance on voluntary cow traffic for milking. Incentives such as shade may be useful in encouraging voluntary traffic, particularly in pasture based systems. The aim of this study was to determine if intermittent shade sources along the laneway between a feed source and the milking facility would impact on voluntary cow movement during hot weather. Twenty four cows with previous AMS experience were divided into four groups of six cows then exposed to two training days each. All cows were fitted with an internal body temperature logger and three cows per group were fitted with a SCR tag to record a range of cow behaviour and activity. Each day for 16 days during December 2015, after being milked by 07:00 one group was taken to the starting point of a 930 m long laneway which led to the milking facility, and locked in with an allocation of lucerne cubes (8 kg/cow). The cows were able to freely traffic from the start point to the milking facility from 10:00 to 17:00 under a partially shaded or non-shaded treatment. During this time, cow behaviour (grazing, ruminating, lying, etc.) was recorded at 10 minute intervals and respiration rate was recorded at 20 minute intervals. When cows had access to shade structures located at approximately 150 m intervals (along the laneway), they travelled further towards the milking facility ($P<0.001$) than when no shade was available. Cows in the shade treatment also had lower average body temperature and respiration rates ($P<0.001$) compared to cows in the no-shade treatment. The results from this investigation show that the provision of intermittent shade sources along a laneway may encourage increased voluntary cow movement during hot weather.

Keywords: heat stress, cow cooling, animal welfare

1. Introduction

Summer weather conditions can challenge the maintenance of production levels on dairy farms as hot weather impacts on the way cows behave and perform. As the environmental temperature humidity index (THI) increases, the performance of cows is negatively affected as they become heat stressed. These conditions deplete the appetite of cows causing reduced feed intake and reduced locomotion (West *et al.*, 2003). Cattle will often spend more time under shade (where possible) or close to a water source with increased water consumption during hot weather (Blackshaw and Blackshaw, 1994). This combination of behavioural changes can lead to significant (often long lasting) declines in production (Ominski *et al.*, 2002).

When an AMS operates close to capacity, it is required that cows voluntarily present for milking throughout the day and night (regardless of climate conditions) to ensure that the cows are milked at a target milking frequency (MF) (Lyons *et al.*, 2014). Feed is the most reliable incentive used to encourage cows to traffic around the farm system (Lyons *et al.*, 2013). However, as the hunger drive of cows typically decreases during hot weather (Kadzere *et al.*, 2002); alternative incentives need to be investigated to enable voluntary cow traffic to be maintained during hot climatic conditions. To compound the problem (of reduced value of feed as an incentive), locomotion typically decreases as THI increases. As a result (during hot weather) it is likely that the prevalence of extended milking intervals (MI) would increase and MF decrease, which can negatively affect both udder health (Hammer *et al.*, 2012) and milk production (Hogeveen *et al.*, 2001).

Whilst a reduction in voluntary cow traffic can be mitigated (to some extent) through increased levels of encouragement by dairy staff, alternative methods of encouragement may be available which could minimise increases in numbers of cows requiring fetching. As feed is no longer highly desired by the cows when THI is high, the encouragement of voluntary movement may be possible through the strategic provision of shade which could result in improved cow comfort and minimise animal welfare concerns. Water consumption (Meyer *et al.*, 2004) and the use of shade (Kendall *et al.*, 2006) increase as THI rises, therefore, the careful manipulation of access to these resources may enable farmers to encourage cow movement throughout the day. It is preferable to avoid limiting or restricting cows' access to water since the welfare implications of reduced water intake would be undesirable (and the impact on milk production could be significant). The careful placement of non-natural shade structures could act as an incentive to encourage cows to traffic between feed sources and the milking facility on pasture based AMS's. It was hypothesised that the provision of intermittent shade sources along a laneway would result in improved voluntary cow traffic between a feed source and an automatic milking facility.

2. Method

2.1 Animals and management

A group of 24 lactating Holstein Friesian dairy cows were recruited for the study. The cows were selected based on their pregnancy status (all pregnant) to ensure that no cows would come into heat, and were further selected to be between 51 and 79 months of age, MF between 1.3 and 2.0 times per day, producing an average of 15.5 to 29.1 kg/milk/day (MY) and were in the range of 186 and 292 days in milk (DIM).

The cows were housed at the University of Sydney's Corstorphine dairy farm in Camden, NSW, Australia. Cows were milked in an automatic milking rotary (DeLaval AMR™, Tumba, Sweden), provided with three pasture allocations per day and managed to voluntarily traffic to the milking facility. The day before the commencement of the training period, the 24 cows were separated from the main herd and housed in a separate paddock with a small batch milked herd where they were fetched for milking in the AMR twice daily for the duration of the trial.

Cows were divided into four groups of six, with each of the groups closely matched for average cow age (63.61±1.525 months), MF (1.72±0.04 milking/cow/day), MY (19.84±0.66 kg milk/cow/day) and DIM (2.31.17±8.5). On the first experimental day of each group, cows were encouraged into a cattle restraint crate where they all had identification numbers painted on their sides and were fitted with an internal temperature logger and three cows per group were fitted with a SCR tag on a neck collar. The temperature logger was mounted to a blank (no hormone) controlled internal drug release (CIDR®) and inserted vaginally into each cow.

2.2 Treatments and design

The cows were exposed to one of two treatments, SHADE or NO SHADE. The SHADE treatment consisted of five individual shade structures which were 3×3×2.4 m (w×l×h) and covered in shade cloth (blocking 90% of UV light). The shade structures were positioned every 150 m (approx.) along the laneway with the first shade structure located 150 m from the starting point and the final shade structure located approximately 180 m from the entrance to the AMR. The shade structures were used to separate the laneway into six separate 'zones'. Zone 1 was the area from the starting point to the first shade structure (inclusive), zone 2 was the area after the first shade to the second shade structure (inclusive), etc.

A pre-existing laneway that the cows had regularly used prior to the experimental period was selected as the test arena. The laneway was approximately 930 m in length and had three 90° corners (approx.). Cows were provided with 8 kg/cow of lucerne cubes in a single (shared) feed trough at the starting point with access to *ad libitum* water and a loafing area. A second water trough was available halfway between the starting point and the milking facility.

One group was observed in the laneway per day, with the groups tested in a consistent order to ensure that they had the same number of days between test runs. The treatments were allocated to the groups so that the cows were not exposed to the same treatment twice in a row. After morning milking (batched), the test group was drafted by electronic drafting gates into a holding yard, where they were manually walked from the dairy to the starting point at approximately 08:00 each morning. Cows were granted immediate access to the feed allocation and restricted to the feeding area until 10:00 when observations commenced. An eight day acclimation period took place to enable the cows to become familiar with the lucerne cubes, shade structures and laneway to ensure that they were aware that they could leave the start area voluntarily. During the acclimation period, each group was exposed to a shade and no-shade treatment (n=2 acclimation runs per group). Following this, 16 experimental days took place where each group was exposed to each treatment twice (n=4 test runs per group).

2.3 Observations

At 10:00 the gate was opened from the feeding area so that the cows could freely leave and walk towards the milking facility. Stance (standing, lying, walking or other), behaviour (ruminating, idle, grazing or other) and location (feeding area, water trough, shade (structure 1-5) or out in the open (zone 1-6)) were recorded every 10 minutes. Respiration rate of each cow was recorded (at 20 minute intervals) by observing the expansion and contraction of the mid-section of the cows from a short distance over 15 seconds. Observations were conducted from 10:00 and ceased when the cows had entered the milking facility, or at 17:00 (whichever occurred first). Environmental temperature was logged every 10 minutes (USB datalogger 98582, OnSolution, NSW, Australia) and solar radiation was measured every 20 minutes during observations using a handheld pyranometer (Ciderhouse tech, Victoria, Australia).

2.4 Analysis

All data were entered and manipulated in Microsoft Excel (Microsoft Office Corporation 2010) and analyses were performed using Genstat® 16th Edition (VSN International, UK) and ASReml (VSN International, UK). Linear mixed models (LMM) were used to identify if treatment had an impact on internal body temperature and breathing rate, and generalized linear mixed models (GLMM) were generated to determine if treatment had an effect on cow behaviour. Environmental temperature and solar radiation were included in these models to correct for weather effects.

Zone was converted into approximate distance from the starting point, based on the midpoint distance of each zone, where the start/feeding area was defined as 0 m. An Ordinal Logistic Regression with random effects was conducted to determine if treatment had any effect on the distance from the starting point the cows had travelled at each observation. Here distance, or zone, was categorised into three groups, 0 m, <375 m (zone 1-3) and >375 m (zone 4-6) along the laneway towards the milking facility.

3. Results

During the experimental period daily maximum temperature ranged from 22.6 to 38.6 °C and daily maximum solar radiation ranged from 340 W/M^2 to 1,365 W/M^2. On the first testing day of each group, all cows had voluntarily entered the milking facility from the feeding area prior to 17:00. On all other testing days none of the cows had voluntarily entered the milking facility within the observation period, however, on eight of these days at least one cow had walked further than zone 1. Cows in the SHADE treatment had significantly lower body temperature and respiration rates than cows in the NO SHADE treatment (Table 1).

Significant treatment differences were detected in grazing and ruminating behaviour. Back transformed means indicate the proportion of time cows spent displaying that behaviour (within treatment; Table 2).

A significant effect of treatment on distance travelled was identified ($P<0.001$), where cows were 1.37 times more likely to walk further when shade was provided (Figure 1).

Table 1. Effect of treatment on internal cow body temperature and respiration rate.

	Predicted means		*P*-values		
Response	Shade	No shade	Treatment	Environment temperature	Solar radiation
Body temperature	38.84	39.04	<0.001	<0.001	<0.001
Respiration	19.91	20.36	<0.001	<0.001	<0.001

Table 2. Effect of treatment on cow behaviour.

	Back transformed means		*P*-values		
Response	Shade	No shade	Treatment	Environment temperature	Solar radiation
Grazing	0.19	0.25	0.017	<0.001	<0.001
Ruminating	0.16	0.10	<0.001	0.428	0.168
Idle	0.64	0.64	0.778	<0.001	<0.001
Lying	0.30	0.27	0.107	0.918	<0.001
Standing	0.65	0.69	0.156	0.847	0.005
Walking	0.03	0.03	0.690	0.388	0.422

4. Discussion

The size and location of the shade structures were selected to reduce the likelihood of the cows spending excessive amounts of time under the shade. These structures did not allow enough room for all of the cows to stand comfortably under a single shaded area. It was hoped that the provision of these shade structures would encourage the cows to leave the feeding area to seek shade, moving

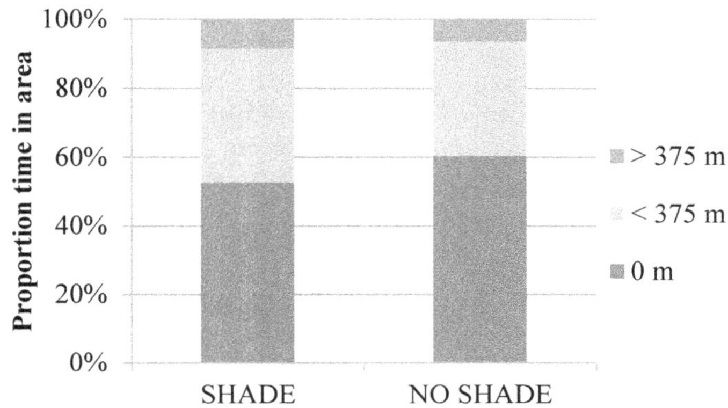

Figure 1. Proportion of observations for each treatment where the cows had travelled 0 m, <375 m or >375 m.

to further shade sources when the number of cows within a given shade area exceeded a comfortable limit. The cooling effect of the shade source was predicted to cool the cows to a level that would enhance the likelihood of cows walking to the milking facility to be milked and receive a feed reward. During the experimental period, environmental temperature and solar radiation were on average higher on NO SHADE days than on SHADE days; as a result, both of these factors were added to the LMM and GLMM to allow for correction of these conditions. When shade was available to the cows, body temperature and respiration rate were significantly reduced (Table 1). Shade minimizes the effects of heat stress by protecting from direct solar radiation (Kendall *et al.*, 2006) and in agreement with the results of the current study, has been shown to reduce respiration rate (Schütz *et al.*, 2014) and body temperature (Tucker *et al.*, 2008).

During the SHADE treatment cows were found to spend 6.0% more time ruminating, and when NO SHADE was available the cows would spend approximately 5.6% more time grazing during the observation period (Table 2). It is not surprising that the cows recorded ruminating more often during the shaded treatment, as hot weather (without the use of cow cooling) is typically associated with reduced rumination time (Soriani *et al.*, 2013). However, it would have been predicted that on non-shaded days the cows would spend less time grazing in line with normally reduced feed intake during hot weather (Kadzere *et al.*, 2002). All efforts were made to keep any grass in the feeding area and alongside the laneway short, however, the cows still picked at the grass in these areas. The increased grazing behaviour (during NO SHADE) was unexpected although with very low levels of pasture mass available it is not likely that total feed consumption would be significantly different. No other behaviors were significantly different between the two treatments.

The main objective of this investigation was to identify if shade on the laneway could positively influence cow traffic during summer. Exploration of the effect of treatment on distance travelled identified that on shaded days the cows were more likely to travel significantly further from the feeding area than on non-shaded days. This shows promise for the use of shade as an incentive to encourage voluntary cow traffic, as even though the cows did not always return to the milking facility before observations ceased at 17:00, when shade was available they had made it significantly closer to the dairy by this time than if shade were not provided. The encouragement of voluntary traffic with methods such as the provision of shade along laneways has the potential to improve farm performance and further investigation into this area is recommended.

5. Conclusions

With the improvement of voluntary cow traffic, particularly during hot weather, the requirement for fetching cows would be reduced and the frequency and distribution of milking events may be improved leading to production benefits. When shade was made available to the cows along the laneway, the cows voluntarily travelled further towards the milking facility than when shade was not available, and had reduced body temperatures and respiration rates.

References

Blackshaw, J. and Blackshaw, A. (1994). Heat stress in cattle and the effect of shade on production and behaviour: a review. Australian Journal of Experimental Agriculture 34: 285-295.

Hammer, J.F., Morton, J.M. and Kerrisk, K.L. (2012). Quarter-milking-, quarter-, udder- and lactation-level risk factors and indicators for clinical mastitis during lactation in pasture-fed dairy cows managed in an automatic milking system. Australian Veterinary Journal 90: 167-174.

Hogeveen, H., Ouweltjes, W., De Koning, C.J.A.M. and Stelwagen, K. (2001). Milking interval, milk production and milk flow-rate in an automatic milking system. Livestock Production Science 72: 157-167.

Kadzere, C.T., Murphy, M.R., Silanikove, N. and Maltz, E. (2002). Heat stress in lactating dairy cows: a review. Livestock Production Science 77: 59-91.

Kendall, P.E., Nielsen, P.P., Webster, J.R., Verkerk, G.A., Littlejohn, R.P. and Matthews, L.R. (2006). The effects of providing shade to lactating dairy cows in a temperate climate. Livestock Science 103: 148-157.

Lyons, N.A., Kerrisk, K.L. and Garcia, S.C. (2013). Effect of pre- versus postmilking supplementation on traffic and performance of cows milked in a pasture-based automatic milking system. Journal of Dairy Science 96: 4397-4405.

Lyons, N.A., Kerrisk, K.L. and Garcia, S.C. (2014). Milking frequency management in pasture-based automatic milking systems: a review. Livestock Science 159: 102-116.

Meyer, U., Everinghoff, M., Gädeken, D. and Flachowsky, G. (2004). Investigations on the water intake of lactating dairy cows. Livestock Production Science 90: 117-121.

Ominski, K.H., Kennedy, A.D., Wittenberg, K.M. and Moshtaghi Nia, S.A. (2002). Physiological and production responses to feeding schedule in lactating dairy cows exposed to short-term, moderate heat stress. Journal of Dairy Science 85: 730-737.

Schütz, K.E., Cox, N.R. and Tucker, C.B. (2014). A field study of the behavioral and physiological effects of varying amounts of shade for lactating cows at pasture. Journal of Dairy Science 97: 3599-3605.

Soriani, N., Panella, G. and Calamari, L. (2013). Rumination time during the summer season and its relationships with metabolic conditions and milk production. Journal of Dairy Science 96: 5082-5094.

Tucker, C.B., Rogers, A.R. and Schütz, K.E. (2008). Effect of solar radiation on dairy cattle behaviour, use of shade and body temperature in a pasture-based system. Applied Animal Behaviour Science 109: 141-154.

West, J.W., Mullinix, B.G. and Bernard, J.K. (2003). Effects of hot, humid weather on milk temperature, dry matter intake, and milk yield of lactating dairy cows. Journal of Dairy Science 86: 232-242.

Automatic milking in large dairy herds

J. Pitkäranta[1*] and J. Rodenburg[2]
[1]4dBarn Consulting, Kauppurienkatu 23, 90100 Oulu, Finland; [2]DairyLogix, 173 Falcon Drive, Woodstock, Ontario, N4T 1W5, Canada; jouni.pitkaranta@netikka.fi

Abstract

To date, the main application of automatic milking is on dairy farms of less than 200 cows where the impact on lifestyle and labour efficiency is greatest. 'Large' dairy herds with 500 or more cows did not initially adopt automatic milking, but in the last few years increasing costs of labour and feed have prompted some large dairies to implement voluntary automatic milking with single box. double box and small multibox systems. This study reports on survey data collected from 9 large dairy farms and 44 smaller farms and compares management and housing practices. Despite the opportunity to do so, we found that with one exception, large farms did not use bigger milking groups than the 2 robots per group commonly used on smaller dairies, but they did house first lactation animals separate from older cows, two herds had a designated group for mastitis cows and one grouped by stage of lactation. Unlike small herds fresh cows were handled as a separate group for 2 to 4 weeks after calving and either milked in a parlor or a dedicated automatic milking system (AMS). Handling systems varied and while all large herds had separation areas many were not well utilized. With staff in the barn for more hours of the day, fetch routines on large farms were more strategic and the number of cows fetched were lower (3.0 vs 4.8 per robot). Large farms reported more milkings (2.85 vs 2.70) and more milk per cow (36.8 vs 31.8 Kg) than small herds. Estimated labor per cow per day varied widely, but was lower on large farms (3.6 minutes vs 5.0 minutes) and milk produced per hour of labor was also higher on large dairies (640 vs 473 litres). Having a clearer understanding of success factors for automatic milking in larger herds will guide future adopters in developing appropriate barn designs and management strategies.

Keywords: robotic milking, big dairies, labor

1. Introduction

Automatic milking has become mainstream technology on family run dairy farms of less than 200 cows where the impact on lifestyle and labour efficiency is greatest. Success factors in management, and barn layout for these herds are relatively well defined (Rodenburg, 2013). 'Large' dairy herds with 500 or more cows that can make very efficient use of automated milking parlors did not initially adopt automatic milking, but in the last few years increasing costs of labor have prompted some large dairies to implement voluntary automatic milking with single box, double box and small multibox systems. We estimate there are less than 60 herds worldwide milking more than 500 cows in this way, but interest in adoption is high. Ideal strategies for group size and grouping by age or stage of lactation, ideal systems for cow handling, the practical organization of labour and ideal barn layouts for these larger herds may be different than for small dairies and are not well defined. This field study reports on survey data collected in 2015 to 2016 from 9 dairies milking more than 500 cows per farm and compares their practices and results to information collected from 44 smaller herds in 2014-2015.

The 9 large robot farms located in Denmark (5), Germany (1), Canada (1) and United States (2) milked with Lely (5 farms) and DeLaval VMS robots (4 farms). These herds milked 360 to 1100 cows (average 644), with 7 to 22 milking boxes (average 12). The number of cows per robot ranged from 40-63 and averaged 54.5. The 44 smaller farms were located in Finland, Sweden, Denmark, Holland and United States. The majority of the small farms had Lely and DeLaval robots, but the group also included GEA, Fullwood and SAC systems. Herd size ranged from 48-433 milking cows (average 128) milked with 1 to 8 robots (average 2.7) with 34-69 cows per milking stall (average 54.5). In the small herds the majority of labor data was collected by direct observation of morning chores and recording the time required for specific tasks. This became impractical in large herds so more of the data was taken from clocked time and time for specific tasks was based on estimates provided. An extensive questionnaire was completed with each herd manager or owner. This questionnaire focussed on less routine work scheduled weekly, handling practices for calving, breeding, reproductive health and treatments as well as opinions about functionality of the barn and perceived strengths and weaknesses in the system and reasons for choosing robotic milking. While the number of herds is too small to attempt any kind of statistical analysis, their experiences provides a starting point for future investigation.

2. Grouping strategies

Seven of the large dairies were newer barns designed for robotic milking and 2 were retrofits. The 5 Lely farms used free traffic and the 4 DeLaval farms used guided traffic, 3 with gating set up for milk first and 1 with feed first. All of the small dairies housed the cows in groups accessing one or two robots. One large herd had 394 cows in a single group accessing 8 robots, but groups accessing two robots were also most common on large dairies, and one worked with one robot groups. Time required for fetching was longer in the very large group and moving through the groups for bedding delivery was more complicated for the herd with single robot groups. While there was very little opportunity to group cows strategically on the small dairies, all of the large herds did reduce competition for first lactation animals by housing them separate from older cows. Two farms had a separate group for contagious mastitis and chronic high somatic cell count (SCC) cows. Only one farm grouped cows by stage of lactation. While on the small dairies fresh cows were added to the main milking groups within a few days of calving, the large herds all applied some form of special management for fresh cows. In the large herds, cows calved in a separate maternity barn. Three of the large dairies had parlor milked herds on the same site and used common dry cow and maternity facilities for automatic milking systems (AMS) and parlor milked herds. In these herds all fresh cows were milked in the parlor and housed in a fresh cow group for 2 to 3 weeks before being allocated to groups in the robotic milking barn. The 6 large dairies that milked exclusively with AMS all had a special group with one or two milking stalls used specifically for fresh cows. These groups had fewer cows per robot and offered more emphasis on cow comfort with larger stalls or a bedding pack. Management protocols and labor records indicated that much more time was spent with these cows for health monitoring, treatments, fetching and training than the rest of the herd. The length of time cows were in the fresh group ranged from 2 to 4 weeks. Group sizes were quite uniform except the number of cows per robot in the fresh group was much lower.

3. Handling and fetching

Handling cows for breeding, reproductive exams, vaccinations and treatments, hoof trimming and removing udder hair, is one of three main areas requiring labor on AMS farms. With one exception the large farms had individual separation areas for cows separated at the exit from the robot or the commitment pen. Typically this space provided room for 6 cows per robot in freestalls or a bedding pack area. This area was used for insemination, herd health, vaccinations, drying off cows and in some cases it was also used for longer term housing of lame cows to permit easy fetched milking and for trimming and treatment. In the best barns the separation area provided a simple, voluntary route back to the robot as well as a resting area, access to feed and water and good access to restaint in a chute or headlocks for handling and treatment. Several of the farms were not making good use of these facilities. Cows were being sorted from the group manually immediately before handling and in some cases cows were examined and inseminated in the freestalls. Based on our observations in these herds the strategic use of properly designed handling areas that rely on automatic post milking sorting, offer a substantial opportunity for additional labor saving with automatic milking. In large herds, separation can be programmed to take place during daytime hours so that a herdsman can rotate from group to group, apply the required treatment and return the separated cows to the groups immediately afterward. Separation space for just 2 to 3 cows per robot would be adequate with this handling strategy.

Fetching cows that do not attend voluntarily is the second major labor component on AMS dairies. In our survey data, large herds were more strategic about fetching than smaller herds, in part because having staff in the barn for most of the day made it possible to vary the fetching interval based on expected production. Large herds fetched fresh cows more often and later lactation cows less often than small herds which typically fetched at two fixed times of the day. Large herds reported leaving late lactation cows for 16 to 18 hours before fetching them. Large herds fetched an average 3.0 cows per robot versus 4.8 cows per robot in the smaller herds. Grouping by age may also contribute to lower fetch rates by improving the voluntary attendance of more timid heifers. It is well known that lameness is a major deterrent to voluntary milking and our survey data indicated that large herds followed more consistent protocols for foot bathing, hoof trimming and treatment of lame cows, so this may also be a contributing factor to the lower reported number of fetch cows. Large barns with alley scrapers removing manure from long alleys present a special challenge with regard to keeping cows feet clean because a large 'bath' of manure forms in front of the scraper sled. The three North American large barns in this study used special scraper systems with a tube below the floor, to alleviate this problem.

4. Bedding and grooming freestalls

Cleaning, grooming and adding bedding to freestalls is the third identified area of labor but we found little difference in how this is done in large and small herds and little difference in the time required on a per cow basis. Bedding was typically added 2 times a week, and stalls were cleaned and raked 2-3 times per day by hand. On small farms this was usually combined with fetching but on large farms these tasks were often done by different employees at different times. Two farms used rail mounted automatic bedding systems but a third farm had abandoned a similar system because of technical problems. On some of the small farms stall cleaning was done strategically just after feeding when most of the stalls were empty but this was more difficult on large dairies.

The technology and methodology used for washing large robotic milking systems varied widely from farm to farm and was viewed by some of the large herds interviewed as an area where improvement was needed.

5. Production and labor

Although average number of cows per robot were identical on small and large farms, large farms reported more milkings per cow than small farms (2.85 milkings per cow per day versus 2.7) At an average of 36.8 litres, milk production per cow per day in large herds was substantially more than in the 31.8 litres reported by the small herds. Grouping by age, more consistent hoof care and more intensive management in the fresh group the first three weeks after calving may be contributing factors.

The primary reason for choosing robotic milking given by large herd owners was to save labor. While variation in the range of animals housed on all the farms and difficulty in separating out the robotic component on farms with both robots and parlors made it difficult to come up with accurate labor measurements we estimated the labor used for all daily work with cows, heifers, calves and feeding and expressed it in hours per milking cow per day and in litres of milk per hour of labor. As illustrated in Figure 1, on small farms labor used ranged from 1.8 minutes to 14.7 minutes per cow per day (average 5.0 minutes) and in large herds it ranged from 2.1 to 4.9 minutes (average 3.6 minutes). As per Figure 2. in the small herds milk per hour of labor ranged from 40 litres to 930 litres (average 473 litres) while in large herds it ranged from 502 to 1,060 litres (average 640 litres).

These numbers suggest that while economies of scale in robotic milking offer an economic advantage to larger herds the gap between large and small herds with this technology is probably substantially smaller than with parlor milking. Fetching and cleaning stalls takes a similar amount of time, and walking distances in large barns can be a challenge if robot locations are not well thought out. One herdsman in a large dairy calculated that in 8 hour shift 1.5 hours were spent walking from task to task.

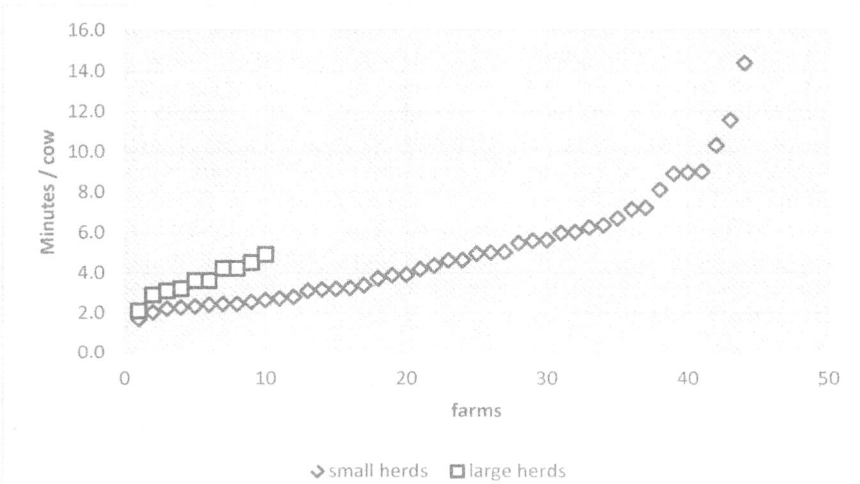

Figure 1. Minutes / cow in large and small herds.

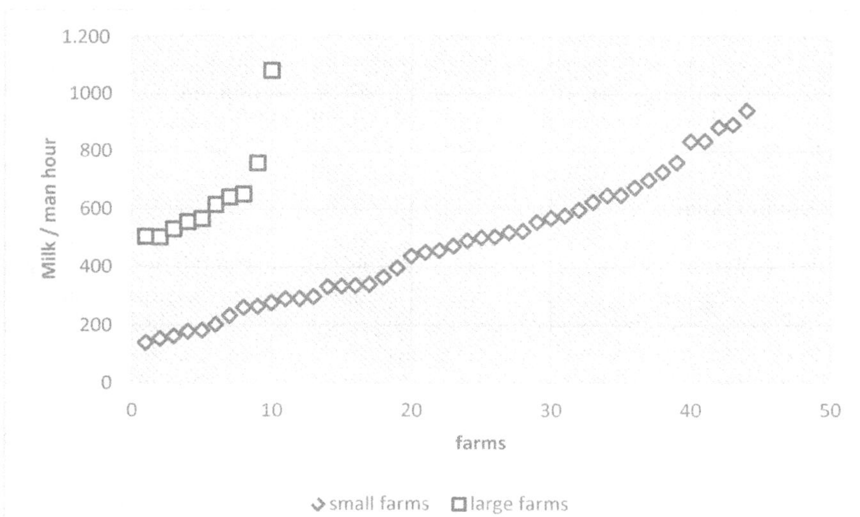

Figure 2. Milk in litres produced/man hour in large and small herds.

In our experience the gap between large and small dairies with milking parlors is substantially larger. Typically farms with 100-200 cows milked in a parlor can reach 150-250 litres milk / man hour, and with 600-800 cows this can be 350-450 litres. It is generally reported that robotic milking reduces dairy farm labor by 20-40% compared to parlor milking and it would seem that small herds are likely to be at the top end of this range and large herds closer to the bottom end.

In addition to labor saving, a benefit of robotic milking that was mentioned by several of the large herd owners is the improved cow comfort and welfare resulting from eliminating the stress of moving cows to the parlor, and crowding them in a holding area. Individual concentrate feeding according to production, stage of lactation and dynamic adjustment and individual control over milking interval all offer opportunities for improved efficiency compared to traditional parlor milking and total mixed ration feeding.

In terms of future direction for robotic milking in large herds, rising labor costs and decreasing cost and increasing capabilities of technology will lead to more rapid adoption of AMS in dairies over 500 cows. Currently the dominant philosophy is to limit group size to 2 robots per group, and while this reduces fetching labor, both handling and bedding labor could be reduced in larger groups. If newer technology reduces the number of fetch cows and sensors that locate fetch cows accurately within the group come into more common usage, larger groups with 3 to 4 robots per group may become more practical.

Having a clearer understanding of success factors for automatic milking in larger herds will guide future adopters in developing appropriate barn designs and management strategies.

Reference

Rodenburg, J. (2013). Success factors for robotic milking. In: Proceedings of the Precision Dairy Conference and Expo: Precision Dairy 2013, University of Minnesota, Rochester, MN, USA, pp. 21-34.

A tool to analyze and optimize milking intervals in robotic milking systems

J. Harms[1*] and R.M. Bruckmaier[2]
[1]Bavarian Research Center for Agriculture, Prof.-Dürwachter-Platz 2, Germany; [2]Veterinary Physiology, University of Bern, Bern, Switzerland; jan.harms@lfl.bayern.de

Abstract

In robotic milking systems, settings for the milking intervals (MI) have to be shorter than the desired average interval as cows visit the system mostly at variable intervals. However, the controlling by the farmer is mainly directed to overdue cows, and often (too) short MI are not recognized. Also most of the installed herd management tools do not consider short MI. Too early milking, especially at low milk yields (MY), should be avoided for physiological reasons but also to increase the capacity of the milking unit and milk quality (FFA). At low udder filling no or only little cisternal milk is present to bridge the lag time until milk ejection, and it can take up to 3 min until the alveolar milk ejection occurs. Until then milk flow is low or even absent. We have, therefore, developed the tool 'MICheck' that simultaneously analyses MI and yield per milking (MYpM). MICheck should show to the farmer how often undesired combinations of MYpM and MI occur on his farm. Based on the collected data, an analysis of all participating farms is performed to evaluate the practical impact of the possible improvements. For the development phase of MICheck, a MI of less than 7 h was defined as 'short' and of more than 16 h as 'long'. MYpM of less than 8 kg were defined as 'low', more than 14 kg as 'high'. Individual milkings were classified in 5 categories, depending on which combination of these criteria was fulfilled. Data from 82 farms were analyzed: 19.3% of all milkings had a MI of less than 7 h of which 37.8% also showed MYpM of less than 8 kg. On the farm with the highest values those percentages were 39.4 and 54.8%, respectively, at a daily MY of 26.4 kg. Farmers and advisers confirmed that short MI were set to avoid overdue cows. They were, however, unaware of the physiological effects and underestimated by far the frequency of their occurrence. Currently an online version of MICheck is implemented that can be used by most of the Bavarian and Austrian farmers.

Keywords: milk yield, milking permission, udder filling, milk quality

1. Background

Automatic milking systems (AMS) are increasingly installed in many countries. In some regions, including Germany, newly installed milking systems are more often AMS than conventional systems. Worldwide an estimated number of 25,000 farms are milking with AMS today.

The goal of AMS milking is that the cows visit the milking box voluntarily. Guided cow traffic or driving the cows into the milking box is necessary for some animals to achieve the desired milking frequency. Farmers as well as milking machine producers try to find solutions for cows that are not willing to visit the milking system, often called 'lazy cows'. Discussions about ways to optimize cow traffic or settings of the herd management systems are aimed to reduce such animals.

To avoid the occurrence of too long intervals between milkings, the setting of the AMS for minimum milking intervals is often shorter than actually desired. However, too short milking intervals remain mostly undetected because current herd management software does not allow a surveillance and evaluation of milking intervals of the herd or of individual animals over a longer time period. Animals which are regularly milked too often, and/or achieve only small milk yields, are not especially considered in the available standard evaluation systems.

A too early milking at a low degree of udder filling (relative to its storage capacity) should be avoided for physiological reasons. At low udder filling, the amount of cisternal milk is too low to bridge the time until milk ejection, and simultaneously the lag time to induce milk ejection is particularly long, and can take up to 3 min if the amount of stored milk is extremely small (Bruckmaier and Hilger, 2001; Dzidic *et al.*, 2004). At very low udder filling, the time requirement for an adequate udder preparation to allow the availability of the alveolar milk fraction at the start of milking may be longer that the actual milking time (Bruckmaier and Hilger, 2001). Although the lag time between teat preparation through the cleaning device and attachment (if not longer than 45-60 s) adds to the required latency period, the efficiency of the AMS may be reduced (Weiss and Bruckmaier, 2005). On the other hand, the milking at full vacuum and regular pulsation should not be started until sufficient milk is available to guarantee a continuous refill of the teats by the positive intramammary pressure during the liner closed phases. Transient milking on empty teats causes an early increase of the vacuum in the liner mouth piece, and hence climbing of the teat cup combined with an increased impact on the teat tissue (Besier and Bruckmaier, 2016; Besier *et al.*, 2016). Besides a prolonged milking time, this can also cause a reduced emptying of the udder, and hence increased risk for intramammary infections. In addition, too early milking on insufficiently filled udders can lead to an increased content of free fatty acids and off-flavors in the harvested milk with the respective effects on milk processing (Klungel *et al.*, 2000).

2. Objectives

The goal of the present investigation was the development of a tool for farmers and farm consultants to evaluate intervals between milkings and the harvested milk yields concomitantly to allow an easy estimate of milking intervals relative to the udder filling. Based on that, we have also analyzed the current frequency of suboptimal milkings with respect to milking interval and udder filling.

3. Materials and methods

During the preparation of the project, the data sets of the DHI systems were found to represent the optimal data base to obtain the most important information on cow identification number, intervals between milkings, and milk yields independent of the brand of the milking system. The prototype of the new tool 'MICheck' is based on Microsoft Excel because it allows easy handling and fast adaptations during the developmental period.
Intervals between milkings were classified as 'short' and 'long' if the time was <7 h and >16 h, respectively. Milk yields were classified as 'low' and 'high' if the harvested milk was <8 kg and >14 kg, respectively. Intervals and milk yields between these thresholds were classified as 'normal'. Individual milkings were classified under consideration of both criteria as shown on Table 1. For future use of MICheck under practical conditions, the thresholds can be adjusted by the farmer or farm consultant to consider the production level of the herd and hence to allow replacement of the degree of udder filling by the milk yield as a rough estimate for udder filling.

4. Results

Data from 82 farms were analyzed with the MICheck tool and discussed with farmers and farm consultants. Figure 1 shows the obtained results. As an average over all farms, 19.3% of intervals after the previous milking were short (<7 h), with simultaneous low milk yields (<8 kg) in 7.3% of these milkings. Among the investigated farms, the highest recorded percentages for short milking intervals and simultaneous low milk yield were 39.4 and 21.6%, respectively, for a farm with an average daily milk production of 26.4 kg.

Table 1. Classification in categories for milking interval and yield per milking.

Color in the diagram	Category	Milking interval (MI)	Milk yield per milking (MYpM)
	A	short (<7 h)	normal (>8 kg)
	B	short (<7 h)	low (<8 kg)
	C	long (>16 h)	normal (<14 kg)
	D	long (>16 h)	high (>14 kg)
	E	normal (7-16 h)	any milk yield

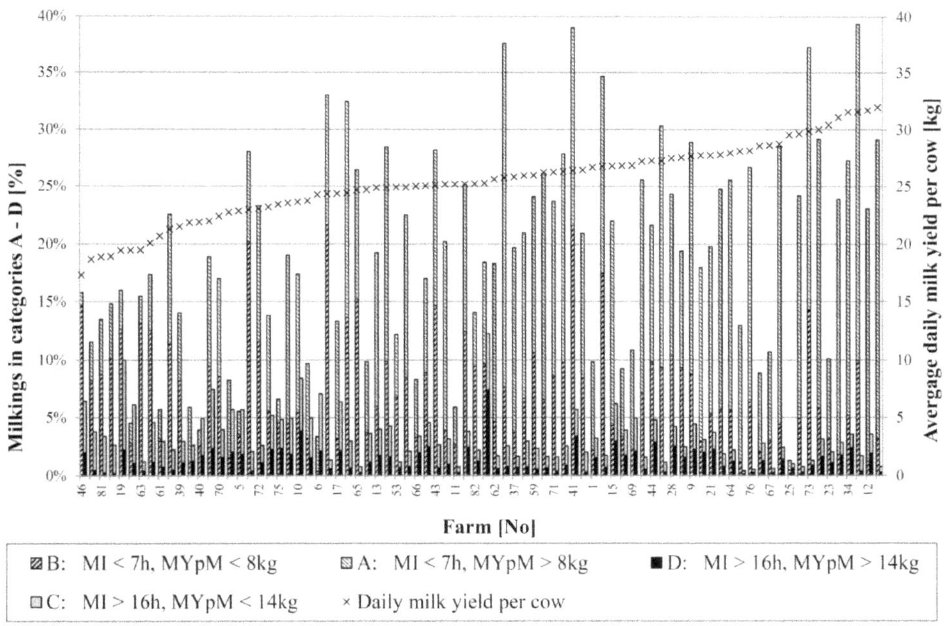

Figure 1. Milkings in the categories for milking interval and yield per milking.

The results of the evaluation were discussed with farmers and farm consultants. Discussions made clear that short milking intervals were principally acceptable for many farmers, whereas they did not expect the frequency of their occurrence. Obviously, farmers were not aware about the negative consequences of milking at low udder filling, in particular about the required lag time of milk ejection and hence requirements for udder preparation. Also the potential impact on the capacity of the milking system had not been taken into consideration. In particular, the farmers had not realized the consequences of a combination of short milking intervals and low milk yields (category B). MICheck can visualize the results graphically over time, and, most importantly, for each individual animal. This allows the farmer to adjust the settings of the milking system to optimize its performance. Figures 2A and 2B show the distribution of the milking intervals of one exemplary farm (average daily milk yield 25 kg) before and after the consulting with the new tool. Before the consulting, milking intervals <7 h occurred at 37% of all milkings, and at 14% of all milkings the harvested milk

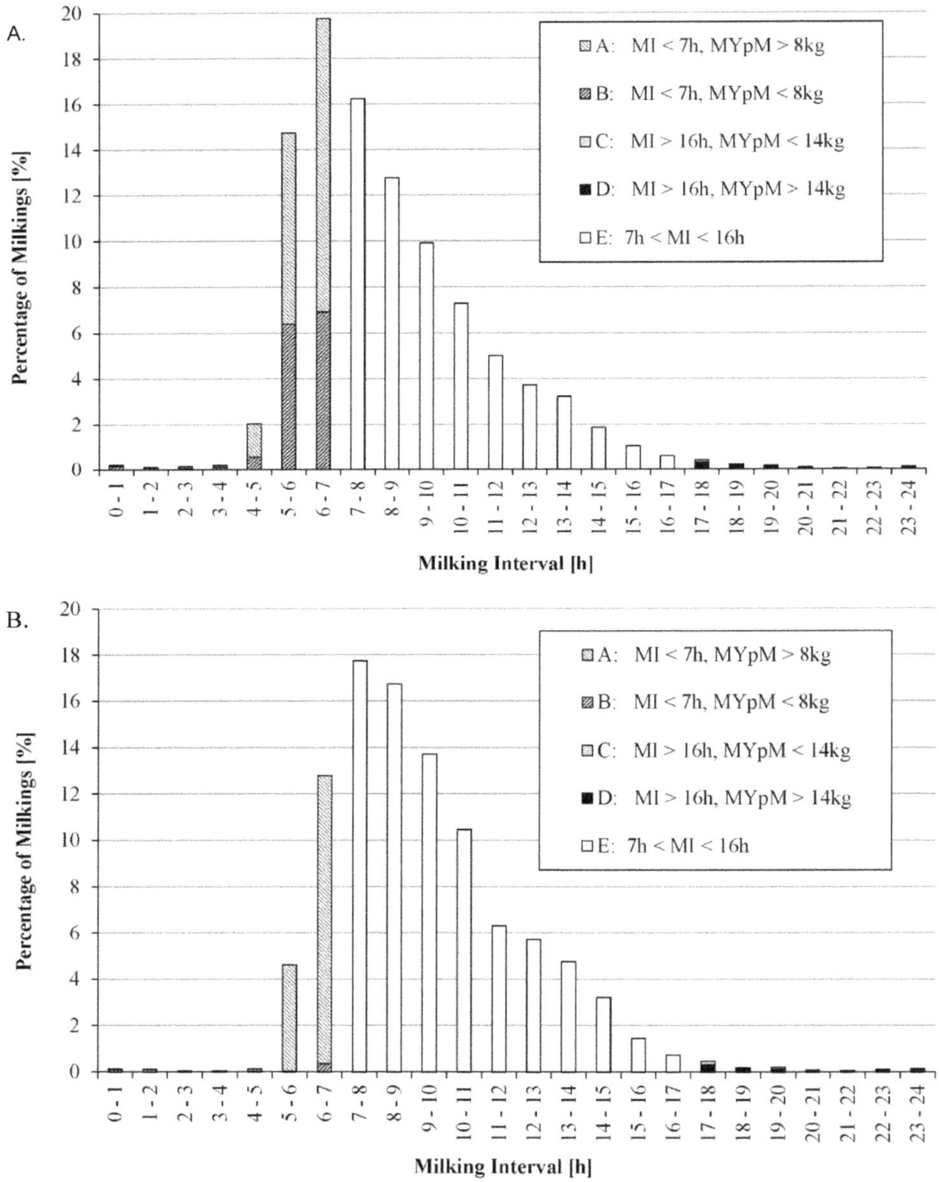

Figure 2. Frequency distribution of the milkings in the categories A-E before (A) and after (B) advice.

was <8 kg. Adjustments of the settings of the milking system reduced the milking intervals <7 h to 18%. These represented almost exclusively animals with high milk production, i.e. milkings with milk yields <8 kg were reduced to 0.6%.

5. Outlook

Our new tool 'MICheck' was adopted by the LKV Bayern e.V., and will be adjusted and further developed in collaboration with several farm consultants. With this cooperation we are able to also reach the AMS farms in Baden-Württemberg and Austria. We will be able to include DIM and maximum milk production during the current lactation in the calculations and thus improve the precision of the udder filling estimate. In addition, MICheck will be optimized to be more user-friendly that it can be used by consultants and farmers without our additional support.

References

Besier, J. and Bruckmaier, R.M. (2016). Vacuum levels and milk-flow-dependent vacuum drops affect machine milking performance and teat condition in dairy cows. Journal of Dairy Science 99(4): 3096-3102.

Besier, J., Lind, O. and Bruckmaier, R.M. (2016). Dynamics of teat-end vacuum during machine milking: types, causes and impact on teat condition and udder health – a literature review. Journal of Applied Animal Research 44(1): 1-10.

Bruckmaier, R.M. and Hilger, M. (2001). Milk ejection in dairy cows at different degrees of udder filling. Journal of Dairy Research 68: 369-376.

Dzidic, A., Macuhova J. and Bruckmaier, R.M. (2004). Effects of cleaning duration and water temperature on oxytocin release and milk removal in an automatic milking system. Journal of Dairy Science 87: 4163-4169.

Klungel, G.H., Slaghuis, B.A. and Hogeveen, H. (2000). The effect of the introduction of automatic milking systems on milk quality. Journal of Dairy Science 83: 1998-2003.

Weiss, D. and Bruckmaier, R.M. (2005). Optimization of individual prestimulation in dairy cows. Journal of Dairy Science 88: 137-147.

Effects of a low moisture block supplement on cow distribution and time budget

A.J. DiGennaro[1], A.R. Lee[1], B.A. Wadsworth[1], S.A. Janse[2], C.L. Wood[2], J.D. Clark[1] and J.M. Bewley[1*]
[1]*Department of Animal and Food Sciences, University of Kentucky, Lexington, 40546 KY, USA;* [2]*Department of Statistics, University of Kentucky, Lexington, 40546 KY, USA; jbewley@uky.edu*

Abstract

The objective of this study was to compare cow visit time and frequency for cows provided access to either a low moisture block supplement barrel (LMB) (Buffer-lyx®, Ridley Block Operations, Flemingsburg, KY) or an empty control barrel (ECB). The study was conducted in two freestall pens (n=40 and n=43, respectively) at the University of Kentucky Coldstream Dairy from 1/21/15 to 3/05/15. A crossover study was performed by exposing all cows to a 7 day acclimation period with an empty control barrel (ECB), a 14 day period with either LMB or ECB (P1), a 7 day washout period with an ECB, and a 14 day period with the opposite supplement (P2). Pen 1 cows were provided access to the LMB then ECB (TC) and pen 2 cows were provided access to the ECB then LMB (CT). Cow visit frequency and time around the barrels was measured using SmartBow (MKW Electronics, Jutogasse, Austria), an eartag based real time location system. SmartBow eartags recorded per second XY coordinates to calculate cow location within the housing facility in min/d. A start or stop time was recorded when a cow entered or exited the 3 m radius around barrel. The MIXED procedure of SAS 9.3 (SAS Institute, Inc., Cary, NC) was used to analyze all data. Cows spent more time in the area around the LMB than the ECB (16.49±1.57 vs 4.98±1.62 min/d, $P<0.01$). A significant difference ($P<0.01$) was observed between total visit time during period (P1: 15.68±1.62 and P2: 5.78±1.57 min) and sequence (TC: 19.79±1.84 and CT: 1.67±1.82 min). Cows with LMB had greater visit frequency/d than cows with ECB (2.01 vs 0.09 visits/d, $P<0.01$). Period and sequence effects may be attributed to inclement weather during P2 and cows becoming conditioned to expect the ECB. Strategic placement of LMB may alter dairy cattle movement and behavior in a freestall setting.

Keywords: cow distribution, automated milking system, SmartBow

1. Introduction

Low moisture block supplements (LMB) are composed of molasses with additional nutrients (McKenzie, 1974) to supplement grazing animals with additional energy, protein, vitamins, and minerals (Bowman and Sowell, 1997). The blocks are formed through a high temperature evaporative process to remove moisture (McKenzie, 1974). Atmospheric moisture is absorbed by the molasses to create a syrup layer on the surface of the block for cows to consume (Benton and Patrick, 1988). Because molasses supply can be limited and expensive, alternative ingredients such as filter muds, or scums, from the sugar processing industry are used (Sansoucy, 1995). Subacute ruminal acidosis challenged cows given unlimited access to buffer blocks experienced a smaller decrease in mean ruminal pH compared to cows without buffer block access (6.00 to 5.87 vs 6.13 to 5.69, $P<0.05$) (Krause *et al.*, 2009). Challenged cows without access to buffer blocks also experienced a greater amount of time (249 vs 83 min/d) below pH 5.6 when compared to cows given access to buffer blocks (Krause *et al.*, 2009).

Cattle may find molasses based supplements more palatable than others supplements. Herds showed greater visiting frequency (2.82 vs 2.47 herd visits/d; $P=0.02$) and increased visit duration to sweet LMB (125.7 vs 54.9 min/herd visit; $P<0.01$) (Aubel *et al.*, 2011). Beef cattle spent on average 66±2 min/d more within 10 m of an LMB than a salt block (Bailey *et al.*, 2008). A greater percentage of cattle were observed visiting an LMB than a conventional dry mix over 4 weeks (74 vs 56%) (Bailey and Welling, 2007). Additionally, LMB have been shown to manipulate grazing patterns by successfully luring cattle toward grazing underutilized rangeland (53% of cows observed within 600 m of supplement when grazing between 245 ha and 330 ha pastures) (Bailey *et al.*, 2001).

With automated milking systems (AMS), increasing visits to the AMS is desired to increase milk yield and AMS efficiency. To entice cow entry and optimal visiting frequency, producers have used concentrate feeding (Madsen *et al.*, 2010). However, supplementation does not guarantee cows will visit the optimal number of times/d (Winter and Hillerton, 1995). Dairy farmers may consider the use of LMB to entice cows towards an AMS. The objective of this study was to compare how frequently and for how long cows visit an area with a LMB (Buffer-lyx®, Ridley Block Operations, Flemingsburg, KY, USA) or an empty control barrel (ECB).

2. Materials and methods

2.1 Low moisture block supplement and control barrel

The study was conducted at the University of Kentucky Coldstream Dairy from January 21st 2015 to March 05th 2015 (IACUC Protocol number 2014-1322). Lactating Holsteins (n=83 parity=2±1.27, DIM=122±110) were housed in two pens and balanced for parity, cow volume, and DIM. The two pens were separated by a shared, raised feedbunk and both pens retained access to freestalls. Pen 1 (n=40) had Dual Chamber Cow Waterbeds (Advanced Comfort Technology, Reedsburg, WI, USA) as the stall base. Pen 2 (n=43) had rubber-filled mattresses (PastureMat; Promat, Ontario, Canada) as the stall base. Stocking density never exceeded 100% in either freestall pen. Sawdust was used to cover stall bases in both pens. All other surfaces of both pens (feed bunk, holding pen, alleys) consisted of grooved concrete. Pens were scraped 2× daily during AM/PM milking. For 1 h/d at 10:00, weather permitting, cows had access to a grass seeded exercise lot. Cows were fed a free-choice total mixed ration (TMR) balanced to meet lactating cow requirements, consisting of grain mix, corn silage, alfalfa silage, whole cottonseed, and alfalfa hay daily at 06:00 and 13:30. Cows were milked twice daily at 04:30 and 15:30.

In this crossover study design, each freestall pen had an LMB for alternating two-week periods. The LMB and ECB were placed in the same position in each pen relative to the freestall and waterer. Novelty effect (Kondo and Hurnik, 1988) was reduced by introducing an ECB 7 days before Period 1 (P1). The study schedule consisted of two 14 day periods with a 7 day washout period with an ECB between. Period 1 consisted of 14 d in which pen 1 had LMB and pen 2 had ECB. Period 2 (P2) consisted of 14 d in which pen 1 had ECB and pen 2 had LMB (P2). Pen 1 was treated with sequence LMB then ECB (TC) and Pen 2 with sequence ECB then LMB (CT).When pens had a LMB supplement, a farm manager monitored the content of the barrel daily. Empty LMB were replaced with full barrels within 24 h. All barrels were secured inside tires to prevent the cows from moving the barrels.

2.2 Cow visits to supplement

Before each observation period, a 3 m radius was approximated. When a cow entered or exited the 3 m radius around either barrel, a start or stop time was recorded, and total visit time was calculated. SmartBow (MKW Electronics, Jutogasse, Austria), an eartag based real time location system, recorded relative location. All cows were fitted in the pinna of the right ear with SmartBow before P1. SmartBow forwarded XY coordinates to multiple receivers throughout the housing facility. Data collected was used to calculate the location of each cow within the housing facility in time interval for min/d.

2.3 Statistical analysis

Visit time was calculated as the daily sum of the time that a cow's coordinates were within the area around the barrel. A visit was first classified for each time that a cow transitioned from another region within the pen to the area around the barrel. Then, if the visit was within five minutes of the previous visit, an assumption was made that this was the same visit, and those visits were removed before calculating visit frequency. The MIXED procedure of SAS (SAS ® 9.3, SAS Institute, Inc., Cary, NC, USA) was used to test the fixed effects of treatment, period, and sequence on visit frequency and visit time. The model accounted for the random effect of cow within sequence. A compound symmetry covariance structure was used and effects with $P<0.05$ were considered significant. Stepwise backward elimination was used to remove non-significant interactions ($P\geq0.05$).

Table 1. LSMeans for visit frequency and total time spent in the area surrounding a barrel in a crossover study using two pens of cows with alternating access to a low moisture block supplement barrel and an empty control barrel.

		Visit frequency (per day)[4]	Visit time (minutes per day)[4]
Treatment[1]	LMB	2.01±0.05[a]	16.49±1.57[a]
	ECB	0.09±0.01[b]	4.92±1.62[b]
Period[2]	1	0.91±0.03	15.68±1.62[a]
	2	0.59±0.03	5.78±1.57[b]
Sequence[3]	TC	0.63±0.03	19.79±1.85[a]
	CT	0.87±0.03	1.67±1.81[b]

[1] LMB (low moisture supplement barrel). ECB (empty control barrel). The LMB and ECB resided in the same position in each pen relative to the freestalls and waterer.

[2] Study schedule consisted of two 14 d periods with a LMB and a 7 d washout period with ECB between. Period 1 consisted of 14 d in which pen 1 had LMB and pen 2 had ECB. Period 2 consisted of 14 d in which pen 1 had ECB and pen 2 had LMB.

[3] Pen 1 was treated with sequence LMB then ECB (TC) and Pen 2 with sequence ECB then LMB (CT).

[4] LSMeans for pairs with different letters are significantly different ($P<0.05$).

3. Results and discussion

3.1 Low moisture block and empty control barrel

Of the 83 cows on study, only 71% (n=59) were recorded visiting the LMB or ECB during P1 or P2. Similar percentages of herd visit, 53 and 88% respectively were reported by Bailey *et al.* (2001, 2008). Data collected with SmartBow showed that cows spent greater total time around the LMB than the ECB (16.49±1.57 and 4.98±1.62 min, $P<0.01$). Bailey *et al.* (2008) also reported similar total LMB visit time (11±2 min/day) for beef cattle when both a LMB and salt block was present. However, when LMB was fed alone, cattle were recorded within 10 m of supplement for 73±9 min/d (Bailey *et al.*, 2008). Aubel *et al.* (2011) reported that beef cattle spent more time at the LMB when grazing quality was poor. During summer months, when grazing quality was typically better, cattle tended to visit the LMB less than during winter months (Aubel *et al.*, 2011). In cows fed a balanced TMR or partial TMR with grazing, visit time may be decreased to the quality and consistency of feed (Bargo *et al.*, 2002). Visit frequency analysis indicated cows visited the LMB on average two more times/d than the ECB (2.01±0.05 and 0.10±0.01 visits per day, $P<0.01$). Similar visit frequency to molasses-based block was observed in beef cattle when compared to a dry, granular form of mineral (2.82 vs 2.47 herd visits/d; P=0.02) (Aubel *et al.*, 2011).

3.2 Period effect

A period (P1: 15.68±1.62 and P2: 5.78±1.57 min, P<0.01) effect was observed between periods. Visit frequency was not different between periods (P1: 0.91±0.01 and P2: 0.60±0.01 min, P=0.08). During P2, reported temperature averaged -3.38 °C (mean max=1.67 °C, mean minimum=-9.44 °C) (Spindletop, Lexington, KY) whereas during period 1 temperature averaged 1.67 °C (mean max=7.22 °C, mean minimum=-3.90 °C). Around 0 °C cows may begin to experience cold stress (Angreka and Herbut, 2015; Brouček *et al.*, 1990). Cold stress is characterized by decreased milk production and decreased activity due to energy diversion for maintenance (Brouček *et al.*, 1990; Collier *et al.*, 1982). Cold temperatures in P2 may have contributed to decreased activity and decreased frequency of visit to the unsheltered area where the LMB and ECB were placed.

3.3 Sequence effect

Visit frequency was not significantly different between the sequences (TC: 0.63±0.01 and CT: 0.87±0.01 min, P=0.20). However, a sequence effect was seen for total visit time (TC: 19.79±1.84 and CT: 1.67±1.82 min, P<0.01). Cows in pen 2 had an empty barrel for four weeks during the acclimation period, P1, and washout period. The low visit time observed in sequence CT may have been caused by cows in pen 2 being conditioned to expect an empty barrel when LMB was available. Kondo and Hurnik (1988) proposed that novel stimulus can arouse the motivation of cattle, but behavioral changes and severity decreases over time. The sequence effect may be attributed to cows in sequence CT continuing to expect an ECB after the addition of the LMB.

4. Conclusions

Cows spent a greater total time and had a greater number of visits at the LMB compared to the ECB. Time spent in the area around the LMB may suggest a LMB could be used to influence cow distribution.

A period effect occurred between time spent in area around LMB or ECB in P1 and P2, which likely resulted due to cold temperatures during P2. A sequence effect occurred between time spent in area around LMB or ECB in sequence TC and CT, which may suggest LMB novelty decreases over time. Additional work must be conducted to determine cow preference for LMB and to explore long term effects in barns that have an AMS.

Acknowledgments

The authors would like to extend their sincerest thank you to Ridley Block Operations, Inc. for the donation of the LMB, Buffer-lyx® to the University of Kentucky Coldstream Dairy and for the financial support needed to complete this study. We also thank Smartbow EN for their donation of the technologies and for their technical support throughout the study.

References

Angrecka, S. and Herbut, P. (2015). Conditions for cold stress development in dairy cattle kept in free stall barn during severe frosts. Czech Journal of Animal Science 60(2): 81-87.

Aubel, N., Jaeger, J., Drouillard, J., Schlegel, M., Pacheco, L., Linden, D., Bolte, J., Higgins, J. and Olson, K. (2011). Effects of mineral-supplement delivery system on frequency, duration, and timing of supplement use by beef cows grazing topographically rugged, native rangeland in the Kansas Flint Hills. Journal of Animal Science 89(11): 3699-3706.

Bailey, D., VanWagoner, H., Weinmeister, R. and Jensen, D. (2008). Comparison of low-moisture blocks and salt for manipulating grazing patterns of beef cows. Journal of Animal Science 86(5): 1271-1277.

Bailey, D.W. and Welling, G.R. (2007). Evaluation of low-moisture blocks and conventional dry mixes for supplementing minerals and modifying cattle grazing patterns. Rangeland Ecology and Management 60(1): 54-64.

Bailey, D.W., Welling, G.R. and Miller, E.T. (2001). Cattle use of foothills rangeland near dehydrated molasses supplement. Journal of Range Management: 338-347.

Bargo, F., Muller, L.D., Delahoy, J.E. and Cassidy, T.W. (2002). Performance of high producing dairy cows with three different feeding systems combining pasture and total mixed rations. Journal of Dairy Science 85(11): 2948-2963.

Benton, A.E. and Patrick, J. (1988). Molasses feed block for animals and method of making the same. Google Patents.

Brouček, J., Letkovičová, M. and Kovalčuj, K. (1991). Estimation of cold stress effect on dairy cows. International Journal of Biometeorology 35(1): 29-32.

Bowman, J. and Sowell, B. (1997). Delivery method and supplement consumption by grazing ruminants: a review. Journal of Animal Science 75(2): 543-550.

Collier, R.J., Beede, D.K., Thatcher, W.W., Israel, L.A. and Wilcox, C.J. (1982). Influences of environment and its modification on dairy animal health and production. Journal of Dairy Science 65(11): 2213-2227.

Krause, K., Dhuyvetter, D. and Oetzel, G. (2009). Effect of a low-moisture buffer block on ruminal pH in lactating dairy cattle induced with subacute ruminal acidosis. Journal of Dairy Science 92(1): 352-364.

Kondo, S. and Hurnik, J.F. (1988). Behavioral and physiological response to spatial novelty in dairy cows. Canadian Journal of Animal Science 68(2): 339-343.

Madsen, J., Weisbjerg, M. and Hvelplund, T. (2010). Concentrate composition for automatic milking systems – effect on milking frequency. Livestock Science 127(1): 45-50.

McKenzie, C.O. (1974). Molasses feed block for animals and method of doing the same. Inventor, Patent No. US3961081 A.

Sansoucy, R. (1995). New developments in the manufacture and utilization of multinutrient blocks – short communications. World Animal Review 82(1).

Winter, A. and Hillerton, J. (1995). Behaviour associated with feeding and milking of early lactation cows housed in an experimental automatic milking system. Applied Animal Behaviour Science 46(1): 1-15.

Individualized recommendations using clustering of robotic milking systems

M. Tremblay[1*], J.P. Hess[1], B.M. Christenson[1], K.K. McIntyre[1], B. Smink[2], A.J. Van der Kamp[3], L.G. De Jong[3] and D. Döpfer[1]

[1]Departments of Medical Sciences, Section of Food Animal Production Medicine, School of Veterinary Medicine, University of Wisconsin-Madison, 2015 Linden Drive, Madison 53706, USA; [2]Lely North America, 775 250th Avenue, Pella, Iowa 50219, USA; [3]Lely International N.V., Cornelis van der Lelylaan 1, 3147 PB, Maassluis, the Netherlands; mtremblay@wisc.edu

Abstract

Automatic milking systems (AMS) are being implemented in a variety of conditions. There is a need to characterize individual farming practices and regional challenges in order to streamline management advice and goals for producers. Benchmarking is often used in the dairy industry to compare farms by computing percentile ranks of the production values of farms in the same geographic region or utilizing the same breed of cattle. We hypothesize that a clustering approach using herds' production data and management information would yield improved peer groups of farms compared to conventional benchmarking methods. A mixed latent-class model-based cluster analysis of 529 North American AMS dairy farms with respect to 18 significant risk factors identified 6 clusters. Each cluster (i.e. peer group) represents unique management styles, challenges and production patterns. When compared to peer groups based on criteria similar to the conventional benchmarking standards, the 6 clusters better predict milk production per robot per day. Each cluster represents a unique management and production pattern that requires specialized advice. For example, cluster 1 farms are farms that recently installed AMS robots. Cluster 3 farms, the most northern farms, feed high amounts of concentrates through the robot to compensate for low energy feed in the bunk. In addition to general recommendations for the farms within a cluster, farms can generate their own specific goals by comparing themselves to farms within their cluster. The improvement that cluster analysis allows for is characterized by the multivariable approach and the fact that comparisons between farms can be accomplished within a cluster and between clusters as a choice.

Keywords: cluster analysis, automatic milking systems, benchmarking, comparisons

1. Introduction

The dairy industry is very familiar with benchmarking as a method of making percentiles of production values from a large number of farms. Usually farms are placed in comparison groups based on a few features such as cattle breed or region. Tremblay *et al.* (in press) used a cluster analysis of 529 North American AMS dairy farms to identify 6 clusters (i.e. peer groups) that represented unique management styles, challenges and production patterns using 18 variables simultaneously. These clusters were found to better predict the farms' milk production compared to usual benchmarking grouping methods and enabled customized advice per cluster (Tremblay *et al.*, in press). Our aim was to show that making comparisons between farms within the same cluster yields a better representation of a farm's strengths and weaknesses compared to when all farms are compared together.

2. Materials and methods

In Tremblay *et al.* (in press) data between 2011 and 2014 were collected from 529 dairy farms throughout North America with Lely Astronaut AMS (Lely Industries N.V., Maassluis, the Netherlands). The final data set with 54,065 observations had previously been analyzed for significant associations between available variables and milk production per cow per day using a generalized linear mixed regression model (Tremblay *et al.*, 2016). Five categorical variables were used (i.e. Traffic_Type, Robots_per_Pen, Breed, Farm_Goal and New_Retro) and thirteen AMS numeric variables were used (i.e. Cows_per_Robot, Average_DIM, Concentrates, Rest_Feed, Milk_Production_per_Cow_per_Day, Refusals, Failures, Milkings, Milk_Speed, Boxtime, Connection_Attempts, Robot_Free_Time, and Days_Since_Installation) (Tremblay *et al.*, in press). Tremblay *et al.* (in press) used the 18 variables to perform a mixed latent-class model-based cluster analysis using the flexmixedruns function in the R package 'fpc'. The final cluster analysis, described in Tremblay *et al.* (in press), resulted in a latent-class model forming 6 clusters of farms with an average of 88 farms per cluster (range: 50-124 farms).
The 10th, 25th, 50th, 75th and 90th percentile of numeric variables were calculated per cluster and for all farms together for comparisons among farms.

3. Results and discussion

3.1 Cluster 2

Many Cluster 2 farms had forced traffic systems in Canada (i.e. quota system) with Jersey cows (Tremblay *et al.*, in press). They had the lowest average values of Milk_Production_per_Cow_per_Day, Milkings, Milk_Speed, and Connection_Attempts. Cluster 2 farms had the highest average value of Rest_Feed, Failures, Days_Since_Installation and Average_DIM (Tremblay *et al.*, in press). These lower producing farms should consider converting to a free traffic system, evaluate their feed allowance settings to make sure cows are allowed enough milkings per day, and make appropriate changes to their robots' settings to reduce failures (Tremblay *et al.*, in press).
Since forced traffic type and Jersey cows are associated with lower milk production, it is intuitive that they should not be held to the same standard of what constitutes 'high production' clusters (Tremblay *et al.*, 2016). For example a cluster 2 farm with an average of 30.2 kg of milk per cow per day would fall in the 32nd percentile when compared to all farms (Table 1). This result could be worrying and could initiate the farmer to focus on increasing this value. When compared to only cluster 2 farms the same farm is in the 90th percentile (Table 1). This more realistic comparison could allow the farmer to switch his or her focus on another aspect of management that could better serve their farm.

Table 1. Milk production per cow per day (kg) percentiles of 529 farms (all) and per clusters 1-6.

Percentile	All	Cluster 1	Cluster 2	Cluster 3	Cluster 4	Cluster 5	Cluster 6
90%	36.9	35.0	30.2	35.6	33.4	39.4	37.1
75%	34.4	33.3	27.6	33.7	31.8	36.9	35.5
50%	32.2	31.0	24.4	32.0	29.9	35.1	33.3
25%	29.3	29.1	23.0	30.4	27.6	33.8	31.1
10%	26.2	26.8	19.8	28.7	26.0	32.5	28.5

3.2 Cluster 5

Cluster 5 was mostly made up of Midwestern Holstein farms in the United States with high average values of Milk_Production_per_Cow_per_Day, Milk_Speed and Cows_per_Robot, and low average values of Concentrates, Rest_Feed, Refusals and Average_DIM.

To continue increasing milk production, these highest producing farms should focus on making small adjustments, such as reducing failures and selecting cows for fast milk speed and cow-robot efficiency (Tremblay *et al.*, in press).

The benefit of the cluster analysis is revealed by the fact that a cluster 5 farm with an average milk speed of 2.94 kg/min would be in the 90th percentile when compared to all farms (Table 2). This might give an exaggerated sense of achievement. When compared to only cluster 5 farms, this farm would be part of the 77th percentile potentially motivating the herd management to set higher goals.

Table 2. Milk speed percentiles of 529 farms (all) and per clusters 1-6.

Percentile	All	Cluster 1	Cluster 2	Cluster 3	Cluster 4	Cluster 5	Cluster 6
90%	2.94	2.92	2.64	2.83	2.76	3.06	2.95
75%	2.78	2.78	2.48	2.67	2.60	2.92	2.79
50%	2.61	2.62	2.33	2.53	2.47	2.77	2.70
25%	2.44	2.48	2.25	2.41	2.27	2.65	2.52
10%	2.28	2.33	1.93	2.33	2.15	2.55	2.36

3.3 Cluster 3

Cluster 3 farms were located the farthest to the north (i.e. all in Canada under the quota system) and had the highest average value of Concentrates (Tremblay *et al.*, in press). These farms could consider adding additional automatic feeding stations outside of the AMS so that cows can reach their concentrate allowance in between milkings (Tremblay *et al.*, in press).

Although increased Refusals is associated with decreased milk production (Tremblay *et al.*, 2016), with so much of feed being consumed in the robot in cluster 3 farms, the robot becomes extra inviting to cows and therefore more refusals will be seen. A cluster 3 farm with 1.66 average refusals per cow per day would be in the 50th percentile compared to all farms (Table 3). The same farm is in the 90th percentile when compared to only cluster 3 farms. It becomes clear that the recommendation of the ideal number of refusals need to be adjusted for Cluster 3 farms, because of their cows' increased motivation to go to the robot.

Table 3. Refusals percentiles of 529 farms (all) and per clusters 1-6.

Percentile	All	Cluster 1	Cluster 2	Cluster 3	Cluster 4	Cluster 5	Cluster 6
90%	0.81	1.07	1.22	1.66	0.71	0.64	1.25
75%	1.10	1.36	1.64	2.01	0.94	0.80	1.60
50%	1.65	1.72	2.15	2.35	1.44	0.99	2.41
25%	2.42	2.29	3.41	3.21	1.73	1.18	3.64
10%	3.43	2.73	3.87	4.00	2.44	1.42	5.01

3.4 Cluster 6

A high percentage of Cluster 6 farms were newly built barns milking Holstein cows in Canada (i.e. quota system) with low ratios of cows per robot and low average Boxtime, Failures, Milkings, Robot_Free_Time, and Refusals (Tremblay *et al.*, in press). Due to their limited available milk quota, these farms should focus on their efficiency (i.e. producing more milk with less input such as feed, and cost of operation and reproduction) (Tremblay *et al.*, in press).

Table 4 illustrates the percentile ranks of the number of failed milkings per robot per day for all 529 farms and for Clusters 1-6. A cluster 6 farm with 3.25 average Failures would have been classified within the 90th percentile of herds when compared to all farms, but will be classified in the 68th percentile when compared to cluster 6 herds only. Since cluster 6 farms need to increase efficiency, this farm should be motivated to lower their number of failures further.

Table 4. Failures percentiles of 529 farms (all) and per clusters 1-6.

Percentile	All	Cluster 1	Cluster 2	Cluster 3	Cluster 4	Cluster 5	Cluster 6
90%	3.25	5.25	5.26	3.57	3.20	3.05	2.59
75%	3.97	5.85	5.99	4.18	3.73	3.71	3.11
50%	5.30	7.41	7.99	4.88	4.71	4.64	3.62
25%	7.06	8.95	11.53	5.81	5.91	6.01	4.92
10%	9.09	10.67	14.11	6.73	7.78	6.82	5.99

3.5 Cluster 4

Cluster 4 had high percentage of farms with one robot per pen and had the highest average Boxtime and lowest average Robot_Free_Time value. These farms should work towards removing cows which are being fetched often or are too low ranking in the herd hierarchy for milking in a one robot pen (Tremblay *et al.*, in press). Having a single robot per pen has been shown to be associated with lower milk production (Tremblay *et al.*, 2016). Therefore, making comparisons amongst other farms with a single robot per pen, such as those in cluster 4, is more informative (Table 1).

3.6 Cluster 1

Cluster 1 had a high percentage of farms with 2 or more robots per pen, farms with 'Other' breeds, farms with most recent AMS installations and the highest average Connection_Attempts (Tremblay *et al.*, in press). As these farms become more established, they need to select cows best suited for AMS milking and examine whether they can better adjust their robots' settings to accommodate their non-Holstein cows (Tremblay *et al.*, in press). Systems that have been in place for greater than 4 years have been shown to significantly increase milk production compared to farms with recently installed AMS (Tremblay *et al.*, 2016); Therefore, these relatively new farms in Cluster 1 should not compare themselves with farms that are well established and past their 4th year with AMS.

4. Conclusions

The 6 clusters, characterized by distinct multivariable production and management patterns, represented more appropriate peer groups for comparisons among farms. These characteristics prompted individualized recommendations per cluster and more realistic and beneficial goals per farm.

References

Tremblay, M., Hess, J.P., Christenson, B.M., McIntyre, K.K., Smink, B., Van der Kamp, A.J., De Jong, L.G. and Döpfer, D. (2016). Factors associated with increased milk production for automatic milking systems. Journal of Dairy Science 99(5): 3824-3837.

Tremblay, M., Hess, J.P., Christenson, B.M., McIntyre, K.K., Smink, B., Van der Kamp, A.J., De Jong, L.G. and Döpfer, D. (in press). Customized recommendations for production management clusters of North American automatic milking systems. Journal of Dairy Science.

KPI project: monitoring 9 Australian automatic milking systems

N.A. Lyons[1*] and K.L. Kerrisk[2]
[1]NSW Department of Primary Industries, EMAI Private Bag 4008, 2567 Narellan, Australia; [2]Dairy Science Group, Faculty of Veterinary Science, School of Life and Environmental Sciences, University of Sydney, Camden 2570, New South Wales, Australia; nicolas.lyons@dpi.nsw.gov.au

Abstract

Farmers considering installing AMS want to know how commercial AMS farms perform. This project will provide the industry with a better understanding of what is achievable on farms with AMS. Nine Australian AMS farms have been identified, which cover a representative range of those farms currently installed. They will be monitored on a monthly basis for a 12 month period (to June 2016). A monthly report is extracted from the system to summarize the main key performance indicators for the previous month relating to milk production, AMS utilization and farm demographics. The majority of participating farms were milking either Holstein cows or crossbreeds (Holstein and Jersey or Brown Swiss). Three farms had seasonal calving, two had split calving and the remainder year-round calving. The farms had either three, four or six single box robots, whilst one farm had a robotic rotary. All farms except two had feed stations (between two and 14 units). All the farms except one were pasture-based managing either three or four allocations per day. On average over the eight months so far, participating farms milked 199 cows (3.1 cows/ha or 52 cows/milking unit). These farms were producing 4,793 kg milk/day with a milk composition of 3.9% fat, 3.3% protein and a somatic cell count of 201,000. Individual cows were milked 2.4 times/day, producing 25.1 kg milk/day and consuming 6.5 kg concentrate/day. Average milk harvested per visit, was 10.7 kg milk and average duration was 6:20 (min:sec) in the robot. Each individual robot performed 123 milkings/day, operating for 14 hours/day and harvesting 1,334 kg milk/day. There is a wide range in farm performance, which highlights that it is the combination of farm system and farm management that determines the whole farm performance. The NSW Department of Primary Industries has committed to increase the awareness and successful adoption of automatic milking systems in Australia. This information might help farmers that are considering investing in this technology.

Keywords: performance, production, utilization, demographics

1. Introduction

Since 1990 the number of dairy farms in Australia has decreased by 18% whilst average herd size has increased by around 160% (Dairy Australia, 2015). Ninety-eight percent of dairy operations are family owned and operate with variable levels of external labor (Dairy Australia, 2013). Farmers see an opportunity in technology to address labor issues and increase production efficiency.

Automatic milking systems (AMS) are one of the most significant and relatively recent technologies that create a significant change to the whole dairy operation and impact on the farm systems and broader dairy industry. These systems were introduced in the early 1990's (De Koning, 2011) but only arrived in pasture-based Australian dairy systems in 2001 (Greenall *et al.*, 2004). Currently in Australia there are 39 farms operating with AMS and at least another three farms being installed. This represents 153 robots, approximately 10,700 milking cows and 60 million liters of milk per year. Every dairy region in Australia has farmers that have decided to commission AMS. These installations cover all the following commercially available types of robot (single box, multi box and robotic rotary). Furthermore, they operate in an array of farming system types: grazing with variable levels of supplementation (85% of farms), corral systems (5% of farms) and indoor systems (10% of farms).

Farmers looking at AMS as an option when either installing or replacing a dairy are interested in gaining an understanding of the farming operation and farm system performance before taking any investment decision. Currently, farmers have access to high quality relevant information captured by research conducted within FutureDairy (Garcia *et al.*, 2007), a national industry driven and funded project which has been running continuously since 2006. However, given the number of AMS installations currently operating with this technology, farmers are interested in understanding what levels of performance and productivity are achievable under commercial conditions.
Previous studies conducted mainly overseas have indicated variable levels of performance for commercial AMS farms (Castro *et al.*, 2012). Similar studies have not been conducted in Australia. This study aims to capture key performance indicators (KPIs) related to milk production, AMS utilization and farm demographic information of commercial Australian AMS farms on a monthly basis. This will help understand how AMS farms perform as well as provide the industry with a better understanding of what is achievable under AMS operation.

2. Materials and methods

An expression of interest was put out to Australian AMS farmers that had been in operation for at least 3 months by July 1st 2015 to participate in the KPI Project. Nine Australian AMS farms were identified, which covered a representative range of those farms currently installed.
Farmers agreed to have monthly (day 1-5 of each month) reports extracted from the herd management system that would allow the generation of summary KPIs for the previous month.
Raw data was extracted for each of the nine farms from July 2016. The information provided as well as the reports obtained, were summarized and collated by the authors of this paper. Descriptive statistics were then calculated to capture average, minimum and maximum for the main KPIs from July 2015 to February 2016. Whilst some general farm demographics will be presented, the main KPIs will be related to cow or robot performance and results have not been adjusted for farm and herd size.

3. Results

3.1 Farm demographics

The participating farms had either three, four or six single box robots, whereas one farm had a robotic rotary. They had been in operation for anywhere between five months and four years. Most of them were milking either Holstein cows or crossbreeds (Holstein and Jersey or Brown Swiss). Three farms had seasonal calving, two had split calving and the remainder operated with year-round calving patterns. All farms except two had feed stations (between two and 14 units). All the farms except one were pasture-based managing either three or four feed allocations per day.
On average over the eight-month period so far, participating farms milked 199 cows (3.1 cows/ha or 52 cows/milking unit). These farms were producing 4,793 kg milk/day with a milk composition of 3.9% fat, 3.3% protein and a somatic cell count of 201,000. Cows per milking unit and somatic cell count varied considerably both within and between months (Figure 1). Over the eight-month period robot stocking rate ranged between 14 and 75 cows/robot, whereas somatic cell count (×1000) ranged between 25 and 481.

3.2 Cow performance

On average over the eight-month period so far, individual cows were milked 2.4 times/day, producing 25.1 kg milk/day and consuming 6.5 kg concentrate/day. In every milking visit, cows produced 10.7 kg milk and spent 6:20 min in the robot.
Milking frequency and daily milk yield on average did not vary much between months but had a great variability within months (Figure 2). Over the eight-month period milking frequency ranged between 1.5 and 3 milkings/cow/day, whereas daily milk yield ranged between 14.8 and 34.4 kg/cow/day.

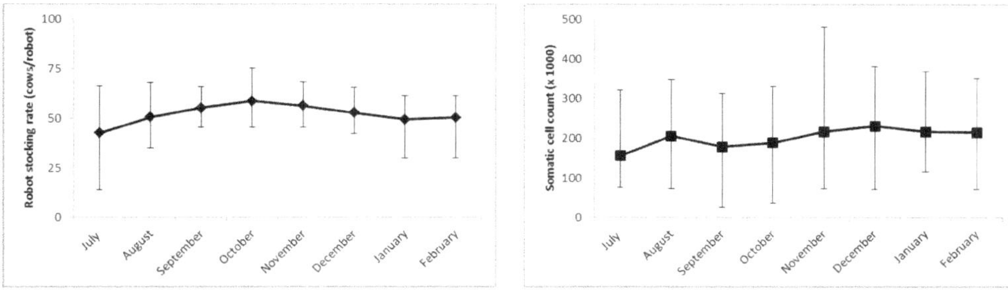

Figure 1. Robot stocking rate (A) and somatic cell count (B) from July to February. Solid dot – connected line indicates average value whereas vertical lines indicate range (min-max) for every month.

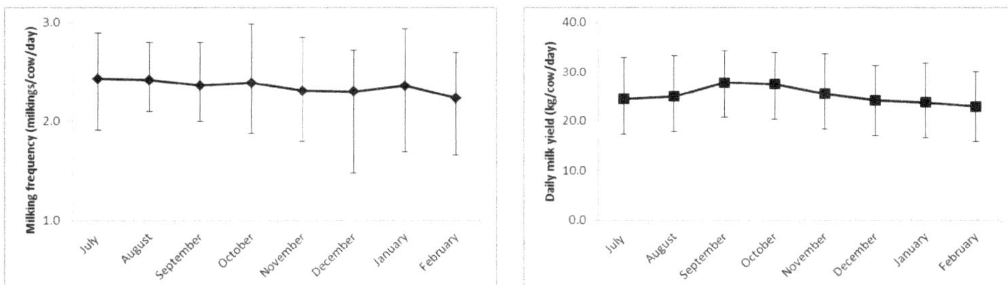

Figure 2. Milking frequency (A) and daily milk yield (B) from July to February. Solid dot – connected line indicates average value whereas vertical lines indicate range (min-max) for every month.

3.3 Robot performance

On average over the eight-month period so far, each individual robot performed 123 milkings/day, operating for 14 hours/day and harvesting 1,334 kg milk/day.

Milkings, milking time and milk harvested were variable within and between months (Figure 3). Over the eight-month period milkings ranged between 39 and 173 milkings/robot/day, milking time ranged between 5:34 and 21:10 and milk harvested ranged between 331 and 2,484 kg milk/robot/day.

4. Discussion

This aim of this study was to capture KPIs related to milk production, AMS utilization and farm demographic information of commercial Australian AMS farms on a monthly basis.

Results presented here show there is a great variability within and between months for most of the KPIs. Higher levels of variability for some months are likely to reflect the different calving patterns implemented across the farms. For example, the high level of variation in July (between farms) of robot stocking rate, milkings/robot/day, milking time/robot/day and milk harvested/robot/day likely reflects the fact that some herds had a large proportion of the herd not lactating (the spring calving cows).

Some of the variables such as robot stocking rate and daily milk yield have an impact on robot performance variables such as milkings, milking time and milk harvested per robot. These variables

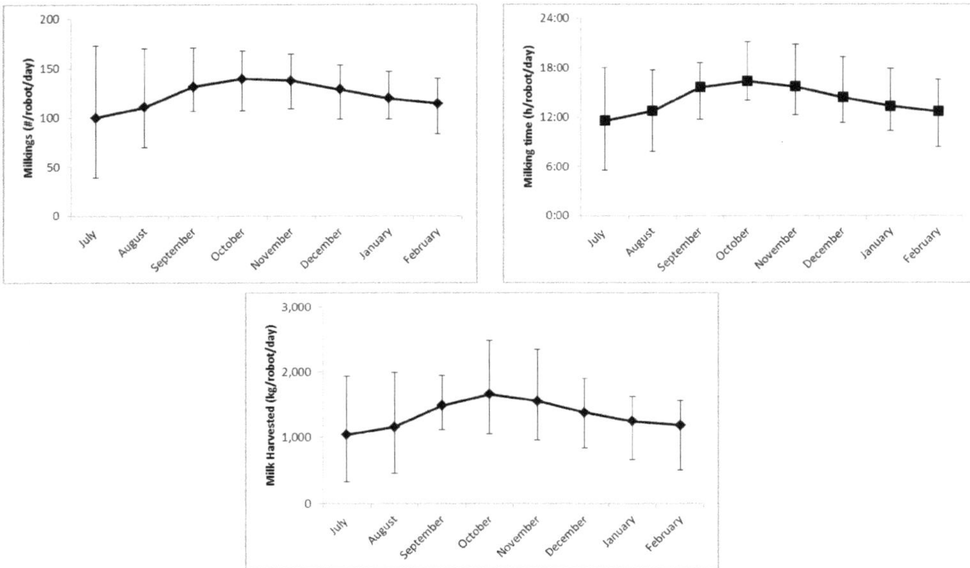

Figure 3. Milkings (A), milking time (B) and milk harvested (C) from July to February. Solid dot – connected line indicates average value whereas vertical lines indicate range (min – max) for every month.

also respond to patterns that belong to seasonal calving, typical of pasture-based dairy systems in which a high proportion of cows calve in a time period that matches pasture growth in order to match feed demand and supply.

Milking robots have a set capacity and it is up to the farmer to decide how to utilize that capacity which is usually measured in milkings or litres of milk harvested per robot and per day. Results presented suggest that it is possible for robots to perform at least up to 173 milkings per day (monthly average on one farm) and harvest almost 2,500 kg milk per day under pasture-based commercial conditions which is similar to the performance reported for indoor systems (Castro *et al.*, 2012). In most months the majority of the farms monitored appeared to have some spare capacity that would enable them to either milk their existing cows more often, and potentially increase daily milk yield (per cow), or increase herd size, or a combination. However, it is recognized that farm size, feed availability and seasonal influences are amongst a number of factors that might limit the ability for a farm to increase the utilization of the milk harvesting equipment.

Traditionally there are two main considerations regarding maximizing system utilization: milk more animals at a lower milking frequency, or milk less animals more frequently (Pettersson *et al.*, 2011). Previous research has reported that under pasture-based conditions, target milking frequency and total daily milk yield per cow, are usually lower than in indoor loose housing systems (Garcia and Fulkerson, 2005). Therefore, some farmers have opted to increase robot stocking rate to maximize utilization (Woolford *et al.*, 2004). In this study, average milking frequency has shown little variation between months, with values similar to those reported from commercial indoor operations (Castro *et al.*, 2012) but higher than that of previous pasture-based operations (Lyons *et al.*, 2013). Furthermore, robot stocking rate has been maintained (on average) at around 50 cows/robot, with some farms able to manage up to 75 cows during certain months.

Somatic cell count as a reference for milk quality and herd health was highly variable within and between months. This could mean that the challenges of producing high quality milk in AMS are similar to those in conventional milking systems, where adequate management and monitoring of cow, environment and milking procedure are key for success. Farmers who are diligent in monitoring the information generated from sensors in the robot and proactively respond whenever it starts sending warning signals can generally manage a consistently low bulk milk cell count. This is one of the competitive advantages of robotic milking machines.

The whole farming system should be taken into consideration recognizing that farm management practices have an impact on several KPIs. In addition to the variables reported in this project, AMS farmers will still monitor the traditional KPIs related to things such as health or reproductive management. Furthermore, this project only captured milk production, AMS utilization and farm demographic information and no economic or farm financial analysis. The authors recognize the importance of farm business management and the role this plays in the success of any dairying operation.

The NSW Department of Primary Industries has committed to increase the awareness and successful adoption of automatic milking systems in Australia. The authors believe that by having access to adequate information, put into the right context, farmers and industry will be able to make well informed decisions in relation to milk harvesting systems as they move into the future. Information about what is achievable under commercial conditions will help farmers that are considering investing in this technology. It will also provide current AMS farmers with a benchmark to monitor certain KPIs and how they track against other commercial farms.

5. Conclusions

Automatic milking systems could be a viable solution to labor issues on Australian dairy farms. The findings contained within this study provide useful information for farmers who will contemplate the installation of AMS in pasture-based systems. Overall there is great variability within and between months for most of the KPIs with some farmers achieving similar results to those observed in more mature indoor markets.

Acknowledgements

The authors would like to thank the participating farmers for providing access to their records on a monthly basis.

References

Castro, A., Pereira, J.M., Amiama, C. and Bueno, J. (2012). Estimating efficiency in automatic milking systems. Journal of Dairy Science 95(2): 929-936.

Dairy Australia (2013). Dairy people factfinder. Victoria, Australia.

Dairy Australia (2015). Australian dairy industry in focus. Victoria, Australia.

De Koning, K. (2011). Automatic milking: common practice on over 10,000 dairy farms worldwide. Paper presented at the Dairy Research Foundation Symposium 2011, University of Sydney, Camden, Australia.

Garcia, S.C. and Fulkerson, W.J. (2005). Opportunities for future Australian dairy systems: a review. Australian Journal of Experimental Agriculture 45(9): 1041-1055.

Garcia, S.C., Fulkerson, W.J., Nettle, R., Kenny, S. and Armstrong, D. (2007). Future dairy: a national, multidisciplinary project to assist dairy farmers to manage future challenges methods and early findings. Australian Journal of Experimental Agriculture 47(9): 1025-1031.

Greenall, R.K., Warren, E. and Warren, M. (2004). Integrating automatic milking installations (AMIs) into grazing systems – lessons from Australia. Paper presented at the Automatic milking: a better understanding, Lelystad, the Netherlands.

Lyons, N.A., Kerrisk, K.L., Dhand, N.K. and Garcia, S.C. (2013). Factors associated with extended milking intervals in a pasture-based automatic milking system. Livestock Science 158(1-3): 179-188.

Pettersson, G., Svennersten-Sjaunja, K. and Knight, C.H. (2011). Relationships between milking frequency, lactation persistency and milk yield in Swedish Red heifers and cows milked in a voluntary attendance automatic milking system. Journal of Dairy Research 78(3): 379-384.

Woolford, M.W., Claycomb, R.W., Jago, J., Davis, K., Ohnstad, I., Wieliczko, R. and Bright, K. (2004). Automatic dairy farming in New Zealand using extensive grazing systems. Paper presented at the Automatic milking: a better understanding, Lelystad, the Netherlands.

Increasing milking frequency and milking yield in robotic milking by using an automatic herding system

U. Drach[1,2], A. Degani[1], T. Pnini[3], I. Izhaki[4] and I. Halachmi[2*]
[1]The Division of Environmental, Water and Agricultural Engineering, Faculty of Civil and Environmental Engineering, Technion – Israel Institute of Technology, Haifa, Israel; [2]Institute of Agricultural Engineering, Agricultural Research Organization (A.R.O.), Volcani Center, Bet Dagan, Israel; [3]Pnini Shiran Milk Corporation Ltd., Tel Adashim, Israel; [4]Department of Evolutionary and Environmental Biology, University of Haifa, Haifa, Israel; halachmi@volcani.agri.gov.il

Abstract

Cow motivation to be milked is a key factor in milking robot utilization, if a cow does not voluntarily attend a robot stall, fetching work is required. Feeding, cooling and behavioral studies have previously tried to solve this problem by willingly attracting the cows to the milking robot. None of them achieved full success since the fetching work was still required. Therefore, this research suggests a new concept, herding all the cows to the milking robot using an Automatic Herding System (AHS). The AHS was built as a system of slow moving mobile fences controlled by an industrial controller. The AHS herds all the cows to the milking robot. The AHS was built in a commercial farm with two milking robots, and the experiment was conducted for three month. The farm was divided into a control group (43 cows) and an experiment group (39 cows). The AHS was installed only in the experiment group. For statistical analysis, two control groups were used: (1) the group without the AHS, (2) the group with AHS before installation of the system. Milking frequency increased in the experiment group by 47% (1.88 vs 2.75 milkings/d), while there was no major change in the milking frequency in the control group 2% (2.38 vs 2.39 milkings/d). Milk yield increased in the experiment group 17% (35 vs 41 kg/d). There was no major change in the yield in control group 4% (31 vs 29.8 kg/d). The results were statistically tested by mixed model ANOVA and were not rejected ($\alpha<0.005$). The system appeared to be a complementary system to the automatic milking system.

Keywords: automatic milking system, milking interval, dairy cow

1. Introduction

Milking robots were developed in order to improve farmer daily routine by reducing labor, and by improving cow welfare (Jacobs and Siegford, 2012). However, the success of the milking robot is based on the cow's good will to voluntarily be milked (John *et al.*, 2016). Therefore, if a cow does not voluntarily attend a robot stall the consequences are economical losses due to: (1) fetching work is required – about 3 times per day, total of 7 hours per 100 cows; (2) interference to cow's routine; (3) loss of potential milk that could have been produced by a higher milking frequency during early and late lactation period (Bar-Peled *et al.*, 1995); (4) low efficiency of an economical investment – due to the robot's idle time, especially at nights. A robot is milking only 45-90% of the Occupation Rate (OR) – defined as percentage of hours the robot is available per day (John *et al.*, 2016). André *et al.* (2010) found that a herd of 62 cows with 64% OR can increase the milk revenue from 498 €/d to 529 €/d at 85% OR. The number of cows that have to be fetched is not consistent, and varies as a function of cow, stage of lactation and the farm. During the first 14 days of lactation, 56-100% of the cows have to be fetched at least once a day. The number is lower in later stages of the lactation, but it still varies between 6-45% of the cows (Jacobs and Siegford, 2012). In order to solve this problem, different types of methods have been proposed to attract the cows to the milking robot.

These methods reduced the number of cows that have to be fetched, but none of them achieved full success, and fetching work was still required. Therefore, we propose a new concept. Instead of trying to attract the cows to voluntarily walk to the milking robot, we suggest to herd all the cows to the milking robot by using an Automatic Herding System (AHS).

In this research we developed and tested a low cost automatic system of mobile fences that herd all the cows to the milking robot. This method assures that every cow is milked in a frequency that is set by the farmer. Our hypothesis was that the AHS can add 0.25 milkings per day (MPD) for each cow. The objective of the research was to maximize the mean number of milkings (per cow, per day), without labor, while maintaining farm profitability and cow welfare.

2. Materials and methods

The experiment was performed at a commercial farm in Tel Adashim (Israel), with 100 high-yielding Holstein-Friesian cows housed in an open cowshed (no cubicle housing) and fed a total mixed ration (TMR) that is commonly used in Israel. The farm was equipped with two voluntary milking robots (VMS) that were operating since 2007. The farm was divided into two cowsheds: primiparous and second lactation cows (control), multiparous cows (experiment). An AHS was installed in the multiparous cows' cowshed. The AHS comprises two mobile fences (Figure 1).

Fence A was slowly guiding the cows along yard A from the robot towards yard B, while fence B was slowly guiding the cows along yard B from yard A towards the robot. In order to reduce the stress that might be caused by the movement speed of the fences, the fence was moving in short time pulses. In each pulse the fence moved 0.8 m in 4 s and then paused for 30 s. In order to minimize the interference for the cow routine and to allow the cows to lie down in the yard, the fences paused for different time periods in the middle of the yards and at the yards ends. The total time for each movement cycle was approximately eight hours. The AHS worked continuously throughout the experiment. In the control treatment was a semi forced cow traffic routine, with manual herding 3 times a day of cows with: long time period since last milk (>12 hour>, high milk yield expected (21 kg). The study was carried out according to the guidelines and under the supervision of the Agricultural Research Organization Animal Care Committee.

Individual milk yield and milkings per day data were recorded automatically by the VMS management program. The data was collected during the period from 1/1/2015 to 14/4/2015. At the beginning, the AHS worked intermittently, however, since 19/02/2015 the system worked continuously. During the experiment period there were 127 cows, 18 cows were sold during the experiment period and had been removed from the database. Artifacts cow milk yield <10 [kg/day], DIM >500 [day], DIM <3 [day] were also removed from the database.

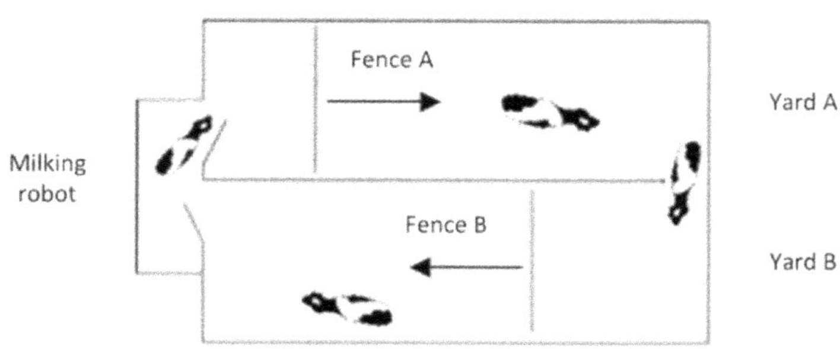

Figure 1. Automatic herding system (AHS) scheme.

A statistical analysis was performed on 10 days, in two different time periods. Period 1, 10-20/01/2015, baseline time, before operating the AHS. Period 2, 09-19/03/2015, experiment time – during operating the AHS. For the analyses, the cows were divided into four treatments. Cows that participated only in one time period were also removed from the database.

Milkings per day and milk yield were used as response variables for mixed model ANOVA analysis in SPSS. Since there is an initial difference between the control and experiment groups, the analysis was constructed for the change in each group between two time periods. Since milk yield and DIM are correlated the mean of DIM in each period was used as a covariate to compensate for their mutual interrelation.

3. Results

The milking frequency of the experimental group that was equipped with the AHS was increased between the two periods by 46% (exp_1 vs exp_2, 1.88 vs 2.75 Table 1, Figure 2) and the milk yield was increased by 17.1% (35.05 vs 41.06 kg/day, Table 1, Figure 2). We observed no significant difference between the treatment groups ($cont_1$ and $cont_2$ Table 1, Figure 2) that were not equipped with the AHS. There were significant interactions between the change in milk yield and the group $F(1,79)=17.592$, $P<0.001$ and between the change in milk yield and the DIM, $F(1,79)=10.656$, $P=0.002$. There was a significant interaction effect between the change in milking frequency and the group, $F(1,79)=69.139$, $P<0.001$ while there was no significant interaction effect between the change in milking frequency and the DIM, $F(1,79)=0.671$, $P=0.415$.

According to the milking frequency histograms (Figure 3) it can be seen that in treatment C most of the cows were milked only twice per day, while in treatment D more than 50% of the cows were milked three times per day.

Table 1. Herd size, characteristics of the focal cows, milk yield and milking frequency of the four treatments.

Treatment	$cont_1$[1]	$cont_2$[2]	exp_1[3]	exp_2[4]
Herd size (n)	52	48	50	47
Focal cows (n)	43	43	38	38
Parity (mean±SE)	1.3±0.02	1.3±0.02	3.36±0.23	3.38±0.23
DIM[5] [days] (mean±SE)	187.1±14.6	245.2±14.6	138.8±15.4	184.7±13.4
Milk yield (kg/day±SE)	31.0±0.89	29.76±1.07	35.05±1.62	41.06±1.66
Milking frequency (milkings/day±SE)	2.38±0.09	2.39±0.08	1.88±0.06	2.75±0.04
Total MPD[5] in AMS[5]	116.4	115.3	91.2	129.1

Treatments:

[1] $cont_1$ – manual herding combined with voluntary cow movement. control group. period 1: 10/1/15-20/01/15.

[2] $cont_2$ – manual herding combined with voluntary cow movement. control group. period 2: 09/3/15-19/03/15.

[3] exp_1 – manual herding combined with voluntary cow movement. experiment group. period 1: 10/1/15-20/01/15.

[4] exp_2 – Automatic herding. experiment group. period 2: 09/3/15-19/03/15.

[5] DIM = Days In Milking; MPD = Milkings Per Day; AMS = Automatic Milking System.

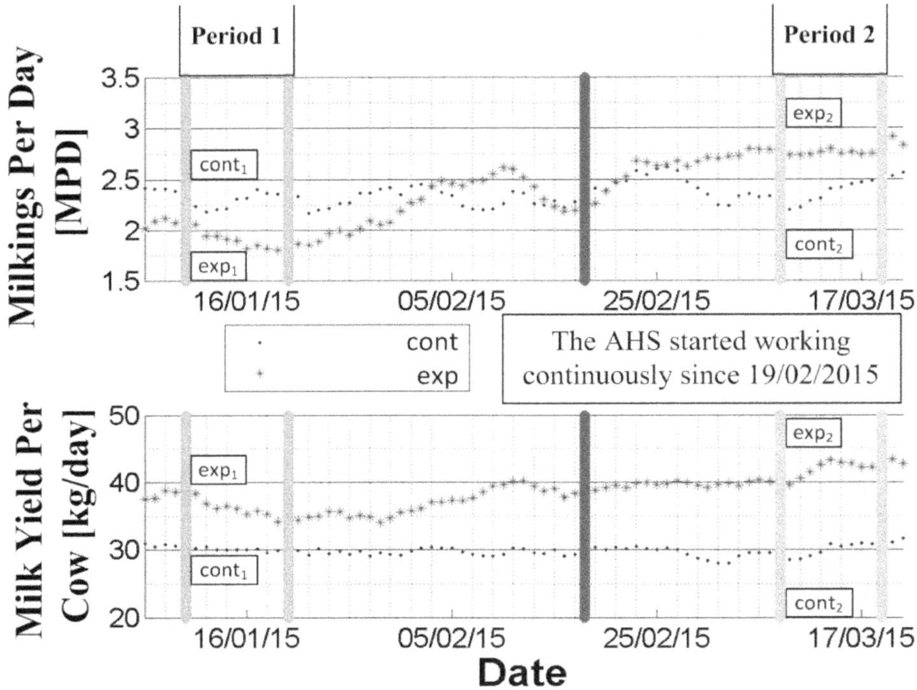

Figure 2. Average milking frequency and milk yield during the experiment periods 01.01.2015-21.03.2015.

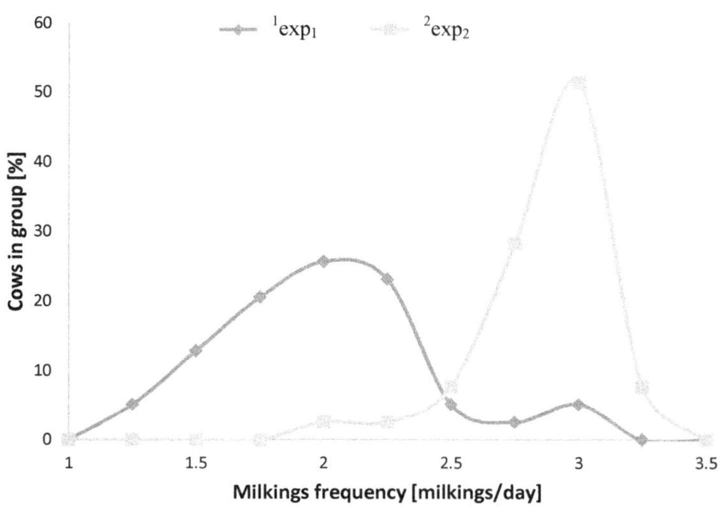

Figure 3. Milking frequency histograms.

[1] exp_1 – manual herding combined with voluntary cow movement. experiment group. period 1: 10/1/15-20/01/15.

[2] exp_2 – automatic herding. experiment group. period 2: 09/3/15-19/03/15.

4. Discussion

The results support the research hypothesis and show the economic benefit of using an AHS as a complementary system to the automatic milking system.

4.1 Milking frequency

Our hypothesis was that an automatic group herding system can add 0.25 MPD for each cow while our results (1.88 vs 2.75 MPD) showed that the tested system actually added 0.88 MPD. There are some possible reasons for this difference. (1) The initial milking frequency in the experiment group was relatively low (1.88) while the Israeli average is 2.76. (2) During the Israeli winter, there is less motivation for the cows to go to the cooling area that are located next to the milking robot. When using the automated herding system, the milking frequency depends on the milking robot availability, and the time of pauses in the fence movement. The average idle robot time that was measured is 07:18 (hour:min), thus it is possible to increase the milking frequency by reducing the fence pausing time. We suggest that in future research there should be an optimization process for finding the ideal pausing time. The ideal pausing time will allow optimal frequency that balance between maximum yield and minimum interruption to cow welfare and routine. The milking frequency can also be increased by using one way gates that will allow motivated cows to walk to the milking robot voluntarily.

4.2 Milk yield

The results show a rise of 17.1% (35.05 vs 41.06 kg/day) in the milk yield when using the AHS. The main reason for this rise is the rise of 46% in milking frequency. According to the milking frequency histograms (Figure 3) it can be seen that in treatment exp_1 most of the cows were milked only twice per day, while in treatment exp_2 more than 50% of the cows were milked three times per day. This coincides with the observations made in other studies regarding to the influence of milking frequency of 2 MPD vs 3 MPD on milk yield of multiparous cows (18.5% (Amos *et al.*, 1985), 15% (DePeters *et al.*, 1985)).

Our research was intended to validate a proof of concept for the new herding method. Therefore, it was limited for a short time, and the system's mechanical durability was low. Future research should be conducted for lactation period of at least one year, and should examine the effect on primiparous cow.

5. Conclusions

In this work, we designed and implemented an AHS. The system was examined and appeared to be a positive complementary system to the automatic milking system. This system was associated with higher milking frequency (1.88 vs 2.75 milkings/day) and a higher milk yield (35 vs 41 kg/day).

References

Amos, H.E., Kiser, T. and Loewenstein, M. (1985). Influence of milking frequency on productive and reproductive efficiencies of dairy cows. Journal of Dairy Science 68: 732-739.

André, G., Berentsen, P.B.M., Engel, B., De Koning, C. and Lansink, A.O. (2010). Increasing the revenues from automatic milking by using individual variation in milking characteristics. Journal of Dairy Science 93: 942-953.

Bar-Peled, U., Maltz, E., Bruckental, I., Folman, Y., Kali, Y., Gacitua, H., Lehrer, A.R., Knight, C.H., Robinzon, B. and Voet, H. (1995). Relationship between frequent milking or suckling in early lactation and milk production of high producing dairy cows. Journal of Dairy Science 78: 2726-2736.

DePeters, E.J., Smith, N.E. and Acedo-Rico, J. (1985). Three or two times daily milking of older cows and first lactation cows for entire lactations. Journal of Dairy Science 68: 123-132.
Jacobs, J.A. and Siegford, J.M. (2012). Invited review: the impact of automatic milking systems on dairy cow management, behavior, health, and welfare. Journal of Dairy Science (95): 2227-2247.
John, A.J., Clark, C.E.F., Freeman, M.J., Kerrisk, K.L., Garcia, S.C. and Halachmi, I. (in press). Review: milking robot utilization – a successful precision livestock farming evolution. Animal.

Part 3. Data management

When cow health counts, probabilities matter

R. Van der Tol*, P. Kool and A.J. Van der Kamp
Lely Industries, Cornelis van der Lelylaan 1, 3147 PB Maassluis, the Netherlands; rvandertol@lely.com

Abstract

The new generation of automatic milking systems (AMS) promises to take more of the toil out of farming by automating the business of the dairy farmer. However, automating factories is easier than automating farms being far less predictable environments: weather constantly changes, food quality alters, and cows are wandering around. Nowadays, AMS become smarter due to improvements in sensing systems in combination with better computational power and artificial intelligence. Instead to work harder, farmers tend to work smarter. On the one side AMS automatically runs in a safe mode using their 'senses', hence increasing the span of control of the farmer. On the other side, such smart systems enable to reduce the wasted time in the working process related to health checks by showing the cows with high probabilities for sickness. A new health module was tested for its ability to reduce the work load on farms. The main changes were the Naive Bayes Classification Technique calculating probabilities for all diseases and the user interface showing the cows with the highest probability on top. The objective was that the module decreased the number of animals to check by 50% without compromising on the sensitivity to find sick animals. Since April 2015 data gathering was started on 4 farms (approximately 670 cows) and still continues. Recall (Sensitivity) and precision (Positive Predictive Value) of health attentions were calculated during one week in January 2016 using health records from farmers. A cow on the in-herd health report was considered true positive when she required some kind of health check as judged by the farmers. Farmers were asked for their opinion and how it affected their daily routines. Preliminary results show shorter attention lists and the PPV increased with 50 up to 200% depending on the farm. It is clear that with the economically imposed increase in farm sizes, for the fast majority of farmers the span of control is a challenge. By using sophisticated data mining techniques with all available data sources, the new module could unburden the farmer. In addition providing a clue what disease it concerns by including the origin of information deduced from sensor data.

Keywords: decision support, easy to use, sensor technology, user interface

1. Introduction

1.1 Automation in dairy industry

A new generation of dairy equipment promises to take more of the toil out of farming by automating the business of the dairy farmer. However, automating factories is easier than automating farms that are far less predictable environments: the weather constantly changes, the light alters, the ground can turn from grass to mud, and there are individuals (animals and people) wandering around.

However, robots become smarter due to improvements in sensing systems in combination with the increase in the power of computing and artificial intelligence. In the field and barn, manual labor was first made mechanical with all kinds of machines, but now those machines can automatically run in a safe mode using their 'senses'. For example crops that used be handpicked, are now harvested by GPS-guided, electronically-controlled tractors. Second, cows were fed with pitch forks and buckets, are now fed with computer controlled rations by feeding robots. Third, hand milking cows in a bucket, has now changed to robotic milking with controlled vacuum characteristics and measuring conductivity and color of milk, concentrate uptake and many more items during each milk visit. In addition, animal records were hand-written on paper, are now digitized using computers providing many more features. So, dairy farming is no longer about just working hard, it's about working smarter!

1.2 Automating decisions

But still, digitizing animal records is not automation! The decisions made by farmers when it comes to detecting disease or judging underperformance is still a 'manual' job in the majority of cases. They wander through numerous attentions lists of all available sensors, basically screening all the data they have available to judge a cow in an attempt to make a decision. And, sometimes an additional manual test is needed for proper diagnosis. It is clear that with the economically imposed increase in farm sizes, for the fast majority of farmers the span of control is a challenge. Moreover, it is questionable whether famers are objective and consistent in their decisions, one can have a strong debate on this topic.

What if all data is transformed to information, using all sensor data available on farm in combination with artificial intelligence? That would provide a route to automatic diagnostics, likely faster than human observation and more consistent as it will be rules based. The objective of this experiment is to show that a Naive Bayes Classification Technique has the potential to show only those cows that are in need the attention of the farmer.

2. Materials and methods

A new report (user interface) was designed for a mobile interface containing the information from the Naive Bayes Classification Technique using all on farm sensor data available (Figure 1). Farmers using the Lely T4C InHerd app now see the cows with highest probabilities for sickness 'Sick Score'on top of this list (Lely T4C InHerd, Lely Industries NV, Maassluis, the Netherlands). In addition it is shown on what data source the information is based, also hinting towards the most likely diagnosis.

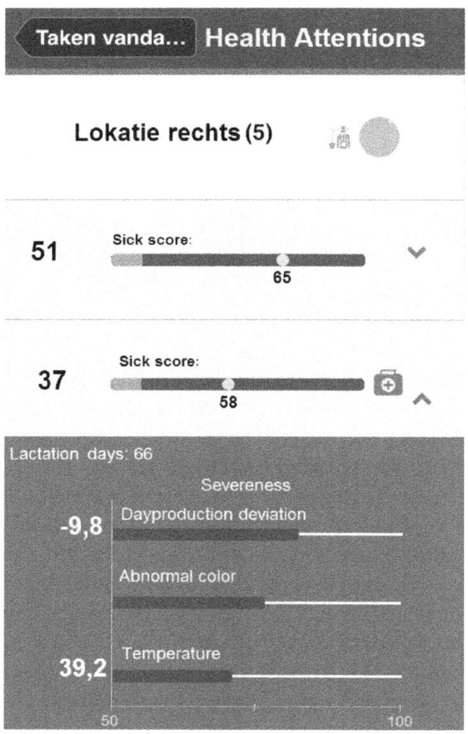

Figure 1. User interface.

Moreover, the new report is now a general health list as it includes udder health related diseases, as well as other production related diseases.
Since April 2015 one by one, four farms (approximately 670 cows in total) were equipped with earlier test-versions of this DSS. Farmers could adopt their methods up to January 2016, than the Beta-version of the software was installed. For one week in January 2016 the farmers kept records of every cow they saw on the new report, whether they checked the cow's 'cowcard' in Lely T4C Office (Lely Industries NV, Maassluis, the Netherlands), 'watch' the cow physically, the cow was diagnosed by a 'veterinarian' and/or was 'treated'. In addition the farmers could report whether it was desired to have another cow listed that did not appear on the new report including some personal comments. For that week in January the Positive Predicted Values (PPV) and Recall were calculated for the new report and the 'udder health attention list' (report 12) and a mastitis analysis report (report 23) from previous versions of T4C using the same underlying data sources.

3. Results and discussion

A total number of 130 cows appeared on the new report during the week of interest, which is a lot less than 294 and 578 attentions in the reports of the previous udder health lists (Table 1). From those 130, 51 cows were diagnosed sick, of which 46 received treatment. Farmers indicated that they missed 31 cows on the reports of which seven cows were diagnosed as sick, hence the Recall was 0.88 with a PPV of 0.39 in terms of sickness. The PPV of the Lely T4C udder health report 12 and 23 were respectively 0.11 and 0.06 with Recalls of 0.57 and 0.60.
One must understand that those cows that the farmers wished to have seen on the reports, are not by definition sick cows. Hence the recall of cows that they desired to check either on the 'cowcard' in Lely T4C Office, examine themselves or by a vet was 0.81. The PPV of the new report was 0.71, which in contrast to the previous versions is relatively high; respectively they were 0.24 and 0.14. When it comes to disease detection, mastitis is the best researched topic. In case cows are milked using an automatic milking system the (fore)milk is not visually judged for the presence of clots to detect mastitis (Rutten *et al.*, 2013). Alternative methods are required to diagnose mastitis and many manufacturers rely on sensor data; mainly conductivity and colour of the milk (Song and Van der Tol, 2010). Separately those sensors perform relatively well on sensitivity, but they lack specificity resulting in attentions lists with a low PPV questioning its trustworthy (Hogeveen *et al.*, 2013). But when combining sensor data and using a Naive Bayes Classification Technique, the PPV can be drastically increased (up to 0.7), a more than threefold increase compared to previous reports. Recently on-line sensors for somatic cell count information became available, adding some new information to the equation. Hence giving a boost to the performance of mastitis detection, potentially allowing automatic decisions taken for (early) treatment.

Table 1. The main data descriptives of the three reports in a week time using similar inputs.

Report	Health attentions	Report 12 (acute)	Report 23 (chronic)
# cows (n-farms=4)	670	670	670
# cows with attention	130	294	578
# cows desire to check	91	71	81
PPV checking	0.70	0.24	0.14
# sick on the report	51	33	35
PPV sick	0.39	0.11	0.06
Recall sick	0.88	0.57	0.60

A wide variety of sensors is already available for dairy farms (Rutten *et al.*, 2013). However a decision support system (DSS) is often lacking. DSS systems are rules-based systems that are most useful for repetitive management problems in which automatically solutions are provided using the electronically available information. By means of the Naive Bayes Classification Technique and an improved user interface the T4C InHerd 'health Attentions' report shows the cows that are most probable to be in need of some attention. Hence, providing a window of opportunities for optimal dairy performance, assisted self-cure and early interventions.

There are numerous articles available showing that prevention of disease is key and early intervention pays off, for example with mastitis treatments (Milner *et al.*, 1997). To conclude, knowing that the fast majority on the attention list are correct allows to be confident in this decision support system and widens the span of control for the modern farmer.

References

Hogeveen, H., Buma, K.J. and Jorritsma, R. (2013). Use and interpretation of mastitis alerts by farmers. Proceedings of the 6th European Conference on Precision Livestock Farming, pp. 313-319.

Milner, P., Page, K.L. and Hillerton, J.E. (1997). The effects of early antibiotic treatment following diagnosis of mastitis detected by a change in the electrical conductivity of milk. Journal of Dairy Science 80: 859-863.

Rutten, C.J., Velthuis, A.G.J., Steeneveld, W. and Hogeveen, H. (2013). Invited review: sensors to support health management on dairy farms. Journal of Dairy Science 96(4): 1928-1952.

Song, X. and Van der Tol, R. (2010). Automatic detection of clinical mastitis in astronaut A3™ milking robot. Proceedings of the 1st North American Conference on Precision Dairy Management.

Semantics for big data applications in the smart dairy farming domain

J.P.C. Verhoosel* and J. Spek
TNO, Kampweg 5, 3769 DE Soesterberg, the Netherlands; jack.verhoosel@tno.nl

Abstract

In the Dutch SmartDairyFarming project, better decision support for the dairy farmer on daily questions around feeding, insemination, calving and milk production processes is made possible using a variety of big data sources containing static and dynamic sensor data of individual cows. This paper is concerned with the inherent semantic interoperability problem between the information in these data sources. Semantic alignment is achieved using ontologies and linked data mechanisms on a large amount of sensor data, such as grazing activity, feed intake, weight, temperature and milk production of individual cows at 7 dairy farms in the Netherlands. A Common Dairy Ontology (CDO) and a specific measurement ontology have been developed and used to transform the sensor data into 310 GB of RDF triples, made accessible via a SPARQL interface on a triplestore. A few example applications have been developed to show how the CDO can be used for decision support and historic analysis. The performance of our linked data semantic solution is acceptable for analysis queries on large sets of data. Without optimization of queries the time for answering queries ranged from a few seconds to a couple of minutes.

Keywords: semantic alignment, ontology, sensor data, triples, sparql

1. Introduction

Dairy farmers are currently in an era of precision livestock farming in which information provisioning for decision support is becoming crucial to maintain a competitive advantage. Therefore, getting access to a variety of data sources on and off the farm that contain static and dynamic individual cow data is necessary in order to provide improved answers on daily questions around feeding, insemination, calving and milk production processes.

In the Dutch SmartDairyFarming project, sensor equipment is used to monitor cows at 7 dairy farms in the Netherlands. Thereby, a large amount of sensor data is generated on grazing activity, feed intake, weight, temperature and milk production of individual cows. A challenge in this sense is how to tackle the problem of semantic interoperability between the concepts present in dairy farming data sources. Semantic alignment of the different meanings of similar concepts in various data sources is therefore necessary for improved decision support and historical analysis.

We have focused on the use of ontologies and linked data mechanisms as a solution direction for this semantic interoperability problem. A Common Dairy Ontology (CDO) has been developed that serves as the main semantic interface to applications for dairy farming decision-making and analysis. Besides semantic alignment, the CDO also enables the reasoning on the dairy concepts and therefore on a large variety of different analysis questions. In addition, all sensor data has been transformed into triples according to the Resource Description Framework (RDF) standard, made available in a triplestore and accessible via a SPARQL engine. With a few example applications we have measured the performance of this solution and assessed the feasibility and performance of our solution direction.

In the remainder of this paper we will describe subsequently our linked dairy sensor setup, the ontologies used for semantic alignment, the test applications and the performance of our setup.

2. Linked dairy sensor data

At each of the 7 dairy farms involved in our SmartDairyFarming project sensor equipment is installed to monitor an average of 400 cows per dairy farm. These cows are continuously monitored since 2014, which has generated a yearly 12 GB of sensor data. This sensor data is made available via a software component called the InfoBroker to be used by various different applications for the dairy farmer. Our approach is to transform this sensor data into linked data in order to make it semantically rich and easily accessible.

To do this in a methodological way, we have developed a Linked Data Roadmap. This roadmap is developed to standardize the process of converting data of different types into linked data. In this roadmap we have defined nine steps containing tools- and best practices to generate high quality linked data. The Linked Data Roadmap is visualized in Figure 1. For the sake of simplicity we will not describe the details of the roadmap here and further details can be found at http://www.pilod.nl/wiki/BoekTNO/stappenplan.

We applied the process defined in this roadmap to make the data that is currently available through the InfoBroker available as linked data. Thereby, it becomes possible to query this data using semantic queries and make use of the advantages that this technology offers.

For the second step of the roadmap on preparing the data, we needed to extract all the data from the InfoBroker. Because the InfoBroker offers its data as a REST-API, we designed a python script that extracts data from the InfoBroker by calling a sequence of API calls, and transforms these JSON responses into CSV files. We then used GoogleRefine to determine the quality of the data to make a clean and consistent dataset.

Based on this clean dataset we applied step three of the roadmap on modeling the data. Modeling data in terms of linked data means defining an ontology or knowledge model for the data. This can be seen as the database schema, it defines the meaning of things in terms of relationships. For this case, we defined a specific measurement ontology based on three basic classes: *Cow, Sensor* and *Measurement*. This ontology is depicted in Figure 2. With this ontology it was possible to express all the data offered by the InfoBroker in terms of triples of the form *<subject, predicate, object>*. We then defined a sensible naming convention for the URI's, so that all URI's would be unique and meaningful.

Finally, we've developed a python script that automates the process of converting the CSV files to RDF files. This script uses the API of GoogleRefine to automatically convert the CSV to RDF triples.

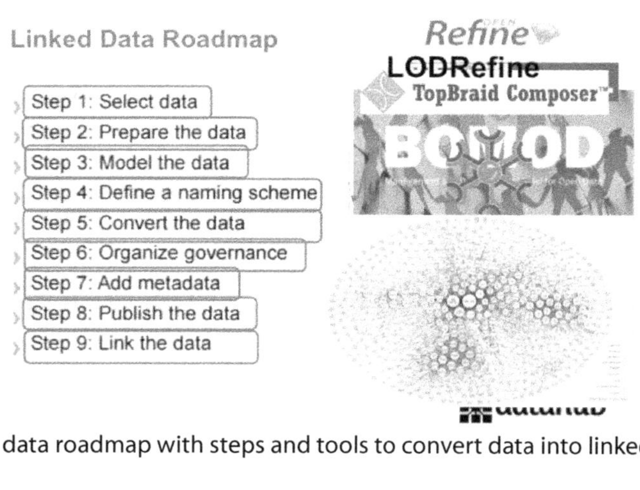

Figure 1. Linked data roadmap with steps and tools to convert data into linked data.

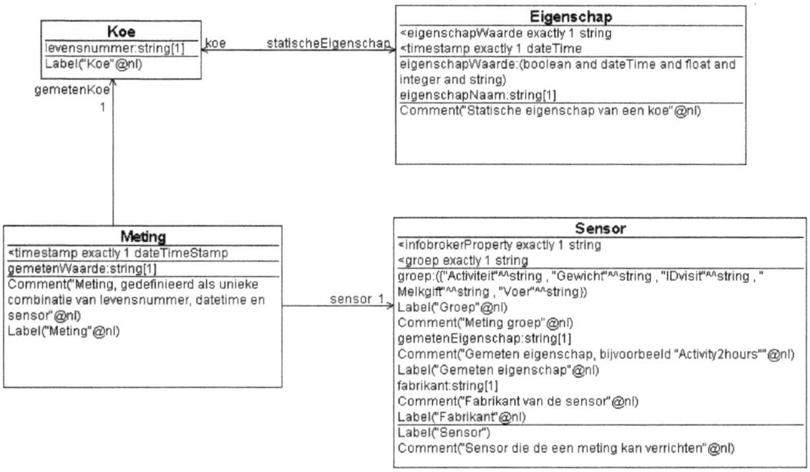

Figure 2. A measurement ontology for modeling measurements of sensor parameters for cows.

This script uses earlier defined operations on the CSV data (including the definition of the RDF structure). This script then exports the GoogleRefine projects as RDF files, which it then uploads to our triplestore, so expose the data and make it queryable using the included SPARQL endpoint. The yearly 12 GB sensor data was transformed into 310 GB of triple data.

We've chosen to use Apache Fuseki as our triplestore and SPARQL endpoint. It uses Apache TDB (Triple DataBase) which is a graph database for storing of triples. Apache Fuseki in combination with the Apache TDB triplestore offers advanced SPARQL support, including federated queries, as well as a very high performance compared to for instance Apache Marmotta. The linked data is

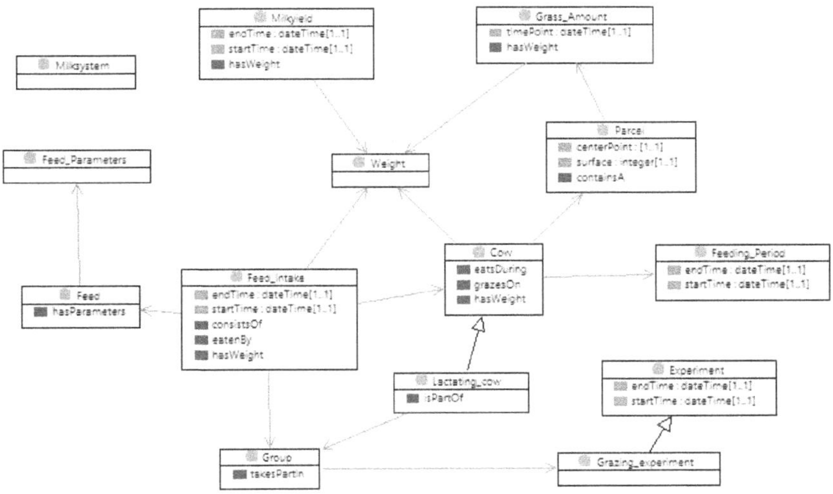

Figure 3. Common Dairy Ontology excerpt with main concepts in dairy farming.

queryable through the Apache Fuseki SPARQL endpoint. This data is kept up-to-date with a fully automated process that daily extracts data from the InfoBroker and stores it as triples in the Apache Fuseki store.

3. Semantic alignment

One of our goals is to enable the answering of analysis questions on the combination of large sources of measured sensor data. One of these questions is for instance 'What is the average weight per day over the last lactation period of a cow of a farmer?' In order to achieve this, we have developed a Common Dairy Ontology (CDO) that is meant to contain the main, common dairy farming concepts. The CDO ontology includes concepts like *Farm, Farmer, Cow, Weight, Milkyield, Activity, Feed, Parcel, Equipment* and so on. In addition, the CDO ontology covers the most important relationships between these concepts. The CDO then functions as the 'semantic interface' to the users of the information that is captured by the sensor data system. The users can be farmers, advisors of farmers or other stakeholders around the farm and they want to express their information needs in terms of these common concepts in the CDO. See Figure 3 with an excerpt of the concepts in the CDO.
The CDO can be used as the knowledge model for accessing the sensor data that has been modeled as triples using the specific measurement ontology. In order to do this, we have made a mapping from the CDO to the specific measurement ontology and the specific parameters that are measured by the sensor equipment. For example, the common concept *Weight* is mapped onto various different parameters for weight measured by different weighing equipment at the farm, such as *Lely. BodyWeight* and *GallagherDairyScale.dsweight*. In addition, the concept Activity is mapped onto various specific parameters for activity, such as *Lely.Activity2hours* and *DeLaval.HoogActNiveau*. We used the rdfs:label mechanism to make the mapping between the classes and properties of the CDO and the measurement ontology.
See Figure 4 with an example of the mapping between the CDO and the measurement ontology. This mapping is being used when big data analysis question are asked to the CDO for which measured sensor data is needed.

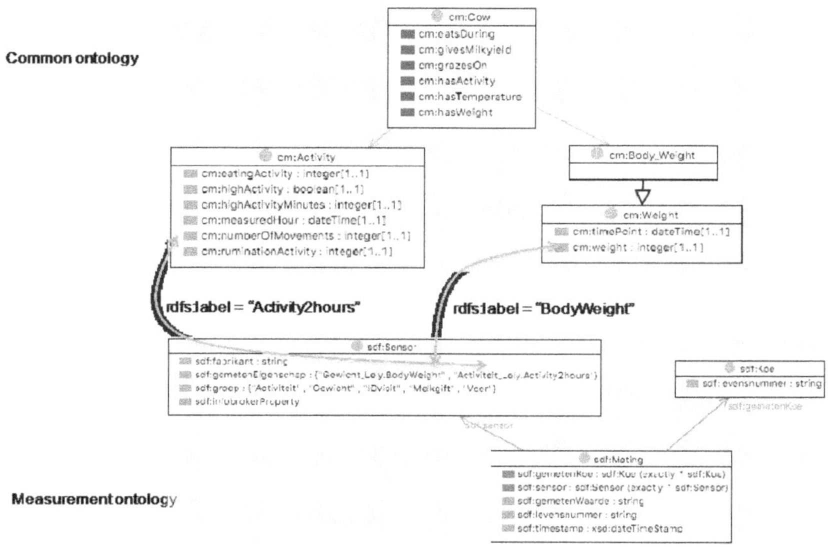

Figure 4. Mapping of classes and properties between the CDO and the measurement ontology.

4. Big data applications

In order to use the large set of historical sensor data and to assess the performance of the triple solution, big data applications have been developed that focus on the analysis of possible patterns in the data of individual cows over the period of one year.

One application looks at the relation between bodyweight and milk yield of individual cows during the lactation period in 2014. For each individual cow the development of bodyweight and milk yield during the lactation period can be drawn in a graph. In addition, the increase/decrease of the bodyweight on a weekly basis during that same period can be drawn as well. Finally, an overall view of the average weight over all the cows of the same parity can be depicted. Using these views, the farmer can derive possible relationships between the bodyweight and the milk yield. Another application tries to find similar relations between the different types of feed and milk yield of individual cows during the lactation period in 2014. For each individual cow the development of total feed intake and milk yield during the lactation period can be drawn in a graph. In addition, the division of the total feed intake over various feed types during that same period can be drawn as well. Finally, an overall view of the total feed intake over all the cows of the farmer can be depicted. Using these views, the farmer can derive possible relationships between the intake of different types of feed and the milk yield.

See Figure 5 for a view on the front end of the applications.

Each of the graphs that can be shown by the applications is build up based on the result of a SPARQL query on the large set of triples. The SPARQL queries use FILTER statements to select those triples that are related to measurements for the specific cow number, the correct time period and the measured parameters. See Figure 6 for the SPARQL query code for two of the queries that we used to select feed and weight data from the Fuseki triplestore.

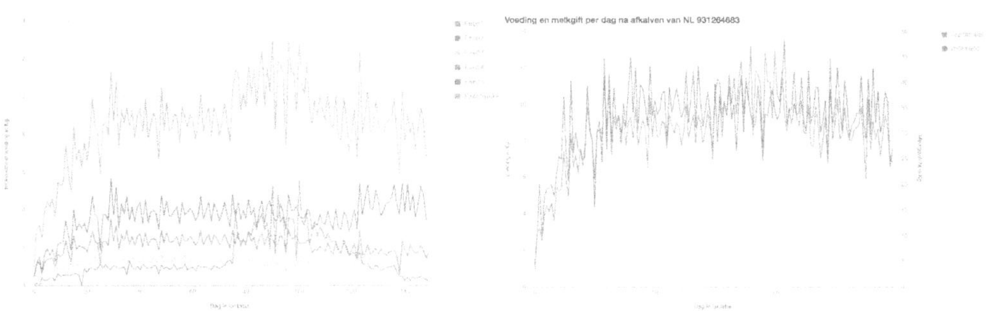

Figure 5. Big data analysis applications on historical weight, feed and milk yield data.

5. Performance

We've started our experiments with the Apache Marmotta triplestore including a SPARQL endpoint. However, it turned out that our large volume of data had a significant impact on the performance of the Marmotta triplestore. This was likely due to the fact that Marmotta stores the data in a relational database, and translates this to triples when it's queried. We tried to optimize the Marmotta and the underlying PostgreSQL database configuration. However, performance remained unacceptable for our purposes. Another drawback of the Marmotta server is the limited support of the full SPARQL1.1 standard and federated queries. This was a functionality we wanted to have to be able to query the data combined with other datasets offered through other SPARQL endpoint. Apache Fuseki

```
# Feed per day per type over all cows of a farmer

PREFIX xsd: <http://www.w3.org/2001/XMLSchema#>
PREFIX minion: <http://minion02.sensorlab.tno.nl:8080/fuseki/SDF/data/>
PREFIX sdf: <http://minion02.sensorlab.tno.nl/ontologies/SDF.ttl#>
SELECT ?type ?dag (SUM(xsd:float(?w)) AS ?waarde)
FROM minion:%s #Machine
WHERE {
  ?sensor sdf:gemetenEigenschap ?type .
  FILTER regex(?type, "Feed") .
  ?meting sdf:sensor ?sensor .
  ?meting sdf:timestamp ?ts .
  FILTER( xsd:dateTime(?ts) >= xsd:dateTime("2014-01-01T00:00:00.0") &&
xsd:dateTime(?ts) < xsd:dateTime("2015-01-01T00:00:00.0"))
  ?meting sdf:gemetenWaarde ?w .

  #calculate JND of ?ts
  BIND(floor((14-month(?ts))/12) as ?a_ts).
  BIND(year(?ts) + 4800 - ?a_ts as ?y_ts).
  BIND(month(?ts) + (12*?a_ts) - 3 as ?m_ts).
  BIND(day(?ts) + floor((153*?m_ts + 2) / 5) + (365*?y_ts) + floor(?
y_ts/4) - floor(?y_ts/100) + floor(?y_ts/400) - 32045 - 2456658 as ?
dag).

}
GROUP BY ?type ?dag
ORDER BY ?type ?dag
```

```
# Average weight of a cow in a lactation period

PREFIX fuseki: <http://minion02.sensorlab.tno.nl:8080/fuseki/SDF/data/>
PREFIX sdf: <http://minion02.sensorlab.tno.nl/ontologies/SDF.ttl#>
PREFIX xsd: <http://www.w3.org/2001/XMLSchema#>
SELECT DISTINCT ?parity ?date (ROUND(AVG(xsd:integer(?waarde))) AS ?gewicht)
FROM fuseki:Antonides_Lely
FROM fuseki:Antonides
WHERE {
        ?koe sdf:levensnummer "%s". #Cow
        ?staticProp sdf:koe ?koe.
        ?staticProp sdf:eigenschapNaam "calvingDate".
        ?staticProp sdf:eigenschapWaarde ?calvingDateStr.
        BIND(xsd:dateTime(?calvingDateStr) as ?calvingDate)
        ?staticProp2 sdf:koe ?koe.
        ?staticProp2 sdf:eigenschapNaam "dryDate".
        ?staticProp2 sdf:eigenschapWaarde ?dryDateStr.
        BIND(xsd:dateTime(?dryDateStr) as ?dryDate).
        ?staticProp3 sdf:koe ?koe.
        ?staticProp3 sdf:eigenschapNaam "parity".
        ?staticProp3 sdf:eigenschapWaarde ?parityStr.
        BIND(xsd:int(?parityStr) as ?parity)
        ?sensor sdf:gemetenEigenschap "BodyWeight" .
        ?meting sdf:sensor ?sensor .
        ?meting sdf:gemetenKoe ?koe .
        ?meting sdf:timestamp ?ts.
        FILTER( ?ts >= xsd:dateTime(?calvingDate)) .
        ?meting sdf:gemetenWaarde ?waarde .

        BIND ( CONCAT( STR(YEAR(?ts)),"-",STR(MONTH(?ts)),"-" , STR(DAY(?
ts)) ) AS ?date).
}
GROUP BY ?parity ?date
ORDER BY ?ts
LIMIT %d #Number_of_days
```

Figure 6. Two SPARQL queries that were used to select feed and weight data from the Fuseki triplestore.

performed a lot better in that sense. This is most likely because Fuseki implements a native triple graph database, instead of a relational database that converts data into triples. We did a number of measurements on the response times for the SPARQL queries executed in our application. In Table 1 we present an overview of the queries, the input parameters (farmer or cow), the graph size in number of triples in which the query is executed, the number of search parameters in the query and the response time in seconds.

An interesting observation is that there is hardly any pattern to be recognized between the graph size, the number of search parameters and the response times. Apparently, the SPARQL search engine has specific ways of indexing to assist the search process through the set of triples. The performance of our linked data semantic solution is acceptable for analysis queries on large sets of data. Without

Table 1. Overview of graph size, selection parameters and response times for the main queries.

Query	Input	Graph size	Search par	Response
Select an overview with the number of cows of a farmer	Stokman	111,604,625	1	0.04s
	Bakker	167,894,559	1	0.03s
	Antonides	79,739,365	1	0.37s
Select the list of cows with number and parity	Stokman	28,704	3	0.934s
	Bakker	9,400	3	15.110s
	Antonides	45,816	3	27.006s
Select feed per type per day over all cows of a farmer	Stokman	66,551,765	3	913.003s
	Bakker	38,034,692	3	350.917s
	Antonides	45,637,592	3	380.470s
Select average weight over all cows per day per parity	Antonides	45,637,592	3	348.704s
Select static info for a cow	NL 715820911	45,816	2	0.094s
Select weight per day in lactation period	NL 715820911	45,683,408	5	5.129s
Select weight and milkyield per day in lactation period	NL 715820911	45,683,408	7	13.714s
Select milkyield per day in lactation period	NL 715820911	45,683,408	3	4.142s

optimization of queries the time for answering queries ranged from a few seconds to a couple of minutes. This is acceptable for analysis purposes, but for analyzing real-time data, this is probably not acceptable. The more complex queries take several minutes to generate a response (this could even take up to 30 minutes for a complex query when the server-load is high). A farmer might not be willing to wait that long for a response.

6. Conclusions and future work

The overall conclusion of our work presented in this paper is that the use of linked data mechanisms on large sets of sensor data is feasible. Even if big data in the order of hundreds of gigabytes is put into RDF triple format, it remains accessible with reasonable response times for analysis purposes. For near real-time purposes the question remains whether this solution approach is still feasible, but this is a topic for further work. Another future work topic is to further extend the Common Dairy Ontology with more dairy concepts and make it available as the semantic interface in conjunction with the InfoBroker or a local relational database. In that setup the sensor data is not being put into RDF triples and the CDO is used as mapping tool towards the specific sensor data sources.

Using big data and advanced analytics to optimise health and fertility

E. Harty* and J. Healy
Dairymaster, Causeway, Tralee, Co Kerry, V92 NWK0, Ireland; eharty@dairymaster.com

Abstract

Megatrends in technology are revolutionising industry. With farms growing bigger and time becoming limited there is a need to see how technology can help on farm. Monitoring the health and fertility status of each individual cow within a herd is one of the key drivers for milk production and farm profitability. The MooMonitor+ is a new type of animal activity and behaviour monitor which continuously monitors each cow in the herd and provides more granular data than heretofore was possible. This device records an animal 24×7 and stores a digital fingerprint of 210,000 data points per animal per year on a cloud-based data analytics platform. In this paper the MooMonitor+ is compared against existing automated technology. A trial was conducted with an existing technology tag and the MooMonitor+. Both tags were fitted on the same collar and worn by cows in a commercial farm for the last six weeks of the breeding season. Heats were recorded for both devices. Milk samples were collected three times per week for progesterone analysis from all cows in the trial. Heat events indicated by both the MooMonitor+ and the existing technology tag were compared with the available progesterone data. The results from this study show high heat detection rates and lower false positives with the new technology (MooMonitor+) versus existing technology. The new device had a detection rate of 94% and an error rate of 9.2% compared to 86% detection rate and 33% error rate for the commercially available existing device. Better heat detection leads to higher submission rates, decreased calving intervals, less use of straws and increased pregnancy rates on farms. On the behavioural side, this new technology has identified various cases of Salmonella, mastitis, abomasal displacement, metabolic upset and lameness. In addition, cohort analytics are available in real-time which compare group analysis with the individual cow in all of the aforementioned categories of behaviour. Cows come into milk a lot faster with better health and fertility management. This in turn makes dairy farming much more profitable and sustainable.

Keywords: oestrus, activity, rumination, resting, progesterone

1. Introduction

Introduction of new technologies has become an integral part of the way we look at farming. The last few decades farms have increased in size and the number of people managing these herds have been reduced. Farm processes take place on a much larger scale than before but the focus now needs to be on each individual animal. Technology in farming is creating a safer work environment, as well as making the farming business more profitable and sustainable. Extra free time and making life easier for farmers is also desirable.

One of the areas technology is emerging rapidly in is the area of heat detection. Dairy cows have a heat cycle that reoccurs approximately every 21 days. A normal heat event should last for 15 to 18 hours. However increased emphasis for breeding on production traits instead of fertility has left its marks. Nowadays a farmer can consider himself lucky when his high producing cows are in heat for six to eight hours every time. Research has shown the majority of cows (60-70%) come in heat at night (Hall, *et al.*, 1959; Hurnik *et al.*, 1975; Stevenson *et al.*, 2014). Also about 15% of all standing heat events have a duration of less than four hours. If these short duration heats occur during the night a large number of them will be missed by farm management. This leads to longer calving intervals and missed profits for the farmers.

At the moment various heat detection devices are available commercially. All devices promise high detection rates for heat detection and many have proven their worth in that area. However accuracy of a device is not only translated in heat detection rate alone. For a device to be more accurate one should also look at the number of false positive alerts are generated. This is also translated in the error rate of a device. The error rate is the inverse value of how precise a device actually is. With other words: whether or not the number of alerts actually correlate with real events. In this study performance of an existing activity monitoring technology was compared with a new activity monitoring technology.

2. Materials and methods

2.1 Cows and data collection

This trial was conducted on a commercial farm. Animals calved >14 days were selected for this trial. In order for a fair comparison to be done between both technologies problem cows with BCS under 2.5 and above 4 were excluded from the study as well as cows with a locomotion score above 3. A total 61 cows were selected for the trial.

Cows were fitted with both a commercially available activity meter and the new MooMonitor+ (Dairymaster, Causeway, Co Kerry, Ireland). The MooMonitor+ is an activity monitor worn around the cow's neck which detects individual cow heats and behavioural changes. It monitors cows 24×7 and identifies specific types of behaviour such as feeding, rumination, resting time and different types of activity intensity. These behaviours can be used for heat detection, monitoring feeding and rumination patterns, cow welfare and managing the health status on farm. The second activity meter is also a neck worn device that focusses on certain behaviours of the cow and can also be used for heat detection. Both devices were fitted on the same collar on the same animal. Heat alerts were extracted from both devices (after a normalisation period had passed). Alerts were then compared to Progesterone profiles taken from milk samples.

2.2 Milk sampling

The trial ran for a duration of six weeks from 07/05/2015 till 12/06/2015. During this timeframe milk samples were taken thrice weekly during the evening milking. A total of 16 samples per cow were taken. Samples were analysed to determine milk progesterone concentrations using an ELISA, supplied by Ridgeway Research Ltd. The method for detecting heats using the progesterone profiles was as follow: Progesterone profiles were created for each individual animal. Activity profiles of both technologies were then overlaid with the progesterone profiles (see Figure 1).

2.3 Performance analysis

Where a heat alert of either two devices coincided with a low progesterone event (<5 ng/ml) it was seen as a true heat event. If a heat alert occurred and progesterone levels were high (>5 ng/ml) the

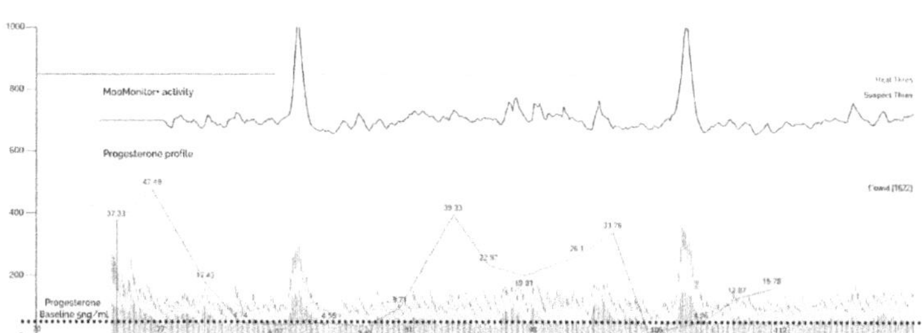

Figure 1. Activity profile cow overlaid with progesterone profile and baseline for heat detection.

heat alert was regarded as false positive. Missed heats were determined when progesterone profiles showed that an ovulation with preceding heat event had occurred and the device had not given a heat alert. Performance of both devices were analysed and compared.

3. Results

Performance analysis of both devices resulted in a heat detection rate of 94% for the MooMonitor+ system compared to an 86% detection rate for the existing technology (P=0.70). Error rate for the MooMonitor+ was 9.2% whereas the existing technology had an error rate of 33.3% (P<0.001). This means that with the existing technology one in three inseminations would have been performed on a cow that was not in heat. Consequences of inseminating at the wrong time are a higher risk of abortion by disturbing the pregnant uterus in early stages, higher risk of introducing infection in the uterus because immune response is not as high with a cow in dioestrus than when a cow is in heat, a higher number of straws used per cow, longer calving intervals and higher 'failure-to-conceive' culling rates since the cow does not seem to conceive on insemination. The difference in error rate was significant meaning that the MooMonitor+ performed better than the existing automated technology in this trial.

4. Practical implications

The MooMonitor+ records an animal 24×7 and stores a digital fingerprint of 210,000 data points per animal per year on a cloud-based data analytics platform and this provides more information than existing technology.

The added benefits of more accurate heat detection using activity sensing technology are early identification of non-cycling cows, repeat breeders and cows that are cystic. Not every heat is similar in strength and duration to the previous one. Both external and internal circumstances can interfere with the way a cow expresses heat. High producing cows tend to have shorter and weaker heats. This because the intense milk production can counteract follicle growth and oestrogen production. Research has shown that inseminating on good strong heats gives higher conception rates and a higher chance of producing fertile offspring (Ismael *et al.*, 2015). It therefore makes sense to monitor heat events from start to end to decide whether to inseminate or not, what quality semen to use and when would be the best time to inseminate after the start of a heat. In this study both devices followed the entire heat event from start to finish. Research has shown that with MooMonitor technology highest conception rates are achieved when inseminating between 9-12 hours from the onset of heat with cows and between 13-16 hours for heifers (Stevenson *et al.*, 2014). Knowing the exact time when a heat event starts can greatly improve accuracy for insemination and increase the possibility of a positive pregnancy check 30 days later.

Heat detection devices have been in existence since the '80s but recent developments in the farming world (abolition of quota, worldwide increased production and ever increasing demand for milk) require more competent and intelligent systems than what we've seen so far. Where it used to be enough to just measure the number of steps a cow took per day we now realise there is a whole lot more to good fertility management than an increase of activity every three weeks. In order to work properly, heat detection devices must be able to pick up heat related behaviour. If a cow is sick, lame or very low in energy heat behaviour is supressed in order to allow the cow to recover from her circumstances. This is why it is important to monitor the entire cow, not just one type of behaviour. The MooMonitor+ can also be used for monitoring other behaviours such as rumination, feeding and resting time of the cow. These behaviours can be used for monitoring the animal's health status. If a cow is getting sick rumination levels will drop and resting time will increase even before a temperature might develop (Dantzler, 2001; Roessen *et al.*, 2015). Monitoring the entire cow will give a farmer extra information about the animals making it easier to detect any underlying health problems when cows are not expressing heats optimally.

Health equals fertility. All production diseases have an effect on the fertility and production status of the animal. A cow with retained foetal membranes could come in heat later and will therefore be inseminated at a later point in time, increasing calving interval. If metritis develops then this further complicates matters, calving to conception interval could be more than 40 days delayed compared to a normal healthy cow. In addition, illness can increase the likelihood of getting another illness. For example, a cow with milk fever is eight times more likely to get mastitis (Curtis *et al.*, 1983). A cow in NEB (negative energy balance) before calving is three times more likely to develop a displaced abomasum (Duffield *et al.*, 2009). And a cow suffering from SARA (sub-acute ruminal acidosis) has an increased risk of getting lame (Plaizier *et al.*, 2008). In commercial applications on dairy farms the MooMonitor+ has identified various cases of Salmonella, mastitis, abomasal displacement, metabolic upset and lameness. These behaviour alerts were acted upon by farm management providing the best chance of recovery for these animals. Using proprietary software group analysis can flag up cows that are in the danger zone of developing a health related problem. Individual cows can be compared to the lactation group they belong to or can be compared to the herd.

Backtracking individual cow behaviour is also a useful resource for vets when visiting a farm. Objective animal history will aid in quicker diagnosis and gives a good perspective of the recovery. In dairy cow herd health programmes data collected by the MooMonitor+ can be used and easily integrated in management programs. Behaviour monitoring of animals and fertility status determination can aid in concluding current herd status and benchmark this against pre-set farm targets.

In this study the MooMonitor+ offers excellent heat detection rates with a low error rate. These results leads to more accurate heat detection which means less wastage on straws and the features of the MooMonitor+ allow for timely insemination on healthier animals. Milk production is driven by the cow's health and fertility status. A cow that calves down healthy and is kept healthy throughout the entire lactation cycle will come in heat a lot faster and milk production will reflect her potential more accurately. This in turn will lead to a more sustainable way of farming that is more profitable for both farm managers and livestock making the practice of farming more enjoyable. This paper shows that using MooMonitor+ provides more data for animal behaviour and easy to use advanced analytics results in significantly better on-farm performance.

References

Curtis, C.R., Erb, H.N., Sniffen, C.J., Smith, R.D., Powers, P.A., Smith, M.C., White, M.E., Hillman, R.B. and Pearson, E.J. (1983). Association of parturient hypocalcemia with eight periparturient disorders in Holstein cows. Journal of the American Veterinary Medical Association 183(5): 559-561.

Dantzler, R. (2001). Cytokine-induced sickness behavior: where do we stand? Brain, Behavior, and Immunity 15(1): 7-24.

Duffield, T., Lissemore, K., McBride, B. and Leslie, K. (2009). Impact of hyperketonemia in early lactation dairy cows on health and production. Journal of Dairy Science 92(2): 571-580.

Hall, J., Branton, C. and Stone, E. (1959). Estrus, estrous cycles, ovulation time, time of service, and fertility of dairy cattle in Louisiana. Journal of Dairy Science 42(6): 1086-1094.

Hurnik, J., King, G. and Robertson, H. (1975). Estrous and related behaviour in postpartum Holstein cows. Applied Animal Ethology 2(1): 55-68.

Ismael, A., Strandberg, E., Kargo, M., Fogh, A. and Lovendahl, P. (2015). Estrus traits derived from activity measurements are heritable and closely related to the time from calving to first insemination. Journal of Dairy Science 98(5): 3470-3477.

Plaizier, J., Krause, D., Gozho, G. and McBride, B. (2008). Subacute ruminal acidosis in dairy cows: the physiological causes, incidence and consequences. The Veterinary Journal 176(1): 21-31.

Roessen, J., Harty, E. and Beirne, C. (2015). MooMonitor + smart sensing technology and big data – resting time as an indicator for welfare status on farms. ICAR Technical Workshop, Krakow.

Stevenson, J., Hill, S., Nebel, R. and DeJarnette, J. (2014). Ovulation timing and conception risk after automated activity monitoring in lactating cows. Journal of Dairy Science 97(7): 4296-4308.

The Ovalert concept: adding value to sensor data

K. Huijps, A. Pasman and H.M. Knijn*
CRV, Wassenaarweg 20, 6843 NW, Arnhem, the Netherlands; hiemke.knijn@crv4all.com

Abstract

Precision Dairy Farming is becoming more important at dairy farms, generating a lot of data on and around the farm. This data has to be transformed to information and be used in decision making processes. Support of farmers using that information is a crucial element in adding value to precision farming. To help dairy farmers with precision farming around one of the top priority issues, fertility, CRV has developed the Ovalert concept. This concept, incorporated in the CRV software's, includes automatic heat detection (SCR or Nedap technology) combined with mating advice, inseminations and advice from fertility experts. The goals of this concept are: (1) improvement of fertility; (2) help farmers managing the fertility on the farm; and (3) simplify their daily fertility management. Data of 715 farmers who started with the Ovalert concept between 2011 and 2014 were used to analyze the effect on calving interval and satisfaction of the farmer. Calving interval data was available from 2010 till 2015. Microsoft Excel was used to calculate the average calving interval per year over farms. From this averages the differences between years were calculated per year of enrollment. The calving interval improved for all farmers when looking at the difference between 2015 and the year of enrollment in the concept. The amount of days differed with the year of enrollment (17±18 for 2011, 15±27 days for 2012, 13±21for 2013, and 4±19 days for 2014). The average improvement of the calving interval in the Netherlands was 4 days from 2010 to 2015. Data about the satisfaction of the farmers was collected by online farmer surveys in 2013. Ovalert users were randomly selected among users enrolled for a minimum of one year. 103 farmers received an online survey (NetQ). Results were analyzed graphically. Farmers indicated the fact that the concept eases their fertility management as main advantage. The Ovalert concept shows how precision dairy farming can be beneficial for the dairy farmers. The concept helps dairy farmers in decision making processes around fertility and results in technical improvement of fertility and it simplifies their fertility management.

Keywords: farmer support, information, combined solutions

1. Background

In today's market, efficient and effective management are the key to a successful dairy farm. Farms face the need to upscale and produce in a sustainable manner, to manage animals on an individual level and to balance between work and private life. This presents farmers with a big challenge.
Parallel to this Precision Dairy Farming is becoming more and more important at dairy farms. Precision farming is generating a lot of data on and around the farm, but this data has to be transformed to information. And that information should be used in decision making processes. Support of the dairy farmers in using that information is a crucial process in adding value to precision farming.
One of the main issues on a dairy farm, is fertility management. But fertility is at the same time a difficult, multifactorial issue. It starts with finding the cows in heat. Approximately 60% of cows show signs of heat during the night. There is a high risk of missing these heats, or being unable to respond in time. Next, with the farms increasing in number of cows, it takes more and more time to visually check all animals for heat multiple times a day. To support farmers with this process, automated heat detection systems are in the market. These systems register heats (mostly based on activity and rumination behaviour) and generate attentions for the farmer. But with the detection of heat, the cycle of fertility is not made yet. After a detected heat, the cow has to be inseminated with

high quality semen at the optimum insemination moment and should conceive. If that is managed, the cow will calve, get in production and generate income for the farmer (Figure 1).
CRV has developed the Ovalert concept to help dairy farmers with precision farming around this top priority issue, fertility. This concept includes automatic heat detection combined with mating advice, inseminations and advice from fertility experts. The main goal of this concept is: (1) to help farmers managing the fertility on the farm; (2) simplify their daily fertility management; and (3) improvement of the technical data (e.g. calving interval). The true added value of this concept lies in the integrated fertility management. The Ovalert package includes: (1) Webbased herd management software; (2) SMS attentions; (3) Mating advice based on farmers breeding goal; (4) AI-service; (5) All-inclusive semen; (6) Pregnancy check; and (7) Fertility consultancy (Figure 2).
Hardware is available from Nedap (Groenlo, the Netherlands) and SCR (SCR Engineers Ltd., Netanya, Israel Netanya, Israel), but other heat detection systems can be linked into Ovalert by public webservices.
Independent on the hardware supplier, farmers have one dashboard including all information of the Ovalert concept (Figure 3).

2. Materials and methods

Data of 715 farmers who started with the Ovalert concept between 2011 and 2014 were used to analyze the effect on calving interval. Calving interval data was available from 2010 till 2015. Microsoft Excel was used to calculate the average calving interval per year over farms. From this averages the differences between years were calculated per year of enrollment.
Data about the satisfaction of the farmers was collected by online farmer surveys in 2013. Ovalert users were randomly selected among users enrolled for a minimum of one year. 103 farmers received an online survey (NetQ). Results were analyzed graphically.

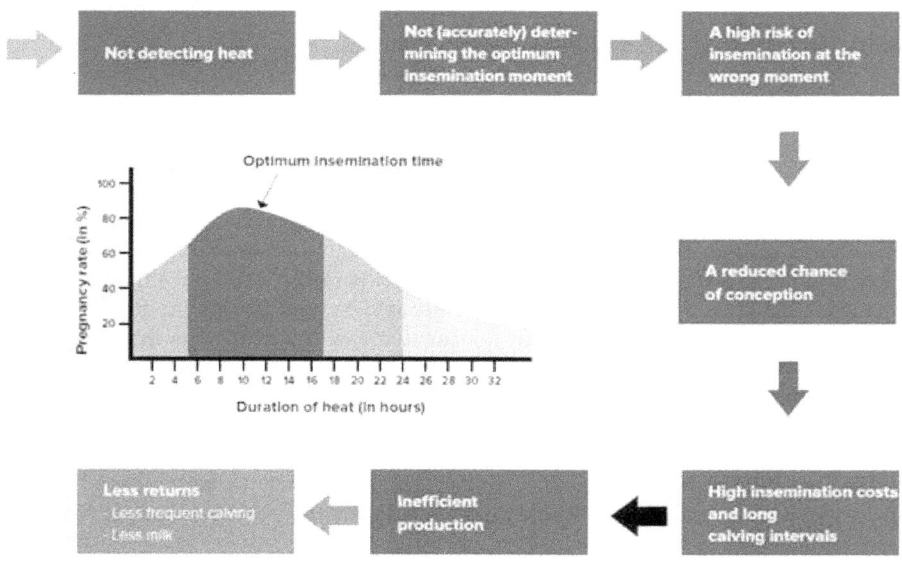

Figure 1. What can go wrong in fertility management.

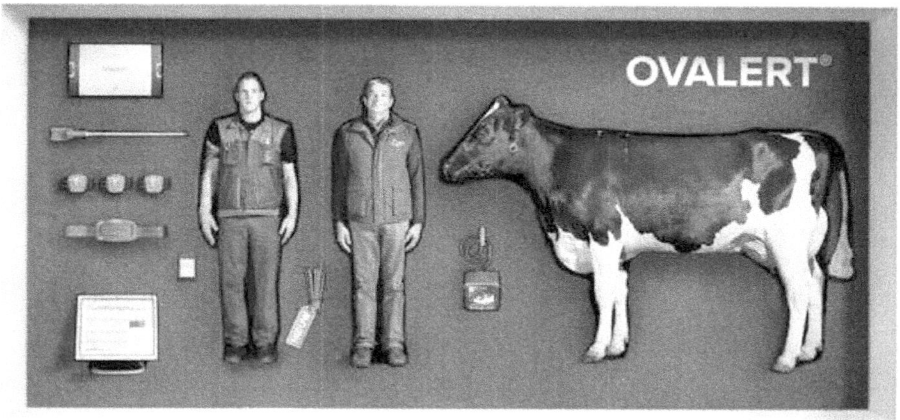

Figure 2. The Ovalert concept.

Figure 3. Printscreen dashboard for farmers from Ovalert concept.

3. Results

3.1 Technical results

The calving interval improved when looking at the difference between 2015 and the year of enrollment in the concept. The decrease of calving interval differed with the year of enrollment, 16 days for 2011, 15 days for 2012, 13 for 2013, and 4 days for 2014. The average improvement of the calving interval in the Netherlands was 4 days in the same period, from 2010 to 2015 (Figure 3).

3.2 Simplify fertility management

From 103 farmers, 60 returned the online survey (Houwelingen *et al.*, 2013). Of these 60, 56 were complete and used for further analyses. Farmers were asked to their experiences with the Ovalert

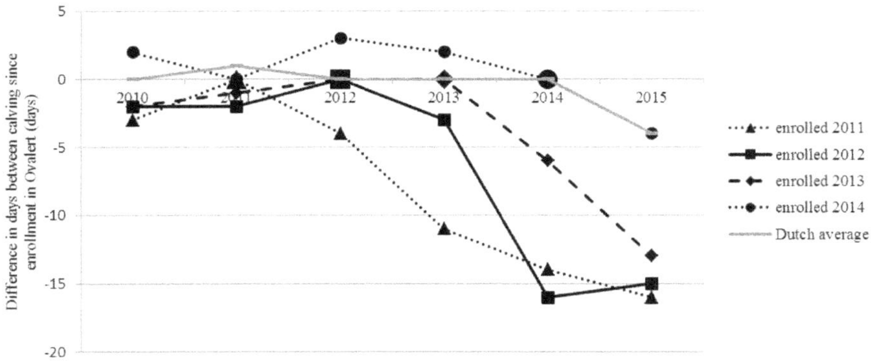

Figure 3. Differences in days between calving since enrollment in the Ovalert concept.

Concept. Two most important advantages of the Ovalert Concept mentioned by farmers were: (1) the way Ovalert supports the farmer in fertility management. An often mentioned reason was the fact that the concept takes away a lot of hassle regarding the whole process of getting cows pregnant; and (2) the positive effect on time between calvings. All farmers perceived an improvement in the time between calvings after they start working with the Ovalert concept (Figure 4).

4. Conclusions and discussion

Combining data and services to support farmers in their management is of crucial importance for Precision Dairy Farming. Combined solutions make sure farmers receive that information that is really needed for their decision making process. Regarding fertility management, Ovalert does exactly this.

The Ovalert concept shows how precision dairy farming can be beneficial for the dairy farmers. The concept helps dairy farmers in decision making processes around fertility and results in technical improvement of fertility. But it also simplifies their fertility management which results in more time to be spend on other issues, in personal life or on the farm.

The introduction of the Ovalert concept added extra improvement in the days between calvings compared with the Dutch average. The Dutch average improved with 4 days, while participants in the concept improved with 4 to 16 days, depending of the year of enrollment. The farmers participating in the Ovalert concept had higher times between calvings when enrolled in the concept then

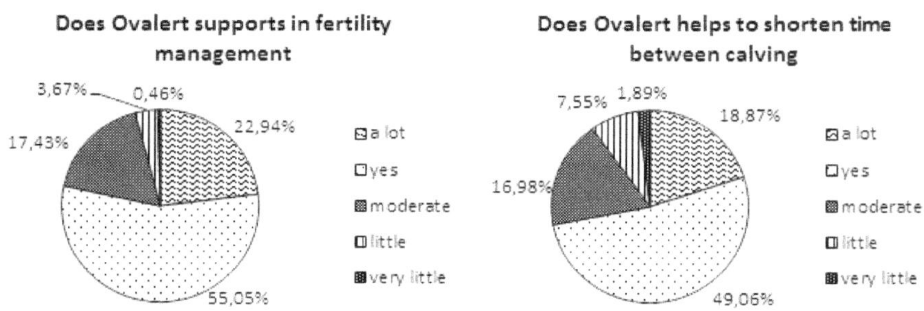

Figure 4. Highest perceived advantages of the Ovalert Concept by dairy farmers.

the Dutch average. This can be caused by the fact that especially farms with problems in fertility management felt the need to invest in a concept improving their fertility. Also the Dutch average improved, which can partly be caused by the fact that in general more heat detection systems are installed (other than Ovalert).
With 60 out of 103 responses on the online survey. This high response indicates the importance of the subject for farmers. From the 56 farmers, 78% indicated that the Ovalert concept supports them a lot in fertility management and 68% indicated that the calving interval decreased significantly. This high satisfaction results in a growing number of farmers using Ovalert. Currently 1,500 dairy farmers are enrolled in Ovalert in the Netherlands and Flanders. The concept is now also available in other countries.

Reference

Van Houwelingen, L., Okkerse, S., Tanis, M. and Westermand, T. (2013). Student report: 'stap vooruit'.

Information and precision technologies used for improving the milking routine performance

C.A.V. Paiva[1*], T.R. Tomich[1], F.S. Machado[1], M.M. Campos[1], G.N. Souza[1], F.O. Velasco[2], F.P. Possas[2], J.A.M. Lima[2], A.L. Ferreira[2] and L.G.R. Pereira[1]

[1]*Brazilian Agricultural Research Corporation, Rod. MG133, km 42, Coronel Pacheco/MG, 36155-000, Brazil;* [2]*Federal University of São João Del Rei, Pç. Frei Orlando 170, São João Del Rei/MG, 36307-352, Brazil; claudio.paiva@embrapa.br*

Abstract

Most of Brazilian farmers are not able to identify the impact of the milking routine procedures on the milking performance, usually because the information required is not ready available. Thus, the information and precision dairy technologies usage can help the farmer to collect, measure and analyze the data to support them on this subject. Fifty seven dairy crossbred Holstein × Gyr cows were milked in a double-4, herringbone parlor, fitted with electronic milk meters, automatic cluster remover and Alpro dairy manager system software (Alpro, DeLaval, Stockholm, Sweden). Data of total milking time, number of cows milked/hour, total milk production, milk yield/milker/hour, milking cluster attachment time, milk yield/cow, and number of automatic milking cluster removal was automatically provided by the milking system software. The experimental was performed during 21 days according to a factorial 2×2 design to evaluate milker (A or B) and milking routine (non-standardized or standardized) effects. Comparisons among treatments were conducted with orthogonal contrasts and statistical significance was considered at $P<0.05$. The results showed interaction between milking routine procedure and milker factors ($P<0.05$). Within both milking routine, the Milker A showed better performance for most milker performance parameters compared to Milker B ($P<0.05$). Within milker, the Milker A improved the milker performance parameters to the standardized compared to non-standardized milking routine procedures ($P<0.05$), whereas no difference were observed within Milker B ($P\geq0.05$). These results highlight the importance of precision dairy technologies usage to identify failures in the milking routine procedures, and hence improve the milking performance once the milker is committed to the procedures improvement.

Keywords: automation, data management, milk quality, sensors system, udder health

1. Introduction

Farmers are becoming more acquainted to adopting the precision dairy farming technologies available in some conventional milking systems (Borchers and Bewley, 2015; Steeneveld and Hogeveen, 2015). The main reasons for its adoption are related to usefulness of the information provided by the management system for reliable decision-making, and also reduction of the milking labor (Butler *et al.*, 2012; Dela Rue, 2008). The use of milking equipment integrated with sensor and herd management systems are increasing on Brazilian dairy farms. According to Rasmussen (1993) and Jago *et al.* (2010), these milking systems allows to verify individual data of milking and milker performances parameters such as total time spent to milking each cow included the beginning and end times, cluster attachment time, use of automatic or manual/forced cluster removal, milk yield/cow, among others. For Borchers and Bewley (2015), the production parameters associated to economic ones can help the farm manager to identify failure points during the milking routine that usually leads to inefficiency in production and labor costs. The present study aimed to evaluate the milking performance parameters of standardized and non-standardized milking routines, comparing the work of two milkers, by usage of the precision technologies available in conventional milking system.

2. Materials and methods

The experiment was conducted at Multi-use Complex on Livestock Bioefficiency and Sustainability, Brazilian Agricultural Research Corporation (Embrapa Dairy Cattle), located in Coronel Pacheco, Minas Gerais. Fifty seven dairy crossbred Holstein × Gyr cows were housed in a free stall barn, grouped by stage of lactation (1 to 3), and were from 100 to 200 days in milk. All cows were milked in the morning between 06:00 and 08:00 and in the afternoon between 14:00 and 16:00. The cows were milked in a double-4, low-line, herringbone parlor, fitted with MM27 electronic milk meters, MPC 580/680 controllers, Automatic Cluster Remover (ACR), and Alpro Manager System software (Alpro, DeLaval, Stockholm, Sweden). The treatments differed in the milking routine steps performed by each milker. The non-standardized milking routine was performed without a repeatable sequence and the use of ACR was determined freely by each milker. The standardized milking routine was implemented the day after the milkers training day on the new milking routine procedure. The standardized milking routine consisted in to follow a fixed steps beginning from first to last cow, where one side of the milking parlor had the routine finished before starting the opposite side as follows: (1) forestripping and pre dipping all teats with chlorine sanitizer moving from first to last cow; (2) backwards to the first cow, wiping and attaching the milking cluster, move forward to next cows; (3) mandatory use of ACR and once the milking cluster was automatic removed the teats were immediately pre dipped in an iodine solution. The informations provided by the milking system software based on the sensor systems were: (1) total milking time/day; (2) number of cows milked/hour; (3) total milk production; (4) kg of milk/milker/hour; (5) milking cluster attachment time; (6) milk yield/cow; and (7) number of automatic/manual milking cluster removal. Each treatment had the data recorded and analyzed for 21 days. Statistical analyses were performed according to a completely randomized 2×2 factorial design including the fixed effect of treatments (milking routine, milker and milking routine × milker) and the random effect of daily data collection. Comparisons among treatments were conducted with orthogonal contrasts, which included an overall comparison between milking routine procedures and milkers. The correlations among variables were evaluated with Pearson correlation coefficients. Statistical significance was considered at $P<0.05$.

3. Results and discussion

The results showed interaction between milking routine procedure and milker factors ($P<0.05$) indicating that the results for each milking routine procedure were influenced by the milker results (Table 1). Within both milking routine, the milker A showed shorter total milking time and milking cluster attachment time and higher number of milked cows/hour and kg of milk/milker/hour ($P<0.001$) compared to milker B (Table 1). The premilking routine presents great influence on the milk ejection reflex (Bruckmaier and Blum, 1996) which in turn has the milking cluster attachment time as an important parameter involved in the measure of the milking routine efficiency (Rasmussen *et al.*, 1990). Additionally, we found a high and positive correlation between the milking cluster attachment time with the total milking time ($r=0.91$, $P<0.01$), inversely to the number of cows milked/hour ($r=-0.94$, $P<0.01$) and to kg of milk/milker/hour ($r=-0.92$, $P<0.01$). The correlation results explain the importance of the milking cluster attachment time to the overall milking performance. Within milker, the Milker A showed better performance to milking cluster attachment time ($P=0.02$) and number of automatic cluster removals ($P=0.04$) for standardized milking routine compared to non-standardized. The milking cluster attachment times for milker A decreased from 162 to 94 seconds comparing the period before and after the milkers training day on the new milking routine procedure, respectively (Table 1). Nielsen and Rasmussen (1987) found an interval of 1.0 to 1.5 min from the beginning of udder preparation until milking cluster attachment to be optimal, which is in accord to our study. It represents a great impact to the farm economics. This type of evaluation is essential to strategic decisions regarding the choice of the milker profile and performance evaluation. Another important milker performance parameter concerns to the number of automatic cluster removal. An ACR device tends to improve the udder health by decreasing the number of over milking (Rasmussen, 1993). The number of automatic cluster removals to the milker A increased from 35 to 109 automatic removals ($P<0.05$) comparing the non-standardized milking routine procedure with standardized, respectively. No improvements were observed within Milker B to the parameters aforementioned ($P\geq0.05$), as the milking cluster attachment times increased from 217 to 225 seconds and the number of automatic cluster removal decreased from106 to 101 automatic removals, respectively (Table 1). Those results showed a strong interaction ($P<0.01$) of the milker factor over the milking routine procedures, in which the willingness to change for a new procedure can be determinant to the success of the milking routine performance. The production parameters as number of milked cows and total milk production did not present difference ($P\geq0.05$). Conversely, Rasmussen *et al.* (1990) reported an increased in the total milk production, which was not observed in our study. Differences between studies may be related to herd size, milking facilities, management practices, labor efficiency, and stage of lactation adopted in each study. The use of precision dairy farming technologies as described in this study turns the identification of problems in the milking routine procedure less laborious and more reliable. Moreover, these technologies allows the farmer to take the control (proactively) of the operation, taking faster decisions (in real time), aimed to increase the bioeconomic performance of the dairy system production (De Koning, 2004) with a great impact observed in the time and costs related to the routine activities like milking cows.

Table 1. Average results of milking and production performance parameters, by milking routine procedure and milker.

Parameters	MR				SEM	*P*-value		
	NSP		SP			MR	M	MR×M
	MA	MB	MA	MB				
Number of milked cows/day [2]	113	115	114	105	0.98	NS	NS	NS
Total milking time/day (min)[1]	235	266	204	279	7	NS	***	*
Number of cows milked/hour[1]	29	26	34	23	0.99	NS	***	**
Total milk production/day (kg)[2]	1,098	1,104	1,104	1,038	12.11	NS	NS	NS
kg of milk/milker/hour[1]	281	249	324	223	9.37	NS	***	*
Milking cluster attachment time (sec)[1]	162	217	94	225	14	*	***	*
Number of automatic cluster removal[1]	35	106	109	101	11	*	NS	*

[1] Milker performance parameters.

[2] Production parameters; MR: milking routine; NSP: non-standard procedure; SP: standard procedure; M: milkers; MA: milker A; MB: milker B; NS: not significant; *$P<0.05$; **$P<0.01$; ***$P<0.001$.

4. Conclusions

The use of information and precision dairy technologies can help farmers to identify deviations in the milking routine procedures and hence improve the milking routine performance through adoption of standardized procedures. However, the success of any new milking routine procedure adoption depends on the milker willingness to change about the old habits. This can be a key factor for the success of the activity once the milk production and udder health parameters can be impacted negatively by a non-standardized milking routine.

Acknowledgement

This research was supported by Embrapa, CNPq and FAPEMIG.

References

Borchers, M.R. and Bewley, J.M. (2015). An assessment of producer precision dairy farming technology use, prepurchase considerations, and usefulness. Journal of Dairy Science 98: 4198-4205.

Bruckmaier, R.M. and Blum, J.W. (1996). Simultaneous recording of oxytocin release, milk ejection and milk flow during milking of dairy cows with and without prestimulation. Journal of Dairy Research 63: 201-208.

Butler, D., Holloway, L. and Bear, C. (2012). The impact of technological change in dairy farming: robotic milking systems and the changing role of the stockperson. Royal Agricultural Society of England 173: 1-6.

Dela Rue, B. (2008). Keeping on top with information. In: Proceedings of the 6th Dairy Conference, Rotorua, New Zealand, pp. 161-165.

De Koning, K. (2004). Present and future. Innovations to improve milk harvesting. International Dairy Federation, Brussels, Belgium.

Jago, J.G., Burke, J.L. and Williamson, J.H. (2010). Effect of automatic cluster remover settings on production, udder health, and milking duration. Journal of Dairy Science 93: 2541-2549.

Nielsen, S.M., Rasmussen, M.D. (1987). Milking routines and equipment for tie up cow sheds. In: Cattle production research. Danish status and perspectives. Landhusholdningsselskabets Forlag, Copenhagen, Denmark, pp. 155.

Rasmussen, M.D., Frimer, E.S., Horvath, Z. and Jensen, N.E. (1990). Comparison of a standardized and variable milking routine. Journal of Dairy Research 73: 3472-3480.

Rasmussen, M.D. (1993). Influence of switch level of automatic cluster removers on milking performance and udder health. Journal of Dairy Research 60: 287-297.

Steeneveld, W. and Hogeveen, H. (2015). Characterization of Dutch dairy farms using sensor systems for cow management. Journal of Dairy Science 98: 709-717.

Tool for assessing the intervention effect on milk production in an evolutionary operation setup

A.H. Stygar[1*], M.A. Krogh[2], S. Østergaard[2] and A.R. Kristensen[1]
[1]*Department of Large Animal Sciences, University of Copenhagen, Grønnegårdsvej 2, 1870 Frederiksberg, Denmark;* [2]*Department of Animal Science, Aarhus University, 8830 Tjele, Denmark; as@sund.ku.dk*

Abstract

Modern dairy herds resemble factories. Cows, organized in production units, are manufacturing milk from many components (e.g. concentrates, silage). However, both production units and components can greatly differ between each other. Therefore, production optimization based on general recommendations might be inefficient. Instead, as in manufacturing industry, decision support could be based on systematic experimentation within ongoing production system. This concept, known as Evolutionary Operations (EVOP), is based on small changes to the production system. However, a challenge here is lack of a tool which would allow a farmer to assess how small changes, for example in feeding, influence productivity. The objective of this study was to construct a multivariate dynamic linear model (DLM) to assess the intervention effect on milk production. The DLM was built to account for intervention at individual and herd level. It consisted of an observation and a system equation. The observation equation links the observations to parameters describing the herd (lactation curve), individual cows and an intervention effect. The system equation expresses how the parameters may change over time. The lactation curve was modeled by two linear expressions and was parameterized using: milk yield 60 days after calving, slope over the first 60 days in milk and slope after 60 days in milk. The variance components of the DLM were estimated using a maximum likelihood method. The application of the model was demonstrated on a field experiment in a commercial herd with 4 automatic milking systems (AMS). The herd was split into 2 groups based on the AMS. The experiment relied on two steps. The first step was to reduce the feed energy given to cows in the AMS and instead supply the feed energy to the cows at the feed bunk. The second step was to reduce the feed energy given in two of the four AMS. The DLM presented here was successful in providing estimates of the effects on milk yield of change in feed energy given to the cows in the AMS. The DLM results support the sequential tactical decisions within EVOP and are readily applicable in other herd experiments.

Keywords: automatic milking system, EVOP, dairy cows

1. Introduction

Proper herd management is demanding. Frequently a manager must decide e.g. how to breed, how to feed or when to terminate the production. Those decisions are very often based on knowledge subjected to considerable uncertainty. Moreover, recommendations given to managers are commonly grounded on data from experimental settings which might be different than those present at a particular farm.

Nowadays, farmers are running big production units, with more and more data collected during the production process. Additionally, like in other industries, farmers are experimenting with production factors in order to reach economic optimum.

A systematic experimentation with Evolutionary Operations (EVOP) during the production process is a method for process-optimization which is finding a way to alter production to be more efficient (Box, 1957). The advantage of this method is that it uses simple statistical ideas and it is run during the normal production process. During EVOP experiment only small changes into the production factors and procedures are made by the personnel themselves. For that reason, the method seems to be well suited to large dairy herds, which nowadays, just like big factories, are equipped with many sources of data collection.

One of the major challenges in practical application of EVOP in dairy production is to estimate the effect of the experimental challenge made to the production system. Verstegen *et al.* (1995) reviewed normative evaluation approaches of management information systems with focus on strengths and weaknesses of the different designs. They found that time series data, which could be described by for example Dynamic Linear Models (DLM), in absence of a control group are the only design that provide reliable results and have internal validity. DLMs have been widely applied in effect evaluation (West and Harrison, 1997). In dairy production DLMs have been applied to milk production (Nielsen *et al.*, 2010) but not with the focus on estimation the effect of an experimental challenge.

The objective of this study was to demonstrate the application of a DLM in effect evaluation on milk yield within an EVOP framework.

2. Material and methods

2.1 Milk yield

The milk yield of a cow in a particular day of lactation was described by four parameters (a herd and a cow level, local production effect and measurement error). The herd level was represented by the average herd milk yield taking into account cow lactation number and number of days in milk. The animal level consisted of a variable representing the production potential of a particular cow and was modeled as autoregressive process of order 1 with mean 0 and known autocorrelation. Additionally, a variable representing a local production effect on milk yield as a result of temporary environmental effects was included to the model. Finally, the model was supplemented by a measurement error. It was assumed that measurement errors were mutually independent.

The herd level lactation curve was modeled using an approach based on two linear functions with two parameters each. Only three possible lactation curves (1, 2 and ≥3) were considered. The first linear function described the milk yield during the first 60 days after calving (slope of lactation function during the first 60 days and milk yield 60 days after calving). The second function described the remaining milk yield (slope of lactation function after the first 60 days and milk yield 60 days after calving).

2.2 Dynamic linear model

Based on the structure for milk yield modeling defined above, a multivariate dynamic linear model (DLM) was build. A DLM consists of an observation and system equation as described by West and Harrison (1997). The role of the observation equation of the form:

$$Y_t = F_t'\theta_t + v_t \quad (1) \qquad (1)$$

is to link individual observations of milk yield with parameters describing the milk yield in the herd and the properties of the cow. The observation vector Y_t consists of observed milk yields of all cows that have been milked at particular day t. θ_t is a vector of parameters composed of 3 sub-vectors so that $\theta_t=(\phi_t,\psi_t,\lambda_t)'$. The very first sub-vector ϕ_t contains information of the herd level milk yield (the nine parameters describing the shape of lactation curve at day t). The second sub-vector ψ_t contains parameters describing the individual cows in the herd (two parameters per cow). The third sub-vector λ_t describes the intervention effect for each cow. For non-treated cows, the intervention effect simply takes the value of 0. From the time when intervention takes place, the effect of intervention is expressed by the parameter λ. The remaining components of the observation equation are: the design matrix (F_t) and measurement error $v_t \sim N(\underline{0}, I\sigma_v^2)$, where 0 is a vector of zeros and I is identity matrix.

The system equation expresses how the parameter θ_t fluctuates over time. The following system equation was developed:

$$\theta_t = G_t \theta_{t-1} + w_t \qquad (2)$$

where G_t is a system matrix, $w_t \sim N(\underline{0}, W_t)$ is the evolution error describing the random fluctuation and W_t is a variance-covariance matrix. The effect of intervention according to the system equation is assumed to be constant over time (expressed by the coefficient in G_t and 0 for corresponding diagonal element of W_t).

The variance components used in the DLM were estimated using a maximum likelihood method.

2.3 Model run

Each day the milk yield was observed, a new distribution of the parameter vector θ_t was obtained from the Kalman filter (the description of the applied algorithm can be found for example in Dethlefsen, 2001). Values obtained from the Kalman filter for the mean and the variance-covariance matrix were used to compute mean and the standard deviation of intervention effect.

The R statistical software (R Core Team, 2014) was used for data analyses and plotting.

2.4 Data from the herd

In a Danish dairy herd with 220 Danish Holsteins and 4 AMS units an EVOP was conducted from March to June 2015. The herd manager wanted to explore the possibility of reducing the amount of concentrate provided to the cows in the AMS and instead supply this amount of energy in the partial mixed ration (PMR) provided to the cows at the feed bunk. This would be economically sound, because energy provided in the PMR is cheaper than energy provided as concentrate fed separately in the AMS. The EVOP consisted of two steps. On the 23th of March 0.85 kg/cow/day of soy-based concentrate was removed from the AMS and the energy concentration in the PMR was increased, so the total average energy intake in MJ/cow/day remained constant. Subsequently the herd was split in 2 groups based on AMS. In the Group 1 (2 AMS) the energy provided to the cows was sequentially reduced (April 21th and May 20th) by 0.2 kg of barley and 0.2 kg of soybean. The Group 2 (2 AMS) served as a control group that was given the same PMR and remained at the original level of concentrate of the AMS. Therefore, cows from the Group 1 would have to increase the intake of PMR to keep the same energy intake as the Group 2. Evaluation of the steps of the EVOP was based on changes in raw milk production (kg). Milk yield was monitored on daily basis, through data from the AMS units.

3. Results and discussion

The constructed DLM was used for monitoring milk production with special emphasis on estimating the effect of planned intervention (change in the feeding strategy) in order to evaluate it and to decide whether the farmer could reduce the energy supplied in the AMS even more.

Given the model formulation (the milk yield of particular cow was defined by 3 sets of parameters: the milk yield at a herd level, the individual cow level and the intervention effect) separate filtered mean values for herd, animal and intervention effect were obtained. In this study we will concentrate on the obtained mean values for the intervention effect together with confidence intervals.

An example from the model run is shown on the Figure 1. In this case only the intervention on the 20th of May was evaluated.

After the amount of concentrate provided from the AMS was reduced (on day 20th of May) the initial value for intervention parameter was introduced. From this point, each the time milk yield was observed the DLM model filtered the collected data and updated the information on the distribution of the intervention parameter.

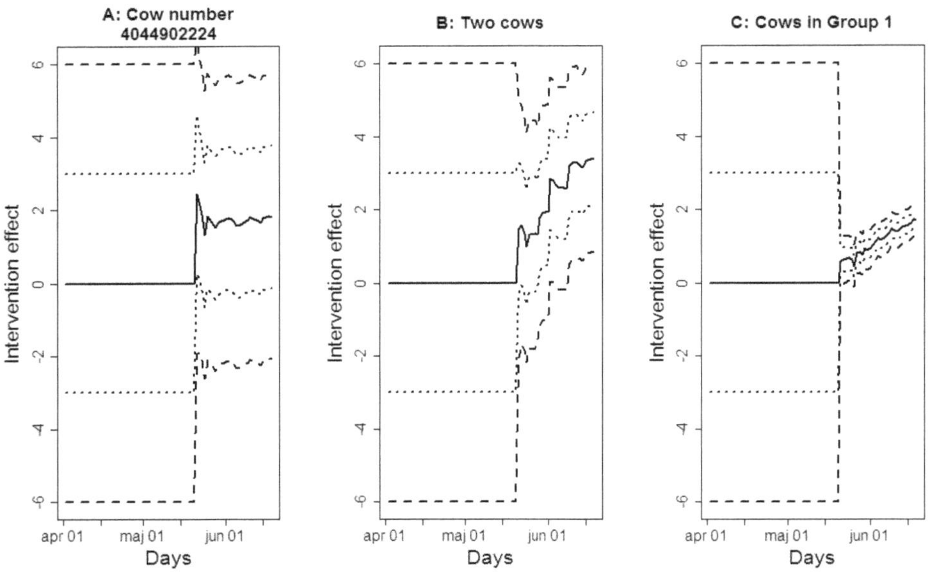

Figure 1. The example of estimated intervention effect (solid line) together with confidence intervals (± standard deviation dotted line and ±2 standard deviation dashed line) for: (A) selected cow; (B) two selected cows; and (C) whole AMS group. In this plot only estimates for cows from Group 1 are presented.

For example for the cow with number 4044902224 the value of intervention parameter for almost whole duration of the experiment oscillated around 2 (Figure 1A). However, acquired confidence limits of intervention effect were very wide. Therefore, information obtained from that particular cow was not sufficient to conclude that the effect of intervention was significant.

With two cows observed (Figure 1B), the mean value of intervention effect was much above 3, and lower confidence limit 19 days after the intervention exceeded the value of 0. Therefore, at the end of experiment we could conclude that the effect of intervention on milk yield was positive and significant at the 5% level.

As expected, when observing more animals (Figure 1C – selected AMS group) the precision of the estimates for the intervention effect increased. In this example, already after 6 days from the intervention it was possible to determine that the changes in feeding had significant positive impact on the milk production.

Based on the obtained results, we concluded that reducing the amount of concentrate provided to the cows in the AMS and instead supplying this amount of energy in the Partial Mixed Ration has positive influence on milk production and can be applied at the studied farm.

As demonstrated on the above example, the constructed tool is very flexible, allowing a user to assess the effect of intervention based on observation of any number of animals. Moreover, the estimates are obtained every time new data on milk yield have been recorded. Therefore, an updated assessment of the intervention is available to the manager on a daily basis. Also the precision of calculated intervention effects is known and reported which allow manager to assess the credibility of mean values.

The presented tool could be used by dairy farm managers on a daily basis to assess what is the effect of small changes in the used factors (like feeds) on milk production. However, to be applicable on other farms, a quantitative modeling of farms specific parameters (e.g. parameters describing a milk yield curve) should be performed.

References

Box, G.E.P. (1957). Evolutionary operation: a method for increasing industrial productivity. Journal of the Royal Statistical Society 6: 81-101.

Dethlefsen, C. (2001). Space time problems and applications. PhD thesis, Aalborg University, Aalborg, Denmark.

Nielsen, L.R., Jørgensen, E., Kristensen, A.R. and Ostergaard, S. (2010). Optimal replacement policies for dairy cows based on daily yield measurements. Journal of Dairy Science 93: 75-92.

R Core Team (2014). R: a language and environment for statistical computing. R Foundation for Statistical Computing, Vienna, Austria.

Verstegen, J.A.A.M., Huirne, R.B.M., Dijkhuizen, A.A. and Kleijnen, J.P.C. (1995). Economic value of management information systems in agriculture: a review of evaluation approaches. Computers and Electronics in Agriculture 13: 273-288.

West, M. and Harrison, J. (1997). Bayesian forecasting and dynamic models. Springer-Verlag, New-York, NY, USA.

Templates to classify progesterone profiles, built using real-time milk progesterone measurements

M. Derks[1,2,3*], P. Blavy[4], J.K. Höglund[5] and N.C. Friggens[4]
[1]*INRA, UMR1213 Herbivores, 63122 Saint-Genès Champanelle, France;* [2]*Clermont Université, VetAgro Sup, BP 10448, 63000 Clermont-Ferrand, France;* [3]*Université de Lyon, VetAgro Sup, UMR1213 Herbivores, 69280 Marcy l'Etoile, France;* [4]*AgroParisTech, UMR Modélisation Systémique Appliquée aux Ruminants, 16 rue Claude Bernard, 75231 Paris, France;* [5]*Aarhus University, Faculty of Science and Technology, Department of Animal Science, Blichers Allé 20, 8830 Tjele, Denmark; marjolein.derks@clermont.inra.fr*

Abstract

This study aims to use data derived from progesterone measurements in milk to build templates for classifying progesterone profiles. Between 2002 and 2006 milk progesterone sample data was collected daily on a Danish research herd. Data was smoothed and 1,418 oestrus cycles were identified. To align oestrus cycles, a model consisting of a base length and a quadratic curve was fitted to each cycle. The cycle was cut at the intersection of base and quadratic curve, leaving the part where progesterone rises and again drops, the luteal like phase (LLP). LLPs were divided in five quantiles based on length (Q1-Q5, in days) and, within each length class, clustered on LLP shape. For each LLP cluster upward slope, downward slope, average of the highest 5 progesterone values, and progesterone concentration on day 5 of the median cluster shape were identified, summarized, and used to build templates. LLPs in Q2 and Q3 have the fastest rise of progesterone, the highest peak levels, and the steepest slopes compared to other quantiles. There were no significant differences found between Q2 and Q3. Compared to Q1, LLPs in Q2 had a shape that is 1.068 times steeper and 1.048 times higher. LLPs in Q3 were 1.053 times steeper and 1.018 times higher. LLPs in Q4 were 0.977 times steeper and 0.973 times higher than LLPs in Q1. These templates can be used to create knowledge about differences in progesterone profiles between farms, breeds, cow phenotypes or management systems. This is a first step towards creating benchmarks (e.g. identify problems just by analyzing the progesterone profiles of a specific farm, or know the implications on progesterone by choosing a certain management system or phenotype). This would greatly support everyone working within the dairy industry. The full results of this study will be published in Theriogenology in 2016.

1. Introduction

Maintaining good fertility in dairy cows is one of the biggest challenges dairy farmers are facing. One way to visualize dairy cow fertility during lactation, is by looking at the progesterone profile. The profile can be used to see whether a cow is in oestrus, or pregnant (Bulman and Lamming, 1978). Earlier studies that examined progesterone profiles discarded abnormal cycles according to a priori set rules (Gorzecka *et al.*, 2011; Royal *et al.*, 2000), that were often based on older research. Given the increase in deviant profiles with increasing milk yield and dry matter intake of dairy cows over the past decades, there is a need to reevaluate the standard progesterone profile. This study aimed to look at progesterone profiles, using a naïve data-driven approach. This allowed us not only to gain a better understanding of the variability in shapes and features of the profile, but also to build new templates, which in the future can be used as a benchmark for other studies.

2. Methods

Between 2002 and 2009 milk progesterone sample data was collected daily on a Danish research herd. Data was smoothed and 1,418 oestrus cycles were identified. To align oestrus cycles, and to be able to discard post-partum anoestrus, a model consisting of a base length and a quadratic curve was fitted to each cycle. The cycle was cut at the intersection of base and quadratic curve, leaving the part where progesterone rises and again drops, the luteal like phase (LLP) (Figure 1).
LLPs were divided into five quantiles based on length (Q1-Q5, in days) and, within each length class, clustered on LLP shape. For each LLP cluster upward slope, downward slope, average of the highest 5 progesterone values, and progesterone concentration on day 5 of the median cluster shape were identified, summarized, and compared using a linear mixed model where quantiles were included as fixed classes, and cow ID and sampling year as a random variable. Results were used to build templates.

3. Results

LLP lengths ranged from 11.2 days to 211.4 days (Q1: 11.2-18.9; Q2: 18.9-21.2; Q3 21.2-23.7; Q4: 23.7-31.2; Q5: 31.3-211.4). The LLPs in Q1-4 showed mostly regular profiles (Figure 2). They show either a continuous rise and fall (Figure 2 Q1 B), or an undulating rise and continuous fall (Figure 2 Q2 B), or a continuous rise and fall with evidence of a plateau (Figure 2 Q3 D). Profiles in Q5 showed a profile shape indicating prolonged luteal phases/embryonic death. These profiles were not used for templates, since they may have been inseminated successfully, after which pregnancy was terminated.
Figure 3 shows the medians of Q1-4 plotted in the same plot. The first plot shows the unscaled medians, plot two shows the same medians after normalizing for LLP height and length. From these plots we can conclude that the LLPs in the different quantiles have the same base shape, and that variance can largely be explained by LLP height and length, not by slope.
Table 1 shows the differences in profile features for LLPs in Q1-4. Table 2 shows the significance of the differences between quantiles. LLPs in Q2 and Q3 had the fastest rise of progesterone, the highest peak levels, and the steepest slopes compared to other quantiles. There were no significant differences found between Q2 and Q3. Compared to Q1, LLPs in Q2 had a shape that is 1.068 times steeper and 1.048 times higher. LLPs in Q3 were 1.053 times steeper and 1.018 times higher. LLPs in Q4 were 0.977 times steeper and 0.973 times higher than LLPs in Q1.

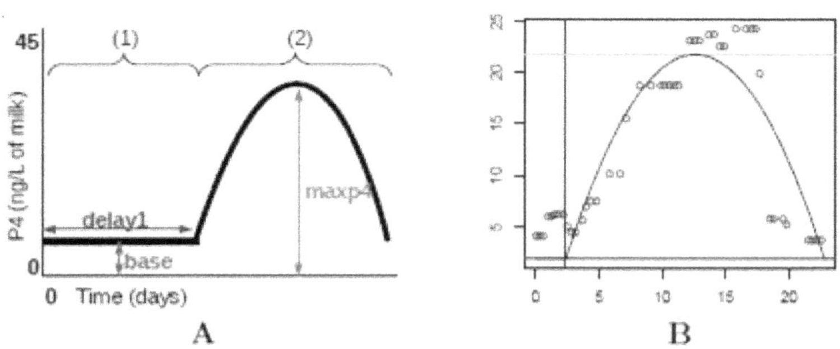

Figure 1. (A) Model used to detect delay before cycle. (B) An example of model adjustment. o = raw progesterone values, — adjusted model.

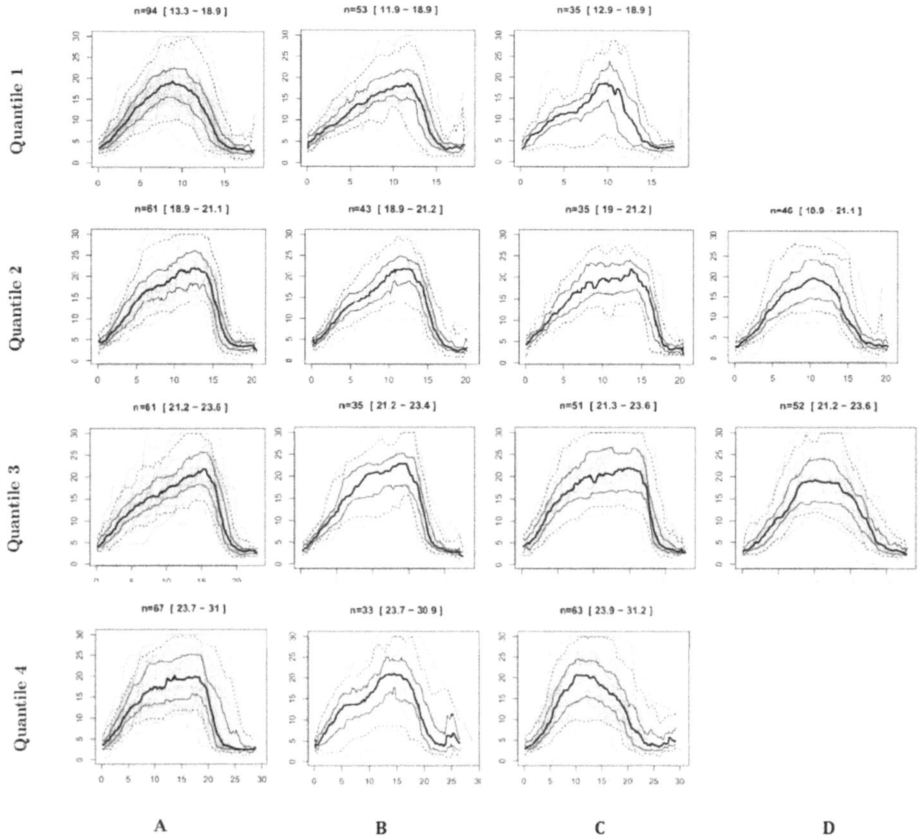

Figure 2. Progesterone profile clustered on cycle shape per length quantile, with days on the X-axis and progesterone in ng ml-1 on the Y-axis. Only clusters where n>30 are shown. Progesterone (ng/ml of milk) is plotted as a function of time (days) with ▬ median, – 25% and 75% percentiles, ····· 5% and 95% percentiles. Plots are titled n=a [b; c] where *a* is the number of LLPs in this cluster, *b* the length of the shortest LLP in this cluster, and *c* the length of the longest one.

4. Discussion

The main results of this paper were that the majority of progesterone profiles have a regular shape, even the ones with a cycle length longer than 24 days, and that variability between profile shapes is mainly explained by profile height and length, so not by a difference in base shape.

Understanding why fertility is failing is difficult, because fertility is a chain of events. If oocyte development is impaired, pregnancy rate will be impaired. Also, factors known to influence fertility (e.g. milk yield and negative energy balance) target fertility at different levels: impaired oocyte development, decreased chance of ovulation, impaired embryo development are examples. Using progesterone profiles to evaluate fertility is interesting because it allows us to detect where the profile is deviant, and to act accordingly. However, in order to be able to do so, we need a clear understanding of what is deviant and what is not. So far, a priori rules have been used which are often based on previous knowledge. These may not always be correct. One example is the optimum

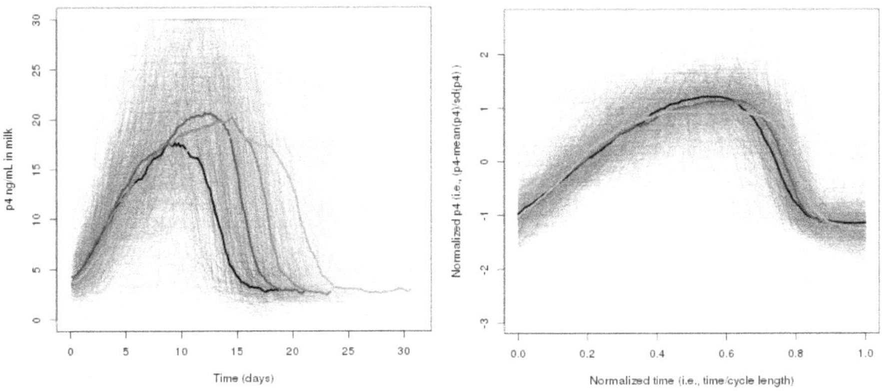

Figure 3. Median plot of Q1-4, unscaled and normalized.

Table 1. Cycle features per quantile.

Statistics	All cycles		Quantile 1		Quantile 2	
	n	median (min-max)[1]	n	median (min-max)[1]	n	median (min-max)[1]
Progesterone slope between d1-10	1,048	1.31 (-4.08-8.51)	212	1.33 (-3.18-5.26)	207	1.42 (-3.58-1.40)
Progesterone concentration d5	1,060	13.39 (0.74-29.5)	212	13.68 (2.96-25.44)	208	14.22 (4.96-14.83)
Progesterone slope end-2d until end-6d	803	-4.55 (-21.27-2.35)	194	-4.59 (-15.57-0.49)	181	-4.85 (-19.16-1.73)
Average 5 highest progesterone values	1,060	21.71 (3.75-30)	212	19.69 (4.66-30)	208	22.76 (7.24-30)

Statistics	Quantile 3		Quantile 4	
	n	median (min-max)[1]	n	median (min-max)[1]
Progesterone slope between d1-10	210	1.38 (-0.76-5.34)	210	1.28 (-1.57-4.67)
Progesterone concentration d5	212	13.64 (4.02-14.38)	216	13.06 (1.47-13.44)
Progesterone slope end-2d until end-6d	171	-4.75 (-16.34-0.96)	141	-3.83 (-13.95-1.83)
Average 5 highest progesterone values	212	22.11 (9.55-30)	216	21.12 (6.48-30)

[1] In order to minimize the effect of outliers, for the min and max the average of the lowest resp. highest five values was used.

Table 2. Linear mixed model estimates (β) of differences in cycle feature between quantiles. Differences are expressed relative to quantile 1 (left hand columns) or relative to quantile 4 (right hand columns). Cow ID and sampling year were included as random effects.[1]

Variable	Quantile	β	SE	DF	t-value	*P*-value
Progesterone slope between d1-10	Q1 (intercept)	1.25	0.08	705.30	16.07	<0.01
	Q2	0.14	0.10	834.20	1.37	
	Q3	0.25	0.10	838.10	2.51	<0.05
	Q4	0.13	0.10	814.10	1.24	
Progesterone concentration d5	Q1 (intercept)	13.74	0.34	733.00	33.87	<0.01
	Q2	1.14	0.44	804.30	2.59	<0.01
	Q3	0.36	0.45	825.90	0.81	
	Q4	-0.47	0.46	848.00	-1.01	
Progesterone slope end-2d until end-6d	Q1 (intercept)	-4.54	0.30	533.40	-15.37	<0.01
	Q2	-0.62	0.35	686.60	-1.79	
	Q3	-0.34	0.36	685.10	-0.96	
	Q4	0.47	0.38	664.90	1.24	
Average 5 highest progesterone values	Q1 (intercept)	18.95	0.70	722.60	26.89	<0.01
	Q2	2.36	0.54	827.90	4.41	<0.01
	Q3	2.40	0.55	842.00	4.41	<0.01
	Q4	1.96	0.56	842.70	3.49	<0.01
Progesterone slope between d1-10	Q4 (intercept)	1.38	0.08	740.40	19.30	<0.01
	Q1	-0.13	0.10	810.10	-1.23	
	Q2	0.01	0.10	830.80	0.14	
	Q3	0.12	0.10	835.00	1.26	
Progesterone concentration d5	Q4 (intercept)	13.27	0.39	733.70	33.86	<0.01
	Q1	0.47	0.46	844.00	1.01	
	Q2	1.61	0.45	832.40	3.57	<0.01
	Q3	0.83	0.44	819.60	1.88	
Progesterone slope end-2d until end-6d	Q4 (intercept)	-4.07	0.32	616.30	-12.66	<0.01
	Q1	-0.47	0.38	661.00	-1.24	
	Q2	-1.10	0.38	678.00	-2.87	<0.01
	Q3	-0.82	0.38	682.30	-2.12	<0.05
Average 5 highest progesterone values	Q4 (intercept)	20.90	0.69	722.60	30.33	<0.01
	Q1	-1.96	0.56	838.70	-3.48	<0.01
	Q2	0.40	0.55	842.50	0.74	
	Q3	0.44	0.53	835.40	0.83	

[1] SE: standard error; DF: degrees of freedom.

for cycle length. In literature, cycle length is reported to last between 18-24 days, with a mean of 21 days and an optimum between 19-24 days (Royal *et al.*, 2000; Savio *et al.*, 1990). An increase in cycle length is assumed to lead to an increase in abnormal progesterone profiles (Royal *et al.*, 2000). In our study, the cycles that exceeded the maximum optimal cycle length of 24 days (Q4) did not show an increase in deviant progesterone profile shapes. When normalized for length and height, they were comparable to the shapes in Q1-3. If a priori rules were used, cycles longer than 24 days may have been discarded. In this study this would have led to discarding a large proportion of cycles which are not classified as deviant.

Meier *et al.* (2009) compared cycle shapes before with a smaller sample (28 cows). They found three shapes: a peaked profile with a distinguishable peak, a profile with a wavelike shape in the upward slope, and a profile with a plateau. This is comparable to the shapes we found. Also, Martin *et al.* (2013) and Friggens *et al.* (2008) identified shapes that were comparable to our normalized reference shapes.

The templates presented here can be used to create knowledge about differences in progesterone profiles between farms, breeds, cow phenotypes or management systems. This is a first step towards creating benchmarks (e.g. identify problems just by analyzing the progesterone profiles of a specific farm, or know the implications on progesterone by choosing a certain management system or phenotype). This would greatly support everyone working within the dairy industry.

References

Bulman, D.C. and Lamming, G.E. (1978). Milk progesterone levels in relation to conception, repeat breeding and factors influencing acyclicity in dairy cows. Journal of Reproduction and Fertility 54: 447-458.

Friggens, N.C., Bjerring, M., Ridder, C., Højsgaard, S. and Larsen, T. (2008). Improved detection of reproductive status in dairy cows using milk progesterone measurements. Reproduction of Domestic Animals 43: 113-121.

Gorzecka, J., Codrea, M.C., Friggens, N.C. and Callesen, H. (2011). Progesterone profiles around the time of insemination do not show clear differences between of pregnant and not pregnant dairy cows. Animal Reproduction Science 123: 14-22.

Martin, O., Friggens, N.C., Dupont, J., Salvetti, P., Freret, S., Rame, C., Gatien, J., Disenhaus, C. and Blanc, F. (2013). Data-derived reference profiles with corepresentation of progesterone, estradiol, LH and FSH dynamics during the bovine estrous cycle. Theriogenology 79: 331-343.

Meier, S., Roche, J.R., Kolver, E.S. and Boston, R.C. (2009). A compartmental model describing changes in progesterone concentrations during the oestrus cycle. Journal of Dairy Research 76: 249-256.

Royal, M.D., Darwash, A.O., Flint, A.P.F., Webb, R., Woolliams, J.A. and Lamming, G.E. (2000). Declining fertility in dairy cattle: changes in traditional and endocrine parameters of fertility. Animal Science 70: 487-501.

Savio, D., Boland, M.P. and Roche, J.F. (1990). Development of dominant follicles and length of ovarian cycles in post-partum dairy cows. Journal of Reproduction and Fertility 88: 581-591.

Enhanced accuracy of cows' retention, breeding, and milking values with precision dairy technologies

A. De Vries
Department of Animal Sciences, IFAS, University of Florida, 2250 Shealy Drive, Gainesville, FL 32608, USA; devries@ufl.edu

Abstract

Accurate future cash flow projections are needed for the decision to keep or cull, continue or stop breeding eligibility, and continue or stop milking. Existing models lack accurate cash flow projections for individual cows because attributes such as milk components, detailed reproductive status, and body weights are often generalized. An improved model is needed. Precision dairy farming technologies such as real time milk analyzers and body weight scales may provide inputs but their importance for decision values is unknown. A Markov chain model was developed to predict cash flow projections for each cow and her replacements. Individual cow attributes include lactation curves for yield, fat and protein; probabilities of conception; current reproductive status; and day of voluntary culling if open (among others). The model was parameterized and run for 498 lactating dairy cows from one Florida farm. Retention, breeding eligibility, and milking values were calculate for each cow. A sensitivity analysis included voluntary culling criteria ($1 or $2 income over feed cost per day), and presence or absence of cow-specific fat percentages and body weights. Median retention, breeding eligible, and milking values were $800, $325, and $1,790 when voluntary culling criterion was $1/day. Spearman rank correlations were >0.950 when culling criterion was $2/day or when cow-specific fat percentages and body weights were absent. Maximum absolute differences for individual cows were $312, $327, and $271 when cow-specific fat percentages and body weights were absent vs present. Practical implications include that voluntary culling criterion and inclusion of individual fat percentages and body weights had generally small effects on the rankings of cows. Effects for the cows of interest may be greater, however. The developed model allows for inclusion of accurate future cash flow projections that may be enhanced by precision dairy technologies. One straightforward extension is the inclusion of disease treatment values.

Keywords: culling, do-not-breed, dry off, decision

1. Introduction

Decisions regarding which cows to cull, put on a do-not-breed list, or dry off are often based on rules of thumb and emotion without a good enumeration of the economic consequences. These decisions are generally not financially optimized because the effects of the various alternatives (keep vs cull, breed vs do not breed, milk vs dry off) on cash flows are not easily determined with accuracy. A comparison of the cash flows from the various alternatives gives a monetary value of one alternative decision over another. The optimization criterion is generally maximized profitability for a slot (place for an animal) on the farm.

A retention value (also known as a keep value, or retention pay off, or cow value or future value) is the difference between the decisions to keep the current cow longer vs culling her followed by replacing her with another cow. Cows with the lowest retention value are the best candidates for culling in the herd. A breeding eligibility value is the difference between the decision to continue trying to breed (inseminate) the current cow vs the decision to not trying to breed her but still keeping her in the herd. Cows with the lowest breeding eligibility values are the best candidates to put on do-not-breed lists. A milking value (also known as a 'wet' value) is the difference between

the decisions to keep milking the cow vs drying her off now. Generally this applies to pregnant cows and the decision is made in the later stage of gestation. Cows with the lowest milking value are the best candidates to dry off.
Of these three related values, calculation of the retention value has received the most attention in the literature with the first publications at least 50 years ago (for older references see for example De Vries (2004) and Kalantari *et al.* (2014)). Because the problem is one of sequential decision making, the problem has been formulated and solved using dynamic programming.
In practice, the assumptions held in the dynamic programming models are seldom met and not all cows of interest are represented. Future decisions may not have to be optimized, however. Without the dynamic programing algorithm to optimize future decisions, the algorithm can be simplified (e.g. Cabrera, 2012) and the relative ranking of cows remains very similar (e.g. Kalantari *et al.*, 2014). Cows with particular long lactations, which are of particular interest for culling, may not be represented in the existing dynamic programming models either. Further, detailed information about cow performance, such as differences in fertility (e.g. Shahinfar *et al.*, 2014) is becoming available and is useful for more accurate cash flow projections and decisions.
It appears, therefore, that practical models do not need to necessarily optimize decisions into the future, but they should be able to accommodate accurate predictions of future cow performance when available, such as those more routinely collected with precision dairy farming technologies. Precision dairy farming technologies such as real time milk analyzers and body weight scales may provide inputs but their importance for retention, breeding eligibility and dry-off decision values is unknown.
The first objective of this paper is, therefore, to describe a model that is flexible enough that it can incorporate detailed cow performance predictions for individual cows, such as individual milk yield and milk component predictions, fertility, body weights and feed intakes. The second objective is to evaluate the effects of precision dairy farming technologies such as real time milk analyzers and body weight scales on the retention, breeding, and dry-off decision.

2. Model for cash flow projections

We modified an existing herd budget model (De Vries *et al.*, 2016) to predict cash flow projections for each cow and her replacements. Briefly, the herd budget model mimics the technical and economic performance of a herd of young stock and cows given a calving heifer. Animal flow through the herd is modelled by Markov chains which determine the daily probability an animal is in a state characterized by parity, days since calving, and days pregnant. Approximately 732,000 states for cows are possible, depending on the length of the insemination period. Transition from state to state is calculated by the probabilities of culling, conception, abortion, and calving. Technical performance in a state is determined by technical inputs such as milk production curves, feed intake functions, body weights, probabilities of insemination, conception, abortion, and culling, and management inputs such as lengths of the insemination period and dry period. Economic performance in a state is calculated from the technical performance and economic inputs such as prices of milk, culled and sold animals, feed, and various other variable costs. Finally, herd outputs are calculated by summations of the multiplication of the technical and economic performance in each state by the fraction of animals in that state.
The herd budget model was originally designed to calculate effects of changes in herd input parameters such as probabilities of conception, lactation curves, or culling risks. At the start of the simulation of a cow was a calving heifer. However, the model was modified (called Calculator now) to start the simulation with an existing cow in the herd and tweak some of the input parameters such that the relevant cash flow for the remainder of her life is calculated. For example, to calculate the cash flow for a cow no longer eligible to get bred, the probability of insemination was set to zero. In addition to determining the cash flow for each individual cow, the cash flow for a replacement heifer was calculated.

Over time, the current cow has a daily probability of getting culled, and the same fraction of the cash flow from a replacement heifer is be added. This method results in a full slot, which consists in a fraction of the current cow and associated cash flow, and a remainder fraction of the replacement heifers with her associated cash flows. This principle is similar to the principle described by Cabrera (2012). Each of the three decision values requires the calculation of two cash flows each (keep vs cull, breed vs do not breed, milk vs dry off). Because the herd budget model has many input parameters which also can be used to define the current cow, the future cow performance and resulting cash flows are simulated with much flexibility.

3. Application and results

The model was parameterized and run for 498 lactating dairy cows from one Florida farm. This farm milked twice per day and had automatic milk recording, Afilab (Afikim, Israel) fat and protein measurements (Kaniyamattam and De Vries, 2014), and Afiweight body weight scales at the exit of the parlor. First, a report for the 498 lactating cows was generated in the Afifarm herd management program with relevant cow attributes (parity, days in milk, days since breeding, known pregnancy status, current production of milk, fat and protein, and body weights among others) and saved as an Excel file. This report was imported into the Calculator model. Additional herd input parameters were added, such as milk prices. In addition, input parameters were added to simulate an average default cow. In the current application, the input parameters of the default cow served to both simulate the cash flows from the replacement heifer, and to simulate the default values for each of the 498 cows if no data were available. For example, if body weigh data were absent, the Calculator would assume the average body weight curves. Fat measurements affected dry matter intakes (and hence feed costs) which was dependent on fat-corrected milk, and on the milk price which in Florida is a function of milk and fat but not protein. Body weights also affected dry matter intakes and cull values. Individual cow performance predictions were kept simple. For example, the last daily milk yield was compared to the daily milk yield predicted for the default cow at the same parity and days in milk. The ratio was then multiplied with the daily milk yield from the default lactation curves. The same principle was used to predict fat and body weight performance for the current cow into the future. The milking values (keep milking until regular dry off at 220 days of gestation, versus dry off earlier) did not include any effects of earlier dry off on cow performance in the current application. Open cows continued to be inseminated into the future (unless the cash flow was for the cull or do not breed decision) until her milk income minus feed cost decreased below $1/day in the default situation at which time she would be culled.

Retention, breeding eligibility, and milking values were calculate for each of the 498 lactating cows (Figure 1). Median retention, breeding eligible, and milking values were $691, $767, and $1,929 when voluntary culling criterion was $1/day. Eight open cows had retention values of -$999 which was assigned because their milk income minus feed costs were < $1. Breeding eligibility values of pregnant cows were low because only in the case of a future abortion in the current lactation is there an opportunity cost of no longer breeding a then open cow. Milking values were generally much greater early in lactation for open cows than for pregnant cows later in lactation.

Figure 2 shows the differences (y-axis) from the retention, breeding eligibility and milking values when individual fat percentage and body weights are included in the cash flow projections (x-axis) versus when default values are used. The differences can be several hundreds of dollars. Especially the breeding eligibility values are sensitive to individual measurements of fat percentage and body weights. The availability of individual fat and body weight measurements had little effect on the ranking of cows for the three values. Spearman rank correlations were >0.950 when culling criterion was $2/day or when cow-specific fat percentages and body weights were absent. Maximum absolute differences for individual cows were $312, $327, and $271 when cow-specific fat percentages and body weights were absent versus present.

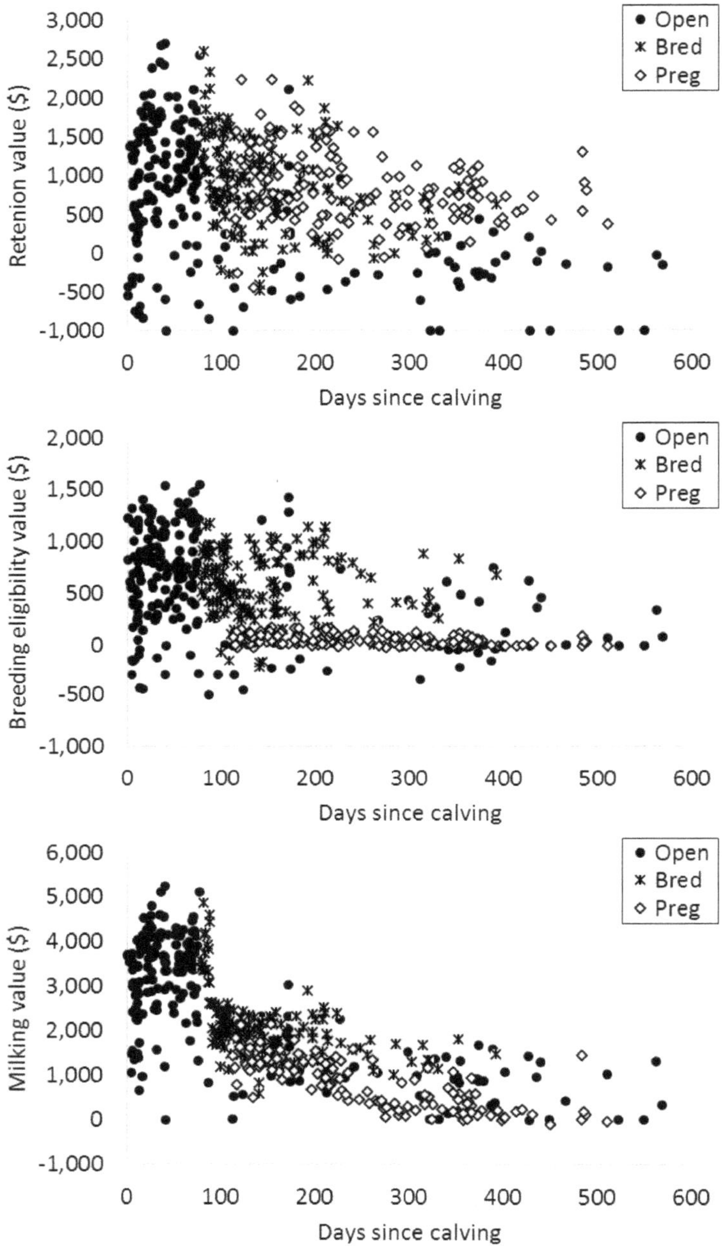

Figure 1. Retention, breeding eligibility, and milking values for 498 cows from one Florida farm. Included in the cash flow projections underlying the three values are individual daily milk yield, fat percentages and body weights obtained with precision dairy farm technologies.

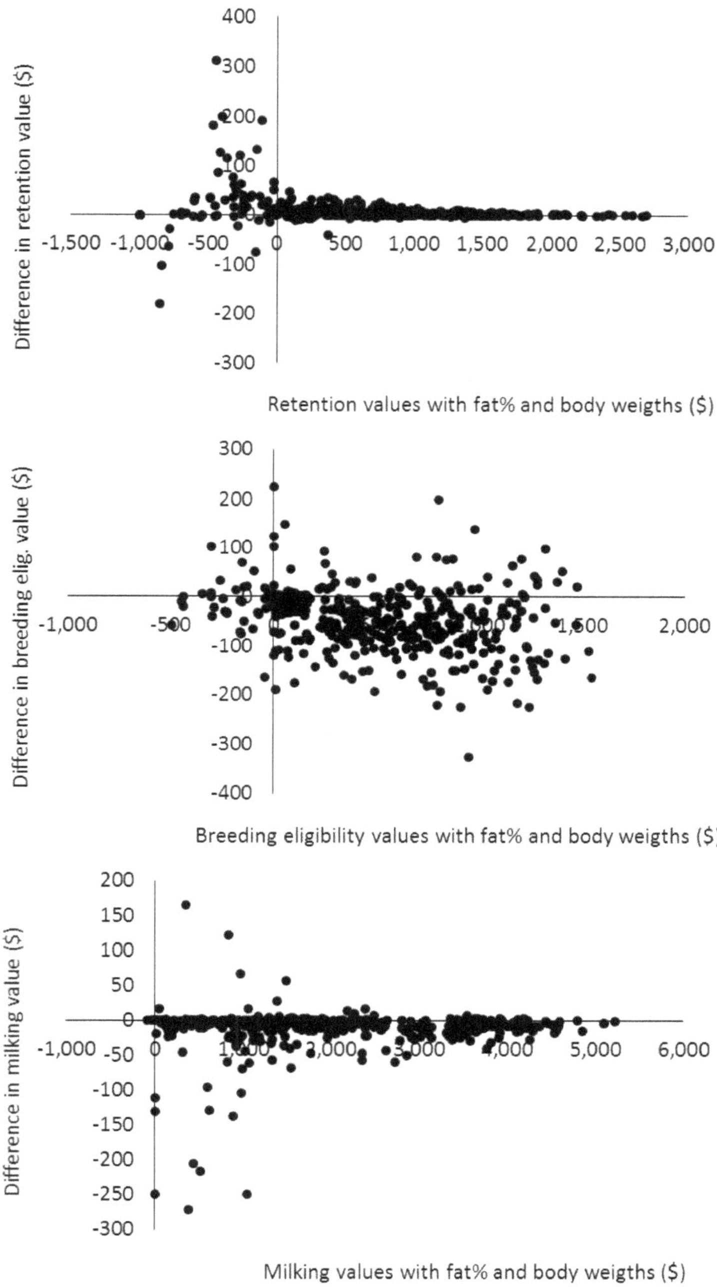

Figure 2. Differences in retention, breeding eligibility, and milking values for 498 cows calculated with and without individual measures of fat percentage and body weight.

4. Implications and conclusions

The Calculator cash flow projector is a flexible tool that allows for accurate predictions of future cash flows for culling, do-not-breed, and dry off decisions. The limited sensitivity analysis shown here revealed that individual fat percentages and body weights affect the decision values but differences in ranking of cows were small. The accuracy of the decision values depends greatly on the accuracy of methods to predict future cow performance. For example, fat and body weight predictions in the current application were made based only on the last fat and bodyweight observations. Fat percentages are quite variable from day to day (Kaniyamattam and De Vries, 2014) and body weights also vary greatly. Missing observations are also common. Future cow performance predictions may be improved when predictions are based on more historical data or correlated variables or both. For example, the accuracy of milk yield predictions may improve when they are based on more past performances such as the Bestprediction model from USDA (Cole *et al.*, 2009). Shahinfar *et al.* (2014) used machine learning techniques to çombine over 20 variables to best predict fertility for individual dairy cows. These authors then used these fertility predictions to calculate the value of delayed inseminations on less fertile cows. The Calculator model is also able to evaluate delaying inseminations (not shown). One straightforward but not very simple extension is the inclusion of disease treatment values. The current Calculator model is used monthly to make culling decisions on a 2,200 cow Florida dairy farm. No novel precision dairy farming technologies are used. In conclusion, the Calculator model allows for inclusion of accurate future cash flow projections that may be enhanced by precision dairy farming technologies.

References

Cabrera, V.E. (2012). A simple formulation and solution to the replacement problem: a practical tool to assess the economic cow value, the value of a new pregnancy, and the cost of pregnancy loss. Journal of Dairy Science 95: 4683-4698.

Cole, J.B., Null, D.J. and Van Raden, P.M. (2009). Best prediction of yields for long lactations. Journal of Dairy Science 92: 1796-1810.

De Vries, A., Dechassa, H. and Hogeveen, H. (2016). Economic evaluation of stall stocking density of lactating dairy cows. Journal of Dairy Science 99: 3848-3857.

Kalantari, A.S., Cabrera, V.E. and Solis, S. (2014). A comparison analysis of two alternative dairy cattle replacement strategies: optimization versus simulation models. Economía Agraria Volumen 18: 12-24.

Kaniyamattam, K. and De Vries, A. (2014). Agreement between milk fat, protein, and lactose observations collected from the Dairy Herd Improvement Association (DHIA) and a real-time milk analyzer. Journal of Dairy Science 97: 2896-2908.

Shahinfar, S., Page, D., Guenther, J., Cabrera, V., Fricke, P. and Weigel, K. (2014). Prediction of insemination outcomes in Holstein dairy cattle using alternative machine learning algorithms. Journal of Dairy Science 97: 731-742.

Shahinfar, S., Guenther, J.N., Page, C.D., Kalantari, A.S., Cabrera, V.E., Fricke, P.M. and Weigel, K.A. (2015). Optimization of reproductive management programs using lift chart analysis and cost-sensitive evaluation of classification errors. Journal of Dairy Science 98: 3717-3728.

Comprehensive analysis of AMS data: a cluster-graph approach tested on a dairy farm

F. Bonora*, S. Benni, A. Barbaresi, P. Tassinari and D. Torreggiani
Department of Agricultural Sciences, University of Bologna, V. le Fanin 48, 40127 Bologna, Italy; filippo.bonora@studio.unibo.it

Abstract

As it is well known, the introduction of AMSs has deeply changed the barn layout and herd management in dairy farms. AMSs measure and record specific data, providing farmers with useful real-time information for each cow. At the same time, the remarkable amount of information stored in an AMS database has a great potential for herd characterization and management optimization. Within this challenging research topic, this study aims at developing and testing an innovative procedure for the comprehensive analysis of AMS-generated multi-variable time series. Cluster analysis can prove useful to examine the state of a dairy farm through a panoramic view, without loss of information about the single cows. In this study, cluster-graph analysis is applied to a dataset containing the time series of the following five parameters recorded for each cow in the summertime in an Italian barn: number of daily milking events, activity, parity, weight, and 'time regularity' of milking. Each variable has been used to cluster the herd using a k-means cluster algorithm. k-values have been selected to avoid outliers and characterize the herd as a whole. Then, this analysis has been joined with the graph theory, where each cow is represented by a node. Weights assigned to each edge have been calculated with a cow-similarity index. Finally, the final clusters have been computed. The combination of cluster analysis and graph theory has allowed to obtain the results automatically with known statistical and mathematical accuracy. The results can improve the characterization of the herd and lend support to cow monitoring, through the comparison between actual and expected cow behavior. Herd partitioning could help to regulate the number of milking or the supplementary feeding of specific groups. Integration with video/RFID position analysis could provide a comprehensive behavioral description. Further studies will focus on benefits for dairy barns design.

Keywords: precision livestock farming, graph theory, cluster analysis, herd characterization

1. Introduction

Information and Communications Technologies (ICT) are becoming increasingly widespread in farming – especially in barns for dairy cattle – and their application in precision livestock farming (PLF) has risen sharply in the last decades. It is well known that the main expected benefits deriving from PLF deal with the real time monitoring of animal welfare and alert to health problems and disease, aiming at an increase in milk production with a reduction in production costs, and an improvement of farmers' work conditions and quality of life.

In particular, the introduction of automatic milking systems (AMS) in the late Nineties has deeply changed barn layouts and herd management in dairy farms. In the last years, many studies have focused on data from AMS-equipped livestock farms (Cook *et al.*, 2005; Fregonesi and Leaver, 2001; Haley *et al.*, 2000).

Within those challenging research topics, the specific goal of the study is to develop and test a mathematical and computational model using AMS-generated data, designed with the aim of leading to an automatic clustering of the cows based on their production and behavioral features, with the definition of the corresponding levels of statistical precision.

2. Materials and methods

The methodology has been calibrated and tested on a dairy farm located in the municipality of Budrio, about 20 km North-East of Bologna (Emilia Romagna Region, Italy).
The barn is a rectangular building 51 m long and 23 m wide, with SW-NE orientation of the longitudinal axis. It consists of a storage area for straw and hay along the SE side, a resting area in the central part of the building, and a feeding area with an external feed delivery lane on the NW side. The areas accessible to cattle have a precast concrete slatted floor.
The resting area hosts 78 cubicles with straw beddings, where about 65 lactating and 13 dry Friesian cows are housed: two blocks of head-to-head rows are located in its central part, and another row runs along the entire length of the resting area near the storage lane. Dry cows are housed in the NE part of the two central rows of cubicles. The milk-room is located on the SW side of the building, where there are also an office and a technical plant room. Ventilation is controlled by three high-volume and low-speed (HVLS) fans with five horizontal blades, which are activated by a temperature-humidity sensor situated in the middle of the barn. Cow milking is performed by means of a Lely 'Astronaut A3 Next' robotic milking system, which is placed in the SW side of the barn.

2.1 Data acquisition

Cow-related and milk production data are recorded by the AMS each time a cow passes through the milking unit.
Based on the database downloaded from the AMS management software, we have created a matrix called 'visit', where each row corresponds to a cow passage, and the columns contain the following parameters:

- cow identification number (*Cin*);
- date and time of the cow passage (*t*);
- milk yield (*My*);
- days of lactation (*Ld*);
- feed intake (*Fi*) (supplementation with additional concentrates). Total daily amount is calculated and provided by the AMS based on milk yield and day of lactation;
- cow weight (*Cw*);
- parity (*Pa*).

Cow behavior data are acquired by means of a cow identification and activity sensor mounted on a collar, which monitors the activity levels (*A*) of a cow by means of an acceleration sensor measuring the duration and intensity of every movement of the animal. Cows' activity is recorded in 2-hour blocks. Cow activity data, used in livestock management as an early indication of cows' heat, are downloaded from the collars into the AMS each time a cow passes through the milking unit. Based on the activity data downloaded by the AMS management software, we have created an 'activity' matrix, where each row corresponds to a two-hour acquisition, and two columns contain the cow identification number and the mean activity of each cow. The 'visit' and the 'activity' matrices have been processed jointly.

2.2 Database creation

Since milking is voluntary with AMS, the frequency and time distribution of milking events and associated data acquisitions are highly variable. Therefore, datasets cannot be compared directly on a regular basis. Firstly, the rows of the 'visit' matrix with null milk yield or feed intake data have been deleted, since they corresponded to unsuccessful milking sessions (for example in case of stressed or uncomfortable cow kicking in the machine, or too high milking frequency) which would have led to an uneven rate distribution. Similarly, milking events closer than 4 hours for the same cow have been aggregated to allow a smoother data distribution. Then, milk production and activity data have been interpolated over 6-hours time steps (0-6 am, 6 am-12 pm, 12 pm-6 pm, 6 pm-12 am), to allow a comparison of synchronous data. This has been achieved by calculating for each cow the average hourly rate both of milk production (Lh) for each passage i (Equation 1), and cow activity (Ah) for each measurement j (Equation 2), and multiplying each hourly rate for the relevant time duration, and summarizing over each time step:

$$Lh_i = \frac{My_i - My_{i-1}}{t_i - t_{i-1}} \qquad (1)$$

$$Ah_j = \frac{A_j - A_{j-1}}{t_j - t_{j-1}} \qquad (2)$$

A further aggregation of the four above-mentioned time steps over each day has provided synchronous daily production and activity data for each cow.
The normality of the distributions of milk production (Figure 1A) and activity data (Figure 1B) for each cow has been tested and confirmed both graphically and by means of the Jarque-Bera test for fixed time steps.

2.3 Clustering

The cow clustering based on production and behavioral features has been focused on data surveyed in summer 2015 (from June 21 to September 30) and applied to the overall number of lactating cows which were bred in the barn in the considered study period, equal to 88 animals.
A k-means algorithm has been used to study the following parameters for each cow:

- number of milking events per day (*#M*);
- parity;
- average daily activity (dA);
- milking regularity, in terms of standard deviation of the time intervals between milking events in the study period (Mr);
- animal weight (W).

This method has allowed us to find k similar subset of the starting set in a statistical way (MacQueen, 1967). For each variable, different k values have been selected a posteriori to highlight some particular trends. By way of example, the results concerning the number of milking events per day and the milking regularity are reported in Table 1 and 2, respectively.
The clusters obtained through the k-means algorithm have been joined in a network graph through the software Gephi (0.9.1 version), an open source and cross-platform exploration tool for networks and complex systems (Bastian *et al.*, 2009).
The network has been designed assigning each cow a node, and linking two nodes if the cows belong to at least one common 'k-means cluster'.

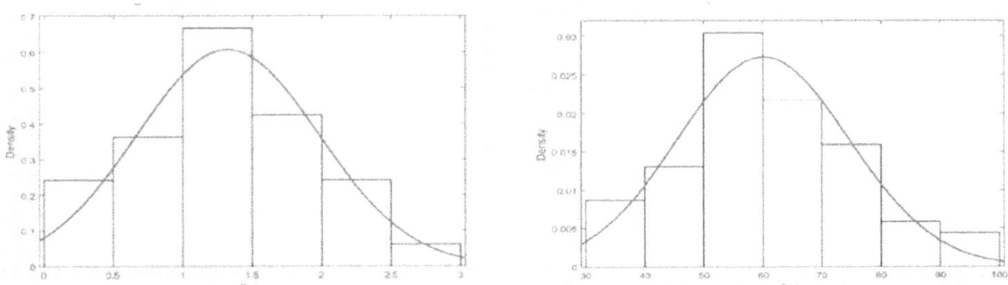

Figure 1. (A) Normality of the average hourly milk production in a generic time step; (B) Normality of activity data in a generic time step.

Table 1. Statistics about #M clusters.

Cluster	Cardinality	Min	Max	Centroid
C1_*#M*	10	1.2	1.7	1.5
C2_*#M*	38	1.8	2.3	2.0
C3_*#M*	30	2.3	2.8	2.6
C4_*#M*	10	2.9	3.7	3.1

Table 2. Statistics about *Mr* clusters.

Cluster	Cardinality	Min (h)	Max (h)	Centroid (h)
C1_*Mr*	4	0	1.3	0.7
C2_*Mr*	39	1.7	3.0	2.5
C3_*Mr*	26	3.2	4.4	3.7
C4_*Mr*	14	4.5	6.3	5.2
C5_*Mr*	5	7.0	9.6	7.7

The weight of the link between two nodes A and B is defined by means of an overall 'similarity index' (S), calculated for each parameter i as follows:

$$S_i = \frac{1\text{-}variable(A)_i\text{-}variable(B)_i}{centroid_i} \quad (3)$$

The graph has been then analyzed by means of the software to find subnetworks, i.e. graphs composed by subsets of nodes and edges from the starting network. This research is based on the modularity, a procedure that minimizes the number of edges from two different clusters (Newman, 2006).

3. Results and discussion

The three clusters resulting from the final modularity process are shown in Figure 2, where the graph is drawn based on the 'Force Atlas' layout (Repulsion strength=10.000, Gravity=400, adjustment by sizes). This is a homebrew layout of Gephi, very similar to Noack's edge-directed force layout. The size of each node has been drawn proportionally to the daily milk yield (dMy): the higher is the milk yield, the longer is the node radius. Parity is indicated as a number inside each circle.
The resulting subdivision of the herd shows a clear differentiation of each cluster from one another for what concerns all the considered descriptive parameters (Table 2).
The main diversification concerns parity as far as cluster 2 shows a mean of one, which indicates that the group includes only first-calf cows. The remaining two groups have similar mean parity, between two and three, and mostly differ in the number of milking events and milking regularity. The milk yield has not been selected as a parameter for clustering because it has been considered as a dependent variable according to which the effectiveness of the clustering procedure and the productive features of each resulting group of cows have been assessed. In fact, the clustering procedure has identified groups of animals with clearly different average productivity (Table 3).

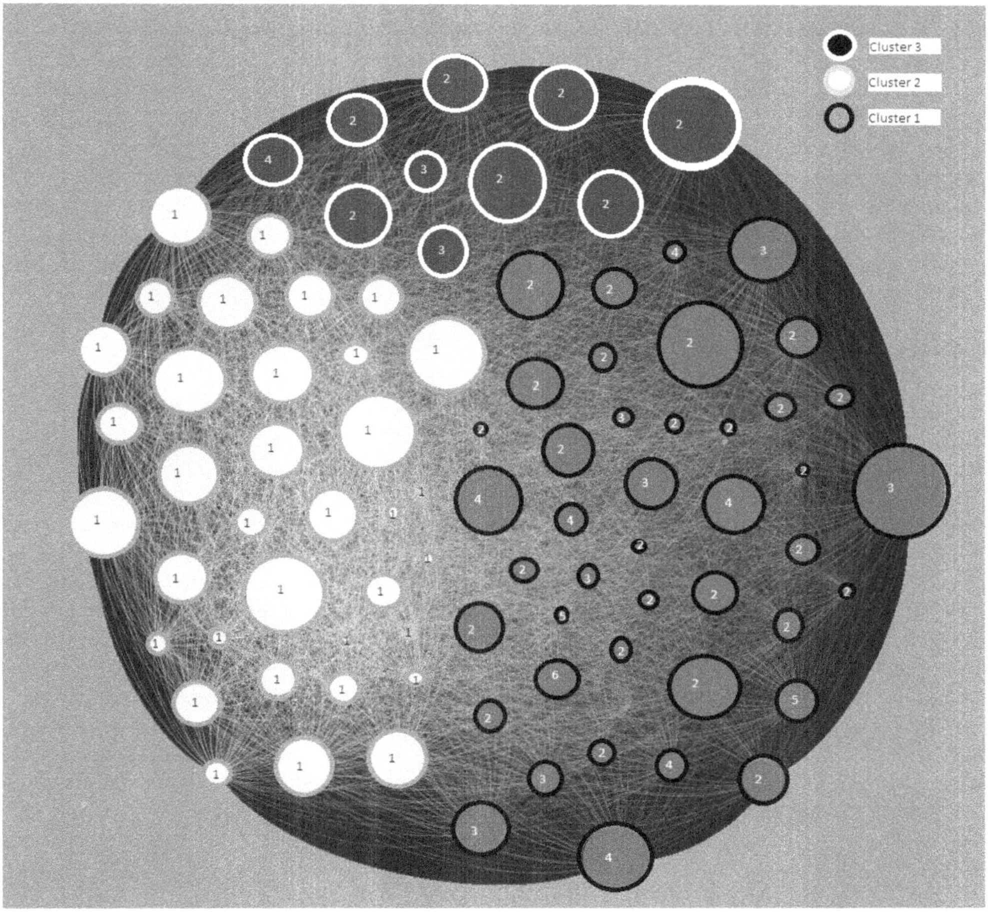

Figure 2. Final clusters obtained with Gephi, showing the herd in a graph representation.

Table 3. Statistics for the final clusters: mean (*m*) and standard deviation (*std*) of each variable.

Cluster	*#M*		*Pa*		*dA*		*dMy*		*Mr*		*W*	
	m	*std*	*m*	*std*	*m*	*std*	*m*	*std*	*m*	*std*	*m*	*std*
1	2.0	0.4	2.7	1.0	53.5	9.8	24.7	10.0	4.3	1.5	674.5	58.0
2	2.3	0.3	1.0	0.0	53.9	12.9	27.0	10.4	2.9	1.3	587.4	33.0
3	3.0	0.3	2.4	0.7	60.2	13.0	36.7	7.0	2.2	0.3	646.4	66.8

The results highlight that lactating cows have very diversified conditions and the cluster analysis is an effective tool to identify the most significant groups, provided that data about animal behavior and conditions are available, as it is the case of AMS equipped barns.

The results suggest also that particular attention should be paid to single animals whose milk yield is poorly consistent with the average value of the cluster they belong to. In particular, cows in cluster 1 showing high milk yield values should be monitored in order to prevent that their production decrease to the more common values of that cluster. In those cases, proper measures should be defined to increase the number of daily visits to AMS and their regularity in time. Another aspect deserving attention is represented by cows of cluster 3 showing milk production significantly below the cluster mean. Proper investigations should be performed on those animals to identify the causes of such a low milk yield and to identify possible corrective strategies. Finally, cows of cluster 2 with the lowest production rates should be checked to verify if it is due to their normal lactating curve pattern or other factors may hamper the expected productivity.

4. Conclusions

A cluster-graph analysis method has been applied to a dataset derived from AMS source containing the time series of the main parameters recorded for each cow in the summertime in an Italian barn. The combination of cluster analysis and graph theory has allowed to obtain objective results on the basis of statistical and mathematical methodologies. The herd has been characterized according to three final clusters, which proved diversified from one another in terms of productivity and animal behavior.

The results can lend support to cow monitoring, through the comparison between the actual and the expected cow behavior. Herd partitioning could help to regulate the number of milking event or the supplementary feeding of specific groups. The identification of the clusters may indeed contribute do define feeding strategies based not only on the milk yield and lactation period, but also on the other descriptive variables, which have proved to be effective in characterizing the herd groups.

The integration of the approach proposed with video or RFID position analysis could provide a comprehensive behavioral description of dairy cattle in a farm. Further developments of the research are focusing on the possible implications of cows clustering for innovations in the definition of spatial layouts, with expected benefits for the design of dairy barns.

References

Bastian, M., Heymann, S. and Jacomy, M. (2009). Gephi: an open source software for exploring and manipulating networks. 3rd International AAAI Conference on Weblogs and Social Media: 361-362.

Cook, N.B., Bennett, T.B. and Nordlund, K.V. (2005). Monitoring indices of cow comfort in free-stall-housed dairy herds. Journal of Dairy Science 88(11): 3876-3885.

Fregonesi, J.A. and Leaver, J.D. (2001). Behaviour, performance and health indicators of welfare for dairy cows housed in strawyard or cubicle systems. Livestock Production Science 68(2-3): 205-216.

Haley, D.B., Rushen, J. and De Passillé, A.M. (2000). Behavioural indicators of cow comfort: activity and resting behaviour of dairy cows in two types of housing. Canadian Journal of Animal Science 80(2): 257-263.

MacQueen, J.B. (1967). Some methods for classification and analysis of multivariate observations. In: Proceedings of 5th Berkeley Symposium on Mathematical Statistics and Probability. University of California Press, Berkeley, CA, USA, pp. 281-297.

Newman, M.E.J. (2006). Modularity and community structure in networks. PNAS 106(23): 8577-8582.

Identifying sensor error and real variations in the bovine rumen pH, using the eCow farmBolus

T.T.F. Mottram[1,2*] and D. Bradley[2]
[1]*Royal Agricultural University, School of Agriculture Food and Environment, Stroud Road, GL7 6SJ Cirencester, United Kingdom;* [2]*eCow Devon Ltd., King St Business Centre, EX1 1BH Exeter, United Kingdom; toby.mottram@rau.ac.uk*

Abstract

Since wireless monitoring of rumen pH began in 2010 thousands of datasets have been collected from the rumen reticulum of cows. Determining the difference between accurate and error data is a challenge without the ability to calibrate a sensor. Maintaining a stable reference electrode within the rumen is a challenge and identifying when the reported pH value is no longer accurate is a key objective for bolus designers. The typical eCow farmBolus now provides data for over 150 days but data is only regarded as accurate to 90 days. Defining when the bolus is no longer accurate will enable more effective bolus usage. This paper reports a method of defining reference junction drift based on analysis to identify drift towards pH 7 with reducing daily amplitude. The algorithm was then tested on 102 rumen pH data sets and identified failures in some boluses before 90 days and failed to be triggered by known disease incidents detected by clinical analysis. The significance of this result is that it shows that internal models of sensor performance and failure can be applied to sensors that are in locations where calibration is not possible and this provides an important method of data filtering and improving bolus performance.

1. Background

In many situations a sensor can be calibrated periodically and accuracy maintained. A rumen bolus is not in a position where it can be recovered and calibrated. However by inspection of the data and an understanding of how sensors fail we can develop mathematical methods to identify from sensor data when the sensor has failed.

A pH sensor is made up of a measurement electrode which measures a voltage potential caused by hydronium ions and reference electrode which is required to remain stable. The measurement electrode and the reference are electrically connected through the liquid medium being measured and the potential difference is measured in volts with a standard value of 59.16 per unit of pH at 25 C. Sensors are calibrated so that at 7 pH voltage output is zero while at 4 pH it is +177.480mV. In laboratory use a sensor is calibrated by placing it into reference buffers at pH 7 and 4 and recording the output values. The offset and slope of the calibration line can then be calculated, in the eCow system this is done automatically by the bolus software.

A pH sensor deployed in the rumen fails in two principle ways. If the measurement bulb is compromised so that there is an open circuit between the measurement and reference electrode which is a large value which is interpreted as a low pH signal. A typical cause of this failure is abrasion or destruction of the measurement electrode surface.

A more subtle failure occurs when the reference junction becomes contaminated and instead of giving a stable reference against which the measurement electrode measures the potential. As the potential difference tends to zero volts the signal is interpreted as nearing 7 pH. As the drift worsens the daily range narrows and oscillates about a value of about 7.

As the low pH signal is outside the physiological range it can be detected with a low filter. The reference junction drift needs a more subtle approach and this paper describes methods of detecting reference junction drift.

2. Materials and methods

A simple machine learning algorithm the kNN algorithm was used for the first analysis, it is non-parametric, and therefore it makes no assumption about the underlying data or its distribution. Two methods were trialed on a small data set to investigate what data should be used. In the first method raw data files were chosen to train the algorithm, this resulted in a poor percentage of accuracy ranging from 30-60% Depending on the training and test files.
In the second method, a data table of values of the raw data was produced of the minimum, maximum, average and range of pH. The accuracy was higher ranging from 60-80%.
The data table approach was selected as a better method to improve the accuracy. The last 2880 lines of data were measured, for minimum, maximum, average and range. There are 96 observations of data per day and therefore the values consist of the last 30 days of the life of the bolus. The values and their classification were collated in to a data table. An Rpart Classification Tree was also completed in order to look at other methods of classification.
To create the data table, bolus pH data files were inspected and cross checked with metadata from the eCow directory, 102 bolus data sets values were put in to three categories (normal, cracked sensor and reference junction drift.

1. *Normal*. All data that maintained bolus integrity and longevity (Figure 1).

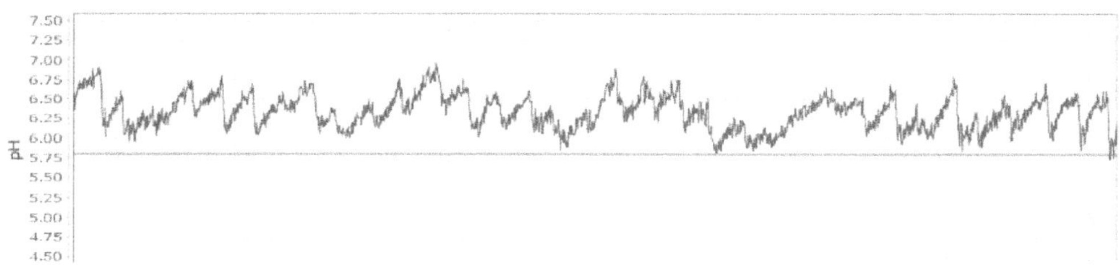

Figure 1. pH data from the rumen of a healthy cow shows variable daily cycles and has to be distinguished from data from a failing bolus.

2. *Reference junction failure*. Boluses that drifted towards a value of 7 with a narrowing range after 90 days of operation. In Figure 2 reference junction drift starts about 25/12/13 about 180 days after insertion into the cow. The warranty of accuracy is shown up to 60 days. Contrast the start of the graph with a high daily range ante-partum going over 7 pH at night with the data from the failing sensor at the end of the graph with a narrow daily range tending toward 7 pH.

The Kmeans method selected was suitable for analysing files retrospectively but we also need a dynamic analysis as data arrives. A simpler method is to normalise the Max pH for each day and the range between the min and max readings for each day so they are both on the same scale. In a normal non-drifting bolus the difference should be between 0.1 or 0.2 pH units. However as reference junction drift starts, the difference between the normalised range and the normalised max pH starts to increase. This algorithm needs further testing but provides a daily warning if the bolus starts to drift.

3. *Cracked sensor*. This is easy to detect as the value given by the sensor drops below 4 pH outside the physiological range as it goes open circuit measuring the potential of rumen liquor.

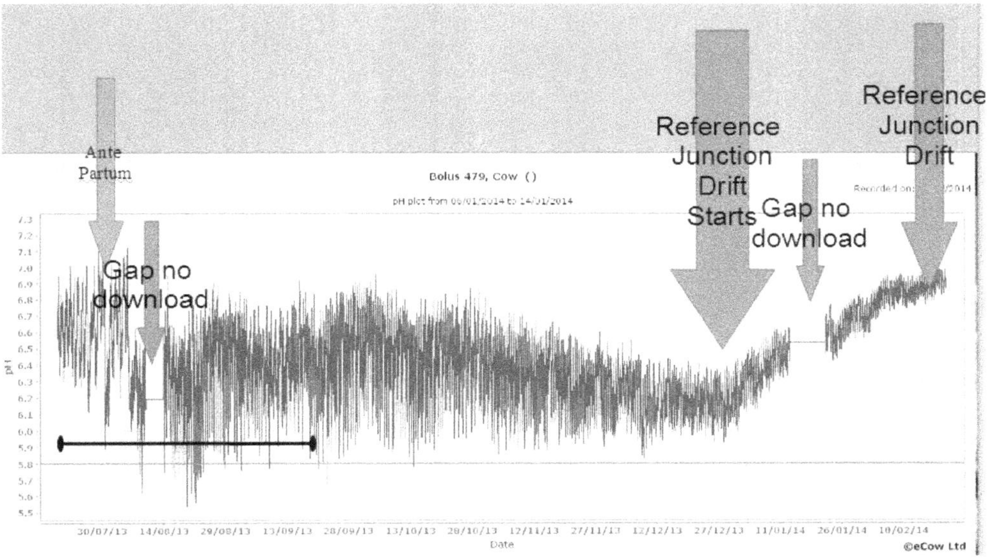

Figure 2. Rumen pH data from a bolus operational for over 200 days.

3. Conclusions

Sensor drift is a serious problem and is non-linear. An algorithm is proposed which classifies a bolus as failing if the following expression is true:

If pH<4 for more than 1 reading
OR
If normalised max pH per day – (difference between min/max pH per day) >0.2
AND
days in operation over 90

These algorithms need checking against clinical signs but they indicate that by using our knowledge of how sensors fail based on the physics of the sensors themselves we can filter out data which is sent by a failing sensor.

Part 4. Animal health, welfare, environment and production

Estimating chances of insemination success using milk progesterone measurements

M. Derks[1,2,3*], P. Blavy[4], J.M. Christensen[5] and N.C. Friggens[4]
[1]INRA, UMR1213 Herbivores, 63122 Saint-Genès Champanelle, France; [2]Clermont Université, VetAgro Sup, BP 10448, 63000 Clermont-Ferrand, France; [3]Université de Lyon, VetAgro Sup, UMR1213 Herbivores, 69280 Marcy l'Etoile, France; [4]AgroParisTech, UMR Modélisation Systémique Appliquée aux Ruminants, 16 rue Claude Bernard, 75231 Paris, France; [5]Lattec I/S, Slangerupgade 6, 3400 Hillerød, Denmark; marjolein.derks@clermont.inra.fr

Abstract

Data derived from progesterone measurements in milk (HerdNavigator©) were used to estimate the chance of insemination success for a given oestrus. During 2010 (January-December) milk progesterone sample data were collected from 16 commercial dairy herds, equipped with a HerdNavigator system. Data was smoothed. Cycle landmarks were identified and used to determine cycle length, height and slope. Inseminations were identified, and deemed successful or unsuccessful based on profile shape post-AI (continuously high P4 in the period 10 to 42 days post insemination: successful). Potential pre-AI predictors (profile slope, cycle length, profile height, milk yield, parity, prolonged anoestrus, and inseminated oestrus no.) were calculated from the dataset. The interval oestrus-insemination was corrected for, to filter out the effect of badly timed inseminations. Predictors were modeled using an a priori curve that described the predictor effect shape observed in visualization plots. The accuracy of the model was estimated by computing D=-log (L), where L is the likelihood of the observed data, according to the model. Model parameters were fitted in order to optimize D. The final variables that were retained in the equation were cycle length, milk yield and parity, leading to the equation: LAIS = LAISmax × f_length × f_milk yield × f_parity, where LAIS = likelihood of artificial insemination success and LAISmax=the maximum likelihood of AI success achievable in the dataset (when cycle length, milk yield and parity are all optimized). In this formula, each function (f_xxx) is scaled to have a positive value and a maximum of 1. By implementing this equation in the Herd Navigator software, using real time data, the chances of an insemination being successful can be estimated at detection of oestrus, supporting the farmer in his decision whether or not to inseminate that cow, at that moment.

1. Introduction

Progesterone is used to assess the reproductive status of the cow, and in-line progesterone measurements have recently become commercially available (HerdNavigator©). Previously, researchers have aimed to predict chances of insemination success of a given oestrus (Friggens and Chagunda, 2005), using the declining slope pre-oestrus and cycle length of the previous oestrus. Chances of insemination success declined when slope and/or cycle length moved away from optimal. However, this decrease in chance of insemination success was based on expert opinion and literature, not on actual studies of progesterone profiles. Also, it did not take into account other influential factors like milk yield. In this study, progesterone profiles from sixteen dairy farms were analyzed to determine actual effects of progesterone slope, cycle length, and peak progesterone values on chance of insemination success. Other factors that were deemed to be important from literature (e.g. milk yield) were included to see if they had any effect on probability of insemination success. The aim of this study was to quantify the effects of progesterone profile features and external factors on insemination success, and to provide a real-time predictor of probability of insemination success.

2. Methods

During 2010 (January-December) milk progesterone sample data were collected from 16 commercial dairy herds, equipped with a HerdNavigator system. Thirteen farms were based in the Netherlands, three in Denmark. The HerdNavigator system takes regular milk samples of cows 20-240 days postpartum. Sampling frequency is dependent on the status of the cow (postpartum anoestrus, cyclic, potentially pregnant) and the cycle itself (days from oestrus, progesterone slope), and is described in detail in the paper by Friggens and Chagunda (2005).

The raw progesterone data was smoothed using a median kernel function. Each kernel was calculated using three consecutive data points (except for the first two and last two data points), with the median values at each time point being used for all subsequent analyses. Cycle landmarks were identified and used to determine cycle length, height and slope. Any increase in progesterone from being <6 ng/ml to being >6 ng/ml was marked as X, representing the start of a new luteal phase. Drop D is the first point where progesterone <6 ng/ml. OS is the oestrus indicator and is the first point where progesterone is <4 ng/ml. Landmarks were used to determine cycle length (OS_n-OS_{n-1}), oestrus to increase (X_n-OS_{n-1}), and oestrus to drop D. Inseminations were identified, and deemed successful, unsuccessful, or unknown as follows: inseminations were said to be successful if the median progesterone value in the period 10 to 42 days post insemination exceeded 12 ng/ml, and there weren't two consecutive progesterone measures below 10 ng/ml or one below 4 ng/ml in this period. Inseminations that were followed by a gap in the data of at least 10 days in the first 42 days were said to be unknown. All other inseminations were classified unsuccessful.

Potential predictors (profile slope, cycle length (days), profile height (progesterone concentration), milk yield (kg), parity, prolonged anoestrus (no rise >35 d post-partum), and inseminated oestrus no.) were calculated from the dataset. Milk yield was calculated by taking the median of the previous eleven highest daily values for this lactation. This secured a smoothed estimation of the maximum milk yield up to that time in the lactation. The interval oestrus-insemination was corrected for, to filter out the effect of badly timed inseminations. Potential predictors were tested in three steps: (1) the proportion of insemination success relative to the time between oestrus and insemination on the x-axis was plotted for each variable (Figure 1A); (2) an a priori curve was fitted to the predictor effect shape observed in the visualization plots (Figure 1B); (3) the predicted and the observed

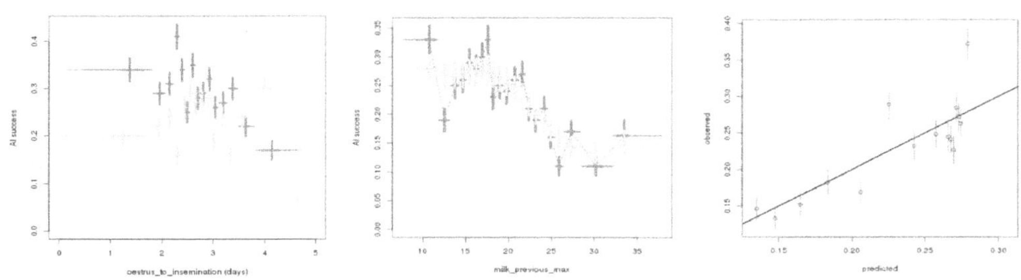

1A 1B 1C

Figure 1. Example of the selection of the appropriate functional form for a potential predictor (milk yield). It was selected in three steps: (1) the proportion of insemination success relative to the interval between oestrus and insemination on the x-axis was plotted for each variable ((A), milk yield divided in 50% highest (light grey) and 50% lowest (dark grey)); (2) an a priori curve was fitted to the predictor effect shape observed in the visualization plots (B); (3) both the predicted and the observed data were plotted to visualize differences between observed and predicted values (C).

data were plotted in the same plot to examine differences between observed and predicted values (Figure 1C).
The model was built starting from a base model containing the effect of oestrus to insemination interval. Predictors were added to the model one by one. The accuracy of the model was estimated by computing D=-log (L), where L is the likelihood of the observed data, according to the model. Model parameters were fitted in order to optimize D.

3. Results

The dataset contained 3,716 lactations from 3,228 cows, from which 6,016 usable oestrus cycles were extracted. Median cycle length was 26 days. Of all cycles, 69.6% were inseminated between days -2 to 5 after oestrus alarm. Cycles were, on average, inseminated 2.73 days after oestrus indicator, with a standard deviation of 0.77 days. Cycles longer than 100 days were assumed to be the result of measurement error and were excluded, as were cycles where the interval oestrus to increase exceeded 50 days, cycles where the interval oestrus to drop D exceeded 70 days, and cycles where the interval oestrus to insemination exceeded or 10 days after or 5 days before ovulation. Distributions of cycle length, oestrus to increase, oestrus to drop D and oestrus to insemination are shown in Figure 2.
This was the optimal equation:

LAIS = LAISmax × f_length × f_milk yield × f_parity
Where:

LAIS=likelihood of artificial insemination success
LAISmax=the maximum likelihood of AI success achievable in the dataset (when cycle length, milk yield and parity are all optimized).

In this formula, each function (f_xxx) is scaled to have a positive value and a maximum of 1
The final variables that were retained in the equation were cycle length, milk yield and parity. They were modeled as follows:

3.1 Cycle length
Optimal cycle length in this dataset was calculated to be 22.03 days. Preceding cycles that were either too short or too long led to a decrease in fertility. A linear relationship between the distance between optimal and actual cycle length and fertility was assumed, with a minimum cycle length of zero and a maximum of 35 days. Cycle lengths that deviate from 22.03 days will have a lower probability of insemination success. When cycle length is optimal, f_length is 1, when it is zero, f_length is 0. If cycles are >35 days the effect is assumed to be the same as when cycle length is 35 days.

f_length = max(min((1-0.03×(previous cycle length-22.03 days)),1),0) × max(min((0.14×(previous cycle length-15 days)),1),0)

3.2 Milk yield
The effect of milk yield was modeled by a sigmoid curve, with an inflection point at milk_i (24.4 kg milk/day), a minimum value of milk_b (0.56 LAIS), a maximum value of 1, and a stiffness of milk_s (0.074 kg/day). An increasing milk yield decreases LAIS.

f_milk yield = 1-(1-0.56LAIS) / (1+exp(-0.074 × (milk yield (kg/day)-24.04)))

3.3 Parity

The effect of parity is modeled by a decreasing function, with a stiffness of parity_s (0,18) a maximum of 1 and a minimum of parity_b.

$f\_parity = 3.01e^{-6}LAIS + (1-3.01e^{-6}LAIS) / (parity \times e^{0.18})$

4. Discussion

This work aimed to investigate the effect of cow traits and progesterone profile features on likelihood of insemination success, and to build an equation that can predict probability of insemination success at the time when an oestrus is detected. Such an equation can be implemented on farms that monitor progesterone, and can support the farmer in deciding when to inseminate an oestrus. This will help move the focus away from the current situation, where each detected oestrus is automatically inseminated, to the situation of near perfect oestrus detection where the question is which oestrus is worth inseminating.

From literature study, different profile features and cow- and environmental traits were found to be important. For profile features they were: cycle length (Gorzecka *et al.*, 2011; Royal *et al.*, 2000), peak progesterone concentration (Diskin *et al.*, 2006), and slope (Friggens and Løvendahl, 2008). For other traits they were: feed regime (Vasconcelos *et al.*, 2003), milk yield (Endo *et al.*, 2013), season (Sartori *et al.*, 2002a,b), calving problems (Fourichon *et al.*, 1999), prolonged anoestrus (Santos *et al.*, 2009), parity (Bagnato and Oltenacu, 1994) and number of previously inseminated oestruses (Lattec). Feed regime, season and milk yield all influence progesterone concentrations, which are all believed to be caused by a shift in metabolism (Sartori *et al.*, 2002b). Given the fact that the background was similar, and no data was available on temperature and nutrition, milk yield was chosen to represent these traits. Prolonged anoestrus is believed to be mostly caused by problems around or after calving (Fourichon *et al.*, 1999). Since there was no data available on calving problems, only prolonged anoestrus was used for further analysis.

Cycle length influences AI success rate (Royal *et al.*, 2000), with optimal cycle length being 18-24 days, and a decline in AI success for both shorter and longer cycles. This is in line with our findings, although we had few cycles <20 d. Also, Gorzecka *et al.* (2011) found that pregnant cows had, on average, shorter follicular phases than non-pregnant cows, indicating a negative effect for longer cycle lengths. They do not mention effects of very short cycles on pregnancy rate. Also the negative relationship between parity and fertility has been described before (Bagnato and Oltenacu, 1994). High milk yield has been found to negatively influence fertility (Kafi *et al.*, 2012), and the results in this study support that. However, the relationship between fertility performance on the farm and level of milk yield is highly influenced by farmers' management skills. More research on the influence of management is needed.

In the present study, variables that were found not to be significant as predictors were cycle height and – slope, inseminated oestrus number and prolonged anoestrus. Both height and slope were found to not significantly improve the predictions of the model. This is expected to be due to data density and –quality. Herd Navigator is meant to detect oestruses; therefore, it takes most samples around the expected time of oestrus, so when progesterone is low (for more information on sampling frequency, see Friggens and Chagunda, 2005). The system is also calibrated to be most accurate when progesterone levels are low. In our dataset, this meant we had fewer data points when progesterone was high, with less accurate values. This makes predicting both height and slope subject to noise. Previous research by Gorzecka *et al.* (2011), who used a more frequently measured dataset, found no differences in oestrus cycle shapes between successful and not successful inseminations. More research is needed to determine the effects of slope and height of progesterone during the luteal phase to be able to answer this question.

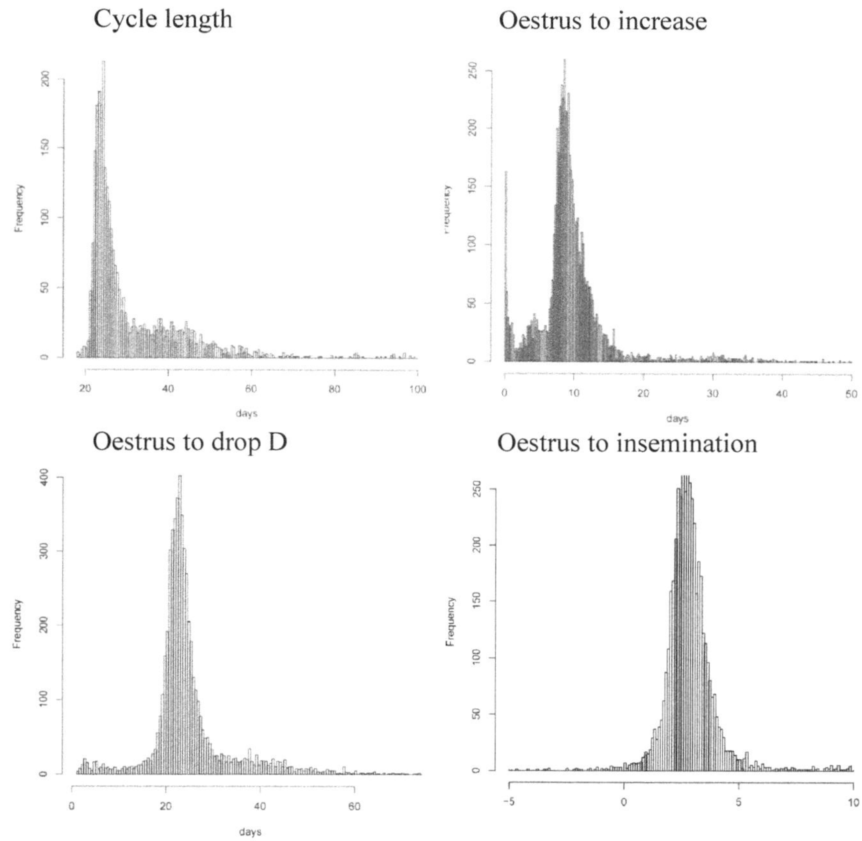

Figure 2. Histograms for the distribution of cycle length, and time from oestrus to increase, oestrus to drop D, and oestrus to insemination for all progesterone profiles in the data set.

By implementing this equation in the Herd Navigator software, using real time data, the chances of an insemination being successful can be estimated at detection of oestrus, supporting the farmer in his decision whether or not to inseminate that cow, at that moment.

References

Bagnato, A. and Oltenacu, P.A. (1994). Phenotypic evaluation of fertility traits and their association with milk production of Italian Friesian cattle. Journal of Dairy Science 77: 874-882.

Diskin, M.G., Murphy, J.J. and Sreenan, J.M. (2006). Embryo survival in dairy cows managed under pastoral conditions. Animal Reproduction Science 96: 297-311.

Endo, N., Nagai, K., Tanaka, T. and Kamomae, H. (2013). Changes in plasma progesterone levels in the caudal vena cava and the jugular vein and luteinizing hormone secretion pattern after feeding in lactating and non-lactating dairy cows. Journal of Reproduction and Development 59: 107-114.

Friggens, N.C. and Chagunda, M.G. (2005). Prediction of the reproductive status of cattle on the basis of milk progesterone measures: model description. Theriogenology 64: 155-190.

Friggens, N.C. and Løvendahl, P. (2008). The potential of on-farm fertility profiles: in-line progesterone and activity measurements. Fertility in dairy cows: bridging the gap, pp. 72-78.

Fourichon, C., Seegers, H., Bareille, N. and Beaudeau, F. (1999). Effects of disease on milk production in the dairy cow: a review. Preventive Veterinary Medicine 41: 1-35.

Gorzecka, J., Codrea, M.C., Friggens, N.C. and Callesen, H. (2011). Progesterone profiles around the time of insemination do not show clear differences between of pregnant and not pregnant dairy cows. Animal Reproduction Science 123: 14-22.

Kafi, M., Mirzaei, A., Tamadon, A. and Saeb, M. (2012). Factors affecting the occurrence of postpartum prolonged luteal activity in clinically healthy high-producing dairy cows. Theriogenology 77: 421-429.

Royal, M.D., Darwash, A.O., Flint, A.P.F., Webb, R., Woolliams, J.A. and Lamming, G.E. (2000). Declining fertility in dairy cattle: changes in traditional and endocrine parameters of fertility. Animal Science 70: 487-501.

Santos, J.E.P., Rutigliano, H.M. and Sá Filho, M.F. (2009). Risk factors for resumption of postpartum estrous cycles and embryonic survival in lactating dairy cows. Animal Reproduction Science 110: 207-221.

Sartori, R., Rosa, G.J.M. and Wiltbank, M.C. (2002a). Ovarian structures and circulating steroids in heifers and lactating cows in summer and lactating and dry cows in winter. Journal of Dairy Science 85: 2813-2822.

Sartori, R., Sartori-Bergfelt, R., Mertens, S.A., Guenther, J.N., Parrish, J.J. and Wiltbank, M.C. (2002b). Fertilization and early embryonic development in heifers and lactating cows in summer and lactating and dry cows in winter. Journal of Dairy Science 85: 2803-2812.

Vasconcelos, J.L.M., Sangsritavong, S., Tsai, S.J. and Wiltbank, M.C. (2003). Acute reduction in serum progesterone concentrations after feed intake in dairy cows. Theriogenology 60: 795-807.

Calving alert system for better working environment and safety

A. Lind[1*], M. Karlsson[2], C. Lindahl[1] and R. Båge[2]
[1]*Swedish Institute of Agricultural and Environmental Engineering, P.O. Box 7033, Uppsala 750 07, Sweden;* [2]*Department of Clinical Sciences, Swedish University of Agricultural Sciences, P.O. Box 7054, Uppsala 750 07, Sweden; ann-kristina.lind@jti.se*

Abstract

In free stall cattle herds it may be difficult to know when it is time to move a cow to a calving pen and to properly monitor and assist the calving. If a cow calves in the free stall, the separation of cow and calf from other animals involves a risk of injury to the handler. The aim of the study was to evaluate a calving alert system, where a vaginal thermometer (Vel'Phone®, Medria, Châteaubourg, France) is used to predict calving time. The thermometer monitors the vaginal temperature, and the farmer is notified via SMS 24-48 hours before calving and again when the thermometer is expelled with the amniotic sac. A clinical observational case-control study was carried out with 9 test and 8 control cows on a dairy farm with Swedish Red breed. Behavioral studies of the cows were carried out during and after the procedure of cleaning, disinfection of the vulva and insertion of the thermometer. Controls were equally prepared but no thermometer was inserted. The behavior of the cows was recorded within 1 to 2 hours and 24 hours after the procedure. Further, the work environment of the handler was investigated by direct observation of the interactions between the cow and the two handlers. The cows were video recorded to investigate both behavior of the cows and exact time the veterinarian was exposed to injury risks. An interview was carried out with the farmer and the veterinarian to address the pros and cons with the technology. On average the veterinarian was in the risk zone (the total time the veterinarian stands in the area around the cow where it was possible to be hit by the hind legs if the cow kicked) for 4.5 minutes per cow during the hygiene and application procedure that on average took 14 minutes per cow. The mean frequencies of back archings ($P<0.01$), tail liftings ($P<0.0001$) and defecation ($P<0.01$) were significantly higher for the test cows compared to the controls. The calving alarm system did alarm 24-48 hours before calving for eight of the nine test cows, and the alarms at calving (when the thermometer was expelled) were accurate and reliable. In conclusion, the studied calving alert system functioned very well technically, but the effects on behavior indicate that the cows experienced discomfort by the vaginal thermometer, which needs to be weighed up against the potential benefits on herd level.

Keywords: animal behavior, calving, dystocia, alarm system, cow, welfare

1. Background

Cows in free-stall housing systems should be separated from other cows before calving and placed in a separate calving pen. However, it happens that cows calve in the free stall. One reason for this is that a few dairy herds and many special beef producers use a bull instead of insemination, which makes it difficult to estimate an accurate calving date. A calving in the free-stall housing is not just an animal welfare problem, but can also be associated with several injury risks for the farmer, when separating the calf and the cow from the other cows (Lindahl and Adolfsson, 2008). Some cows have a very strong maternal protective behavior, which can lead to a cow attacking a person who comes near her calf. It is also difficult to keep track of the other cows in a free-stall housing system when you are focused on helping a cow during calving. Farmers themselves have indicated that calving is one of the most risky and stressful work operations on the farm (Geng *et al.*, 2013). Thus, there is an obvious need to be able to predict calving in time and thereby reduce the risk of injuries.

There are various types of calving alert systems available on the market, e.g. detectors that measure the cow's activity level and the rumination frequency, contractions of the uterus, tail lifts and detectors that measure temperature vaginally (Ouellet *et al.*, 2016; Saint-Dizier and Chastant-Maillard, 2015). Vaginal temperature is considered a reliable predictor of calving (Ouellet *et al.*, 2016). The objective of the study was to evaluate how a calving alert system with a vaginal thermometer (Vel'Phone®, Medria, Châteaubourg, France) affects the cow and the work environment of the handler. The system monitors vaginal temperature and the Medria box sends daily reports by SMS to the farmer. It is possible for the user to decide how many and the time for daily reports. Further, the farmer is notified via text messages 24-48 h before calving when body temperature drops and again when the thermometer is expelled with the amniotic sac.

2. Materials and methods

The study was a clinical observational case-control study, conducted on an organic dairy farm with Swedish Red breed (SRB). The farm had an open-sided group calving pen with deep bedding and free access to pasture.

Ten cows in late pregnancy were selected in the first study period and in the second study period two cows and five heifers were selected. Hereafter, 'cows' are used for both heifers and cows. The cows were selected by calving date, so they would calve as close in time as possible. All cows included in the study were fixed in the cubicles before the start of the test. The practical procedure, including the insertion of the vaginal thermometer, was performed by the same experienced veterinarian in both study periods.

Cow behavior was observed during and after the procedure of cleaning and disinfection of vulva and insertion of the vaginal thermometer. Controls were equally prepared, but no thermometer was inserted. The ethogram protocol contained the behaviors back arching, tail wagging, bellowing, fidgeting, kicking, tail lifting and defecating. Behavior observations were repeated 1-2 hour and 24 hour after the insertion.

The risk zone was defined as the total time the veterinarian stands in the area around the cow where it was possible to be hit by the hind legs if the cow kicked. Direct observations of the interactions between the test cow and the two persons handling the test cow during the clinical procedure (the veterinarian and the farmer that holds the tail) were conducted. In other words, the procedure did not include the time for sorting and fixation of the cows in the cubicles. Furthermore, the total number of minutes the veterinarian was exposed to injury risks was calculated from the video recordings. After the clinical study, the farmer and the veterinarian performing the insertions were interviewed about their subjective opinions on the system. Data was analyzed using GraphPad QuickCalcs (GraphPad Software Inc., San Diego, CA, USA). Means were compared by t-test.

3. Results

3.1 The calving alarm system

Once or twice a day, the farmer could receive an SMS from the Vel'Phone® that reported the cow's temperature. Besides this daily report, the calving alarm system gave an alarm that calving was expected within 24-48 hours, gave an expulsion alarm and an alarm if the Medria box lost contact with the cow's thermometer. Table 1 shows the date and time for the expected calving alarm and the Expulsion alarm has given an alarm on the exact calving date reported by the farmer except for one cow, where the expulsion alarm was in the evening. Further, the expected calving alarm within 24-48 hours is one day prior to the expulsion alarm and the calving date.

Table 1. The date and time for the expected calving within 24-48 hours alert and expulsion alarm by the Medria Vel'Phone, and the reported calving date recorded by the farmer.

	Expected calving 24-48		Expulsion alarm		Reported calving
Cow no.	Date	Time	Date	Time	Date
1	25-10-2015	16:00	26-10-2015	05:57	26-10-2015
2	03-11-2015	18:00	04-11-2015	21:21	05-11-2015
3	-[1]	-	11-11-2015	11:25	11-11-2015
4	30-10-2015	18:00	31-10-2015	11:08	31-10-2015
5	28-10-2015	18:00	29-10-2015	17:07	29-10-2015
6	18-11-2015	15:00	19-11-2015	18:32	19-11-2015
7	16-11-2015	12:00	17-11-2015	00:35	17-11-2015
8	24-11-2015	21:00	25-11-2015	17:10	25-11-2015
9	21-11-2015	03:00	21-11-2015	22:04	22-11-2015

[1] Missing value means that an alarm of Alert absent was recorded.

3.2 Cow behavior

Fidgeting was the most prevalent behavior in all 17 cows, followed by back arching and tail wagging (both present in 14/17 cows). The mean frequencies of back arching (P<0.01), tail lifting (P<0.0001) and defecation (P<0.01) were significantly higher for the test cows compared to the controls. Of all behaviors expressed at insertion, back arching, tail wagging and tail lifting were the predominant ones still remaining 1-2 hour and 24 hour after insertion.

3.3 Interactions and injury risk

On average the time the veterinarian was in the risk zone was 4.5 minutes per cow during the procedure which on average took 14 minutes per cow, and included preparation, washing, disinfection and insertion of the vaginal thermometer as well as washing and disinfection of the application after insertion.

The observed interactions during the procedure were 'the cow kicks without hitting the handler (11 observations)', 'the handler talks softly to the cow (22 observations)', 'gentle pressure on the cow, for her to stand still (21 observations)' and 'pressure on the cow with moderate force to get the cow to move (5 observations)'. In all, 59 interactions were observed between the test cows and the two people who handled the test cows during the preparation procedures, which resulted in 6.56 interactions per cow in 14 minutes.

4. Discussion

4.1 The calving alarm system

Logging cow vaginal temperature is a reliable way to reveal calving's (Aoki *et al.*, 2005; Burfeind *et al.*, 2011; Noakes *et al.*, 2009; Ouellet *et al.*, 2016). The calving detector from Medria Vel'Phone is a thermometer that was deposited in the vagina over a week before the expected calving. The technology worked well and the expulsion alarm and expected calving alerts within 24-48 hours were reliable.

Despite the hygiene procedure with cleaning and disinfection of the applications instrument and the thermometers before the insertion of the calving detector there is a risk of vaginal contamination by infectious agents such as bacteria from the environment. This can lead to infection of the animal, which was highlighted by the study's veterinarian. The risk of infection as a result of the internal deposition was also reported by the farmer as one of the disadvantages of the calving detector. When measuring the body temperature continuously the calving alarm system can immediately after insertion analyze the body temperature of the cow to determine if the cow as a result of an infection associated with the deposit has been hit by a fever (Pohl *et al.*, 2014).

The farmer emphasized the benefits of knowing when to move a cow to a calving pen and being alerted to the onset of calving. The farmer found the technology easy to understand and handle, although the insertion procedure was time consuming. He was reluctant to use such an invasive device affecting the cows' wellbeing. The veterinarian also commented on the invasive nature of the device, potential risks for infection, and the necessity of secure restraint and good knowledge of the cow's anatomy. Despite that, the farmer would recommend the calving system to other farmers, particularly if the farm has calving problems.

It was not possible to see if the Expected calving alarm within 24-48 hours was exact within this timespan since it was not possible for the farmer to give the exact time of birth. However, the results indicate that the vaginal thermometer is very precise in predicting calving.

4.2 Cow behavior

The results indicate that the cows experienced some discomfort both at insertion of the thermometer and the day after. Tail lift is a common behavior around calving (Miedema *et al.*, 2011; Saint-Dizier and Chastant-Maillard, 2015). Tendencies to lift the tail can be seen several days before calving (Saint-Dizier and Chastant-Maillard, 2015), however given that the behavior occurred abruptly, subsequently to the insertion procedure, and that there were significant difference in the behavior between the test group and the control group, indicates that the calving detectors have influenced the test cows negatively. Furthermore, significant differences were found for the behaviors back archings and defecation between the test and control group.

4.3 Interactions and injury risk

The veterinarian stated that the technique was easy to use and the instruction was easy to follow. It felt safe to perform the procedure, however, 11 kicks were observed even though the veterinarian was never hit. It is important that users of the Vel'phone system have good knowledge about cow's anatomy and is properly trained, e.g. by the herd veterinarian, to perform the clinical intervention while ensuring both hygiene and animal welfare. Fixation of the cows is an important aspect to make management safe for the handler. In our study, there were two people at the deposit, which we would also recommend, because it is impossible to keep the cow's tail away while simultaneously inserting the vaginal thermometer.

5. Conclusions

In conclusion, the studied calving alert system functioned well technically, but the effects on behavior indicate that the cows experienced discomfort by the vaginal thermometer, which needs to be weighed up against the potential benefits on herd level, since the vaginal thermometers do affect the cow's welfare negatively.

References

Aoki, M., Kimura, K. and Suzuki, O. (2005). Predicting time of parturition from changing vaginal temperature measured by data-logging apparatus in beef cows with twin fetuses. Animal Reproduction Science 86: 1-12.

Burfeind, O., Suthar, V.S., Voigtsberger, R., Bonk, S. and Heuwieser, W. (2011). Validity of prepartum changes in vaginal and rectal temperature to predict calving in dairy cows. Journal of Dairy Science 94: 5053-5061.

Geng, Q. Field, W.E. and Salomon, E. (2013). Risk assessment of working environment during cattle handling on pasture using WEST-AG. ISASH Technical Papers 13-TP2. The Conference of International Society for Agricultural Safety and Health. June 23-27, 2013. Sawmill Creek Resort, Sandusky, OH, USA.
Lindahl, C. and Adolfsson, N. (2008). En olycka händer så lätt – exempel från jordbruk och ridskolor. JTI-rapport nr 119. JTI – Institutet för jordbruks- och miljöteknik, Uppsala, Sweden.
Medria (2010). Calving detector via SMS. Available at: http://www.medria.fr/en_GB/produits/vel-phone.html.
Miedema, H., Cockramb, M.S., Dwyer, C.M. and Macraea, A.I. (2011). Behavioural predictors of the start of normal and dystocic calving in dairy cows and heifers. Applied Animal Behaviour Science 132: 14-19.
Ouellet, V., Vasseur, E., Heuwieser, W., Burfeind, O., Maldague, X. and Charbonneau, É. (2016). Evaluation of calving indicators measured by automated monitoring devices to predict the onset of calving in Holstein dairy cows. Journal of Dairy Science 99: 1539-1548.
Pohl, A., Heuwieser, W. and Burfeind, O. (2014). Technical note: assessment of milk temperature measured by automatic milking systems as an indicator of body temperature and fever in dairy cows. Journal of Dairy Science 97: 4333-4339.
Saint-Dizier, M. and Chastant-Maillard, S. (2015). Methods and on-farm devices to predict calving time in cattle. The Veterinary Journal 205: 349-356.

Prediction of calving in dairy cows using tail-mounted accelerometers

S. Krieger[1,2*], G. Sattlecker[2], F. Kickinger[2], W. Auer[2], M. Iwersen[1] and M. Drillich[1]
[1]*Clinical Unit for Herd Health Management in Ruminants, Department for Farm Animals and Veterinary Public Health, University of Veterinary Medicine, Vienna 1210, Austria;* [2]*Smartbow GmbH, Jutogasse 3, 4675 Weibern, Austria; stefanie.krieger@vetmeduni.ac.at*

Abstract

Frequent problems on dairy farms are dystocia and stillborn calves, resulting in direct losses due to calf mortality, dam mortality and premature culling, as well as indirect costs due to additional veterinary services, labor and reduced fertility. A real-time monitoring system could support farmers to detect onset and course of parturition and to intervene in cases of dystocia. Thus, an evaluated, accurate and sensitive device is needed. The hypothesis of this study was that the onset of parturition can be predicted by movements of the cows' tail. The study was conducted on the Teaching and Research Farm Kremesberg, University of Veterinary Medicine Vienna, Austria. Approximately one week before the expected calving, cows (n=5) were housed in a straw-bedded single calving box, equipped with a digital video recording system for 24h/d observation. 3D-accelerometers (Smartbow®) were fixed on the upper part of the tail with tape. The video recordings were analysed by encoding the events 'frequency and duration of tail-raising five hours prepartum', 'rupture of the amniotic sac' and 'expulsion of the calf'. The accelerometer data was analyzed retrospectively for the two days before calving. An algorithm was developed to perceive the tail raising by using the encoded videos. In addition, a decision function based on the frequency and duration of the tail raising was created. Exceeding a threshold leads to the decision making and consequent signaling of a birth alarm. In each of the five evaluated cases, the alarm was triggered a short time before the expulsion of the calves at 16, 34, 56, 71, and 160 minutes, respectively. These preliminary results indicate that an accelerometer to detect tail-movement patterns may be useful to predict parturition. Further research is required to specify the algorithm and the decision function, to analyse predictability of dystocia and to develop a real-time alarm system under field conditions.

Keywords: sensor, monitoring, cattle, birth

1. Problem statement

Current socioeconomic issues are leading to new challenges for modern dairy farms. Structural changes and a shift towards intensive livestock farming are necessary to ensure an optimized usage of limited resources, but lead to less time per animal. Nevertheless, maintaining a healthy livestock is the basis for a productive farm, high standards of animal welfare and consumers' demands. Precision dairy farming technology offers the opportunity to focus on individual animal needs and challenges by using e.g. sensors and wireless, real-time technologies.
In the last two decades, the perinatal mortality of dairy calves increased and is ranging between 2 and 10% (Mee, 2008a). Ninety per cent of calves that died during parturition are alive at the beginning of parturition (Mee, 2008b,c). Approximately 50% of stillborn calves are born without any calving difficulties (Philipsson, 1996), whereas dystocia contributes to up to the other half of the perinatal mortalities in cattle (Meyer *et al.*, 2001). Environment and management are important factors. Lack of supervision and inadequate as well as falsely terminated obstetrical intervention were discussed to be decisive management factors for an increase in stillbirth rates (Leister, 2009). Implementing an optimized calving management particularly by improving supervision during parturition (Mee, 2004), reduced the stillbirth rate by 60% under field conditions (Kausch, 2009).

Numerous behavioral changes can be observed in the dam around parturition. Four systems for calving detection are currently available: inclinometers and accelerometers detecting tail raising and changes in activity; abdominal belts monitoring uterine contractions; vaginal devices detecting vaginal temperature and expulsion of the calf or allantochorion (Saint-Dizier and Chastant-Maillard, 2015). However, the high individual variability of these changes is still limiting the sensitivity and specificity of automatic assistance systems predicting or monitoring parturitions in dairy cows.
The Smartbow® System (Smartbow GmbH, Weibern, Austria) is a device that collects accelerometer data. It consists of the Smartbow Tags, Smartbow Wallpoints and the Smartbow Station. The system was developed to collect activity and rumination data of cows to provide a real-time monitoring tool. The general aim of this project was to evaluate the system for analyzing movements of cows to predict and monitor parturitions. The objective of this study was to determine in particular the feasibility of tail movements for the prediction of calving.

2. Methods

The study was conducted on the Teaching and Research Farm Kremesberg, University of Veterinary Medicine Vienna, Austria. One week before the expected calving, cows (n=5) were moved from a group pen and housed in a 17 m^2 straw-bedded single calving box. The dams had visual contact to the rest of the herd at all times. Aside from the daily stable routine unnecessary distraction or examinations were avoided. The box was equipped with a camera system (Hikvision VF IR Bullet Network Camera, Hangzhou, China). The two infrared cameras enabled 24 h observation and the recordings were stored for further analyzes.
Table 1 provides an overview of the cows used in the experiment. Immediately after moving into the calving box, 3D-accelerometers (Smartbow) were fixed on the dams tail. The modified ear-tag sensor was originally designed to measure acceleration during the head movement of the cow in three axis (xyz). The measured values ranged from -2 to +2 g and were recorded and stored 10 times per second (10 Hz). A 1000 mAh battery provided the energy for each tag. Figure 1 demonstrates the attachment of the sensor. For this study, the tags were mounted on the upper part of the tail at the level corresponding with the ventral vulva commissure. They were fixed with tape (Leukoplast® 2.5 cm × 5 m, BSN medical GmbH Hamburg, Germany) providing a secure fit, but still ensuring sufficient blood circulation by applying it as a figure eight bandage. A self-adhesive bandage (HS Kohäsivbinde 10 cm × 4.5 m, Henry Schein Medical Austria GmbH, Vienna, Austria) was applied as an external protection.
For the five hour period preceding birth, the evaluation of the recorded videos was performed using a behavior coding software (Interact, Mangold International GmbH, Arnstorf, Germany). Therefore, the events 'frequency and duration of tail-raising', 'rupture of the amniotic sac' and 'expulsion of the calf' were encoded. This behavior data provided the gold standard. The accelerometer data was analyzed retrospectively for the two days before calving until parturition and compared with the gold standard.

Table 1. Specification of the cows used in the experiment.

Number	Breed	Lactation number	Gestation length
1	Brown Swiss	8	282
2	Simmental	1	286
3	Simmental	2	278
4	Simmental	6	283
5	Simmental	1	285

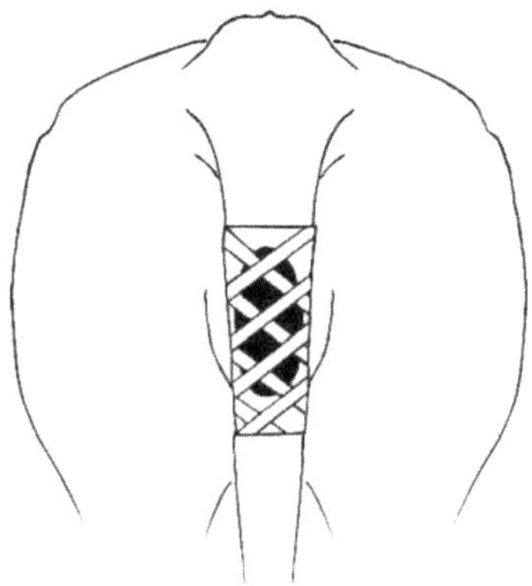

Figure 1. Fixation of the accelerometer at the tail of the cows.

3. Results

In total, 85, 110, 93, 76 and 118 tail-raise events were encoded in the videos during the standing bouts for the five dams. An algorithm was developed to perceive this tail raising. In addition, a decision function based on the frequency and duration of the tail raising was created. Exceeding a threshold led to the decision making and consequent signaling of a birth alarm as shown in Figure 2. In each of the five cases, the alarm was triggered a short time before the expulsion of the calves, at 16, 34, 56, 71 and 160 minutes, respectively.

4. Practical implication

Further research is required to specify the algorithm and the decision function. Additional movement elements will be analyzed and added to an extended feature pool. In following steps, this multi-parameter system will be evaluated for predicting parturition, detecting dystocia and testing its usage as a real-time alarm system under field conditions to reduce calf mortality by providing an optimized calving management.

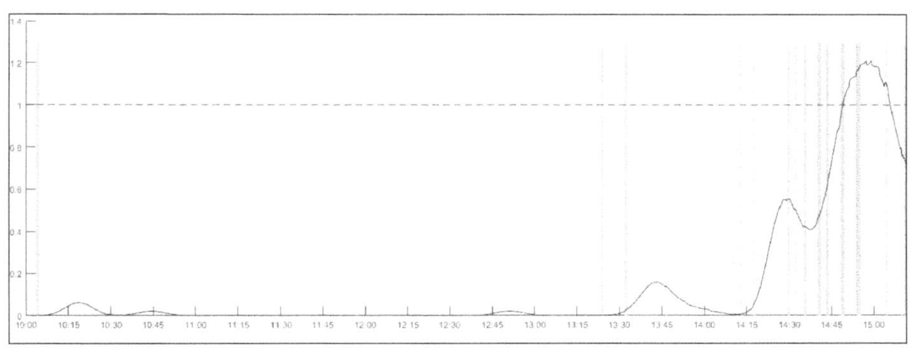

Figure 2. Example of a decision function (—) exceeding the threshoåld (----) for signaling the birth, based on the frequency and duration of the tail raising ( ).

Acknowledgments

The authors are grateful to 'The Austrian Research Promotion Agency' (FFG) of the Austrian Government for financial support of this study.

References

Kausch, M. (2009). Inzidenz und ursachen von totgeburten in einer milchviehanlage in brandenburg bei optimiertem geburtsmanagement. Klinik für Rinder. Doctoral thesis, Tierärztliche Hochschule Hannover, Germany.

Leister, T.M. (2009). Untersuchungen zur vitalität neugeborener kälber in einer milchviehanlage in brandenburg bei optimiertem geburtsmanagement. Doctoral thesis, Tierärztliche Hochschule Hannover, Germany.

Mee, J.F. (2004). Managing the dairy cow at calving time. Veterinary Clinics of North America Food Animal Practice 20: 521-546.

Mee, J.F. (2008a). Prevalence and risk factors for dystocia in dairy cattle: a review. Veterinary Journal 176: 93-101.

Mee, J.F. (2008b). Newborn dairy calf management. Veterinary Clinics of North America Food Animal Practice 24: 1-17.

Mee, J.F. (2008c). Perinatal mortality in heifers – an emerging problem. Cattle Practice 16: 166-173.

Meyer, C.L., Berger, P.J., Koehler, K.J., Thompson, J.R. and Sattler, C.G. (2001). Phenotypic trends in incidence of stillbirth for Holsteins in the United States. Journal of Dairy Science 84: 515-523.

Philipsson, J. (1996). Strategies to reduce problems in calving performance and stillbirth. In: Proceedings of the International workshop on genetic improvements of functional traits in cattle. Interbull Bulletin, Gemloux, Belgium, pp. 65-71.

Saint-Dizier, M. and Chastant-Maillard, S. (2015). Methods and on-farm devices to predict calving time in cattle. Veterinary Journal 205(3): 349-356.

Predicting calving in dairy cattle using machine-learning methods applied to automatically collected activity, rumination, and lying behaviors

M.R. Borchers[1], Y.M. Chang[2], K.L. Proudfoot[3], B.A. Wadsworth[1], A.E. Stone[1] and J.M. Bewley[1*]
[1]*Department of Animal and Food Sciences, University of Kentucky, Lexington, KY 40546, USA;* [2]*Research Support Office, Royal Veterinary College, University of London, NW1 0TU London, United Kingdom;* [3]*Department of Veterinary Preventive Medicine, Ohio State University, Columbus, OH 43210, USA; jbewley@uky.edu*

Abstract

The objective of this study was to use automated activity, lying, and rumination monitors to characterize prepartum behavior and predict calving in dairy cattle. Data was collected from 20 primiparous and 31 multiparous Holstein dairy cattle beginning September 2011, through May 2013 at the University of Kentucky Coldstream Dairy. The HR Tag (SCR Engineers, Ltd., Netanya, Israel) automatically collected neck activity and rumination data in 2-h increments. The IceQube (IceRobotics, Ltd., Scotland) automatically collected number of steps, lying time, standing time, the number of transitions from a standing position to a lying position (lying bouts), and total motion summed in 15-min increments. IceQube data were summed in 2-h increments to match HR Tag data. All behavioral data was collected for 14 d before predicted calving date. To determine if the technologies were useful for predicting calving, random forest, linear discriminant analysis, and neural network machine-learning techniques were constructed and implemented using R version 3.1.0 (R Foundation for Statistical Computing, Vienna, Austria). These methods were used to predict calving events using 14 d of behavioral data. Two machine-learning analyses were combined to predict calving; a daily analysis was used to identify the day of calving and a bihourly analysis identifying the 8-h period before calving (the time period in which calving began). This analysis was performed on individual parameters from each technology, and a combination of parameters from both technologies. A neural network analysis combining parameters from both technologies yielded the best prediction of the day before calving (sensitivity, 100.0%; specificity, 86.8%). The same analysis of bihourly increments identified the 8-h period before calving with 86% sensitivity and 80% specificity. Changes in behavior and machine-learning alerts indicate commercial precision dairy technologies may have potential for calving prediction.

Keywords: cattle behavior, calving prediction, precision dairy monitoring technology

1. Introduction

Parturition is considered one of the most important periods for both cows and their calves. Dystocia and calf mortality in this period can negatively impact farm economics and animal welfare (Mee, 2004). Providing timely calving assistance may reduce the risk of dystocia, reduce the pain associated with assisted labor (Mainau and Manteca, 2011), and improve reproductive performance in the subsequent lactation (Bellows *et al.*, 1988).

Before calving, a cow's udder will engorge with milk, the vulva will swell and loosen, pelvic ligaments will relax (Hulsen, 2006), and the tail will raise more frequently and for a longer duration (Miedema *et al.*, 2011). These cues are used to estimate calving times, but they do not always occur and even experienced personnel cannot accurately predict calving events from these observations (Hofmann *et al.*, 2006; Sendag *et al.*, 2008). Precision dairy monitoring technologies provide alternatives to the subjective observation and assessment of calving behaviors. Evidence exists that dairy cows change feeding and rumination behavior (Huzzey *et al.*, 2005; Schirmann *et al.*, 2013), and lying behavior (Huzzey *et al.*, 2005; Jensen, 2012; Miedema *et al.*, 2011) as calving approaches, making technologies that measure these behaviors potentially useful calving monitors.

Before these technologies can become useful as predictors of calving, research is needed to determine if the behaviors measured by the technologies (e.g. activity, rumination, and lying behavior) are highly sensitive and specific in detecting imminent calving. The objectives of this study were to: (1) quantify activity, rumination, and lying behaviors before calving using two commercially available technologies; and (2) determine how well the parameters measured by these technologies, either individually or in combination, predict calving time.

2. Materials and methods

2.1. Data collection

Data was collected using 20 primiparous and 33 multiparous prepartum Holstein dairy cattle from September 2011 through May 2013 at the University of Kentucky Coldstream Dairy Facility (IACUC Protocol Number: 2010-0776). At least 30 d before expected calving date, cows were moved to dry cow facilities and housed in a 9.15×21.34 m sawdust pack with constant access to pasture (3.64 hectares). A total mixed ration was delivered to the pen once daily. Two technologies were fitted to each cow by 28 d before predicted calving and 2 weeks of data was collected from each cow, retrospectively. The HR Tag (SCR Engineers, Ltd., Netanya, Israel) was placed on the left side of the neck and used to automatically collect neck activity and rumination data in 2 h time increments using a 3-axis accelerometer and a microphone with microprocessor, respectively. The IceQube (IceRobotics, Ltd., Scotland) was attached to the left leg and automatically collected number of steps, time spent lying, time spent standing (inverse of time spent lying), the number of transitions from a standing position to a lying position (lying bouts), and a proprietary total motion variable in 15 min time blocks using a 3-axis accelerometer.

Cows in the dry pen were monitored for signs of calving every 3 h. Individual cows were monitored every 15 min after the first sign of labor was detected (e.g. the amniotic sac or calf feet became visible outside of the vulva). For each calving, farm staff were instructed to record the calving date, the cow's parity, the time calving began, and the approximate time calves were fully outside the cow.

2.2 Prediction algorithm development

Machine-learning techniques were applied to the data sets to predict calving. The three machine learning analysis techniques used for calving prediction were random forest, linear discriminant analysis, and neural network analyses. The random forest method is based on decision tree classification methods and develops a group of tree-structured classification algorithms. Each tree contributes an opinion of how the data should be classified (Bishop, 2006; Breiman, 2001; Shahinfar *et al.*, 2014). Linear discriminant analysis is similar to analysis of variance and regression methods, but uses a categorical dependent variable, and several continuous independent variables (McLachlan, 2004; Wetcher-Hendricks, 2011). Neural networks imitate the structure and function of the human brain, simulating human intelligence, learning independently and quickly, adapting continuously, and applying inductive reasoning to process knowledge (Krieter *et al.*, 2006; Zahedi, 1991). In animal sciences, neural networks are the most frequently used machine-learning method (Shahinfar *et al.*, 2014).

Only cows contributing data from the desired technology, or all data from both technologies (for the combination analysis) for each time interval of interest were considered in prediction models for individual days. Parity, days until predicted calving date, neck activity, rumination time, number of steps, time spent lying, number of lying bouts, and total motion data were used to predict calving events. The IceQube also monitored standing time (the inverse of lying time). Standing time was a parameter supplied by the technology and all available parameters were used in prediction models to simulate actual conditions. For daily calving prediction, the variable of interest for prediction was Day -1 from the 14 days of data. Data was summed from 00:00 through 23:59 for each day. Day 0 was not considered in daily prediction models to exclude calving times.

A 24-h backwards moving average was calculated for each bihourly interval. Machine-learning techniques were applied to 22 h of data before calving. The bihourly block just before calving was excluded from analysis because the alert would come after calving was completed, and thus would not be obtained in a timely fashion for a producer to execute any meaningful interventions before calving. The variable of interest was the 2-h period before calving (bihourly blocks from -2 h to -22 h before calving; 11 total time points). Parity, time of day, neck activity, rumination time, number of steps, time spent lying, number of lying bouts, and total motion data (with the addition of standing behavior, similarly to the daily analysis) were used to predict calving events.

Separate random forest, linear discriminant analysis, and neural network analyses were performed for the IceQube, the HR Tag, and a combination of parameters from both technologies (9 daily and 9 bihourly analyses). A data subset consisting of 80% of the observations was used as a training set to generate prediction algorithms. The remaining 20% were used to evaluate the performance of the algorithms. A leave-one-out cross-validation method was performed for each machine-learning method to develop training phase algorithms. During the testing phase, trained algorithms were used to predict daily or 2-h periods before calving. To make the prediction methods as applicable in actual calving situations as possible, two algorithms performed were performed sequentially. The sequential, two-step algorithm first identified the day of calving. Next, the bihourly data from the day of calving was used to identify the 8 h period immediately preceding calving. True positives, false positives, true negatives, and false negatives were calculated for each daily and 2-h period. For 2-h periods, true positive, false positive, true negative and false negative definitions were modified to predict an 8-h block before calving in order to identify the window before calving where managers could observe and intervene if necessary. Using these values, the sensitivity, specificity, positive predictive value, and negative predictive values were calculated to evaluate the performance of different machine-learning techniques and technology. All analyses were constructed and implemented using the <caret> package in R version 3.1.0 (R Foundation for Statistical Computing, Vienna, Austria).

3. Results and discussion

Prediction performance for daily methods is shown in Table 1. Methods were best able to predict the day before calving using a combination of parameters from the HR Tag and IceQube. The best calving prediction results were obtained in the combined parameter neural network analysis with a sensitivity of 100.0%, a specificity of 86.8%, a positive predictive value of 40.0%, and a negative predictive value >99.9%. The bihourly prediction method was developed under the assumption that following identification of the day before calving, a bihourly analysis could commence. A flaw with this approach would be if the daily analysis failed to identify the day before calving, the bihourly analysis would never commence. In the bihourly analysis, alerts generated in the 8 h time block before calving were classified as true positives. Sensitivity, specificity, positive predictive value, and negative predictive value were calculated for this period, and results are presented in Table 1. Similar to the daily analysis, neural network results of the combination analysis were the most favorable with a sensitivity of 86.2%, a specificity of 80.4%, a positive predictive value of 69.4%, and a negative predictive value of 91.8%.

Predicting calving using a combination of automatically collected behavioral parameters was previously attempted by Maltz and Antler (2007). Their study described calving prediction methods using changes in daily step number, lying behavior, and number of times passing into a feeding area for 12 cows over 7 d. Their method achieved a sensitivity of 83.3% and a specificity of 95.2% in predicting the day of calving. Similarly, this study also indicates a combination of parameters to be most useful in calving prediction.

Table 1. Results of leave one out cross-validated machine-learning analysis for predicting the day and 8-h period before calving[1] using daily and bihourly behavior data from the HR Tag[2] and IceQube[3] for the 14 days before calving (n=53).[4]

Analysis	Technology	Sensitivity (%)	Specificity (%)	Positive predictive value (%)	Negative predictive value (%)
Daily					
Random forest	HR Tag	12.5	95.6	20.0	92.6
	IceQube	37.5	89.0	23.1	94.2
	Combination[5]	25.0	89.0	16.7	93.1
Linear discriminant analysis	HR Tag	25.0	96.7	40.0	93.6
	IceQube	75.0	91.2	42.9	97.6
	Combination[5]	75.0	93.4	50.0	97.7
Neural network	HR Tag	0.0	98.9	0.0	91.8
	IceQube	50.0	87.9	26.7	95.2
	Combination[5]	100.0	86.8	40.0	100
Bihourly					
Random forest	HR Tag	61.5	88.5	72.7	82.1
	IceQube	54.5	88.7	66.7	82.5
	Combination[3]	68.0	92.5	81.0	86.0
Linear discriminant analysis	HR Tag	83.3	83.0	71.4	90.7
	IceQube	78.3	88.5	75.0	90.2
	Combination[3]	75.0	85.2	69.2	88.5
Neural network	HR Tag	81.0	84.9	68.0	91.8
	IceQube	70.8	86.8	70.8	86.8
	Combination[3]	86.2	80.4	69.4	91.8

[1] The day or bihourly period immediate preceding calving was excluded from machine learning analyses.
[2] The HR Tag (SCR Engineers, Ltd., Netanya, Israel) measured neck activity and rumination.
[3] The IceQube (IceRobotics, Ltd., Scotland) measured lying bouts, lying time, standing time, step number, and total motion.
[4] Sensitivity = TP / (TP + FN) × 100, specificity = TN / (TN + FP) × 100, positive predictive value = TP / (TP + FP) × 100, negative predictive value = TN / (TN + FN) × 100; where TP = true positive, TN = true negative, FP = false positive, and FN = false negative.
[5] Parameters from both the HR Tag and the IceQube were used in combination analyses.

To the knowledge of the authors, no technologies monitor rumination, lying behavior, and activity in unison. A two-technology approach, such as that used in this study, would be useful in calving prediction, but potentially not economically justifiable on commercial farms. In the absence of a two-technology calving prediction approach, results indicate machine-learning techniques applied to single technologies similar to the IceQube may be the best option in behavior-based calving prediction at the daily level, while technologies such as the HR Tag may be more suited for the bihourly level. At the daily level, the IceQube sensor alone effectively predicted the day of calvings in the random forest analysis with a sensitivity of 75.0%, a specificity of 91.2%, a positive predictive value of 42.9%, and a negative predictive value of 97.6%. At the bihourly level, the HR Tag alone identified the 8 h time block before calving with a sensitivity of 83.3%, a specificity of 83.0%, a positive predictive value of 75.9%, and a negative predictive value of 90.2%.

True positive, false positive, true negative and false negative definitions were modified to predict an 8-h block before calving. This was necessary because sensitivity and specificity are inversely related and if an alert threshold is increased or decreased to make a respectively more specific or sensitive test, the specificity and sensitivity will proportionally and inversely change (Hogeveen *et al.*, 2010). Prediction methods predicting the 2-h period before calving produced large specificity values and low sensitivity values. Larger specificity values have traditionally been more valued in estrus and health because of the cost associated with missed events (Hogeveen *et al.*, 2010; ISO, 2007; Rutten *et al.*, 2013). In animal illness detection, false positives (type I errors) can cause financial losses through unnecessary treatment (Burfeind *et al.*, 2010). This may not be as applicable in calving prediction. False negatives in calving prediction would be instances where systems do not detect actual calving events. The consequences of missed calving events could be extremely detrimental (dystocia, stillbirth, cow death, etc.) and may outweigh the comparative increase in farm labor from incorrectly identified calving events. Future research in calving prediction and economic modeling may need to further explore this relationship.

4. Conclusions

Automatically monitoring the behavior of prepartum dairy cows can provide additional uses for commercially available technologies already used to generate health and estrus alerts. A combination of parameters from the HR Tag and IceQube in machine-learning techniques yielded promising calving prediction results. In the absence of rumination behavior, lying behavior data could accurately predict the day before calving with machine-learning prediction methods. In the absence of lying behaviors, rumination and activity behaviors could accurately predict the 8-h period before calving using machine-learning prediction methods. To maximize calving prediction alert usefulness, future studies will need to focus on the implementation of these measures in real-time on dairy farms.

Acknowledgements

The authors would like to thank Joey Clark, Denise Ray, and the UK Coldstream Dairy Staff for collecting calving data.

References

Bellows, R.A., Short, R.E., Staigmiller, R.B. and Milmine, W.L. (1988). Effects of induced parturition and early obstetrical assistance in beef cattle. Journal of Animal Science 66: 1073-1080.

Bishop, C.M. (2006). Pattern recognition and machine learning. Springer, New York, NY, USA.

Breiman, L. (2001). Random forests. Machine learning 45: 5-32.

Hofmann, E., Failing, K. and Wehrend, A. (2006). Changes of the vulva and the vestibulum in suckler cows and heifers during the last seven days ante partum. Tieraerztliche Praxis Ausgabe Grosstiere Nutztiere 34: 15-19.

Hogeveen, H., Kamphuis, C., Steeneveld, W. and Mollenhorst, H. (2010). Sensors and clinical mastitis – the quest for the perfect alert. Sensors 10: 7991-8009.

Hulsen, J. (2006). Cow signals: the practical guide for dairy cow management. Roodbont Publishers, Zuthpen, the Netherlands.

Huzzey, J.M., Von Keyserlingk, M.A. and Weary, D.M. (2005). Changes in feeding, drinking, and standing behavior of dairy cows during the transition period. Journal of Dairy Science 88: 2454-2461.

International Standards Organization (2007). Automatic milking systems – requirements and testing. Annex C: example of methods of evaluating detection systems for milk deemed as abnormal due to blood or changes in homogeneity. International Standards Organization, Geneva, Switzerland.

Jensen, M.B. (2012). Behaviour around the time of calving in dairy cows. Applied Animal Behaviour Science 139: 195-202.

Krieter, J., Stamer, E. and Junge, W. (2006). Control charts and neural networks for oestrus dectection in dairy cows. GIL Jahrestagung, pp. 133-136.
Maltz, E. and Antler, A. (2007). A practical way to detect approaching calving of the dairy cow by a behaviour sensor. Precision Livestock Farming 7: 141-146.
McLachlan, G. (2004). Discriminant analysis and statistical pattern recognition. John Wiley and Sons, New York, NY, USA.
Mee, J.F. (2004). Managing the dairy cow at calving time. Veterinary Clinics of North America: Food Animal Practice 20: 521-546.
Miedema, H.M., Cockram, M.S., Dwyer, C.M. and Macrae, A.I. (2011). Changes in the behaviour of dairy cows during the 24h before normal calving compared with behaviour during late pregnancy. Applied Animal Behaviour Science 131: 8-14.
Rutten, C.J., Velthuis, A.G., Steeneveld, W. and Hogeveen, H. (2013). Invited review: sensors to support health management on dairy farms. Journal of Dairy Science 96: 1928-1952.
Schirmann, K., Chapinal, N., Weary, D.M., Vickers, L. and Von Keyserlingk, M.A.G. (2013). Short communication: rumination and feeding behavior before and after calving in dairy cows. Journal of Dairy Science 96: 7088-7092.
Sendag, S., Hofmann, E. and Wehrend, A. (2008). Study about the occurrence of visible signs of approaching parturition in suckler cows and heifers: changes of the udder. Deutsche Tierarztliche Wochenschrift 115: 66-70.
Shahinfar, S., Page, D., Guenther, J., Cabrera, V., Fricke, P. and Weigel, K. (2014). Prediction of insemination outcomes in holstein dairy cattle using alternative machine learning algorithms. Journal of Dairy Science 97: 731-742.
Wetcher-Hendricks, D. (2011). Analyzing quantitative data: an introduction for social researchers. John Wiley and Sons, New York, NY, USA.
Zahedi, F. (1991). An introduction to neural networks and a comparison with artificial-intelligence and expert systems. Interfaces 21: 25-38.

Consistency of measurements from an automatic body condition scoring camera

C. Hallén Sandgren[1*] and U. Emanuelson[2]
[1]*DeLaval International AB, P.O. Box 39, 147 21 Tumba, Sweden;* [2]*Department of Clinical Sciences, Swedish University of Agricultural Sciences, P.O. Box 7054, 750 07 Uppsala, Sweden; charlotte.hallensandgren@delaval.com*

Abstract

The aim of this study was to determine repeatability of body condition score (BCS 1-5) recorded in a technology based on a continuously running 3D camera, linked to a RFID system (DeLaval BCS camera). The study was performed on data over 11-25 months from 4 herds in 3 countries. The breed composition of the herds was two with pure Holstein Friesian (HF), one with Simmental and one with 75% Norwegian Red and 25% HF crosses. The average daily milk yield varied between 27.1 and 29.6 kg. The monitoring period had a median number BCS scores per day per herd ranging from 57 to 548. An analysis of intra-class correlation (ICC) during week 1-8 (excluding day of calving) between repeated measurements of BCS was conducted with a linear mixed model, where herd and cow within herd were the only variables and included as random effects. Estimated variance components were used to calculate the ICC. A total of 8,045 scores were retrieved from the herds and used in the calculations. LS-means for BCS with farm as fixed effect adjusting for season and lactation number in week 1 and week 8 were 3.3/3.0, 3.8/3.5, 3.7/3.4 and 3.2/2.8 respectively in the four herds. The ICC was 0.86 and 0.94 respectively for week 1-4 and 5-8. About 42% of the total variation was explained by cow and 43% by herd. BCS patterns between herds and lactation numbers during the first 8 weeks confirm earlier findings from literature and the reliability and robustness of the tested technology. The technology is reported to be useful to follow up on dry cow feeding and management, selection of inefficient cows for possible culling and monitoring of finishing of cows for slaughter.

Keywords: automatic, body condition scoring, camera

1. Introduction

Cows but also herds with unfavorable body condition scores (BCS) over the lactation cycle will display reduced milk production, decreased feed efficiency, impaired reproduction performance and an increased number of post partum diseases with significant negative impact on farm profitability (Roche *et al.*, 2009). To mention only one example; milk yield is greatly influenced by the body condition score, and differences in milk yield between a too lean cow compared to one in an optimal body condition score could be as high as 545 kg during the first 120 days of lactation (Domecq *et al.*, 1997).
Although it has been evident for centuries that cows lose and gain condition during the lactation cycle, there was no simple measure of a cow's stored energy reserves until the 1970s (Stockdale, 2001). Body weight (BW) alone is not a good indicator of body reserves (Enevoldsen and Kristensen, 1997). Recent work (Thorup *et al.*, 2012) indicates that providing very frequent BW measurements may help overcome some of these issues, but for accurate assessment of adipose and lean tissue, measurements of body condition are needed.
With this as a background, body condition scoring has become the most accepted and used method for assessment of the nutritional status of dairy cows (Roche *et al.*, 2009). However, despite the huge potential in an improved profitability as a result of close management of body condition scores,

few farmers score their cows frequently enough, if at all. This is mainly due to lack of either specific knowledge or time, or alternatively, hesitance to spend money on a professional scorer. Further on, the subjective nature of manual body condition may result in a too low precision of the method (Ferguson *et al.*, 1997; Kristensen *et al.*, 2006).
The purpose of this paper is to describe the recently launched body condition scoring system (DeLaval BCS camera, Tumba Sweden), which overcomes the hurdles hampering a widespread implementation and systematic use of body condition scoring for optimizing herd performance, and to determine the repeatability of the scores during the early part of the lactation.

2. Materials and methods

2.1 Description of the body condition score recording system

The technology is based on a 3D camera, linked to an RFID system which is continuously recording, from above, the rear part of the back from the short ribs to the tail end. Anytime a cow passes under the camera, the system recognizes the movement and selects the best image of the cow in the video sequence. The 3D camera uses light coding technology, which works by projecting a pattern of IR dots on the back of the cow. In a following step the distance between these dots are measured to create an accurate 3D image of the back and an algorithm converting that image information into a body condition score is applied. Depending on the camera location, cows are scored 2-10 times per day. As golden standard the scale used to develop algorithms was visual scoring based on the 1 to 5 scale as described in Elanco Animal Health (2009) supplemented with assessing the spinous to transverse processes (Edmonson *et al.*, 1989) where 1 corresponds to the lowest and 5 to the highest condition score. The system provides daily 7-days rolling average scores, for all cows with at least one, by the system, approved image (the 20% highest and lowest values per cow removed). Currently there are algorithms available for scoring Holsteins including similar breeds, Simmental and Norwegian Red.
The individual BCS scores, comprising 7-day average scores, has been shown to have an average standard deviation (STD) of 0.04, with STD's ranging from 0.0 to 0.12, and with 99% of the 7-day scores having a STD of the individual scores of ≤0.1 (Forsén, personal information). The accuracy of the 3D camera has been verified showing that the BCS camera scores 95% of the cows within a ± 0.25 interval compared to BCS assessment by a professional scorer (Mazeris, 2015).

2.2 Herds and data

Individual cow data on lactation number, daily milk yield and BCS, and dates of calving and BCS recording, was retrieved from the software program Delpro from 4 herds from 3 countries (2 DE, 1 NL, 1 NO). The breed composition of the herds was two with pure Holstein Friesian (HF), one with Simmental and one with 75% Norwegian Red and 25% HF crosses. The monitoring period ranged from between January 2014 to April 2015 until February 2016, with a median number BCS scores per day per herd ranging from 57 to 548.

2.3 Statistical methods

In order to avoid autocorrelation only one 7-day rolling average BCS per week per animal was used in the statistical analyses. Observations on calving day (days in milk at BCS recording is 0) were excluded and only observations during week 1-8 after calving were included. A linear mixed model, with the random effects of herd and cow within herd included as the only explanatory variables, was used to study the repeatability of BCS. The estimated variance components were used to calculate the intra-class correlation (ICC), as a measure of the repeatability of the observations. The ICC was calculated as:

$$ICC = (\sigma^2_{herd} + \sigma^2_{cow})/(\sigma^2_{herd} + \sigma^2_{cow} + \sigma^2_{residual})$$

In addition, a linear mixed model, with the effect of cow included as a random variable, and with herd, week in milk (1-8), lactation number (1, 2, 3+) and season of calving (April-September vs October-March) as fixed explanatory variables, was used to study the patterns of BCS. The same model was used to study patterns in daily milk yield (kg).

3. Results and discussion

3.1 Consistency of measurements

Altogether, 8,045 weekly scorings from the BCS camera, in lactation weeks 1-8 were retrieved from the 4 herds as presented in Table 1. The estimated variance components are presented in Table 2 and shows that, for the whole period, 42% of the variation was explained by differences between individual cows and 43% variation between herds. The ICC was generally high, showing a high repeatability of BCS observations within cow, but the ICC was slightly higher in week 5-8 (94%) as compared with week 1-4 (86%). The ICC increased further in the following 4 week period (data not shown).

3.2 Patterns for body condition score and milk yield

Least-square means from the linear mixed model showed different BCS patterns in the study herds. Mean BCS in the study herds at the first week after calving varied between 3.2 and 3.8 with the highest in the NRF and Simmental herds and lowest in the HF-herds. In herd B nadir BCS was reached in week 7 whereas steady state during the study period was never reached in the remaining three herds. Moreover, the drop in BCS from week 1 to 8 was 0.41 in herd D as compared to 0.25 in

Table 1. Number of observations per herd and lactation week.

	Herd			
Week	A	B	C	D
1	441	73	189	186
2	546	79	201	185
3	576	78	195	177
4	577	75	201	170
5	577	70	199	188
6	578	72	205	175
7	571	71	200	176
8	565	69	203	177
Total	4,431	587	1,593	1,434

Table 2. Variance components and intra class correlations (ICC) for body condition scores in lactation week 1-8.

	Week 1-8	Week 1-4	Week 5-8
Cow	0.089	0.081	0.102
Herd	0.091	0.080	0.099
Residual	0.032	0.026	0.014
ICC	0.85	0.86	0.94

herds B (Figure 1A). The BSC loss during the study period was 0.39 highest for cows in lactation ≥3 and 0.28 for first and second lactation cows that showed almost identical BCS loss. Cows in second lactation started the lactation with a lower BSC compared to cows in 1st and ≥3rd lactation (Figure 1B). Moreover, the linear mixed model was also used to study patterns in daily milk yield (kg). Herd D showed a yield curve for the first 8 weeks clearly distinguished from the remaining 3 herds reaching peak yield in week 4 and then starting to decline in week 5 (Figure 2). Difference in milk yield between lactations was 9.1 and 11.0 kg higher per day for 2nd and ≥3rd lactation cows respectively compared to 1st lactation. Hence, the difference in daily milk yield between lactation 2 and 3 was only 1.9 kg. Lactation curves from cows in lactation 2 showed the weakest persistency starting to decline already in week 8 (data not shown).

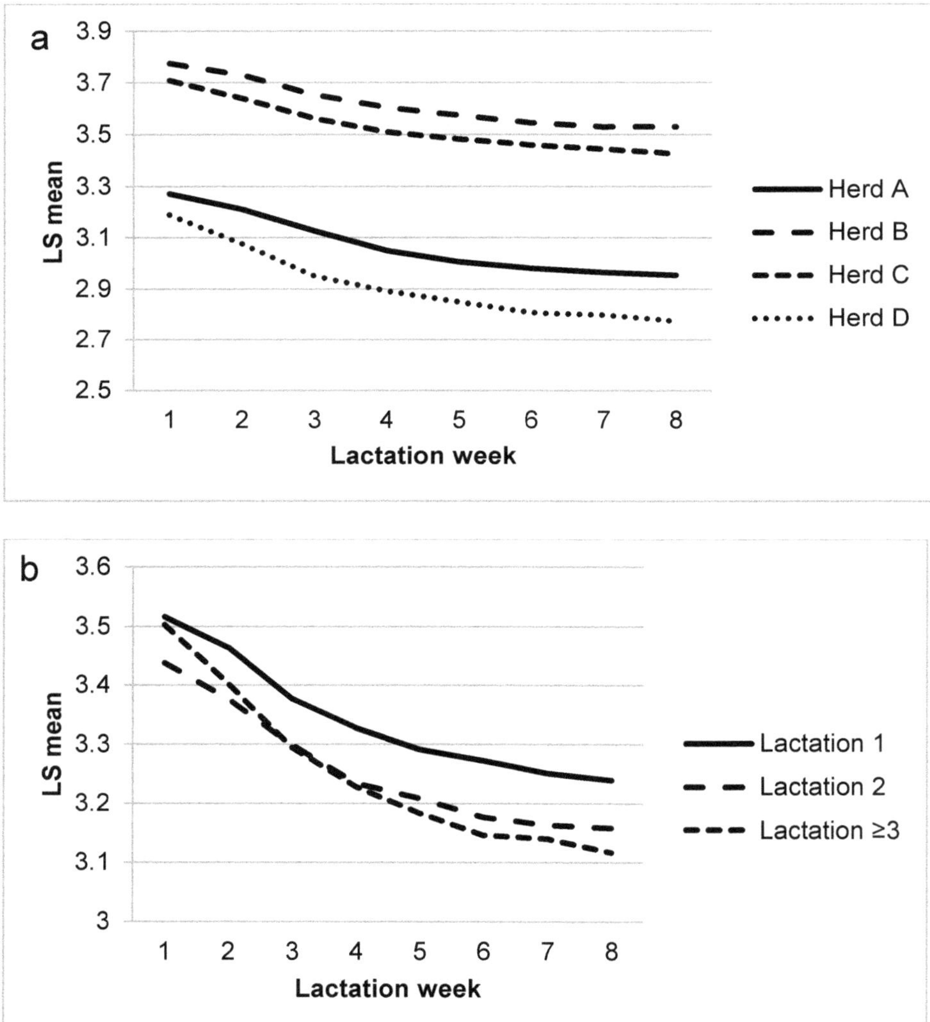

Figure 1. Least-square (LS) means for the effects of the interaction between herd and lactation week (A) and between lactation number and lactation week (B) on body condition score.

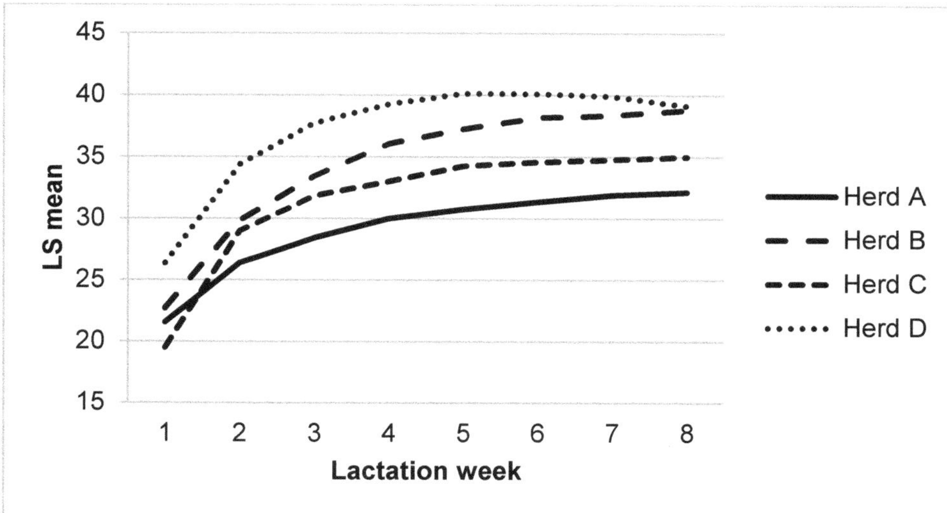

Figure 2. Least-square (LS) means for the effects of the interaction between herd and lactation week milk yield.

We chose to study lactation week 1-8 as this is the time period in the lactation cycle when the body condition score is known to change rapidly and hence the results would be indicative of the validity of the output from the BCS camera. And indeed, we can conclude the technology provides consistent and relevant BCS results also during this period, i.e. a high intra class correlation. The somewhat lower intra class correlation during week 1-4, as compared with 5-8, is likely to be related to the rapid change in BCS during this period. Other indications of the validity of the scoring system was the clear difference in BCS level and pattern between herds consisting of pure HF cows, A and D, as compared with herd B and C with HF/NRF crossings and Simmental, respectively.
The BCS pattern in different lactations compared with milk yield confirms the result from others (Roche *et al.*, 2009) such as the BCS in second lactation cows was lower at the beginning of lactation and 1st lactation cows starting at a high BCS and showing lower loss. The combined evaluation of BCS and milk yield in Herd A indicates a general feeding problem in that herd, which also was confirmed Based on the results and the experience from the farmers and feed advisors using the technology it reported to be very useful in order to make adjustments of dry cow feeding and management. Moreover, BCS camera has been used with great success to select inefficient cows for possible culling, typically cows building up BCS and not responding to feed 60-90 days in milk. Also, larger farms have used the technology for efficient monitoring of finishing of cows for slaughter. The technology is still young and it is obvious that an extended number of applications will develop over time.

4. Conclusions

It is clear that the automatic body condition scoring camera tested in this study will be a valuable tool to support evaluations of feed efficiency and health.
Based on the obtained ICC values it can be concluded that the measurements conducted with the DeLaval BCS camera show a very high consistency within cow and herd during the most dynamic period of the lactation cycle with respect to BCS and daily milk yield. Also, the presented BCS patterns are in line with earlier reported findings from literature which confirm that the technology provide relevant and robust BCS values.

References

Domecq, J.J., Skidmore, A.L., Lloyd, J.W. and Kaneene, J.B. (1997). Relationship between body condition scores and milk yield in a large dairy herd of high yielding Holstein cows. Journal of Dairy Science 80: 101-112.

Edmonson, J., Lean, I.J., Weawer, L.D., Farver, T. and Webster, G. (1989). A body condition scoring chart for Holstein dairy cows. Journal of Dairy Science 72: 68-78.

Elanco Animal Health (2009). The 5-point body condition scoring system.

Enevoldsen, C. and Kristensen, T. (1997). Estimation of body weight from body size measurements and body condition scores in dairy cows. Journal of Dairy Science 80: 1988-1995.

Kristensen, E., Dueholm, L., Vink, D., Andersen, J.E., Jakobsen, E.B., Illum-Nielsen, S., Petersen, F.A. and Enevoldsen, C. (2006). Within and across-person uniformity of body condition scoring in Danish Holstein cattle. Journal of Dairy Science 89: 3721-3728.

Mazeris, F. (2015). DeLaval body condition scoring BCS: daily, automatic and consistent scoring of cows. Precision Dairy Conference, June 24-25, 2015. Rochester, MN, USA, pp. 47-50.

Roche, J.R., Friggens, N.C., Kay, J.K., Fisher, M.W., Stafford, K.J. and Berry, D.P. (2009). Invited review: body condition score and its association with dairy cow productivity, health, and welfare. Journal of Dairy Science 92: 5769-5801.

Stockdale, C.R. (2001). Body condition at calving and the performance of dairy cows in early lactation under Australian conditions: a review. Australian Journal of Experimental Agriculture 41: 823-829.

Thorup, V.M., Edwards, D. and Friggens, N.C. (2012). On-farm estimation of energy balance in dairy cows using only frequent body weight measurements and body condition score. Journal of Dairy Science 95: 1784-1793.

Association between behavior and subclinical ketosis in transition dairy cows

E.I. Kaufman[1], S.J. LeBlanc[2], B.W. McBride[1], T.F. Duffield[2] and T.J. Devries[1*]
[1]*Department of Animal Biosciences, University of Guelph, 50 Stone Road East, Guelph, ON, N1G 2W1, Canada;* [2]*Department of Population Medicine, University of Guelph, 50 Stone Road East, Guelph, ON, N1G 2W1, Canada; tdevries@uoguelph.ca*

Abstract

Automated monitoring of cow behavior may be useful in the early detection of health problems. The objective of this study was to characterize the relationship between rumination behavior, lying behavior, milk production, and subclinical ketosis (SCK) in transition dairy cows. A total of 339 Holstein dairy cows (107 primiparous and 232 multiparous; MP) were monitored for rumination and lying activity and SCK from 14 d prior to calving until 28 d after calving. Rumination and lying behavior were recorded daily using automated systems. A blood sample was taken once weekly from the coccygeal vein of each cow for measurement of beta-hydroxybutyrate (BHB). Cows with BHB ≥1.2 mmol/l in any of the 4 postpartum samples were considered to have SCK. Cases of retained placenta, metritis, milk fever, or mastitis were recorded. Comparisons were made between cows categorized into 1 of 3 groups: healthy cows (HLT) that had no SCK or any other health problem (n=139); cows with SCK (HYK) with no other health problems (n=97); or cows (HYK+) that had SCK and one or more other health problems (n=53). Multiparous HYK cows ruminated 25±12.8 min/d less ($P<0.05$) than HLT cows, while HYK+ cows ruminated 44±15.6 min/d less ($P<0.05$) than HLT cows. Lower rumination times in MP cows during wk -1 and wk +1 were associated with increased odds of HYK only (OR=1.1 per 20 min/d decrease; $P=0.04$) and HYK combined with another health problem (OR=1.2 per 20 min/d decrease; $P=0.006$), respectively. MP HYK+ cows spent 92±24.0 min/d more ($P< 0.001$) time lying than HLT cows in wk +1. Longer lying time in MP cows during that wk was associated with increased risk of SCK combined with another health problem (OR=1.15 per 20 min/d increase; $P=0.05$). A positive association was found between weekly summarized RT and milk yield in 1st ($P=0.04$) and 2nd lactation ($P<0.01$) cows from 4 to 28 DIM. A positive association ($P<0.01$) was also seen in parity 3+ cows, however, the predicted relationship between RT and milk yield differed across weeks in these cows. The results suggest monitoring behavior may be useful in identifying MP cows with SCK and another health problem in the wk after calving; monitoring rumination behavior may aid in the early identification of MP cows at higher risk for developing SCK post-calving. Finally, these results suggest that rumination time may be predictive of milk yield in early lactation cows.

Keywords: illness, subclinical hyperketonemia, automation

1. Introduction

The transition period has been defined as 3 wk before calving until 3 wk after calving (Drackley, 1999). This is a period of high energy demand where cows are going through many physiological and hormonal changes. Not only must cows support their calf during the final stages of development, but their bodies are getting ready to initiate lactation. While these cows are pressured with increased energy requirements, they have a depressed dry matter intake (DMI) in the days leading up to calving (Herdt, 2000). As milk production peaks earlier in lactation than DMI, essentially all cows experience negative energy balance (NEB) at the beginning of their lactation (Herdt, 2000). As milk production increases and cows continue in a state of NEB, they will begin to rely on fat stores to support their energy needs. Cows with subclinical ketosis (SCK) mobilize body reserves releasing ketone bodies in the blood (LeBlanc, 2010). This metabolic disorder is very prevalent in high-producing dairy herds, affecting an average of 43% of cows during the first 2 weeks of lactation (McArt *et al.*, 2012). Many risk factors for SCK have been identified, and there is much information on management approaches to help mitigate this illness. To reduce the risk of SCK cows must be provided with the proper nutrition during the close-up period (Overton and Waldron, 2004), competition at the feed bunk should be reduced, management should aim to reduce the amount of over conditioning during the prepartum period (Ingvartsen, 2006), and producers should implement monitoring programs to identify subclinically sick cows.

There are many methods available to diagnose SCK including milk and urine tests, and blood tests. Some methods are more subjective than others, testing must be done on a regular basis during the first few weeks after calving, and testing may become quite costly for producers. Even with these methods available, it is very difficult to diagnose ketosis in the very early stages. In recent years, there is a growing amount of information of monitoring animal behavior for the detecting illnesses. The primary objective of this study was to characterize the relationship between rumination behavior, lying behavior, and SCK in transition dairy cows. A secondary objective was to association rumination behavior with milk production in early lactation cows.

2. Materials and methods

This prospective observational study was conducted on 4 commercial dairy farms located in Eastern Ontario, Canada between March and October 2014. Herds were selected as a convenience sample according to proximity to the University of Guelph, Kemptville Campus (Kemptville, Ontario, Canada). Participating dairies milked between 125 and 400 Holstein cows, housed cows in free stall facility, fed a total mixed ration 1×/d, and milked in a parlor 3×/d. Animal use, data collection, and study design were approved by the University of Guelph's Animal Care Committee (AUP#2518) and Research Ethics Board (REB#14JA015).

We aimed to study individual cows from 2 weeks before calving until 4 weeks after calving. In total, 346 cows were monitored from an average of -16±5.4 d (mean±standard deviation; min=-34 d, max=-2 d) until +28 d relative to calving. An automated rumination monitoring system (Hi-Tag, SCR Engineers Ltd., Netanya, Israel) was installed at each participating dairy farm. Each cow was fitted with a SCR rumination collar at enrollment, which monitored rumination time 24 h/d, as validated for dairy cows by Schirmann *et al.* (2009), over the 6-wk study period. All enrolled cows were also affixed, weekly, with a data logger (Onset HOBO Pendant G data loggers; Onset Computer Corporation, Bourne, MA, USA) to record standing and lying behavior for 7 d, as validated by Ledgerwood *et al.* (2010). Loggers were replaced weekly and data were downloaded; recordings were used to calculate daily measurements of lying time (min/d), frequency of lying bouts (no. of bouts/d), and average lying bout length (min/bout). Two farms recorded milk weights at each milking to determine daily milk yield (n=170).

A blood sample was taken from the coccygeal vein of each cow for measurement of BHB 1×/wk; BHB was measured using an electronic hand-held device (Precision Xtra meter, Abbott Diabetes Care, Saint Laurent, QC, Canada), as validated by Iwersen *et al.* (2009). Cows with BHB ≥1.2mmol/l at one or more weekly sample postpartum were considered to have SCK (McArt *et al.*, 2012). Producer-reported cases of retained placenta, metritis, milk fever, or mastitis during the study period were also recorded.

Cows that had aborted (n=2), were culled (n=2), or diagnosed ketotic (n=3) before calving were not included in the statistical analysis. A total of 339 cows (107 primiparous and 232 multiparous) were categorized into 1 of 4 groups: healthy (HLT) cows had no SCK or any other recorded clinical disease (n=139); healthy plus (HLT+) cows were not diagnosed with SCK but were treated for at least one other health problem (n=50); SCK (HYK) cows with no other health problems during transition (n=97);

or ketotic plus (HYK+) cows that had SCK and one or more other clinical diseases (n=53). Linear and logistic multivariable mixed-effect models were then constructed to compared the measures of rumination behavior and lying behavior between HYK cows and HLT cows, and HYK+ cows and HLT cows, respectively. Further, linear multivariable mixed-effect models were constructed to determine the associated between rumination activity and milk production during the first 4 weeks of lactation.

3. Results

From 2 weeks before calving (wk -2) to 4 weeks after calving (wk +4), there was no difference (P=0.5) in rumination time (409±9.8 min/d) among HLT, HYK and HYK+ cows in their first lactation. Multiparous (MP) cows in HLT spent an average of 459±11.3 min/d ruminating from wk -2 to wk +4. Multiparous HYK cows ruminated 25±12.8 min/d less (P<0.05) than HLT cows, while HYK+ cows ruminated 44±15.6 min/d less (P<0.05) than HLT cows (Figure 1A). The largest differences

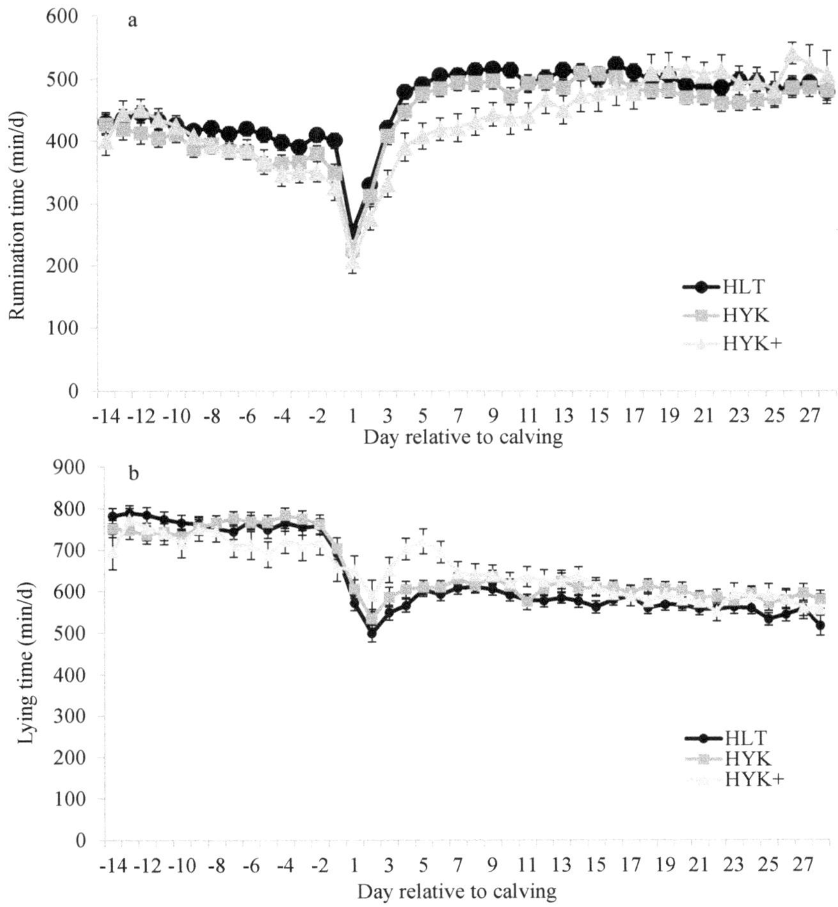

Figure 1. Average (mean±standard error; min/d) daily (A) rumination time and (B) lying time over the transition period (-14 to 28 d) for healthy multiparous (MP) cows with no other illnesses (HLT; n=87), subclinically ketotic MP cows with no other health issues (HYK; n=76) and subclinically ketotic MP cows with other health issues (HYK+; n=39).

between HLT and HYK+ cows were seen during wk -1, +1 and +2, when HYK+ cows ruminated 48±17.2 min/d (P<0.001), 73±16.0 min/d (P<0.001), and 65±19.4 min/d (P<0.01) less than HLT cows, respectively. In MP cows, increased odds of SCK were associated with greater milk yield in the previous lactation, greater loss of BCS across transition, greater stall stocking density in wk -1, and reduced rumination time in wk -1. For every 20 min/d decrease in rumination time during the week prior to calving, the odds of becoming SCK post-calving increased 1.1-fold (P=0.04). Increased odds of SCK and another clinical disease were associated with higher parity, longer dry period, greater stall stocking density in wk -1, and reduced rumination time in wk +1. The odds of developing SCK and another clinical disease were 1.2× greater (P=0.006) for every 20 min/d decrease in rumination time during wk +.

There was no difference among health categories in lying time, bout frequency, or bout length from 2 weeks before (-2) to 4 weeks after calving (+4) for primiparous (PP) cows. Differences in lying time for MP cows were seen in wk +1, when HYK+ cows spent 92±24.0 min/d more (P<0.001) time lying down than HLT cows, and during wk +3 and +4 when HYK cows spent more time (P≤0.03) lying down per day (19±20.7 min/d and 41±18.9 min/d, respectively) than HLT cows (Figure 1B). Increased odds of SCK and another clinical disease were associated with higher parity, longer dry period, greater stall stocking density in wk -1 and longer lying time during wk +1. The odds of developing SCK and another clinical disease were 1.15× greater (P=0.05) for every 20 min/d increase in lying time during wk +1. When comparing HYK to HLT cows, the same variables were associated with odds of SCK, however, lying time was not retained in the final model.

A positive association was found between weekly summarized RT and milk yield in 1st (+0.01±0.003 kg milk/min RT, P=0.04) and 2nd lactation (+0.01±0.004 kg milk/min RT, P<0.01) cows from 4 to 28 DIM. A positive association was also seen in parity 3+ cows (Figure 2), however, the relationship between RT and milk yield differed (P<0.01) across weeks (wk +1: +0.05±0.007 kg/min RT, Figure 2A; wk +2: +0.06±0.008 kg/min RT, Figure 2B; wk +3: +0.05±0.011 kg/min RT, Figure 2C; wk +4: +0.04±0.014 kg/min RT, Figure 2D).

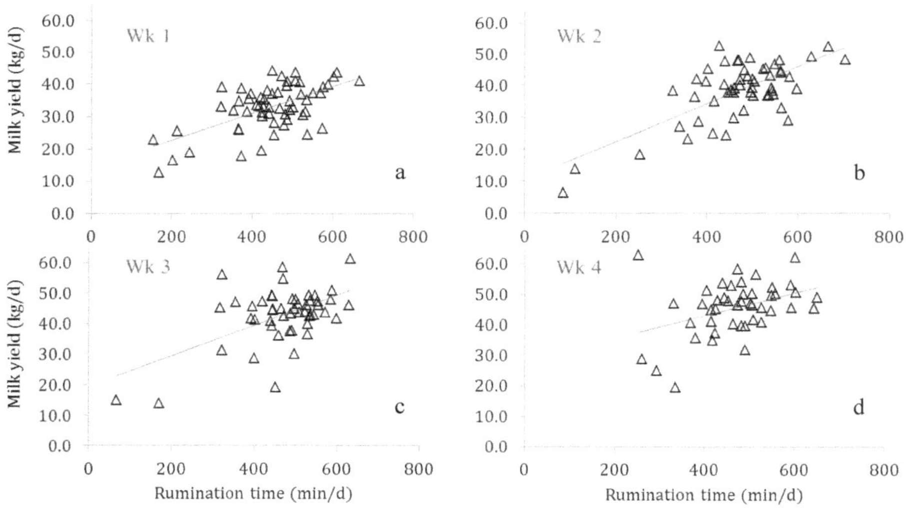

Figure 2. Scatter plot of daily milk yield and daily rumination time summarized by wk for 3rd + lactation dairy (r, n=67) cows during (A) wk +1 (7±2.5 DIM, mean±standard deviation), (B) wk +2 (14±2.4 DIM), (C) wk +3 (21±2.3 DIM), (D) wk +4 (28±2.1 DIM).

4. Implications

The results suggest that rumination monitoring across the transition period may contribute to the identification of MP cows either at risk for developing SCK or those that have SCK in combination with another health problem. To employ rumination data to help identify MP cows at risk for developing ketosis after calving, it is important to begin monitoring during the dry period to establish a baseline for each cow. These results also suggest that monitoring lying behavior across the transition period may not be useful for the early identification of SCK, but may contribute to the identification of MP cows that have SCK in combination with another health issue. Finally, these results suggest that rumination time may be predictive of milk yield in early lactation cows. Further work is needed to determine how changes in rumination time may be related to changes in milk production, whether due to onset of disease or otherwise.

Acknowledgements

We thank all participating farms. We are grateful to Robin Crossley, Lisa Gordon, Morgan Overvest, Caylie Corvinelli, and Hannah Gillespie of the University of Guelph, Kemptville Campus for all their technical help during data collection. Financial support for this research was received from the Natural Sciences and Engineering Research, as well as from the Ontario Ministry of Agriculture Food University of Guelph Research Partnership. We thank Dr. Karen Beauchemin of Agriculture and Agri-Food Canada for providing rumination monitoring equipment, as well as Eastgen, particularly Mark Carson, for contributions towards the rumination monitoring equipment and technical support.

References

Drackley, J.K. (1999). Biology of dairy cows during the transition period: The final frontier? Journal of Dairy Science 82: 2259-2273.

Duffield, T. (2000). Subclinical ketosis in lactating dairy cattle. Veterinary Clinics of North America: Food Animal Practice 16: 231-253.

Herdt, T.H. (2000). Ruminant adaptation to negative energy balance. Influences on the etiology of ketosis and fatty liver. Veterinary Clinics of North America: Food Animal Practice 16: 215-230.

Ingvartsen, K.L. (2006). Feeding and management related diseases in the transition cow. Animal Feed Science Technology 126: 175-213.

Iwersen, M., Falkenberg, U., Voigtsberger, R., Forderung, D. and Heuwieser, W. (2009). Evaluation of an electronic cowside test to detect subclinical ketosis in dairy cows. Journal of Dairy Science 92: 2618-2624.

LeBlanc, S. (2010). Monitoring metabolic health of dairy cattle in the transition period. Journal of Reproduction and Development 56: S29-S35.

Ledgerwood, D.N., Winckler, C. and Tucker, C.B. (2010). Evaluation of data loggers, sampling intervals, and editing techniques for measuring the lying behavior of dairy cattle. Journal of Dairy Science 93: 5129-5130.

McArt, J.A.A., Nydam, D.V. and Oetzel, G.R. (2012). Epidemiology of subclinical ketosis in early lactation dairy cattle. Journal of Dairy Science 95: 5056-5066.

Overton, T.R. and Waldron, M.R. (2004). Nutritional management of transition dairy cows: strategies to optimize metabolic health. Journal of Dairy Science 87: e105-e119.

Schirmann, K., Von Keyserlingk, M.A.G., Weary, D.M., Veira, D.M. and Heuwieser, W. (2009). Technical note: validation of a system for monitoring rumination in dairy cows. Journal of Dairy Science 92: 6052-6055.

Treating cows for ketosis by milk fat-to-protein ratio levels to improve production

T. Schcolnik[1*] and H. Honig[2]
[1]*Applied Research Department, Afimilk, Kibbutz Afikim, Israel;* [2]*Volcani Agricultural Research Center, ARO, Bet Dagan, Israel; tal.wise@afimilk.co.il*

Abstract

Negative energy balance (NEB) and its result, ketosis, increase the likelihood of infectious disease incidence, which adversely affects productivity and reproductive performance. Continuous automatic milk components monitoring enables farmers to be aware of metabolic, invisible (sub-clinical) issues and improves animal productivity, health and wellbeing. The objective of this study was to detect ketosis according to inline automatic measurements of milk fat-to-protein ratio (FPR) in comparison with discrete ketone testes detection alternatives and to establish the optimal schedule for detecting ketosis under field conditions. Milk production data were collected every milking for every cow at a large commercial farm in Northeast Israel in 2012. This herd is a 700 milking-cow herd in a loose-housing system, producing 40 kg milk on daily average and about 12,000 kg of milk per 305 days of lactation. This dairy is equipped with an in-line milk spectrometer (AfiLab™) at each milking station. The milk analyzers provide information on milk components recorded each milking and analyzed by specialized software. Blood Beta-hydroxy-butyric-acid (BHBA) of same 18 cows was measured three times a day, after every milking session. Cows with BHBA above 1.4mmol/l were considered ketotic. In the morning, two cows (11%), at noon five (28%) and in the evening six (33%) were found ketotic. A total of seven cows out of the tested 18 were found ketotic along that same day, in at least one of the three BHBA test sessions. No single measurement session identified all seven ketotic cows together. Looking at the fat-to-protein ratio, we found six of the seven cows (86%) that were ketotic by blood tests.

Keywords: negative energy balance, in-line milk components

1. Introduction

Calving and the onset of lactation impose tremendous physiological challenges to the homeostatic mechanisms of the cow (Goff and Horst, 1997). Therefore, at the onset of lactation, a cow's energy requirements for production cannot be met by its feed consumption alone. Maladaptive response to a peripartum negative energy balance and other stressors may contribute to impaired immune function, resulting in retained placenta or metritis (Kehrli *et al.*, 1999) damaging reproductive performances. A depression of important immune functions has been reported around parturition and hence there is a higher prevalence of clinical mastitis and other diseases during this period (Waller, 2000) damaging productive performances. Deprived appetite, physical and physiological limitations (i.e. rumen volume/surface and microflora adaptation) and ration's composition, often lead to negative energy balance (NEB) with ketosis.

Ketosis was found to affect 26 to 55% of North American, dairy cows and it conservatively estimated to cost dairy producers $78 per case. Ketosis is also correlated with increased risk for other transition cow disorders that decrease dairy farm profitability (Sawall and Litherland, 2014). Due to the inadequate dietary energy supply during freshening, a cow's body mobilizes fat to provide the needed energy, thus leading to increased concentration of non-esterified fatty acids (NEFA) in blood. Body fats are funnelled through blood stream, mainly, to the liver and mammary gland (Heuer and Schukken, 1999). In the liver, it is thrusting ketones production, whereas in the udder, it is accumulating in milk, raising butterfat content (Figure 1). Because of the same lack of energy, also, milk protein production is inhibited, and that is why milk fat to protein ratio is a reliable indicator for NEB and ketosis (Grieve *et al.*, 1986).

In our studies, we were looking for ketosis because it is defined as an event indicating an energy imbalance that causes tremendous economic losses. This can be seen as well in Figure 1, which includes the differences in milk fat and protein contents between ketotic [according to acetoacetate urine tests, performed by a veterinarian at 5-15 days in milk (DIM), with keto-sticks of Bayer®] and non-ketotic cows, in 180 days yield. This data, taken from retrospective health analysis of Israeli herds, includes almost 10,000 ketosis events out of 42,000 lactations (Nir-Markusfeld *et al.*, 2012, unpublished data). Impaired energy balance and ketosis have a significant negative impact on dairy cows all over the world. Excess of circulating ketone bodies in the blood with and mostly without clinical signs was previously associated with calving diseases, delayed ovulations or reduced conception rates and reduced milk production (McDougall *et al.*, 2015). Severe negative energy balance during transition may also increase the risk of retained placenta, metritis, and mastitis through impaired immune function (Duffield *et al.*, 2009). Ketone bodies released from the liver in to the blood stream in pulses affected from circadian rhythm and feed intake.

In the recent past, in order to monitor and manage energy balance status and ketosis, a postpartum clinical examination (with a cow-side urine, milk or blood ketone test) was performed usually once, if at all, a few days after calving (5-12 DIM or so). Sometimes cows were also tested for ketosis according to risk factors, as part of that same routine. Our goal is to improve disease diagnosis and to optimize treatment and prevention of calving diseases complex. The objective of the presented study was to compare real-time, in-line automatic measurements of milk components data, using AfiLab™, with a discrete cow-side test of blood β-hydroxybutyrate (BHBA) concentrations, which considered the 'gold standard' for ketosis event.

2. Materials and methods

The study took place on a northeast Israeli dairy farm with a herd of 700 Israeli-Holstein cows with an average 305 days milk production of 12,690 kg, through 27 January 2012. Cows were fed with total mixed ration (TMR) which conforms with NRC recommendations (NRC, 2001) twice daily (05:00 and 10:00 am). Housing system was roofed loose sheds with dry manure bedding. Blood samples were drawn with a sterile disposable needle from the coccygeal vein, of eighteen postpartum cows (5-40 days in milk, Lactations 2-8). Tests were done 3 times per day at 05:30-07:30, 12:00-14:00

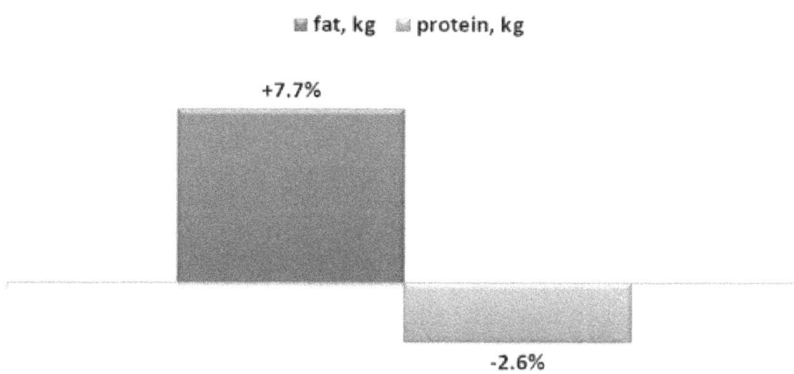

Figure 1. The differences in milk fat and milk protein content of 180 days yield between 9,679 ketotic and 32,676 non-ketotic cows.

and 19:00-21:00, resulting in 54 observations. Blood BHBA concentration was measured using Optium Xceed™ (by Abbott). Cows with BHBA values higher than 1.4 mmol/l in one of the tests were considered to have ketosis. Blood BHBA (as a gold standard for ketosis) results were compared with AfiLab fat to protein ratio values of three milking sessions for each cow. AfiLab™ (by Afimilk) is an electro optic sensor, fixed on the milk line after the milk meter, installed in every stall at the milking parlor. It measures milk components (fat, protein and lactose) and indicates the presence of blood and somatic cell counts, during each milking session, for every cow. Statistical analysis of success rates were conducted with JMP® (by SAS).

3. Results

The BHBA concentrations were elevated along the day. Variations were shown between the three sets of measurements: two cows (11%) had ketosis in the morning, five cows (28%) had ketosis at noon and six cows (33%) had ketosis in the evening (Figure 2). The mean and standard deviation of BHBA was significantly higher at noon and evening than in the morning of the same day (Table 1). Seven (39%) out of eighteen cows were found ketotic in at least one of the three measurement sessions on that day. No single measurement time included all seven cows that were found with ketosis. Looking at the weighted daily fat to protein ratio (from 2-3 milking session measurements, determined on-line by AfiLab), six out of the seven cows (86%) was found to be ketotic by BHBA that day.

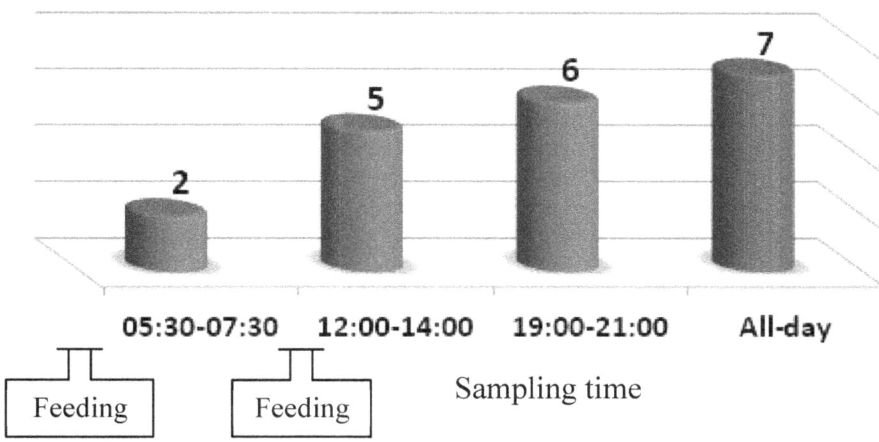

Figure 2. Blood ketone changes during one day in 18 cows, 5-40 days in milk. The different number of cows diagnosed during the day reflects blood ketone changes. It can be assumed that these changes are related to timing of feeding.

Table 1. The mean and standard deviation of β-hydroxybutyrate (BHBA) have shown significantly higher values at noon and evening than in the morning of the same day.

BHBA (mmol/l) / time	05:30-07:30	12:00-14:00	19:00-21:00
Mean	0.9	1.1	1.2
Standard deviation	0.4	0.6	0.7

4. Discussion and conclusions

This study show that ketone test results are inconsistent; therefore, a negative result does not necessarily mean that the cow does not suffer from NEB and ketosis. These results show that random checks during the day might completely miss the problem. Therefore daily variation in blood, urine or milk ketones concentrations makes a single daily ketone measurement less effective in monitoring energy status to prevent NEB and ketosis.

Similar past studies clearly show that in more than 85% of the cases, constant testing of cows identified by fat-to-protein ratio (FPR), for ketosis, will at some point present positive results for each identified animal, either a day before, the same day or a day after the positive test result. However, no single recommended point in time could be set to perform a single test that will identify all cows with ketosis. These findings reinforced our previous knowledge about different NEB profiles in different herds and within herds, between different lactations and days in milk (Schcolnik *et al.*, 2014).

The results of the present study are supported by those of another trial we carried out in December 2014 in a commercial, 400 Israeli-Holstein cows dairy herd, with an average 305 days of milk production totalling 11,412 kg. Cows were fed with TMR every 5 hours. Afimilk system identified four ketotic cows (based on fat to protein ratio values) requiring treatment on that day, out of the four hundred milking cows. In this experiment, all four cows were tested for both blood and urine ketone levels, three times in two hours intervals (at 07:00, 09:00 and 11:00 am). Blood BHBA concentration was measured using Optium Xceed™ (by Abbott). Cows with BHBA values higher than 1.4 mmol/l in one of the tests were considered to have ketosis. Urine aceto-acetate concentration was measured with a reagent strip, Ketostix™ (by Bayer). Cows with urine aceto-acetate concentrations above 1.5 mmol/l were considered to have ketosis.

In the first measurement, we found no signs of ketosis in the urine of any of the four cows, while one cow was diagnosed with ketosis by blood test. The second measurement session did not detect any ketotic cow by both blood and urine tests. In the third measurement, another cow was diagnosed with ketosis by blood test. Still there was no evidence of ketosis in the urine. Using blood tests, two cases of ketosis were diagnosed, but only one animal was identified in each test session (Table 2). Using the urine tests, no cow would have received the necessary treatment. Knowing all four cows identified needed the treatment against ketosis (according to our previous research in 2006 in four commercial Israeli dairy farms), we found cow-side discrete tests insufficient when it comes to monitoring and treating NEB and ketosis.

A view of one, sample herd late NEB profiles (Figure 3), where more than half of the cows (of every lactation) are affected later than 15 days in milk, once again emphasizes the fact that a herd constant monitoring is necessary to confront these challenges of transition period. However, we can safely

Table 2. Every two hours ketones urine and blood test results of four cows identified by daily fat to protein ratio. Using blood tests, two cases of ketosis were diagnosed, but only one animal was identified in each test session. This time by using the urine tests, no cow would have received the necessary treatment.

Cow no./time	07:30		09:30		11:30	
	Blood (BHBA)	Urine (acetoacetate)	Blood	Urine	Blood	Urine
8811	0.8	0	0.7	0	0.6	0
8362	1.8	0	0.9	0	1.3	0
8412	0.7	0	1.0	0	1.7	0
8995	0.6	0.5	0.4	0	0.4	0

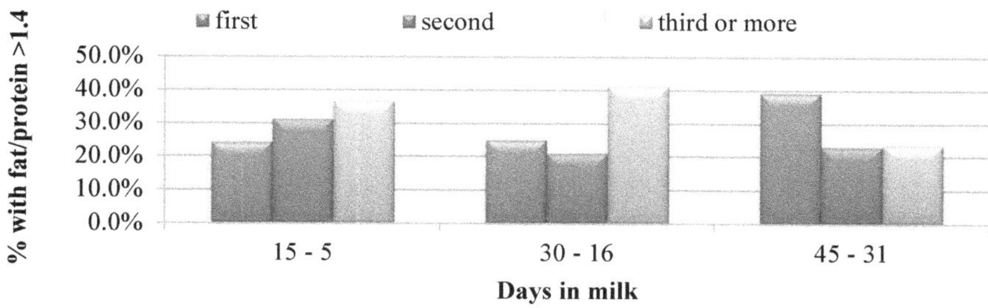

Figure 3. Herd postpartum energy profile (by days after calving and lactation no.) of an Italian herd of 220 Holstein cows, adapted from herd health report of 11/2010-10/2011 (Nir-Markusfeld, 2012).

assume that one or two post-partum urine or blood tests for ketones can miss more ketotic cows than a continuous on line FPR measurements.

Wide scope retrospective analysis was conducted based on the Israeli herd book data, covered 300,000 second or higher lactations in eight years (Nir-Markusfeld and Ezra, 2013, personal communication). This study shows milk yield loss, during 180 days of lactation, for cows with ketosis and FPR above 1.4 compared to cows with 'no factor' (Figure 4). Both cows diagnosed with ketosis and cows with FPR higher than 1.4 had the same lower conception rates, higher open days and higher culling rates. The conclusion from this extensive research is that FPR by itself is an adequate parameter for evaluating energy balance status and its consequences.

As anyone dealing with productive animals knows, controlled energy balance ensures long-term optimal production capabilities. Furthermore, the tests were only performed on the four cows identified. If herd-wide analysis had been performed, many more animals would have been tested in vain in order to locate the affected ones. This also means that other real-time testing solutions, consisting of a single ketone body measurement, suffer from the same two basic problems observed: which cows should be set aside for testing and when should this test be performed. These two dilemmas are automatically and effectively solved by continues daily monitoring of FPR data.

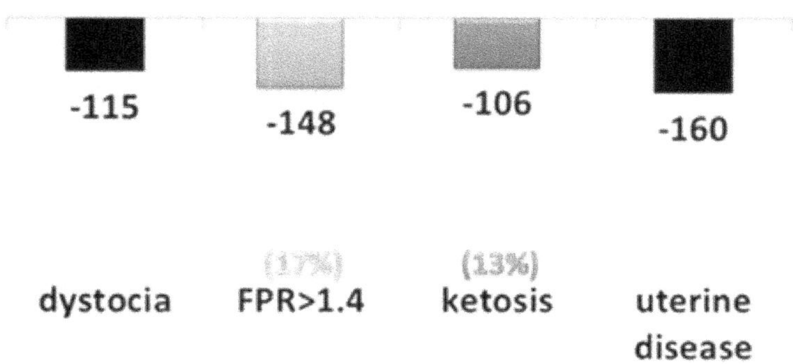

Figure 4. Milk yield loss (kg) of 180 days in milk for cows with ketosis and milk fat-to-protein ratio (FPR)>1.4, compared to healthy cows with no diagnosed postpartum disease. Based on ICBA herd book data including 300,000 (≥2nd) lactations of 160,000 cows (2002-2010).

The bottom line is clear: continuous milk data analysis, correctly analysed, is the most practical and effective tool for identifying cows needing treatment for ketosis and balancing energy resources. Continuous automatic monitoring also enables dairy farmers to minimize abrupt feed management changes and therefore improves animal health and wellbeing.

References

Duffield, T., Lissemore, K., McBride, B. and Leslie, K. (2009). Impact of hyperketonemia in early lactation dairy cows on health and production. Journal of Dairy Science 92: 571-580.

Goff, J.P. and Horst, R.L. (1997). Physiological changes at parturition and their relationship to metabolic disorders. Journal of Dairy Science 80: 1260-1268.

Grieve, D., Korver, S., Rijpkema, Y. and Hof, G. (1986). Relationship between milk composition and some nutritional parameters in early lactation. Livestock Production Science 14: 239-254.

Heuer, C., Schukken, Y.H. and Dobbelaar, P. (1999). Postpartum body condition score and results from the first test day milk as predictors of disease, fertility, yield, and culling in commercial dairy herds. Journal of Dairy Science 82: 295.

Kehrli, E., Burton, L., Nonnecke, J., Lee, K. (1999). Effects of stress on leukocyte trafficking and immune responses: implications for vaccination. Advances in Veterinary Medicine 41: 61-81.

Raboisson, D., Mounié, M. and Maigné, E. (2014). Diseases, reproductive performance, and changes in milk production associated with subclinical ketosis in dairy cows: a meta-analysis and review. Journal of Dairy Science 97: 7547-7563.

Sawall, Z. and Litherland, N. (2014). Fresh cow milk fat to protein ratio (fat:protein) can be used to identify cows experiencing ketosis. Dairy cattle nutrition extension, University of Minnesota, Minneapolis, MN, USA.

Schcolnik, T., Arazi, A. and Nir-Markusfeld, O. (2014). Daily monitoring of energy balance to improve decision-making on a dairy farm. Progressive Dairyman Magazine.

Waller, K.P. (2000). Mammary gland immunology around parturition. Influence of stress, nutrition and genetics. Advances in Experimental Medical Biology 480: 231-245.

Young, L. and McDougall, S. (2015). Subclinical ketosis in post-partum dairy cows fed a predominantly pasture-based diet: defining cut-points for diagnosis using concentrations of beta-hydroxybutyrate in blood and determining prevalence. New Zealand Veterinary Journal 63: 241-248.

Redefining SARA in dairy cows with reticulum pH measurements

T.T.F. Mottram[1,2]
[1]*Royal Agricultural University, School of Agriculture Food and Environment, Stroud Road, GL7 6SJ, United Kingdom;* [2]*eCow Devon Ltd., King St Business Centre, EX1 1BH Exeter, United Kingdom; toby.mottram@rau.ac.uk*

Abstract

The invention and use of rumen pH telemetry now provides large data sets of continuous measurements of pH and temperature from the rumeno-reticulum. In contrast the historic definitions of the rumen pH associated with SARA were based on single and possibly contaminated samples taken by rumenocentesis or oral pump taken from the main ventral sac of the rumen. Readings of reticuloruminal pH are the primary method of diagnosing subacute ruminal acidosis (SARA) in dairy cows, with a pH of 5.5 in the ventral sac of the rumen generally accepted as the threshold for acidosis. The contents of the reticulorumen are not uniform and therefore values of pH can vary greatly depending on where the reading is taken. Readings of reticular pH have been shown to fluctuate less and consistently remain higher than values from the ventral sac. Studies appeared to show the reticulum to be 0.2-0.3 pH units higher than the ventral sac. This paper models the effect of different SARA thresholds which are determined by the dynamic relationship of reticulum and rumen, on the incidence of SARA. Literature reveals a number of different values of association between pH values in the ventral sac and reticulum. Ranging from 0.25 pH units offset to greater than 1 pH unit. Routine reticulum pH measures show fewer than 5% of commercial dairy cows have SARA whereas field studies by clinicians (Atkinson, 2014) indicate that 30% of dairy cows in the UK have SARA. We modelled the effects of various definitions of SARA (mean pH, minimal pH, daily range) on the population prevalence of SARA. The effect applying different definitions of SARA using reticulum pH on 4,000 daily rumen pH profiles will be reported. The variation between clinical diagnosis with limited data and continuous reticular data demonstrates the need for better characterisation of the metadata associated with SARA.

Keywords: bovine, rumen, telemetry, wireless, sensor

1. Background

As dairy cow diets have changed and milk yields have risen cows have become increasingly at risk of sub-acute rumen acidosis SARA. SARA is defined as extended periods with rumen pH below a threshold. Indicators are reduced feed intake. Cows will start eating again when ruminal pH rises and there may be erratic feed intake that often goes unnoticed in individual. Clinical signs often observed during SARA include reduced rumination, mild diarrhoea, foamy feces containing gas bubbles and the appearance of undigested grain. Three to six months dairy herds experiencing SARA usually exhibit secondary signs of the disease. These secondary signs include episodes of laminitis, weight loss and poor body condition despite adequate energy intake, and unexplained abscesses (Mutsvangwa and Wright, 2012).

The diagnosis has traditionally been confirmed by sampling cows via rumenocentesis which requires a sample of rumen liquor through a long vacuumised needle thrust into the rumen through a gap in the ribs (Garrett *et al.*, 1994). Alternatively a 1.2 m long sample tube may be pushed into the rumen via the mouth of the cow and a sample of fluid extracted via ororuminal probes (Tajik and Nazifi, 2011). Neither method can be sure that the sample is taken from the same location or the same time within the daily cycle of rumen change. The withdrawing of the sample from the anaerobic rumen to atmosphere will also change the chemical structure of the sample and introduce another source of error.

There is dynamic change in rumen pH during the day due to timing of periods of eating and ruminating so the pH of any sample will vary. A typical rumen profile might vary as much as 1.5 pH units between the maximum and minimum measures during the day.

In unpublished trials with two boluses inside a cow one tethered to the rumen fistula in mid-ventral sac and the other in the reticulo-rumen a regular offset and a dynamic difference also apparent. For this reason we have used 5.8 pH as a threshold below which a cow will be assumed to be at risk of SARA (Figure 1). The ventral sac pH is far more dynamic in its response to changes in nutrient intake than the reticulo-rumen which is buffered by saliva and is generally flooded and probably better mixed (Figure 2).

Research reports vary about the threshold of rumen pH at which SARA is indicated and about the offset between reticulo-rumen and ventral sac. Readings of reticuloruminal pH are the primary method of diagnosing subacute ruminal acidosis (SARA) in dairy cows, with a pH of 5.5 in the ventral sac of the rumen generally accepted as the threshold for acidosis. The literature reveals a number of different values of association between pH values in the ventral sac and reticulum. Ranging from 0.25 pH units offset to greater than 1 pH unit (Sato, 2016).

In this paper we accept the hypothesis is that clinician is accurate as he assesses the condition of cows and uses an invasive technique consistently and that 30% of cows have SARA (Atkinson, 2013) we can then compute the threshold for reticulo-rumen pH in terms of time below that give us the population statistic.

2. Materials and methods

The boluses used in this study the farmBolus from eCow Ltd. The boluses were 125 mm long by 27 mm diameter weighing 200 g. The sensor end was made of stainless steel which inverts the bolus into a normally sensor down position in cows with a normal shaped reticulum. The electronics were encapsulated with a cold poured resin coat that has proved resilient against rumen liquor in trials and obviates the need for vulnerable seals. The sensor was a combined electrode pH probe

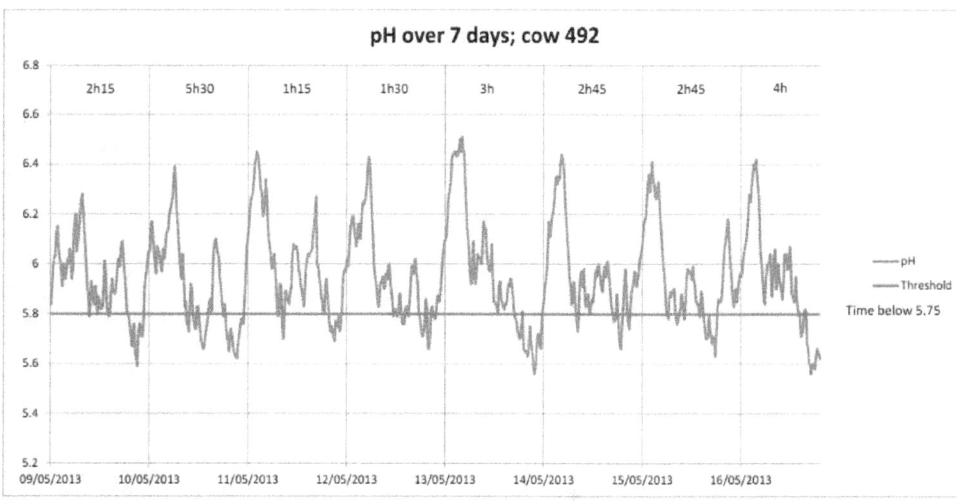

Figure 1. Reticulo-rumen pH from a cow believed to have SARA, it drops below 5.8 after a morning and rises above 6.4 during the night.

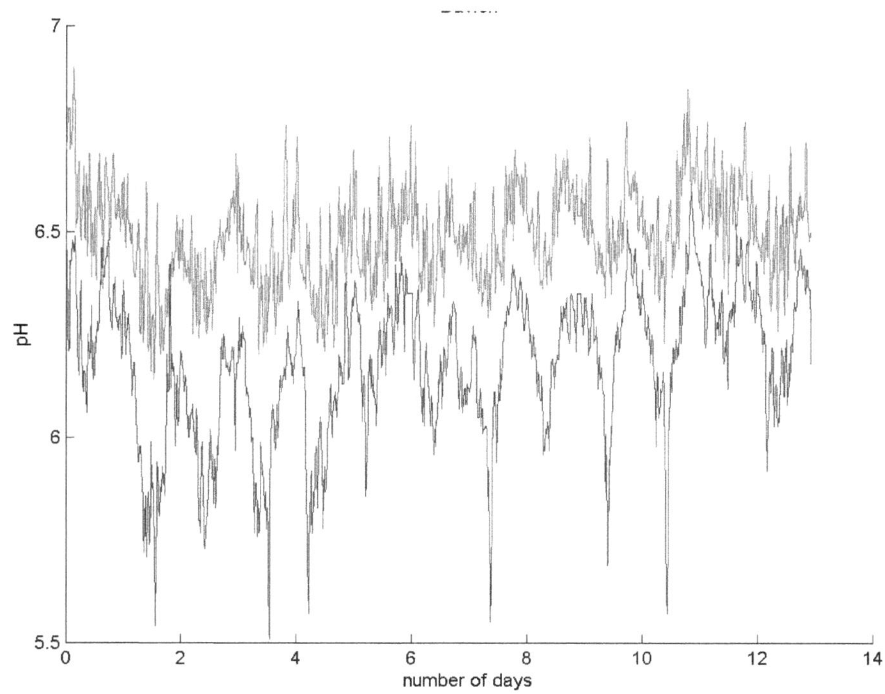

Figure 2. Comparison between reticulo-rumen pH (upper line) and ventral sac in a cow on a nutrition trial.

routinely used in applications in industry. The temperature probe was embedded in the stainless steel end cap, which has machined holes to allow rumen liquor to flow past the sensor tangentially without permitting direct impact of stones or grit on the glass sensing bulb.

The density of the bolus (specific gravity greater than 2) allowed it to remain in the reticulum. The bolus measured pH and temperature every 60 s and took an average value every 15 minutes and stored up to 2,700 lines of data in a text file format of date, time, pH, temp, battery V. If data was not collected within 28 days the file on the bolus was overwritten starting from the beginning.

The boluses were administered to the cow by mouth with a standard bolling gun. The device was calibrated at the factory before use and the calibration was accurate for four weeks in normal storage. The radio frequency used was in the 433 MHz ISM band.

Data was collected wirelessly using an antenna connected to an adapted mobile phone via micro USB. The customer collected the data with the adapted mobile phone handset by standing near the cow usually on the left front side. The customer was advised to administer 3 boluses to 3 normal healthy cows for a feeding group of 100 animals.

In order to create the table of 4,000 days of bolus data, 60 bolus files were analysed. The data files were chosen from boluses that had been categorised as accurate recordings taken in the first 60 days of operation to exclude any possibility of sensor drift.

The raw bolus files were trimmed at the beginning and end of each file so that they contained only full days, they were then loaded into a data table in the statistical program R and the mean, min, and max values for each day was created in to a new data frame. Once all the data files had been collated into the data frame, the daily range between minimum and maximum was calculated. In addition to a new table of values to correspond with the ventral sac pH using the offset of 0.25 and 1 pH.

R was then used to investigate the number of days in association with three pH thresholds:
1. Permanently below 5.2 – 'severe acidosis'
2. Under the SARA threshold 5.5 – 'SARA'
3. Never below 6.0 – 'not subclinical'
Data tables were produced to look at the total number of days under the thresholds in addition their percentage within the thresholds. These tables and the scripts used to analyse them to publish here were too big and will be available from www.researchgate.net after June 20th 2016.

3. Results

Tables show the total number of days under the thresholds in addition to the percentage of time within the thresholds.

Table 1. 'Severe acidosis' with an offset of 0.25.

Values	No. days	Percentage
Minimum pH	289	7.26
Mean pH	10	0.25
Maximum pH	0	0

Table 2. 'Severe acidosis' with an offset of 1.0.

Values	No. days	Percentage
Minimum pH	3,158	78.95
Mean pH	1,460	36.5
Maximum pH	140	3.5

Table 3. 'SARA' using an offset of 0.25.

Values	No. days	Percentage
Minimum pH	1,227	30.68
Mean pH	111	2.8
Maximum pH	4	0.1

Table 4. 'SARA' with an offset of 1.0.

Values	No. days	Percentage
Minimum pH	3,738	93.5
Mean pH	3,093	77.32
Maximum pH	932	23.3

Table 5. 'Never below 6.0' with an offset of 0.25.[1]

Values	No. days	Percentage
Minimum pH	738	18.5
Mean pH	2,290	57.3
Maximum pH	3,816	79.7

[1] Table 5 shows the total number of bolus days that didn't drop below 6.0 pH = 738 of 4,000 days remained over 6.0 for the 24 hours period.

Table 6. 'Never below 6.0' with an offset of 1.0.[1]

Values	No. days	Percentage
Minimum pH	0	0
Mean pH	127	3.2
Maximum pH	558	13.9

[1] Table 6 shows the total number of bolus days that didn't drop below 6.0pH = a Minimum of zero days indicates that all 4,000 days (60 Bolus's/Cows) minimum value did not exceed 6.0pH.

4. Conclusions

We accept the hypothesis of Atkinson that 30% of cows have SARA when we use an offset of 0.25 pH between reticulo-rumen and ventral sac because this value gave a percentage 30.68% of 4,000 days of cows randomly selected pH data below the minimum threshold of 5.5 pH in the ventral sac. We have to reject the conclusion of Sato that the offset between ventral sac and reticulum is 1 pH unit since this would mean that 93% of cows had ventral sac pH values in the SARA or acute acidosis range.

We conclude that wireless telemetry boluses measuring pH can be used to diagnose SARA without invasive sampling procedures.

References

Atkinson, O. (2013). Prevalence of subacute ruminal acidosis (SARA) on UK dairy farms. Proceedings of the BCVA.

Garrett, E.F., Norlund, K.V. (1994). Rumenocentesis: a technique for the diagnosis of subacute rumen acidosis in dairy herds. Bovine Practitioner 28: 104.

Mutsvangwa, T. and Wright, T. (2003). Sub-acute ruminal acidosis (SARA) in dairy cows. Available at: http://tinyurl.com/jqzwdgy.

Sato, S. (2016). Pathophysiological evaluation of subacute ruminal acidosis (SARA) by continuous ruminal pH monitoring. Animal Science Journal 87: 168-177.

Tajik, J. and Nazifi, S. (2011). Diagnosis of subacute ruminal acidosis: a review. Asian Journal of Animal Sciences 5: 80-90.

Using classifiers based on 3D-head acceleration for lameness detection in dairy cows

Y. Link[1*], J. Salau[1], S. Karsten[2] and J. Krieter[1]
[1]*Institute of Animal Breeding and Husbandry, Christian-Albrechts-University, Olshausenstrasse 40, 24098 Kiel, Germany;* [2]*TiDaTier und Daten GmbH, Bosser Strasse 4c, 24259 Westensee, Germany; ylink@tierzucht.uni-kiel.de*

Abstract

Lameness is a great welfare and health problem in dairy cow husbandry and time consuming to detect. An automated monitoring system is of interest for scientists and farmers. Using 3D-head acceleration sensors is a new approach to detect lameness in dairy cows. Between February 2014 and March 2015, 70 cows at a German research farm were equipped with 10 Hz or 1 Hz ear tags. As a reference dataset locomotion scores were recorded three times a week. Sensor data encompassing three weeks prior to each lameness was used to create classifiers for lameness detection by means of Area Under ROC (receiver operating characteristic) Curve (AUC) analysis. For pre-processing, each day was divided into 100 equally long time sections, and nine different methods (e.g. sum or mean) were applied to condense the data within these sections. Further on, 19 features (e.g. mean, variance or skewness) were calculated from the condensed vectors. For each combination of pre-processing method and feature, the data of the first 10 days was used to define ranges for cow individual thresholds. The features' values of the 10 days prior lameness were used to classify cows as lame using the predefined individual thresholds. ROC curves for all classifiers were plotted and AUC values calculated as measure for classifier performance. The best classifier for the 10 Hz dataset was based on the pre-processing method 25% quantile and the feature mean-median with an AUC value of 0.71. The same combination of pre-processing method and feature for the 1 Hz dataset yielded an AUC value of 0.74. However, the highest AUC value (0.88) for this dataset was observed for the pre-processing method standard deviation and the feature 75% quantile. These results indicate that 3D-head acceleration data can be used to detect lameness in dairy cows.

Keywords: acceleration sensor, features, ROC, AUC

1. Introduction

Lameness is a great health and welfare problem in dairy cow husbandry. Early lameness detection reduces loss of milk yield, prevents loss of reproductive efficiency and reduces veterinary fees (Green *et al.*, 2002; Somers *et al.*, 2015). With an occurring lameness the behavior and movement patterns change. Lame cows lie more (Ito *et al.*, 2010; Yunta *et al.*, 2012), which leads to an overall lower activity level. This decrease of activity level can be measured by pedometers (Alsaaod *et al.*, 2012; Mazrier *et al.*, 2006). Besides pedometers and other activity meters, accelerometers are one of the most discussed methods of automatic lameness detection using activity. The differences between pedometers and accelerometers are the place of sensor fixation; whereas activity meters / pedometers are mostly attached to the legs or the neck, accelerometers were attached to legs (Pastell *et al.*, 2009; Thorup *et al.*, 2015), to legs and back (Chapinal *et al.*, 2011), solely to the back (Kofler *et al.*, 2012; Mangweth *et al.*, 2012) or to the neck (Martiskainen *et al.*, 2009). Martiskainen *et al.* (2009) did not describe differences between the gait of lame and non-lame cows in the raw data. Instead they used support vector machines to identify different behavioral patterns, including moderately lame cows described by Sprecher *et al.* (1997). The other authors described the differences in the acceleration data of lame and non-lame cows. For example, vertical acceleration of the legs were lower in lame cows (Pastell *et al.*, 2009) and the acceleration data of the back showed more asymmetries (Kofler *et al.*, 2012). These results did not validate how the head movement or its acceleration alters during lameness, as these studies mainly focused on body and legs. Nevertheless, it is also known that the position and the movement of the head during lameness change, which was used for lameness scoring systems (Flower and Weary, 2006). This compensatory head movement was further discussed within lame horses (Buchner *et al.*, 1996; Krüger, 1941; Leach, 1986) and describes a nodding movement which ensures a positional change of the gravity center. Using an ear tag as sensor has the benefit that it can also be used for monitoring feeding behavior (feeding and rumination) and heat detection. The aim of this study was to identify lame cows using classifiers based on 3D-head acceleration data. To verify these classifiers Area Under ROC (receiver operating characteristic) Curve (AUC) analysis was used.

2. Material and methods

2.1 Animals

During a 13 month period (February 2014 to March 2015), 70 Holstein Friesian cows at Futterkamp, the research farm of the Chamber of Agriculture in Schleswig-Holstein, Germany were used for data acquisition. These dairy cows were in the second to fifth lactation, were milked two times a day, and were provide with a fresh Total-Mixed-Ration in individual feeding places right after each milking time. They had no access to pasture during lactation.

2.2 Ear tags

All animals used for the research were equipped with an ear tag provided by Smartbow GmbH, Weibern, Austria (Smartbow®, 2015). These ear tags had a 3D-acceleration sensor (x, y, z). Acceleration was recorded at a data frequency of either 10 Hz (864,000 values per cow and day) or 1 Hz (86,400 values per cow and day) depending on the sensor. For this study, the three axis of the acceleration sensor were aggregated to a total acceleration (a_t) with the theorem of Pythagoras: $a_t = \sqrt{a_x^2 + a_y^2 + a_z^2}$; where a_t is the total acceleration and a_x, a_y, and a_z are the acceleration data of the three recorded axes (x, y, z). The acceleration data according to the axis can have a maximum acceleration value from -2.04 to 2.03 m/s^2. In consequence, the total acceleration ranges between 0 and 3.54 m/s^2.

2.3 Lameness definition

Locomotion scoring, according to Sprecher *et al.* (1997), was done by one trained observer three times a week after the cows left the milking parlor in the morning. Each cow received a locomotion score (1-5), whereas cows with a score $\leq$2 were classified as non-lame and with a score $\geq$3 as lame. For the 1 Hz dataset, 6 lamenesses of 5 animals entered the analysis. Furthermore, 17 lamenesses of 10 animals represented the 10 Hz dataset.

2.4 Dataset

Two datasets for each sensor type (1 Hz and 10 Hz) were designated. The first one included lame cows and the second one included non-lame cows. The lame dataset included the day of detected lameness (locomotion score >2) and twenty-one days prior. This required no further detected lameness during these twenty-one days. The non-lame dataset included only animals which had no lameness during data collection. For each lame cow a non-lame cow within the same time frame and close to the lactation number was chosen.

2.5 Classifiers

The steps shown in Figure 1 were performed on both datasets (1 Hz and 10 Hz). All steps were done for each cow individually.

The data of one day was divided in 100 equal time sections. Within these sections nine different parameters were calculated. These pre-processing methods were: sum, mean, standard deviation, median, 25% quantile, 75% quantile, skewness, kurtosis and the 1st p-variation. Further on nineteen features were derived from the pre-processing methods on a daily basis. As features were calculated: sum, minimum, maximum, absolute minimum, absolute maximum, range, mean, median, 25% quantile,75% quantile, inter-quartile range, mean-median, standard deviation, variance, skewness, kurtosis, 1st p-variation, 2nd p-variation, 3rd p-variation.

2.6 Receiver operator characteristic / area under the curve

The first ten days (learning phase) were used to identify the span in which each feature ranges. Therefore, the minimum and maximum of each calculated feature were determined. Upper and lower limits were then set as an extension of the values assumed in the learning phase. This span was divided in 100,000 sections. For each threshold-step (0 to 100,000), a point on the ROC curve was calculated. A detected lameness within a four-day-period (day of detected lameness +3 days prior) was classified as true positive, whereas a detected lameness before this four-day-period was classified as false positive. For all ROC curves AUC values were calculated as a measure of classifier performance. AUC values range between 0 and 1. The closer the AUC value is to 1, the better the classifier performs. AUC value=0.5 corresponds to a random decision.

3. Results

The classifiers with the pre-processing methods standard deviation, kurtosis and 25% quantile received in general higher AUC values than classifiers based on other pre-processing methods.
The best classifier for the 1 Hz dataset with an AUC value of 0.88 was based on the pre-processing method standard deviation and the feature 75% quantile (Figure 2). The second best classifier received an AUC value of 0.80 and was based on the pre-processing method kurtosis and the feature standard deviation and variance, respectively.
The best classifier for the 10 Hz dataset was based on the pre-processing method 25% quantile and the feature mean-median with an AUC value of 0.71 (Figure 2). In general, the AUC values for the 10 Hz dataset were lower compared to the 1 Hz dataset.

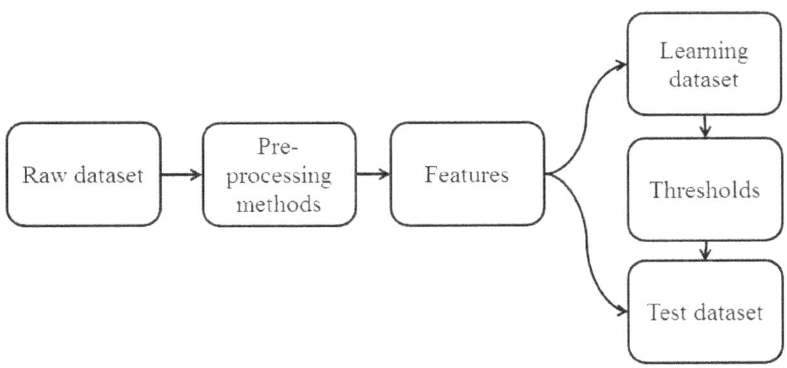

Figure 1. A process outline of establishing a classifier for lameness detection.

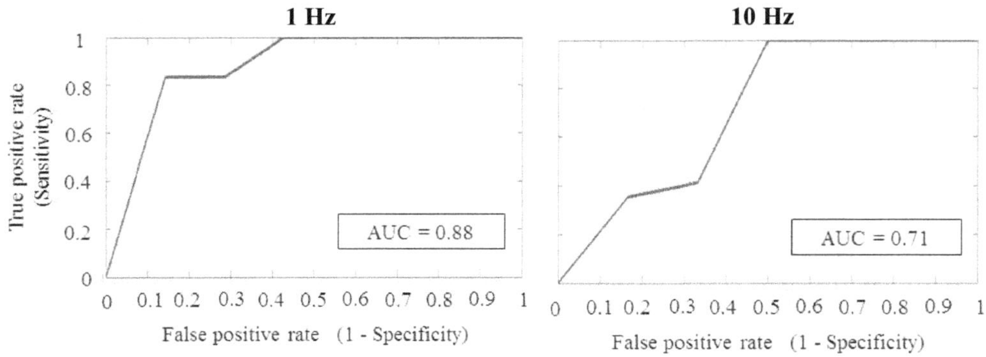

Figure 2. Receiver operator characteristic curve and the corresponding area under the curve value for lameness detection for the 1 Hz (preprocessing method: standard deviation, feature: 75% quantile) and 10 Hz (preprocessing method: 25% quantile, feature: mean-median) dataset.

4. Discussion

The aim of this study was to detect lame cows using 3D-head acceleration data. Several attempts to detect lameness using different types of sensors, like force plates, pedometers and accelerometers were published in previous studies (Mazrier *et al.*, 2006; Pastell *et al.*, 2009; Rajkondawar *et al.*, 2002). In general, the detection of lameness with pedometers or accelerometers are based on a change in the activity level of the animal (Alsaaod *et al.*, 2012; Mazrier *et al.*, 2006; Thorup *et al.*, 2015). The advantage of using ear tags with acceleration sensors is that they offer other possibilities in addition to lameness detection, e.g. monitoring feeding and heat detection. Furthermore, they can be combined with the official ear tag.

For evaluating an automatic lameness detection system a reference is needed, which provides data of the real situation. Therefore, it was chosen to monitor lameness directly by visual examination by one trained observer recording the locomotion score of each cow. Scoring for locomotion was done after milking due to the fact, that a full udder effects the gait (Chapinal *et al.*, 2009). The locomotion scoring system of Sprecher *et al.* (1997) was used. Due to its common use in research (Schlageter-Tello *et al.*, 2014) it allows a good comparability with others studies. Another possibility for a more objective definition of lameness could be the day of treatment (Kramer *et al.*, 2009; Miekley *et al.*, 2013a,b; Van Hertem *et al.*, 2013). However, this possibility was not chosen due to the fact that the aim of this study was to find differences as early as possible in lame cows, before treatment becomes necessary.

The best results for both datasets were received with the pre-processing methods standard deviation, 25% quantile and kurtosis. However, due to the fact of using two different methods to condense the data (pre-processing method and feature) it is more difficult to interpret the single results regarding the here found best fitting classifiers. Due to this fact, a more general point has to be taken in consideration. However, the pre-processing methods standard deviation, kurtosis and 25% quantile showed in a previous study that they are able to indicate lameness in dairy cows (Link *et al.*, 2015). Link *et al.* (2015) found that kurtosis and 25% quantile increased with an occurring lameness and standard deviation decreased, which indicated a lower activity level. Therefore, the well-known change in activity due to lameness (Alsaaod *et al.*, 2012; Mazrier *et al.*, 2006), can be detected in 3D-head acceleration data and can be used to detect lameness on a cow individual basis.

The present results of the ROC curves and the AUC values show that classifiers were able to distinguish between lame animals and non-lame animals. Using a locomotion scoring system to identify lameness holds the possibility of being too early to detect changes in activity level. A more reliable detection of lameness might be possible in a later stage of lameness since the 'stoical nature of cattle' can lead to a reduced pain response (O'Callaghan *et al.*, 2003). Therefore, further studies have to show if the day of treatment provides a more solid reference for lameness detection via accelerometer ear tag than locomotion scoring.

5. Conclusions

The aim of this study was to detect lame cows using classifiers based on 3D-head acceleration data. In both datasets, classifiers with the pre-processing methods standard deviation, 25% quantile and kurtosis detected lameness in dairy cows the best, which indicated a decrease of the activity level in lame cows. Moreover, to establish a lameness detection system with a locomotion scoring system holds the possibility of being too early in the lameness. Therefore, further studies should test the day of treatment for lameness identification. Furthermore, classifiers with fixed thresholds, established out of a learning dataset, have to be tested.

References

Alsaaod, M., Romer, C., Kleinmanns, J., Hendriksen, K., Rose-Meierhofer, S., Plumer, L. and Buscher, W. (2012). Electronic detection of lameness in dairy cows through measuring pedometric activity and lying behavior. Applied Animal Behaviour Science 142: 134-141.

Buchner, H.H., Savelberg, H.H., Schamhardt, H.C. and Barneveld, A. (1996). Head and trunk movement adaptations in horses with experimentally induced fore- or hindlimb lameness. Equine Veterinary Journal 28: 71-76.

Chapinal, N., De Passillé, A.M., Pastell, M., Hänninen, L., Munksgaard, L. and Rushen, J. (2011). Measurement of acceleration while walking as an automated method for gait assessment in dairy cattle. Journal of Dairy Science 94: 2895-2901.

Chapinal, N., De Passillé, A.M. and Rushen, J. (2009). Weight distribution and gait in dairy cattle are affected by milking and late pregnancy. Journal of Dairy Science 92: 581-588.

Flower, F.C. and Weary, D.M. (2006). Effect of hoof pathologies on subjective assessments of dairy cow gait. Journal of Dairy Science 89: 139-146.

Green, L.E., Hedges, V.J., Schukken, Y.H., Blowey, R.W. and Packington, A.J. (2002). The impact of clinical lameness on the milk yield of dairy cows. Journal of Dairy Science 85: 2250-2256.

Ito, K., Von Keyserlingk, M.A.G., LeBlanc, S.J. and Weary, D.M. (2010). Lying behavior as an indicator of lameness in dairy cows. Journal of Dairy Science 93: 3553-3560.

Kofler, J., Mangweth, G., Altenhofer, C., Weber, A., Gasser, C., Schramel, J.P., Tichy, A. and Peham, C. (2012). Messung der bewegung lahmheitsfreier kühe mittels accelerometer im schritt und vergleich der beschleunigungswerte nach kleben eines klotzes. Wiener Tierärztliche Monatsschrift 99: 179-188.

Kramer, E., Cavero, D., Stamer, E. and Krieter, J. (2009). Mastitis and lameness detection in dairy cows by application of fuzzy logic. Livestock Science 125: 92-96.

Krüger, W. (1941). Über das verhalten des schwerpunktes bei der normalen fortbewegung des pferdes. Tierärztliche Rundschau 47: 147-151, 162-166.

Leach, D. (1986). Centre of gravity and the analysis of lame gaits. Equine Veterinary Journal 18: 2-3.

Link, Y., Auer, W., Karsten, S. and Krieter, J. (2015). 3D-head acceleration used for lameness detection in dairy cows. In: 66th EAAP Book of Abstracts,. Presented at the EAAP, Warswaw, Poland, pp. 323.

Mangweth, G., Schramel, J.P., Peham, C., Gasser, C., Tichy, A., Altenhofer, C., Weber, A. and Kofler, J. (2012). Lameness detection in cows by accelerometric measurement of motion at walk. Berliner und Münchener tierärztliche Wochenschrift Journal 125: 386-396.

Martiskainen, P., Järvinen, M., Skön, J.-P., Tiirikainen, J., Kolehmainen, M. and Mononen, J. (2009). Cow behaviour pattern recognition using a three-dimensional accelerometer and support vector machines. Applied Animal Behaviour Science 119: 32-38.

Mazrier, H., Tal, S., Aizinbud, E. and Bargai, U. (2006). A field investigation of the use of the pedometer for the early detection of lameness in cattle. Canadian Veterinary Journal 47: 883-886.

Miekley, B., Stamer, E., Traulsen, I. and Krieter, J. (2013a). Implementation of multivariate cumulative sum control charts in mastitis and lameness monitoring. Journal of Dairy Science 96: 5723-5733.

Miekley, B., Traulsen, I. and Krieter, J. (2013b). Principal component analysis for the early detection of mastitis and lameness in dairy cows. Journal of Dairy Research 80: 335-343.

O'Callaghan, K.A., Cripps, P.J., Downham, D.Y. and Murray, R.D. (2003). Subjective and objective assessment of pain and discomfort due to lameness in dairy cattle. Animal Welfare 12: 605-610.

Pastell, M., Tiusanen, J., Hakojärvi, M. and Hänninen, L. (2009). A wireless accelerometer system with wavelet analysis for assessing lameness in cattle. Biosystems Engineering 104: 545-551.

Rajkondawar, P.G., Lefcourt, A.M., Neerchal, N.K., Dyer, R.M., Varner, M.A., Erez, B. and Tasch, U. (2002). The development of an objective lameness scoring system for dairy herds: pilot study. Trans. ASAE 45.

Schlageter-Tello, A., Bokkers, E.A.M., Groot Koerkamp, P.W.G., Van Hertem, T., Viazzi, S., Romanini, C.E.B., Halachmi, I., Bahr, C., Berckmans, D. and Lokhorst, K. (2014). Effect of merging levels of locomotion scores for dairy cows on intra- and interrater reliability and agreement. Journal of Dairy Science 97: 5533-5542.

SMARTBOW®, 2015. Smartbow. Available at: http://www.smartbow.at/de.

Somers, J.R., Huxley, J., Lorenz, I., Doherty, M.L. and O'Grady, L. (2015). The effect of Lameness before and during the breeding season on fertility in 10 pasture-based Irish dairy herds. Irish Veterinary Journal 68: 14.

Sprecher, D.J., Hostetler, D.E. and Kaneene, J.B. (1997). A lameness scoring system that uses posture and gait to predict dairy cattle reproductive performance. Theriogenology 47: 1179-1187.

Thorup, V.M., Munksgaard, L., Robert, P.-E., Erhard, H.W., Thomsen, P.T. and Friggens, N.C. (2015). Lameness detection via leg-mounted accelerometers on dairy cows on four commercial farms. Animal Journal 9: 1704-1712.

Van Hertem, T., Maltz, E., Antler, A., Romanini, C.E.B., Viazzi, S., Bahr, C., Schlageter-Tello, A., Lokhorst, C., Berckmans, D. and Halachmi, I. (2013). Lameness detection based on multivariate continuous sensing of milk yield, rumination, and neck activity. Journal of Dairy Science 96: 4286-4298.

Yunta, C., Guasch, I. and Bach, A. (2012). Lying behavior of lactating dairy cows is influenced by lameness especially around feeding time. Journal of Dairy Science 95: 6546-6549.

Behavioral comparisons between lame and sound cattle using precision monitoring technologies

B.A. Wadsworth*, L.M. Mayo, I.C. Tsai, A.E. Stone, J.D. Clark and J.M. Bewley
University of Kentucky, 413 W.P. Garrigus Building, Lexington, KY 40546, USA; barbara.wadsworth@uky.edu

Abstract

The study objective was to compare behavior and production variables in lame and sound cows. The study was conducted at the University of Kentucky from June 08, 2014 to July 09, 2015. Holstein (n=126) cows were equipped with multiple precision monitoring technologies measuring daily lying times, lying bouts, steps, rest time, rest bouts, milk yield, body weight, rumination time, activity, no activity, high activity, eating time, time at the bunk, and feed bunk visits. Cow gait was assessed weekly using a one (sound cow) to five (severely lame cow) multi-parameter scale. Final gait score was calculated as a weighted average of general symmetry (25.41%), tracking (19.97%), spine curvature (18.88%), head bobbing (14.88%), speed (12.34%), and abduction/adduction (8.53%). Cows with a weighted score <1.5 overall were classified as sound, cows with a weighted score ≥1.5 and <2.5 were classified as moderately lame, and cows with a weighted score ≥2.5 were classified as lame. The MIXED procedure of SAS was used to evaluate the effects of gait score, parity, milk yield, days in milk, period, and their two-way interactions on lying times, lying bouts, steps, rest time, rest bouts, milk yield, body weight, rumination time, activity, no activity, high activity, eating time, time at the bunk, and feed bunk visits. Stepwise backward elimination was used to remove non-significant interactions ($P \geq 0.05$). Main effects were kept in the model regardless of significance. Primiparous lame cows lay down longer than primiparous sound cows (10.24±0.13 and 9.42±0.06, respectively). Multiparous lame cows weighed less than multiparous sound cows (750.41±3.22 and 756.56±3.10, respectively). Multiparous and primiparous lame cows both had lesser feedbunk time, 2.84±0.05 and 2.93±0.05, than multiparous and primiparous sound cows, 3.01±0.04 and 3.38±0.04, respectively. Multiparous lame cows had less rumination time (7.55±0.04) than multiparous sound cows (7.65±0.03). Using precision dairy monitoring technologies may be useful in lameness detection.

Keywords: lameness, precision dairy monitoring technologies, behavior

1. Introduction

Lameness is a painful disease that decreases performance (Cutler *et al.*, 2013; Whay *et al.*, 1997) and is highly recognized as one of the most important health and welfare concerns for dairy cattle (Ito *et al.*, 2010; Olmos *et al.*, 2009). Many researchers have reported behavioral and production differences in lame versus sound cows (Blackie *et al.*, 2011a; Warnick *et al.*, 2001). Palmer *et al.* (2012) showed that lame cows ate fewer meals compared to sound cows. Warnick *et al.* (2001) found that cows diagnosed lame produced 2.6 kg less milk compared to their sound counterparts. Blackie *et al.* (2011b) discovered that lame cows lay down significantly less than sound cows, lame cows lay down for 11.1 hours per day and sound cows lay down for 9.0 hours per day ($P<0.01$). Van Hertem *et al.* (2013) showed that lame cows had a lesser rumination time at night, 208 minutes, compared to sound cows, 221 minutes, ($P<0.01$).
Visual locomotion scoring is the most common way to detect lameness in dairy cattle (Flower and Weary, 2009). However, this is not only subjective (Flower and Weary, 2009), but is also time consuming and costly (Thomsen, 2009). Therefore, automatic lameness detection for continuous monitoring is warranted. Nevertheless, debate about which behavior and production parameters

to evaluate exists. The study objective was to compare behavior and production variables in lame versus sound cows to evaluate potential usefulness in lameness detection.

2. Materials and methods

This study was conducted at the University of Kentucky Coldstream Dairy from June 08, 2014 to July 09, 2015. Holstein (n=126) cows were equipped with multiple precision monitoring technologies (Table 1). A weather station (HOBO U23 Pro v2 External Temperature/Relative Humidity Data Logger - U23-002, Onset, Bourne, MA, USA) measured daily relative humidity and temperature. Temperature humidity index (THI) was computed using the following formula (NOAA and Administration, 1976): THI=temperature (F)-[0.55-(0.55×relative humidity/100)]×[temperature (F)-58.8]. Days when THI was <68 were classified as cold periods and ≥68 were classified as hot periods. Cow gait was assessed using a one (sound cow) to five (severely lame cow) multi-parameter scale. General symmetry, tracking, spine curvature, head bobbing, speed, and abduction/adduction were scored individually weekly. Final gait score was calculated as a weighted average of general symmetry (25.41%), tracking (19.97%), spine curvature (18.88%), head bobbing (14.88%), speed

Table 1. List of all precision dairy monitoring technologies and parameters measured by each technology

Precision dairy monitoring technology	Parameters measured	Abbreviation
IceQube, IceRobotics Ltd., Edinburgh, Scotland	Lying time (hours/day)	IQLT
	Lying bouts (bouts/day)	IQLB
	Steps (number/day)	IQSTEP
AfiAct Pedometer Plus, Afimilk, Kibbutz Afikim, Israel	Rest time (hours/day)	AFIREST
	Rest bouts (number/day)	AFIRB
	Steps (number/day)	AFISTEP
Track a Cow, ENGS Systems Innovative Dairy Solutions, Israel	Time at the feedbunk (hours/day)	TACET
	Feedbunk visits (bouts/day)	TACFB
	Lying time (hours/day)	TACLT
	Lying bouts (bouts/day)	TACLB
	Steps (number/day)	TACSTEP
CowScout S Leg, GEA Farm Technologies GmbH, Bönen, Germany	Activity (steps/day)	GEAACT
HR Tag, SCR Engineers Ltd.,Netanya, Israel	Rumination time (hours/day)	HRRUM
	Neck activity (units/day)	HRACT
CowManager SensoOr, Agis Automatisering, Harmelen, the Netherlands	Feeding time (hours/day)	SENET
	Rumination time (hours/day)	SENRUM
	Time not active (hours/day)	SENNOACT
	Time active (hours/day)	SENACT
	Time high active (hours/day)	SENHIGHACT
Smartbow (Smartbow GmbH, Jutogasse, Austria)	Lying time (hours/day)	SBLT
	Rumination time (hours/day)	SBRUM
	Time not active (hours/day)	SBNOACT
	Time active (hours/day)	SBACT
	Time high active (hours/day)	SBHIGHACT
Afimilk MPC Analyzer Afimilk, Kibbutz Afikim, Israel	Milk yield (kg/day)	AFIMY
AfiWeigh, Afimilk, Kibbutz Afikim, Israel	Boyd weight (kg/day)	AFIBW

(12.34%), and abduction/adduction (8.53%). Cows with a weighted score <1.5 overall were classified as sound, cows with a weighted score ≥1.5 and <2.5 were classified as moderately lame, and cows with a weighted score ≥2.5 were classified as lame.

The MIXED procedure of SAS (SAS Institute, Inc., Cary, NC, USA) was used to evaluate the effects of gait score, parity, milk yield (MY), days in milk (DIM), period, and their two-way interactions on IQLT, LQLB, IQSTEP, AFIREST, AFIRB, AFISTEP, TACET, TACFB, TACLT, TACLB, TACSTEP, GEAACT, HRRUM, HRACT, SENET, SENRUM, SENNOACT, SENACT, SENIGHACT, SBLT, SBRUM, SBNOACT, SBACT, SBHIGHACT, AFIMY, and AFIBW. Stepwise backward elimination was used to remove non-significant interactions (*P*≥0.05). Main effects were kept in the model regardless of significance.

3. Results and discussion

Multiple parameters showed multiple interactions. However, due to space only main effects are shown here. Significant predictors of IQLT were gait, parity, and dim (Table 2). Lame cows IQLT was longer than sound cows (*P*<0.01). Significant predictors of IQLB were parity and period. Significant predictors of IQSTEP were parity and period.

Significant predictors of AFIREST were gait, parity, MY, dim, and period (Table 2). Lame cows had a longer AFIREST than sound cows (*P*<0.01). Significant predictors of AFIRB were gait, parity, and dim (Table 2). Significant predictors of AFISTEP were gait, parity, and dim (Table 2). Lame cows had less AFISTEP than sound cows (*P*<0.01). Significant predictors of AFIMY were parity, dim, and period. Significant predictors of AFIBW were gait, parity, MY, and dim (Table 2).

Significant predictors of HRRUM were parity, MY, dim, and period. Significant predictors of HRACT were parity, dim, and period. Significant predictors of GEAACT were gait, parity, dim, and period (Table 2). Lame cows had a lesser GEAACT compared to sounds cows (*P*<0.01).

Significant predictors of SENET were gait, parity, MY, dim, and period (Table 2). Lame cows had a lesser SENET compared to sound cows (*P*<0.01). Significant predictors of SENRUM were parity and MY. Significant predictors of SENNOACT were gait, parity, MY, dim, and season (Table 2). Lame cows had a greater SENNOACT than sound cows (*P*<0.01). Significant predictors of SENACT were gait, MY, dim, and period (Table 2). Lame cows had a lesser SENACT than sound cows (*P*<0.01). Significant predictors of SENHIGHACT were gait, MY, and period (Table 2). Lame cows had a lesser SENHIGHACT than sound cows (*P*<0.01).

Significant predictors for TACET were parity, MY, dim, and period (Table 2). Lame cows had lesser TACET than sound cows (*P*<0.01). Significant predictors of TACFB were gait and period (Table 2). Significant predictors of TACLT were gait, parity, MY, dim, and season (Table 2). Lame cows TACLT was greater than sound cows (*P*<0.01). Significant predictors of TACLB were parity, MY, dim, and period (Table 2). Significant predictors of TACSTEP were gait, parity, MY, dim, and period (Table 2). Lame cows has less TACSTEP than sound cows (*P*<0.01). Significant predictors of SBLT were parity, MY, and period.

Significant predictors for SBRUM were gait, MY, dim, and period (Table 2). Lame cows SBRUM was lesser than sound cows (*P*<0.01). Significant predictors of SBNOACT were gait, parity, MY, and period (Table 2). Significant predictors of SBACT were gait, parity, MY, dim, and period (Table 2). Lame cows SBACT was lesser than sound cows (*P*<0.01). Significant predictors of SBHIGHACT were gait, MY, dim, and period (Table 2). Lame cows SBHIGHACT was lesser than sound cows (*P*<0.01). Similar to this study, Blackie *et al.* (2011b) also discovered that hours lying per day were longer for lame cows compared to sound cows. However, in this study, lying times in lame versus sound cows were not as prominent as other studies and in general are lower than other studies. In this study, the IQLT difference for cows was 36 minutes compared to the two-hour difference in Blackie *et al.* (2011b). Evaluation of the eating times between SENET and TACET show differences between the two technologies. The SENET displays head based eating time, where the TACET displays time

Table 2. Least squares means of gait as a main effects by each variable as measured by precision dairy farming technologies.

Variable	Gait		
	Sound	Moderately lame	Lame
IceQube hours lying/day[1]	9.92±0.04[b]	9.97±0.05[b]	10.32±0.07[a]
IceQube steps/day[1]	1,234±8[a]	1,220±9[a]	1,142±12[b]
IceQube lying bouts/day[1]	17.08±0.12[a]	16.92±0.14[a]	17.12±0.19[a]
Milk yield (kg/day)[2]	33.47±0.28[a]	33.51±0.28[a]	33.41±0.31[a]
Afi steps/day[2]	3,621±18[a]	3,600±20[a]	3,463±27[b]
Afi rest time (hours/day)[2]	9.38±0.04[b]	9.44±0.05[b]	9.86±0.07[a]
Afi rest bouts/day[2]	10.01±0.07[b]	10.01±0.09[b]	10.37±0.11[a]
Body weight (kg)[3]	720.91±2.43[a]	720.03±2.50[a]	720.66±2.83[a]
HR rumination (hours/day)[4]	7.75±0.24[a]	7.56±0.28[a]	7.59±0.04[a]
HR activity (units/day)[4]	462.50±4.60[a]	465.72±4.65[a]	464.56±4.93[a]
GEA activity (steps/day)[5]	5,237±47[a]	5,195±49[a]	5,038±60[b]
Sensoor time not active (hours/day)[6]	7.94±0.12[b]	7.90±0.12[b]	8.41±0.16[a]
Sensoor eating (hours/day)[6]	4.04±0.06[a]	4.02±0.06[a]	3.82±0.07[b]
Sensoor ruminating (hours/day)[6]	9.70±0.07[a]	9.71±0.08[a]	9.47±0.11[b]
Sensoor time active (hours/day)[6]	1.13±0.01[b]	1.16±0.01[ab]	1.19±0.02[a]
Sensoor time high active (hours/day)[6]	1.33±0.03[a]	1.32±0.03[a]	1.24±0.04[b]
Track a cow time at feedbunk (hours/day)[7]	3.19±0.27[a]	3.21±0.03[a]	2.89±0.05[b]
Track a cow feedbunk visits per day[7]	8.00±0.05[a]	7.94±0.06[a]	7.49±0.07[b]
Track a cow hours lying/day[7]	10.12±0.05[b]	10.05±0.06[b]	10.46±0.09[a]
Track a cow lying bouts/day[7]	13.64±0.12[b]	13.90±0.13a[b]	14.09±0.17[a]
Track a cow steps/day[7]	2,099±15[a]	2,116±16[a]	2,045±21[b]
Smartbow rumination (hours/day)[8]	8.18±0.05[a]	8.00±0.06[b]	7.79±0.08[c]
Smartbow hours lying/day[8]	9.05±0.10[a]	9.09±0.12[a]	9.19±0.16[a]
Smartbow time not active (hours/day)[8]	5.52±0.06[b]	5.51±0.07[b]	6.03±0.09[a]
Smartbow time active (hours/day)[8]	13.02±0.64[a]	12.87±0.09[a]	12.41±0.12[b]
Smartbow time high active (hours/day)[8]	2.86±0.04[a]	2.87±0.04[a]	2.67±0.06[b]

[a,b] Pairs with different letters are significantly different (*P*<0.05).

[1] Hours lying, steps, and bouts/day were measured by IceQube (IceRobotics, Scotland).

[2] Milk yield (kg/day) steps, rest time and bouts/day were measured by the AfiMilk MPC Analyzer (Afimilk, Kibbutz Afikim, Israel).

[3] Body weight (kg) was measured by AfiWeigh (Afimilk, Kibbutz Afikim, Israel).

[4] HR rumination and activity (units/day) were measured by a HR tag (SCR Engineers ltd., Israel).

[5] GEA activity (steps/day) was measured by a CowScout S Leg (Gea Farm Technologies GmbH, Bönen, Germany).

[6] Sensoor eating, ruminating, no activity, activity, and high activity (hours/day) were measured by a CowManager SensoOr tag (Agis Automatisering, the Netherlands).

[7] Track a cow time at feedbunk, feedbunk visists, hours lying, lying bouts, and steps /day were measured by a Track a Cow (ENGS, Hampshire, UK).

[8] Smartbow rumination, lying time, no activity, activity, and high activity (hours/day) were measured by Smartbow (Smartbow GmbH, Jutogasse, Austria).

at the feedbunk. The TACET shows lesser time eating compared to the SENET. However, Palmer *et al.* (2012) speculated that lame cows may increase meal size to avoid a reduction in dry matter intake. The cows may not want to eat as many meals to reduce time on their feet and to minimize the pain of lameness. Therefore, the dry matter intake may have been the same for the cows in this study, but total time at the feedbunk would be less compared to total time eating. Unlike this study, others researchers have highlighted differences in milk yield in lame and sound cows (Warnick *et al.*, 2001). This study failed to recognize that difference. Evaluation of high activity times between SENHIGHACT and SBHIGHACT shows differences between the two technologies. The SENHIGHACT shows when cows are overly active (i.e. in heat), where the SBHIGHACT shows when cows are overly active, but also when the cows head may be moving during a feeding period. These differences may explain the dissimilar times.
The researchers in this study only evaluated each technology and parameters from each technology separately. Other researchers have evaluated multiple technologies and multiple parameters from each technology (Van Hertem *et al.*, 2013). Future studies may evaluate multiple technologies and multiple parameters from each technology.

4. Conclusions

Differences between lame and sound cows were observed in behavior parameters, as measured by precision dairy monitoring technologies. Lame cows lay down longer, had a lesser body weight, ate for lesser time, had lesser number of steps, and had reduced rumination compared to sound cows. Using precision dairy monitoring technologies may be useful in lameness detection.

References

Blackie, N., Amory, J., Bleach, E. and Scaife, J. (2011a). The effect of lameness on lying behaviour of zero grazed Holstein dairy cattle. Applied Animal Behaviour Science 134(3-4): 85-91.

Blackie, N., Bleach, E., Amory, J. and Scaife, J. (2011b). Impact of lameness on gait characteristics and lying behaviour of zero grazed dairy cattle in early lactation. Applied Animal Behaviour Science 129(2-4): 67-73.

Cutler, J.H.H., Cramer, G., Walter, J.J., Millman, S.T. and Kelton, D.F. (2013). Randomized clinical trial of tetracycline hydrochloride bandage and paste treatments for resolution of lesions and pain associated with digital dermatitis in dairy cattle. Journal of Dairy Science 96(12): 7550-7557.

Flower, F.C. and Weary, D.M. (2009). Gait assessment in dairy cattle. Animal 3(1): 87-95.

Ito, K., Von Keyserlingk, M.A.G., LeBlanc, S.J. and Weary, D.M. (2010). Lying behavior as an indicator of lameness in dairy cows. Journal of Dairy Science 93(8): 3553-3560.

NAHMS (2007). Facility characteristics an cow comfort on U.S. dairy operations. USDA:APHIS:VS, CEAH, National Animal Health Monitoring Systems, Fort Collins, CO, USA.

NOAA and N.O. a. A. Administration (1976). Livestock hot weather stress. In: Proc. US Dept. Commerce, Natl. Weather Serv. Central Reg. Operations Manual Lett. Natl. Oceanic Atmospheric Admin., Kansas City, MO, USA, pp. C31-C76.

Olmos, G., Boyle, L., Horan, B., Berry, D.P., O'Connor, P., Mee, J.F. and Hanlon, A. (2009). Effect of genetic group and feed system on locomotion score, clinical lameness and hoof disorders of pasture-based Holstein – Friesian cows. Animal 3(1): 96-107.

Palmer, M.A., Law, R. and O'Connell, N.E. (2012). Relationships between lameness and feeding behaviour in cubicle-housed Holstein – Friesian dairy cows. Applied Animal Behaviour Science 140(3-4): 121-127.

Thomsen, P.T. (2009). Rapid screening method for lameness in dairy cows. Veterinary Record 164(22): 689-690.

Van Hertem, T., Maltz, E., Antler, A., Romanini, C.E.B., Viazzi, S., Bahr, C., Schlageter-Tello, A., Lokhorst, C., Berckmans, D. and Halachmi, I. (2013). Lameness detection based on multivariate continuous sensing of milk yield, rumination, and neck activity. Journal of Dairy Science 96(7): 4286-4298.

Warnick, L.D., Janssen, D., Guard, C.L. and Gröhn, Y.T. (2001). The effect of lameness on milk production in dairy cows. Journal of Dairy Science 84(9): 1988-1997.

Whay, H.R., Waterman, A.E. and Webster, A.J. (1997). Associations between locomotion, claw lesions and nociceptive threshold in dairy heifers during the peri-partum period. Veterinary Journal 154(2): 155-161.

Validation of Nedap Smarttag Leg and Neck to assess behavioural activity level in dairy cattle

E. Van Erp-Van der Kooij*, M. Van de Brug and J.B. Roelofs
HAS University of Applied Sciences, Onderwijsboulevard 221, Den Bosch, the Netherlands; l.verp@has.nl

Abstract

The last decades, the number of dairy cows per farm has increased and the time spent on individual cows by the farmer has been reduced. To help the farmer detect changes in activity of cows associated with health issues, activity monitoring systems are being developed. The systems can help with daily farm management decisions, thus increasing farm profitability. Besides this economic benefit there is a social benefit: farmers highly value the herd being under continuous surveillance. Nedap Livestock Management (the Netherlands) introduced a leg activity meter and a neck activity collar: Smarttag Leg and Smarttag Neck. By registering activity of individual cows, the Smarttags help the farmer to detect oestrus and give alerts when activity deviates from normal patterns. The objective of this study was to validate results from the Smarttag Leg and Smarttag Neck, by comparing them with results from live observations and video recordings. Eight lactating dairy cows were observed for 22.5 hours each and video recordings were made of six dry dairy cows for 5.5 hours each. Lying, standing, walking, eating, standing up and ruminating were recorded. Data were compared with results from the Smarttag Leg and Smarttag Neck by calculating Cohen's Kappa, univariate linear regression analysis, Pearson's correlation and concordance correlation coefficient. Visual observations and video observations show correlation coefficients of >0.85 with the results of the Smarttags for all behaviours except walking. Correlations between visual and video observations and Smarttag results for walking were 0.45 and 0.50 respectively, possibly due to low incidence and difficulties in observing this behaviour. These results provide strong evidence that the Smarttag Leg and Smarttag Neck can reliably be used to monitor specific behaviours. With this system, the farmer can monitor behaviour and detect behavioural changes.

Keywords: dairy cows, pedometers, locomotion, lying behaviour, sensor technology

1. Introduction

In time the number of cows per dairy farm has increased and, as a consequence, the time spent on individual cows has been reduced. Driven by reducing workload and increasing efficiency and profitability, interest of farmers and industry for precision dairy farming has increased (Bikker, 2014). To support decision making and improve farm productivity and profitability, precision dairy farming technologies are developed (Asseldonk *et al.*, 1999). An example is the development of oestrus detection tools. Several studies (Hockey *et al.*, 2010; Müller *et al.*, 2003; Roelofs *et al.*, 2005) showed that pedometers and neck mounted collars can be used to predict the time of ovulation, showing better results than visual observation (Roelofs and Van Erp-Van der Kooij, 2015) and increasing pregnancy rates (Roelofs, 2013).
Changes in behavioural activity of farm animals are widely used as an indicator of welfare, health and comfort (Mattachini *et al.*, 2013; Müller *et al.*, 2003). To help the farmer detect changes in activity, activity monitoring systems are developed. The systems can help improve the daily farm management decisions and, thus, farm profitability. In addition to this, farmers highly value the knowledge that the herd is under continuous surveillance (Kamphuis, 2015).
Nedap Livestock Management (the Netherlands) introduced a leg activity meter and a neck activity collar: Nedap Smarttag Leg and Nedap Smarttag Neck. These systems measure activity of cows

continuously and help the farmer to detect oestrus and abnormalities in behaviour that may refer to health problems. Besides number of steps, the Nedap Smarttag Leg activity meters measure walking, standing, lying time and stand-ups. Activity recording based on accelerometers is commonly used and is easy to apply, detecting quantity and intensity of activities (Patterson *et al.*, 1993; Tyron, 1991). Apart from oestrus detection, they can be used for health indication. The (increased or decreased) duration and frequency of lying and standing time is directly related to clinical health, especially to identify lameness (Chapinal *et al.*, 2010; Garbarino *et al.*, 2004; Juarez *et al.*, 2003). Cow comfort can be indicated by lying behaviour (Haley *et al.*, 2000). Behaviours are interconnected: if standing and walking time decrease, for example by lameness, lying time will increase. The Smarttag Neck measures eating and resting, as well as rumination time. These behaviours can provide relevant information on food and water intake (Chapinal *et al.*, 2007) and are important in understanding the nutrition, production, social welfare, and overall health of dairy cattle (Krawczell, 2012). On average, a dairy cow ruminates about 35-40% of the day, mostly when lying down but sometimes also when standing, walking or during other activities (Bar *et al.*, 2010; Beauchemin, 1991). Time spent ruminating is an important health indicator, with stress, anxiety or disease decreasing rumination time (Beauchemin, 1991; Bristow *et al.*, 2007; Büchel *et al.*, 2014; Hansen *et al.*, 2003; Herskin *et al.*, 2004; Schirmann *et al.*, 2009; Welch, 1982). Digestibility, forage quality, NDF intake and dietary composition also influence rumination (Beauchemin, 1991; Welch *et al.*, 1970).
The overall objective of this study was to validate the measures generated by the Nedap Smarttag Leg and Nedap Smarttag Neck systems, by comparing these results with measures from live observations and time-lapse video recording.

2. Material and methods

2.1 Animals, feed and housing

To validate results of the Nedap Smarttag Leg and Nedap Smarttag Neck systems, cows were observed visually and using video recordings at a dairy farm in the Netherlands. Visual data were collected from eight lactating Holstein Friesian cows, and video data from six dry Holstein-Friesian cows. Cows were housed in a free stall with slatted floor and cubicles, with *ad libitum* access to roughage and water.

2.2 Measurement of activity

The cows were equipped with activity meters on the front leg and on the neck (Smarttag Leg and Smarttag Neck, Nedap Livestock Management, Groenlo, The Netherlands). Leg meters measured number of steps, walking, lying, standing time and number of stand-ups. Neck collars measured eating, resting and ruminating time. Smarttags have a G-sensor which measures accelerated movements in a particular direction, based on a three-dimensional accelerometer. The tags distinguish in motions for- and backwards, left and right and up and down.

2.3 Observation of behaviour

Five trained observers independently recorded behaviour of eight cows by direct visual observation for a period of three days, eight hours/day. All behaviours measured by the activity meters were recorded. Also the behaviour of six cows was recorded with a webcam (Logitech Webcame Software Inc.), resolution 1290×702 and 15 frames/second, for 12 hours/cow. Camera images were imported to a computer and observed and analysed by two trained observers with ObserverXT12 software (Noldus 2015). All behaviours measured by the activity meters were recorded. Smarttag data were recorded in 15-minute periods.

2.4 Statistics

Results from video and behavioural observations were converted to behaviours per quarter, corresponding to the data from the Smarttags. This was done by finding the main behaviour per minute, and converting this to behaviours per quarter. Data were analysed using SPSS (SPSS21.0.0.0, IBM Company Inc., USA). To evaluate the correspondence between individual observers and between visual or video observations and activity monitor sensors recordings, measures of interrater agreement (Cohen's Kappa statistic), univariate linear regression analysis and Pearson's correlation with concordance correlation analysis (CCC) were used. Kappa is expressed as a number between zero and one, and validated according to Landis and Koch (1977): poor (<0.00), slight (0.00-0.20), fair (0.21-0.40), moderate (0.41-0.60), substantial (0.61-0.80), and almost perfect (0.81-1.00). Pearson's correlation and CCC were determined to the criteria of Hinkle *et al.* (2003): negligible (0.00-0.30), low (0.30-0.50), moderate (0.50-0.70), high (0.70-0.90), and very high (0.90-1.00).

3. Results

Visual observations of eight cows for 180 hours were available, and a total of 720×15 minutes of data (appr. 90 quarters/cow) for both sensors. Video data of six cows for 63 hours were available, 252 quarters of data for the Smarttag Leg and 108 quarters for the Smarttag Neck.
Table 1 shows the results of the correlation analyses: the Kappa value, the univariate linear regression analysis and the concordance correlation coefficients, per Smarttag.
The kappa values for resting, eating and ruminating (Smarttag Neck) were fair to moderate. For the Smarttag Neck, the kappa values for walking were fair, for standing moderate, for lying and stand-ups substantial to almost perfect. Kappa for the agreement of observers showed an almost perfect agreement. In the linear regression analysis, no effect of observers was found. For the visual and video observations, R^2 of resting, eating and ruminating (Smarttag Neck) showed high correlations, and high correlations for lying and standing, a low to moderate correlation for stand-

Table 1. Statistical results of Kappa, univariate linear regression analyse and concordance correlation coefficient for visual (Exp 1.) and video (Exp. 2) observations.[1]

Tag	Behaviour		K	R	C	RC	C_b	CCC
Smarttag Neck	Eating	Exp. 1	0.403	0.801	0.585	0.930	0.998	0.892
		Exp. 2	0.562	0.940	0.315	0.982	0.999	0.969
	Ruminating	Exp. 1	0.470	0.884	0.867	0.920	0.997	0.937
		Exp. 2	0.543	0.807	0.807	0.871	0.996	0.895
	Resting	Exp. 1	0.331	0.803	0.037	0.881	0.994	0.890
		Exp. 2	0.489	0.820	0.537	0.906	1.000	0.905
Smarttag Leg	Lying	Exp. 1	0.838	0.986	0.059	0.992	1.000	0.993
		Exp. 2	0.748	0.940	-0.004	0.962	0.999	0.969
	Standing	Exp. 1	0.560	0.981	0.063	0.960	0.999	0.989
		Exp. 2	0.526	0.930	0.611	0.950	0.999	0.963
	Walking	Exp. 1	0.258	0.256	0.321	0.624	0.901	0.445
		Exp. 2	0.372	0.265	0.179	0.430	0.977	0.503
	Stand-Up	Exp. 1	0.851	0.726	0.015	0.870	1.000	0.851
		Exp. 2	0.817	0.413	0.044	0.866	0.612	0.953

[1] K=Kappa; R^2=Pearson's correlation; C=Constante; RC=Regression coefficient; C_b=Correlation Bias; CCC=Concordance correlation coefficient.

ups and very low correlations for walking (Smarttag Leg) were found. Gradients are mostly ~1 (very high agreement). The concordance correlation coefficient (CCC) was high or very high for all behaviours except walking.

4. Discussion and conclusions

Data of cow behaviour is commonly recorded on video for subsequent analysis, or collected directly by a human observer (O'Driscoll, 2008), which is time consuming, labour intensive, and one may influence behaviour while observing (Krawczell, 2012; Müller, 2003). Only one cow can be observed simultaneously when observing visually (Schirmann, 2009). In this study, one cow was observed at the time. Inter-observer differences can be a disadvantage when more than one observer is used (Martin and Bateson, 1993). In this study, observers studied different cows and no effect of observer was found. Videos were analysed by two researcher. To check for reliability, both researchers analysed two identical videos of two cows, one hour per cow. This showed an almost perfect agreement.
Müller *et al.* (2003) and Munksgaard *et al.* (2006) reported in their study high levels of correspondence between video recording and automatic devices when considering the total duration of behavioural activities. The advantages of video recording as a control to validation studies are that video can be stopped, rewinded and showed slowly so behaviours could be observed easier and reassessed. The camera was placed above the cows, this was an excellent position for all behaviours except ruminating. Most cows ruminate while lying down (Beauchemin, 1991), which was difficult to see. Bikker *et al.* (2014) and Beauchemin *et al.* (1989) found a lower correlation between visual and sensor observation for eating than for ruminating, similar to our findings. However, this is not found in the CCC. Eating shows a higher correlating than ruminating in the video analysis. It was easier to see a cow eating on video than with live observations. Ruminating was always difficult to observe. A side-effect is the tuning of the Smarttag Neck: short periods of resting during ruminating are measured as ruminating, while observers record this as resting. Bikker *et al.* (2014) present in their study the evaluation of a resting sensor with a high correlation (R^1=0.98). We did not find such high correlations. Resting while lying was difficult to spot. Another explanation for the lower result is that we recorded head movements (head bumping, grooming, scratching) as resting, while the Smarttag Neck records it as remaining behaviour.
Observed lying behaviour corresponds almost perfectly with the Smarttag Leg results. Cows lie 12-14 hours/day (Fregonesi *et al.*, 2007; Van't Hoff, 2010). Those lying bouts are easy to observe. Mattachini *et al.* (2013) and O'Driscoll (2008) showed in their study similar results. Standing measured by the Smarttag Leg shows also good results. Walking is a difficult behaviour to measure and the results affirm this. A standing cow may make sporadic movements such as short stops or sudden direction changes, without having ended the sequence of body movements (Trénel *et al.*, 2009), for example to change the eating place. This behaviour was not classified by the observers, while it seems that the Smarttag Leg does record this. Kajava *et al.* (2014) also concluded that lying and standing behaviour have a high agreement (R^2=0.99) and walking a lower agreement (R^2=0.69). When short movements of the leg were added to the walking observations, agreement was higher (R^2=0,89). Another explanation for the difference in standing and walking time is lameness. When the leg equipped with the Smarttag Leg is painful, the cow will stand less on this leg, even lifting it up or putting it in a different direction. Smarttag Leg sensors record this differently than the observers. Furthermore, lameness can influence walking behaviour, putting more pressure on the non-lame leg or switching her legs more quickly (Hassal *et al.*, 1993).
Results for stand-ups were corrected, because of a time lag in the Smarttag results. The agreement is almost perfect, but the gradient is 0.73. This shows that the number of stand-ups measured with the Smarttag Leg was correct, but the timing was incorrect.
We conclude that the Nedap Smarttag Leg and Nedap Smarttag Neck are useful systems to help the farmer monitor behaviour and detect behavioural changes in cattle.

References

Bar, D. and Solomon, R. (2010). Rumination collars: what can they tell us. The first North American conference on precision dairy management.

Beauchemin, K.A., Zelin, S., Genner, D. and Buchanan-Smith, J.G. (1989). An automatic system for quintification of eating and ruminating activities of dairy cattle housed in stalls. Journal of Dairy Science 72: 2746-2759.

Beauchemin, K.A. (1991). Ingestion and mastication of feed by dairy cattle. Veterinary Clinics of North America: Food Animal Practice 7: 439-462.

Bikker, J.P., Van Laar, H., Rump, P., Doorenbos, J., Van Meurs, K., Griffioen, G.M. and Dijkstra, J. (2014). Technical note: evaluation of an ear-attached movement sensor to record cow feeding behaviour and activity. Journal of Dairy Science 97: 2971-2979.

Bristow, D.J. and Holmes, D.S. (2007). Cortisol level and anxiety-related behaviours in cattle. Physiology and Behavior 90: 626-628.

Büchel, S. and Sundrum, A. (2014). Technical note: evaluation of a new system for measuring feeding behaviour of dairy cows. Computers and Electronics in Agriculture 108: 12-16.

Chapinal, N., Veira, D.M., Weary, D.M. and Von Keyserlingk, M.A.G. (2007). Technical note: validation of a sustem for monitoring individual feeding and drinking behaviour and intake in grouphoused cattle. Journal of Dairy Science 90: 5732-5736.

Chapinal, N., De Passillé, A.M. and Rushen, J. (2010). Correlated changes in behavioural indicators of lameness in dairy cows following hoof trimming. Journal of Dairy Science 93: 5758-5783.

Fregonesi, J.A., Tucker, C.B. and Weary, D.M. (2007). Overstoking reduces lying time in dairy cows. Journal of Dairy Science 90: 3349-3354.

Garbarino, E.J., Hernandez, J.A., Shearer, J.K., Risco, C.A. and Thatcher, W.W. (2004). Effect of lameness on ovarian activity in portpartum Holstein cows. Journal of Dairy Science 87: 4123-4131.

Haley, D.B., Rushen, J. and De Pasille, A.M. (2000). Behavioural indicators of cow comfort: activity and resting behaviour of dairy cows in two types of housing. Canadian Journal of Animal Science 80: 257-263.

Hansen, S.S., Norgaard, P., Pedersen, C., Jorgensen, R.J., Mellau, L.S.B. and Enemark, J.D. (2003). The effect of subclinical hypocalcaemia induced by Na_2EDTA on the feed intake and chewing activity of dairy cows. Veterinary Research Communications 27: 193-205.

Hassal, S., Ward, W. and Murray, R. (1993). Effects of lameness on the behaviour of cows during the summer. The Veterinary Record 132(23): 578-580.

Herskin, M.S., Munksgaard, L. and Ladewig, J. (2004). Effects of accute stressors on nociception, adrenocorical responses and behaviour of dairy cows. Physiology and Behaviour 83: 411-420.

Hinkle, D.E., Wiersma, W. and Jurs, S.G. (2003). Applied statistics for the behavioural sciences. Houghton Mifflin, Boston, MA, USA.

Hockey, C.D., Morton, J.M., Norman, S.T. and McGowan, M.R. (2010). Evaluation of a neck mounted 2-hourly activity meter system for detecting cows about to ovulate in two paddock-based Australian dairy herds. Reproduction in Domestic Animals 45: E107-E117.

Juarez, S.T., Robinson, P.H., DePeters, E.J. and Price, E.O. (2003). Impact of lameness on behaviour and productivity of lactating Holstein cows. Applied Animal Behaviour Science 83: 1-14.

Kajava, S., Frondelius, L., Mononen, J., Mughal, M., Ruuska, S. and Zehner, N. (2014). Validation of RumiWatch pedometers measuring lying, standing and walking of cattle. Proceedings of the International Conference of Agricultural Engineering, Zurich, Switzerland.

Kamphuis, C. (2015). Using activity meters to monitor cow health. Proceedings of the 2nd Dairycare Conference, Cordoba, Spain.

Krawczell, P.D., Klaiber, L.M., Thibeau, S.S. and Dann, H.M. (2012). Technical note: data loggers are a valid method for assessing the feeding behaviour of dairy cows using the Calan Broadbent Feeding System. Journal of Dairy Science 95: 4452-4456.

Landis, J.R. and Koch, G.G. (1977). The measurement of observer agreement for categorical data. Biometrics 33: 159-174.

Martin, P. and Bateson, P. (1993). Measuring behaviour. Cambridge University Press, Cambridge, UK.

Mattachini, G., Antler, A., Riva, E., Arbel, A. and Provolo, G. (2013). Automated measurment of lying behaviour for monitoring the comfort and welfare of lactating dairy cows. Livestock Science 158: 145-150.

Müller, R. and Schrader, L. (2003). A new method to measure behavioural activity levels in dairy cows. Applied Animal Behaviour Science 83: 247-258.

Munksgaard, L., Reenen, C.G. and Boyce, R. (2006). Automatic monitoring of lying, standing and walking behaviour in dairy cattle. Journal of Animal Science 84: 304.

O'Driscoll, K., Boyle, L. and Hanlon, A. (2008). A brief note on the validation of a system for recording lying behaviour in dairy cows. Applied Animal Behaviour Science 111: 195-200.

Patterson, S.M., Krantz, D.S., Montgomery, L.C., Deuster, P.A., Hedges, S.M. and Nebel, L.E. (1993). Automated physical activity monitoring: validation and comparison with psychological and self-report measures. Psychophysiology 30: 296-305.

Roelofs, J.B., Van Eerdenburg, F.J.C.M., Soede, N.M. and Kemp, B. (2005). Pedometer readings for oestrus detection and as predictor for time of ovulation in dairy cattle. Theriogenology 64: 1690-1703.

Roelofs, J.B. (2013). Estrus detection technologies to improve pregnancy rates. Cattle Practice 21: 205-208.

Roelofs, J.B. and Van Erp-Van der Kooij, E. (2015). Estrus detection tools and their applicability in cattle: recent and perspectival situation. Animal Reproductive Science 12: 498-504.

Schirmann, K., Von Keyserlingk, M.A.G., Weary, D.M., Veira, D.M. and Heuwieser, W. (2009). Technical note: validation of a system for monitoring rumination in dairy cows. Journal of Dairy Science 92: 6052-6055.

Trénel, P., Jensen, M.B., Decker, E.L. and Skjøth, F. (2009). Technical note: quantifying and characterizing behaviour in dairy calves using the IceTag automatic recording device. Journal of Dairy Science 92: 3397-3401.

Tyron, W.W. (1991). Activity measurement in psychology and medicine. Plenum Press, New York, NY, USA.

Van Asseldonk, M.A.P.M., Jalvingh, A.W., Huirne, R.B.M. and Dijkhuizen, A.A. (1999). Potential economic benefits from changes in management via information technology applications on Dutch dairy farms: a simulation study. Livestock Production Science 60: 33-44.

Van 't Hoff, M. (2010). Individual cattle lying behaviour. Veterinary Medicine University of Utrecht, Utrecht, the Netherlands.

Welch, J.G. and Smith, A.M. (1970). Forage quality and rumination time in cattle. Journal of Dairy Science 53: 797-800.

Welch, J.G. (1982). Rumination, particle size and passage from the rumen. Journal of Animal Science 54: 885-894.

Optimizing dry cow therapy by using frequent SCC monitoring

A. Gouw*, L.G. De Jong, R. Van der Tol and A.J. Van der Kamp
Lely Innovations, Cornelis van der Lelylaan 1, 3147 PB, Maassluis, the Netherlands; agouw@lely.com

Abstract

The growing social pressure to reduce antibiotic use forces farmers to apply antibiotics more prudently. During the first 100 days in lactation (DIL) about half of clinical mastitis cases originate from the same bacteria as the last infection in the previous lactation. This indicates bacterial resistance to antibiotics used during dry-off. To reduce antibiotic use and bacterial resistance, selective dry cow therapy (SDCT) is being adopted by farmers as a new method. Selective treatment requires accurate information about recent cow health history at the moment of dry-off to make a correct decision. Nowadays the decision is often based on low frequent historical somatic cell count (SCC) measurements according to the milk production registration (MPR). With MPR, the time delay between SCC measurements and dry-off can be up to 6 weeks. Instead of using MPR data, another option is to use inline SCC information from an automatic milking system (AMS) that measures cows' SCC continuously up to each milking, depending on udder health status. The objectives of the study were (1) to determine if, for monitoring purposes, daily inline SCC data is preferred over infrequent MPR SCC data, and (2) to determine if individual cow SCC profiles, constructed from daily inline SCC, show a better efficacy of SDCT compared to low-frequency MPR SCC measurements. On five commercial Dutch farms, including 1,697 dry-offs, it was determined whether inline SCC measurements were available, the cow was dried off with or without antibiotics, and whether the cow became sick in the first 30 days of the next lactation. The results showed that median values of sensor data correlate (R=0.58) with laboratory SCC data. The results also showed that inline monitoring of SCC is a better way to detect clinical mastitis and consequently support the process for applying SDCT. Measuring SCC on a daily basis enables the construction of individual cow SCC profiles, taking (natural) daily SCC variations into account. In addition, self-curing effects, treatment efficacy and reoccurrence rates are visualized, enabling the farmer to make more profound decisions when it comes to mastitis management and reduction of antibiotics usage.

Keywords: antibiotics, somatic cell count, mastitis

1. Introduction

Mastitis is one of the most common health problems in dairy farming. It causes decreased animal welfare (Schukken *et al.*, 2003) and is a costly disease because of decreased milk production, discarded milk, treatment costs, increased labour, milk quality penalties and the early culling of cows (Bradley, 2002; Hammer *et al.*, 2012; Huijps *et al.*, 2008). In the Netherlands, the losses due to clinical and subclinical mastitis are estimated to vary between €17 and €198 per cow per year, with an average of €78 per cow per year (Hogeveen *et al.*, 2011). A way to prevent mastitis in early lactation is to apply dry cow therapy (DCT). Giving a long-acting intramammary infusion at the end of lactation has two functions. Firstly, to eliminate current infections at the moment of drying-off and, secondly, to prevent new infections in the dry-period and early lactation (Cameron *et al.*, 2014; Halasa *et al.*, 2009). Until recently, most cows were dried off using antibiotics. However, antibiotic use creates selective pressure on bacterial populations and contributes to the development of antimicrobial resistance (Landers *et al.*, 2012). This emerging problem has led to guidelines for applying selective dry cow therapy (SDCT), where the decision to provide DCT is based on cow characteristics. A meta-analysis conducted by Halasa *et al.* (2009) showed that SDCT provides higher protection against new infections compared with no DCT but lower protection compared with all cows on

DCT. Huijps and Hogeveen (2007) indicated that applying SDCT could improve economic benefits even further, depending on the selection procedure. The directive of the professional organization of veterinarians in the Netherlands (KNMvD) and the Dutch Animal Health Service subscribes to the use of antibiotics at drying off only for primiparous cows with a somatic cell count (SCC) >150,000 cells/ml and for multiparous cows with a SCC >50,000 cells/ml during the last milk recording (MPR). These results can be six weeks old at the time a cow is dried off. However, there are suggestions that the six weekly milk recording test does not provide the most accurate information with which to decide whether or not to apply SDCT. First of all, because of the time delay between SCC measurement during milk recordings and the actual moment of drying off. It is known that SCC increases at the end of the lactation (KNMvD, 2013). This increase is probably caused by a lower milk production, resulting in a higher concentration of somatic cells (Green *et al.*, 2006). In addition, cow health could change significantly just before dry off, affecting the decision for the right treatment. A recent development in automatic milking systems (AMS) is the option to perform inline SCC measurements, making it possible to measure SCC up to every milking. High frequent SCC measuring enables the construction of individual cow SCC profiles taking (natural) daily SCC variations into account. Deviating patterns in the SCC profile enable an earlier and better-supported diagnosis of mastitis episodes. This offers opportunities for using frequent SCC measurements for SDCT. The objectives of this study, therefore, were 1) to determine if, for monitoring purposes, daily inline SCC data is preferred over infrequent MPR SCC data, and 2) to determine if SCC profiles from daily milkings show better efficacy of SDCT compared to MPR results.

2. Materials and methods

2.1 Data collection and selection

Since 2013, udder level SCC data can be routinely recorded with automatic milking systems. In this study, information was used from five commercial Dutch dairy farms that use the in-line SCC sensor (Lely Industries NV, Maassluis, the Netherlands). The in-line SCC algorithm is developed on and calibrated against raw milk lab values, which has two accepted 'gold standards'; the Fossomatic™ flow cytometry and the DNA staining method for counting cells. These two measurement methods use different techniques. The inline SCC method measures viscosity changes, which are directly proportional to the number of whole viable cells. This is effectively an automated CMT test. The Fossomatic method counts the number of stained DNA clumps, which could be from whole cells and any cellular DNA present or released from broken cells. The total measurement time period differed per farm, depending on the installation date of SCC sensors. Eligible farms needed to have at least half a year robotic milking data available before starting with the in-line SCC measurements and half a year with in-line SCC measurements. Data about the application of DCT has been collected through management systems, such as T4C (Lely Industries NV, Maassluis). The dry-off injectors that were used by the farmers included Orbenin dry cow, Orbenin extra dry cow, Prevaclox, Super mastidol and Nafpenzal. Teat sealer information is known but excluded from this research. For all cows in the dataset the value of the last milk recording test day up to six weeks before drying off was collected and compared with the KNMvD-directive. Intramammary infections in the next lactation were defined as cows having a milk separation in the first 30 days of lactation. It is considered to be a consequence of treatment with antibiotics for mastitis. Days of milk separation due to colostrum were excluded.

2.2 Statistical analysis

Data was analysed with statistical software package R (R 3.2.4, R Core Team (2016), Austria). First the correlation of laboratory SCC data with the in-line SCC measurements on the day of milk recording was determined. This was done by comparing the daily average geometric cell count from the in-line sensor with the laboratory SCC data from the corresponding day. In total, 809 laboratory SCC values were compared with the in-line SCC measurements of that day using a Pearson product-moment correlation coefficient. Secondly, the compliance of the farmer with the guidelines of KNMvD was analysed. A total of 294 dry-offs were taken into account. When SCC on the last milk recording exceeded 150,000 cells/ml for primiparous cows or 50,000 cells/ml for multiparous cows, it was allowed to apply antibiotics at dry-off. This was compared with the information whether the cows were indeed treated with antibiotics at dry off. Finally, the efficacy of dry-cow therapy was evaluated. For 1,697 dry-offs it was determined if: (1) inline SCC measurement was available; (2) the cow was dried off with or without antibiotics; and (3) if the cow became sick in the first 30 days of the next lactation. The results were presented in a cross table. A difference of proportions test was used to determine whether the differences between un/availability of in-line SCC measurements and the differences between dry-off with/without antibiotics were significant.

3. Results

To confirm the overall performance of the inline SCC sensor, 294 indications were compared with milk recordings. The correlation between inline geometric SCC and laboratory milk recording SCC data is 0.58.

For proper mastitis management, monitoring udder health during lactation and before dry off is key to determining the dry off strategy. Based on this data and the directives, farmers decide on their dry off management. Table 1 summarizes the number and percentage of cows that received DCT based on the selection criteria from the KNMvD.

In 65% of the dry off treatments the treatment was in line with KNMvD milk recording directives. No DCT advised and not applied happened in 25.5% of the cases and DCT allowed and applied in 39.5% of the cases. In 14.6% of cases the farmer had given treatment while based on the directive no dry off treatment with antibiotics was expected. In the other 20.4% of cases, treatment was allowed but not given. Inline availability of SCC did not affect the percentage of cows treated with DCT (55.6 and 58.9%, respectively), Table 2.

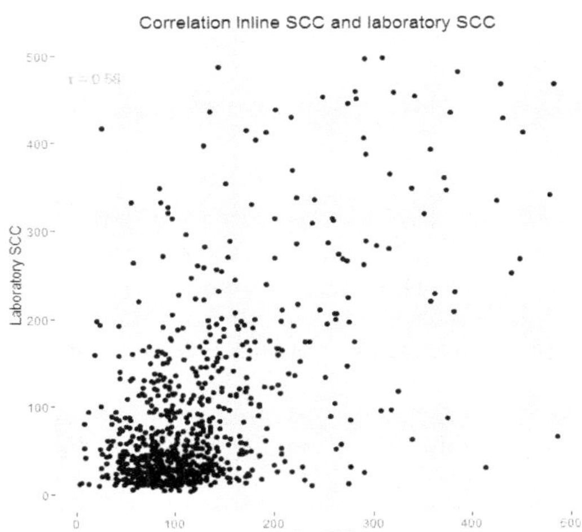

Figure 1. Correlation inline SCC sensor with milk recording test data.

Table 1. Treatment decisions (n=294) compared with selection criteria KNMvD.

DCT advised based on directive KNMvD	Applied DCT	Animals (n)	Animals (%)
No	No	75	25.5
No	Yes	43	14.6
Yes	No	60	20.4
Yes	Yes	116	39.5

Table 2. The number of cows (n=1,697) and the percentage of the total cows that were treated with DCT in the period before inline SCC and with the availability of inline SCC.

Availability inline SCC	Animals (n)	Treated with DCT (n)	Treated with DCT (%)	*P*-value
No	1,031	573	55.6	
Yes	666	392	58.9	0.200

The efficacy of DCT is analysed based on the periods before and after the availability of inline SCC measurement, Table 3. Having inline SCC available significantly supports the efficiency. The number of mastitis cases is almost halved in the first 30 days of lactation: it is reduced from 9.2 to 4.7%. Inline SCC supports the choices made for treatment both with and without antibiotics at dry off, Table 4. Since a significant reduction was observed with the use of inline SCC, mastitis cases in the first 30 days of lactation decreased from 8.3 to 3.6%. Moreover, with DCT a significant reduction from 9.9 to 5.4% was seen.

Table 3. The number of cows (n=1,697) and the percentage of the total cows that became sick in the consequent lactation with or without the availability of inline SCC.

Availability inline SCC	Animals (n)	Mastitis cases (n)	Mastitis cases (%)	*P*-value
No	1,031	95	9.2	0.001
Yes	666	31	4.7	

Table 4. The number of cows (n=1,697) and the percentage of the total cows that were treated with DCT in the period before inline SCC and with the availability of inline SCC in comparison with the development of mastitis in the first 30 days of lactation.

DCT	Availability inline SCC	Animals (n)	Mastitis cases (n)	Mastitis cases (%)	*P*-value
No	No	458	38	8.3%	0.021*
	Yes	274	10	3.6%	
Yes	No	573	57	9.9%	0.014*
	Yes	392	21	5.4%	

4. Discussion

Even though the inline SCC algorithm is developed on and calibrated against raw milk laboratory values, a relatively low correlation (R=0.58) with the data available for farmers is inevitable and can be easily explained. Milk recording data represent a SCC value of a single milking, while inline SCC data is deliberately presented as a geometrical average in order to provide reliable indications that take daily variation into account. It is known that the infection status of a quarter is a major factor affecting SCC. Other non-infectious factors as parity, breed, lactation stage, measurement equipment, fraction of milk sample, stress, season, diurnal variation and prolonged milk interval contribute to SCC (Djabri *et al.*, 2002; Pettersson *et al.*, 2002, Ruegg and Pantoja, 2013; Schukken *et al.*, 2003; Sharma *et al.*,2011). Besides, milk recordings use a different measuring principle as laboratories often count the number of stained DNA clumps from whole cells and some cellular DNA released from broken cells, while the inline SCC indications are based on the proportional number of whole viable cells. For dry cows treatments it was assumed that the milk recording data was the main argument for the farmer to make a treatment decision. However, this was only the case in 65% of the cases. Other than milk recording, bacteriological research (BO), conductivity, LDH assay (lactate dehydrogenase) and some other recorded provisions that determine whether or not a cow has an intramammary infection are allowed (KNMvD, 2013). However, in this research the data on which the farmer based his final decision was unknown. Knowing the reason for treatment is useful for evaluating the efficiency of SDCT or subsequent decision support systems. Before the availability of inline SCC, 8.3% of the animals without DCT and 9.9% of the animals with DCT became sick at the beginning of the subsequent lactation. Nowadays, decisions for not using dry cow therapy are made more effectively; on average 3.6% of the animals became sick during the subsequent lactation, compared with 5.4% of the cows that did receive treatment with antibiotics (having inline SCC available). Selection based on SCC availability can be risky, as no dry cow treatment resulted in a clinical mastitis incidence rate that was 1.7 times higher for quarters dried off without antibiotics (Scherpenzeel *et al.*, 2014). Random selection resulted in an incidence rate that was 1.4 to 2 times higher for quarters dried off without antibiotics as compared with quarters dried off with antibiotics (Hassan *et al.*, 1999; Schukken *et al.*, 1993). While using inline SCC measurements, the percentage of mastitis cases after calving dropped significantly, whereas the percentage of cows that are dried of with antibiotics remained equal. This indicates that inline measurement of SCC provides the farmers with better control of udder health. Hogeveen (*et al.*, 2010) showed that combining SCC and EC increased the probability of a positive sensor outcome in the event of a mastitis case to 33 compared to the probability of 13% when using only the SCC. In this research, farms with the availability of inline SCC also had the availability of e.g. conductivity, giving them better tools with which to control udder health, resulting in a drop from 9.2 to 4.7% of mastitis cases after calving. Inline monitoring of SCC does not lead to a reduction in the use of antibiotics at the moment of dry off, but it does in the first 30 days of the next lactation. To estimate the effect of overall usage, the complete lactation period prior to dry off and consequent lactation should be included. The frequent SCC availability during lactation can lead to more or quicker responses from the farmer in comparison with low frequent milk recording, possibly affecting the use of medicines as well.

5. Conclusions

Using inline SCC helps to better detect mastitis episodes and consequently supports the decision process for applying SDCT. The number of mastitis cases is halved to only 4.7% of the cows in the first 30 days in the new lactation, when using inline SCC. This means that the allocation of antibiotics in DCT has become more effective in automatic milking systems that have available inline SCC. For 65% of the cases farmers still rely on low frequent MPR SCC data. Frequent measurements and the availability of udder health profiles, including all relevant udder health parameters, support a transition towards effective selective dry cow therapy and reduce the infection pressure in our dairy industry.

References

Bradley, A.J. (2002). Bovine mastitis: an evolving disease. The Veterinary Journal 164: 116-128.

Cameron, M., McKenna, S.L., MacDonald, K.A., Dohoo, I.R., Roy, J.P. and Keefe, G.P. (2014). Evaluating of selective dry cow treatment following on-farm culture: risk of postcalving intramammary infection and clinical mastitis in the subsequent lactation. Journal of Dairy Science 97(1): 270-284.

Djabri, B., Bareille, N., Beaudeau, F. and Seegers, H. (2002). Quarter milk somatic cell count in infected dairy cows: a meta-analysis. Veterinary Research 33(4): 335-357.

Godden, S., Rapnicki, P., Stewart, S., Fetrow, J., Johnson, A., Bey, R. and Farnsworth, R. (2003). Effectiveness of an internal teat seal in the prevention of new intramammary infections during the dry and early-lactation periods in dairy cows when used with a dry cow intramammary antibiotic. Journal of Dairy Science 86(12): 3899-3911.

Green, L.E., Schukken, Y.H. and Green, M.J. (2006). On distinguishing cause an consequence: do high somatic cell counts lead to lower milk yield or does high milk yield lead to lower somatic cell count? Preventive Veterinary Medicine 76(1-2): 74-89.

Halasa, T., Nielen, M., De Roos, P.W., Van Hoorne, R., De Jong, G., Lam, T.J.G.M., Van Werven, T. and Hogeveen, H. (2009). Production loss due to new subclinical mastitis in Dutch dairy cows estimated with a test-day model. Journal of Dairy Science 92(2): 599-606.

Hammer, J.F., Morton, J.M., and Kerrisk, K.L. (2012). Quarter- milking-, quarter-, udder- and lactation-level risk factors and indications for clinical mastitis during lactation in pasture-fed dairy cows managed in an automatic milking system. Australian Veterinary Journal 90(5): 167-174.

Hassan, Z., Daniel, R.C., O'Boyle, D. and Frost, A.J. (1999). Effects of dry cow intramammary therapy on quarter infections in the dry period. Veterinary Record 145: 635-639.

Hogeveen, H., Huijps, K. and Lam, T.J.G.M. (2011). Economic aspects of mastitis: new developments. New Zealand Veterinary Journal 59(1): 16-23.

Hogeveen, H., Kamphuis, C., Steeneveld, W. and Mollenhorst, H. (2010). sensors and milk quality – the quest for the perfect alert. The First North American Conference on Precision Dairy Management.

Huijps, K. and Hogeveen, H. (2007). Stochastic modeling to determine the economic effects of blanket, selective, and no dry cow therapy. Journal of Dairy Science 90(3): 1225-1234.

Huijps, K., Lam, T.J. and Hogeveen, H. (2008). Costs of mastitis: facts and perception. The Journal of Dairy Research 75(1): 113-120.

Koninklijke Nederlandse Maatschappij voor Diergeneeskunde (KNMvD) (2013). Richtlijn antimicrobiële middelen bij het droogzetten van melkkoeien.

Landers, T.F., Cohen, B., Wittum, T.E. and Larson, E.L. (2012). A review of antibiotic use in food animals: perspective, policy, and potential. Public Health Reports 127: 4-22.

Lim, G.H., Leslie, K.E., Morgan, J., Dow, B., Kelton, D., Duffield, T.F. and TenHag, J. (2000). An evaluation of the factors affecting the efficacy of a dry cow teat seal. Proceedings of the 39th Annual Meeting Of the National Mastitis Council, pp. 245-246.

Pettersson, G., Berglund, I. and Svennersten-Sjaunja, K. (2002). Effects of temporal technical stoppages in an AMS on bulk milk SCC and number of positive bacterial tests on udder quarter level. In: Proceedings of the NJF Seminar: Technology for Milking and Housing of Dairy Cows, February 11-13, Hamar, Norway.

Ruegg, P.L. and Pantoja, J.C.F. (2013). Understanding and using somatic cell counts to improve milk quality. Irish Journal of Agricultural and Food Research 52: 101-117.

Scherpenzeel, C.G.M., Den Uijl, I.E.M., Van Schaik, G., Olde Riekerink, R.G.M., Keurentjes, J.M. and Lam, T.J.G.M. (2014). Evaluation of the use of dry cow antibiotics in low somatic cell count cows. Journal of Dairy Science 97(6): 3606-3614.

Sharma, N., Singh, N.K. and Bhadwal, M.S. (2011). Relationship of somatic cell count and mastitis: an overview. Asian-Australasian Journal of Animal Sciences 24(3): 429-438.

Schukken, Y.H., Vanvliet, J., Vandegeer, D. and Grommers, F.J. (1993). A randomized blind trial on dry cow antibiotic infusion in a low somatic cell count herd. Journal of Dairy Science 76: 2925-2930.

Schukken, Y.H., Wilson, D.J., Welcome, F., Garrison-Tikofsky, L. and Gonzalez, R. (2003). Monitoring udder health and milk quality using somatic cell counts. Veterinary Research 34: 579-596.

Progesterone evolution modelling for improved reproduction monitoring in dairy cows

I. Adriaens[1*], T. Huybrechts[1], B. Aernouts[1], K. Piccart[2], K. Geerinckx[3], D. Daems[1], J. Lammertyn[1], B. De Ketelaere[1] and W. Saeys[1]

[1]*MeBioS, KU Leuven, Kasteelpark Arenberg 30, 3001 Leuven, Belgium;* [2]*ILVO, Technologie en Voeding, Agrotechniek, Burgemeester van Gansberghelaan 115 bus 1, 9820 Merelbeke, Belgium;* [3]*Hooibeekhoeve, Hooibeeksedijk 1, 2440 Geel, Belgium; ines.adriaens@biw.kuleuven.be*

Abstract

Monitoring milk progesterone (P4) provides the possibility to estimate the cow's reproduction status directly on farm. Usually, fixed thresholds are used to extract information from these measurements. However, as the P4 profiles vary largely between and within cows, fixed thresholds typically lead to sub-optimal performance in practice. To overcome this limitation we developed a physiology-based mathematical model to describe milk P4 profiles of cycling dairy cows based on the growing and degenerating *corpus luteum*, the main effector of serum and milk P4. Cycle-specific parameters can be calculated from the model which allows for identification of the correct reproduction status of the cow based on individual thresholds. For example, time between inflection points of the increasing and decreasing non-linear function is an indication for luteal cysts, whereas the slope and inflection point of the decreasing function are possibly linked with ovulation time. After insemination, an unexpected drop in P4 can indicate unsuccessful insemination or embryonal death. In this way, current algorithms can possibly be complemented and the automated monitoring of reproduction status improved. This model was fitted to 32 (uncomplete) progesterone profiles. An average pseudo coefficient of determination (pseudo-R^2) of 0.84 was obtained when the P4 concentration of all milkings was used, and a pseudo-R^2 of 0.86 was obtained when only one milking per day was included. We observed a large variation between model parameters and characteristics, demonstrating the necessity of an individual approach. Finally, the of potential covariates such as fat content and milking interval on the P4 level was tested by regressing the residuals of the model against these factors. No significant effects were observed. Therefore, it can be concluded that at the current progesterone measurement accuracies the methodology cannot be improved by taking these potential covariates into account.

Keywords: progesterone p4, reproduction monitoring, non-linear model, individual approach

1. Introduction

Milk progesterone (P4) measurements offer the possibility to monitor a direct physiologic parameter representative for the cow's reproduction status (Posthuma-Trumpie *et al.*, 2009). Progesterone profiles give indication of estrus, pregnancy and even of abnormalities (e.g. acyclicity, cysts). Still, the difficulty in the automatic extraction of the reproductive information from progesterone measurements lies in the inter- and intra-cow variability of the profiles. The inter-cow variation is particularly the result of genetic variation, diseases, individual stress factors and so on, where the intra cow variation can also be caused by a large range of factors. One of them, the fat percentage of milk, was already investigated by Ginther *et al.* (1976) and Pennington *et al.* (1981), but their results were contradictory. Moreover, automated monitoring systems are mainly developed for automated milking systems (AMS), in which milking intervals are irregular. To our knowledge no reports are available on the possible effect of the milking interval on milk P4, e.g. due to the link between milking interval and fat content. The algorithms used today based on fixed thresholds can extract valuable information out of the data. However, it should be investigated whether more accurate modelling of the individual progesterone profiles with respect to the influential factors on the intra-cow variability can have a positive impact on the performance of the current monitoring systems. As such, two main goals were set in this research. The first aim was to develop a method to describe the P4 profile of cycling dairy cows with a flexible, but robust mathematical model. This model is based on the growing and degenerating *corpus luteum* (CL). In the growing phase, the CL produces P4 proportional to its size. Later, during metestrus, the P4 production is high, but not directly proportional to luteal size anymore, and the number of follicular waves influences the length of this phase, causing more variation (Mann, 2009). After the peak of $PGF_{2\alpha}$, luteolysis causes the P4 concentration to drop. The individual levels of P4 concentration are both cow and cycle dependent and influenced by e.g. stress and cortisol. By means of the model, cycle characteristics can be calculated from the model parameters and these can be used to monitor reproduction status. In this way, fixed thresholds are avoided and monitoring reproduction can be supported without excessive observation of the animals is necessary. The second aim was to investigate whether certain factors influencing the P4 measured in milk in an AMS setting, and thus should be taken into account in the modelling.

2. Materials and methods

2.1 Cows and samples

Twenty-two Holstein-Friesian dairy cows with parities between one and five were selected from a fifty headed herd on an experimental dairy farm in Geel, Belgium. The cows were milked with an automatic milking system of Delaval (Tumba, Sweden) and mixed milk samples were taken every milking with the sampling unit (VMX). A total of 2,460 samples were useful for the data analysis. The reproduction status of each cow was categorized as cycling, pregnant or post-partum anestrus based on ultrasound and visual observation of heat. Three cows were investigated as pregnant. Two cows were not cycling yet (29 and 34 days in milk (DIM) at the start). The other 17 cows were cycling, from which one was culled during the trial. The data of this cow were not included in the analysis.

2.2 Progesterone assay and milk constituent analysis

Three times a week, analysis of the raw milk samples was done at the certified lab of MCC Flanders in Lier, Belgium. Every milk sample was examined for pH, milk fat, crude protein, lactose and urea content by the use of Milkoscan FT+ (Foss A/S, Hillerød, Denmark) according to ISP 9622:2000 (ISO, 2000). The Somatic Cell Count (SCC) was determined by flow cytometry (Fossomatic, Foss A/S, Hillerød, Denmark) as an indicator for udder health. Two ml of each sample was frozen at -19 °C for the P4 analysis. The P4 concentration in the whole milk samples was measured with a direct ELISA at 405 nm, using a Ridgeway milk P4 kit (Ridgeway, Gloucester, UK). The milk standards used were provided by Ridgeway and contained 0, 1, 2, 5, 10, 20 and 50 ng/ml P4. The intra-assay variation specified by the producer at high and low P4 concentrations was 15 and 5% coefficient of variation (CV), respectively.

2.3 Model description

The model was based on the physiology of the growing and degenerating CL. After a period of acyclicity after the partus, where P4 concentration is low and no pattern can be distinguished, follicular activity restarts and a first CL is formed. This CL grows and produces P4 until luteolysis occurs, followed by a drop in P4. These P4 cycles succeed until pregnancy. After successful insemination, a pregnancy CL is formed and the P4 production remains high. In the case of a luteal cyst, the P4 concentration remains high for an unexpectedly long period. However, when a follicular cyst is present, the opposite is true and an extra-long period of low P4 is observed. The method we present is founded on this physiological P4 pattern and can be divided in three parts:

1. Postpartum anestrus can be described as a constant progesterone level and the standard deviation (SD) can be calculated. By distinguishing between normal and abnormal variation, positive deviations can be detected, which is the sign that luteal activity has started.
2. A combination of non-linear growth curves, one increasing and one decreasing, based on the work of Boer (2012) were combined to describe the growth and degradation of the cyclic CL.
3. After insemination, the same growth curve described the P4 production of the pregnancy CL. During pregnancy, P4 concentration remains high, but when the embryo dies, P4 drops to unexpectedly low levels again.

The non-linear growth functions used to describe the P4 patterns were two Gompertz functions ($f(t)=A\times\exp(-\exp(B-C\times t))$), with A the horizontal asymptote and B, C parameters determining the slope and displacement over the x-axis. One determines the increase of P4 (positive A parameter), another the decreasing phase (negative A parameter).

2.4 Data analysis and factors influencing the milk P4 concentration

All data analysis was done in Matlab 2013a (Mathworks Inc., Natick, MA, USA) and JMP Pro 11 (SAS Institute Inc., Cary, NC, USA). The model was fitted using the Gauss-Newton algorithm and goodness-of-fit was assessed using the pseudo-R^2, calculated as pseudo-$R^2=1-SSEw/SSTO$ with SSE=error sum of squares, weighted for the fitted P4 value to correct for the proportional error of the ELISA essay. Due to the resulting heteroscedasticity, pseudo-R^2 is a better measure for goodness-of-fit than the unweighted R^2 (Christopoulos and Motulsky, 2003). $SSTO$ is the total sum of squares. To investigate if fat or milking interval affect the P4 content of each sample, whether or not corrected for the effect of cycle and the heteroscedasticity of the measurements, P4 was tested for correlation with milking interval and fat content. This was done by regressing the P4 variables against milking interval and (corrected) fat percentage, with corrected fat percentage defined as the fat content after subtraction of the baseline fat content for that cow. The regression coefficients were assessed to determine the effect of these parameters on P4. For the sake of completeness, the effect of the other milk constituents was assessed in the same way, although no effect was expected based on physiology and literature.

3. Results and discussion

The observed milk P4 concentrations varied between 0 and 10 ng/ml, while in many studies milk P4 concentrations between 0 and 40 ng/ml have been reported. This may be explained by the used ELISA kit, as the results are of the same order of magnitude as those reported by Chanvallon *et al.* (2014), who used a similar methodology.

Three examples of the available data and their corresponding model is shown in Figure 1. The dots represent the P4 measurements, the thick line the model. The model fitted all available cycles and cycle parts (increasing or decreasing) well, independently of the cycle-specific minimal and maximal values, the length of the luteal and follicular phases and the slopes of the rising and decreasing phases. The average pseudo-R^2 was 0.84 and varied between 0.77 and 0.96. When the sampling frequency used in the model was lowered to one measure the pseudo-R^2 increased and varied between 0.83 and 0.96. Model characteristics differ between cycles. For example, the time between the inflection point of the rising and that of the decreasing function varies from 10 to 19 days. This supports the observation that the variation between cycles and cows is large and an individual approach could provide added value. However, future validation of this algorithm is required to verify if the model characteristics (cycle length, slope of the growth functions, ...) provide added value in monitoring the reproduction status. It is hypothesized here that the slope and inflection point of the decreasing function can provide information on the ovulation time, while the time between the start and end of each function might be used to calculate cycle length. Embryonic death or the success of insemination may be detected by monitoring P4 during the period after insemination.

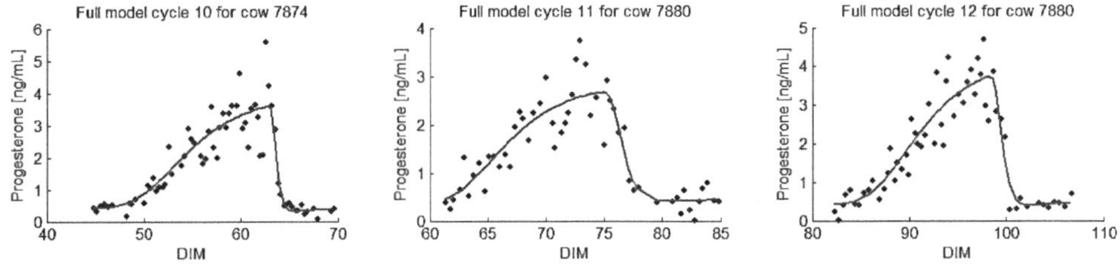

Figure 1. Progesterone model based on a combination of two non-linear Gompertz function fitted on three available P4 profiles.

The second research question in this study concerned the effect of fat percentage and milking interval on the P4 concentration in milk, for example by influencing the diffusion process in the udder, or the P4 measurement with ELISA. No significant correlations could be observed between the P4 and the milking interval or fat content of a particular sample, nor did correlations between P4 and other milk constituents exist. One possible explanation for this observation is that the effect of variable P4 production (e.g. during midestrus) is much larger than the effect of altering diffusion of P4 to the milk. A second possible explanation is the fact that the samples are automatically taken mixed milk samples, and the variation in milk constituents in these samples is too large to detect a significant effect. For example Pennington *et al.* (1981) also found a low correlation between fat in mixed milk and P4 concentration, where Ginther *et al.* (1976) reported a very high correlation when foremilk samples were used. This suggests that the sampling technique could be a very important factor influencing the effect of fat content on the milk P4 concentration. A third possible explanation may be found in the large proportional error of the ELISA kit. This might hide the smaller effect of fat or milking interval. If this is the case, it is recommended to repeat this study when a more precise measurement technique for milk P4 becomes available. Until then, it is not necessary to take milking interval or fat percentage into account when considering individual mathematical models for milk P4 based on mixed milk samples.

References

Boer, M. (2012). All in good time: dynamics of the bovine estrous cycle investigated with a mathematical model, 192 pp.

Chanvallon, A., Coyral-Castel, S., Gatien, J., Lamy, J.-M., Ribaud, D., Allain, C., Clément, P. and Salvetti, P. (2014). Comparison of three devices for the automated detection of estrus in dairy cows. Theriogenology 82: 734-741.

Christopoulos, H. and Motulsky, A. (2003). Fitting models to biological data using linear and nonlinear regression: a practical guide to curve fitting. GraphPad Software Inc., San Diego CA, USA, pp. 351.

Ginther, O.J., Nuti, L.C., Garcia, M.C. and Wentworth, B.C. (1976). Factors affecting progesterone concentration in cow ' s milk and dairy products. Jounal of Animal Sciences 42: 155-159.

Mann, G.E. (2009). Corpus luteum size and plasma progesterone concentration in cows. Animal Reproduction Science 115: 296-299.

Pennington, J., Spahr, S.L. and Lodge, J.R. (1981). Influences on progesterone concentration in bovine milk. Journal of Dairy Science 64: 259-266.

Posthuma-Trumpie, G.A., Van Amerongen, A., Korf, J. and Van Berkel, W.J.H. (2009). Perspectives for on-site monitoring of progesterone. Trends in Biotechnology 27: 652-660.

Dairy cow welfare: effects of a virtual fence on social behavior

P. Koene*, P. Hogewerf and B. Ipema
Department of Animal Welfare, Wageningen Livestock Research, De Elst 1, 6708 WD, Wageningen, the Netherlands; paul.koene@wur.nl

Abstract

Dairy cows may graze more efficiently when allowed grazing in predefined areas with predictable sward height and grass quality. Moving the animals daily or preferably more frequent is needed. Using traditional fencing so-called strip grazing causes high labor requirements. Undesirable effects of electric fencing might be prevented by the use of virtual fencing. We used a non-GPS system with underground wires signaling a boundary that could be partly and differentially activated by remote control together with a collar in with the boundary signal is received. A warning signal could be given to the cow when approaching the boundary and a correction signal when crossing the boundary. One of the questions is whether the virtual fence influences the social behavior and welfare of cows. Dairy cows – grazing in a group of four – in the virtual fenced area were compared with cows in an electric fenced area. This was replicated once with two different groups (n=16). During 12 days the groups were compared in several aspects. One of the aspects was grazing on a single strip (60×10 m) compared to a double strip (60×20 meter). The behaviour, position and activity were monitored during 5 hours per day. During the 12 day period all cows stayed in the strips with only few temporary boundary crossings. The cows in the virtual fence system showed lower nearest neighbor distances than the cows in the electric fence system. The social activity of the cows was significantly influenced by strip size. In a 1-strip area the virtual fence cows were standing closer together. Also behavioral synchronization of the group in the virtual fence condition measured during standing and lying was weaker. Our preliminary conclusion is that a virtual fence has some effects on the social behavior of cows different from social behavior of electric fenced cows. More detailed analysis will show whether the social organization of the cows is also affected.

Keywords: grazing behavior, strip grazing, fence, animal behaviour, animal welfare

1. Introduction

In this project, the question is examined whether it is possible to train cattle using pre-defined signals that indicate boundaries of – for example, pastures – instead of physical and visible fences. Such a 'virtual fence' could make welfare-friendly and grassland-use-efficient grazing possible and are still in development (Umstatter, 2011). There are systems developed and made available in the market to keep cattle with signals within a fixed predefined area (Monod *et al.*, 2009; Umstatter *et al.*, 2015). In this pilot the focus is on strip-grazing. Each cow gets a collar with a box and sensor to determine the position of the animal with respect to an underground wire (a 'virtual fence') in the pasture and can, moreover, produce warning or correction signals when approaching the ring wire. The warning signal (sound) is intended to alert the cows on the approaching a virtual boundary; this signal will have little impact on welfare. The correction signal (an electric signal) is only given if the animal persists in further approaching the virtual fence. This may have a greater effect on well-being of the cow. By learning the animals to associate the warning signal with the correction signal it is expected that the corrective signal is not given often. The feasibility of the development of a method which is based on an area that can be adjusted over time (instead of a fixed area) is estimated to be high and will facilitate the development of efficient strip-grazing methods. This pilot experiment will be reviewed as being successful when the main part of the animals learn the association between the warning signal and location or behavior and when the experimental –virtual fence – group in terms of well-being scored not significantly lower than the control – electric fence – group.

2. Material and methods

2.1 Experimental setup

Animals were heifers without grazing experience and had always been stabled. They had to learn to graze and learned to deal with a fence (barbed wire) in a separate grazing area. The experiment is done with small groups of animals (4) that are allowed to graze on strips of grass which during the test are shifted every day. We used the Boviguard system, described in detail elsewhere (Monod *et al.*, 2009). In short, when approaching a virtual boundary, the animals receive a first warning signal. When they do not stop or return the warning signal is followed by a correcting signal. The research focuses primarily on whether and how fast a cow learns to avoid the correcting signal in small areas of pasture. The animals learn in trials and get in the course of the trials more freedom of movement. This experiment will be conducted in 2 replications with an experimental group and a control group (a total of 4 groups of 4). Groups differ in the manner in which the border of their grazing area is indicated, either by virtual fence (experimental group with warning and correction signal via the collar), or by the traditional electric fence (control group with exposed wires and correction in contact with the wire). Both groups are treated in the same way and the same observations took place in both groups. The experiment is carried out in four phases, which are identical in setup. Every day 2 strips are made available (each 10×60 meter) – an old (grazed the day before) and a new (with fresh grass were available) one. In the first phase of 3 days (three repeats to allow trial learning) an electric wire was placed above the underground hidden virtual fence. In the second phase of three days the electric fence was placed 2 meter past the virtual fence. After these 2 training phases the final test was done, i.e. the electric fence was removed and in fact placed at the actual boundary of the field. In the control group the conventional electric fence, of course, is always on the borders of the area, i.e. the border of the two strips. In the phases 1-3 the animals are from 9:00 am until 15:00 in the test strips. In the remainder of the time the animals are as a group (experimental and control group) kept in separate pastures with fixed fence (not virtual or electric fence) under conventional husbandry conditions. After each trial day after 15:00 the conditions for the next day are prepared. Before 9:00 on test days the animals are prepared to enter the test area. GPS sensors are placed in the collar and observers are preparing the behavioural observations. Observers are present during the test days in all phases during the period of 9:00-15:00. The observers take care of the welfare of the animals and will intervene when an animal is showing obvious signs of stress. Per animal responses will be recorded and correction signals – if possible – are registered. An animal that continuously ignores signals, and wherein no learning effect can be observed will be removed from the test, if more than light distress is detected. In this paper we focus mainly on the following 4th phase. This phase was somewhat different from the first 3 in that only one strip was available on day 1, two strips on day 2 and one strip on day 3 between 9:00 and 15:00. Furthermore, the animals stayed on the strip for 24 hours. However, we report the observations and GPS recordings of the animals between 15:00 and 9:00.

2.2 Measurements and statistics

The animals were monitored individually and as a group. Daily in each group each animal was observed with continuous protocols (5 protocols with a duration of 10 min per animal) and the group is scanned behaviorally (25 times per day). The ethogram included the behaviors walking, lying, standing, grazing, social positive, social negative behaviors and the reactions of the animal on the signals recorded at the borders of the grazing area using the Observer (Noldus, 1991). The experimental area was divided in several subareas (Figure 1) and also the locations are recorded by the observers. The locations and movements of the cows (x, y co-ordinates) are recorded by GPS (Extreme 66 GPS travel recorder attached to the collar of the animal). This GPS system gives no inconvenience for the animal.

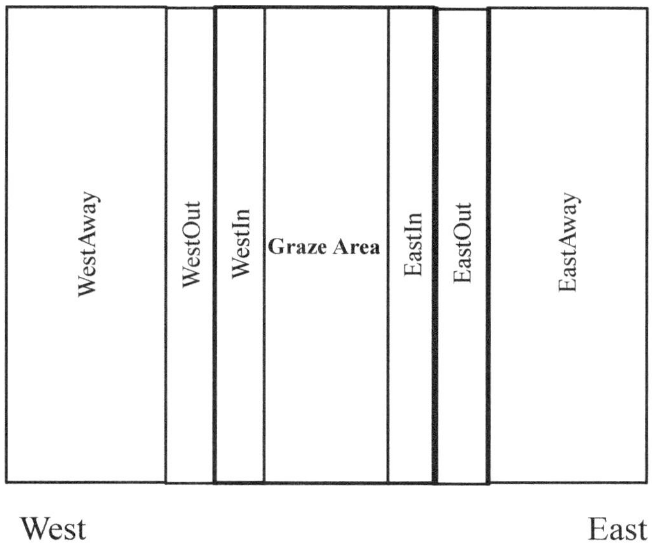

Figure 1. The black line indicates the fenced area (20×60 meter). The presence near the fence (WestIn, EastIn), over the fence (WestOut, EastOut) or escaped (WestAway, EastAway) in measured in the observations and by GPS.

Primary outcome parameters from the observations (visual observations and automated) were derived per cow per day. The behavioral observations provided insight into the nature, intensity and duration of the responses to the signals. Performance of the experimental group (virtual fence) is compared with that of the control group (electric fence) using Linear Mixed Models (LMM) of SPSS (IBM version 2016). In this paper special attention is given to social aspects of the cows behavior, while in a twin paper the focus is on individual aspects of behavior and welfare. Analyzed are nearest neighbor distance, walked distance and synchronization of the behaviors grazing and standing.

3. Results

The cows in the virtual fence system showed lower nearest neighbor distances than the cows in the electric fence system (LMM; $P<0.001$). The nearest neighbor distances of the cows were also significantly influenced by strip size (LMM; $P<0.01$). In a 1-strip area the virtual fence cows were behaving (standing, grazing, lying?) closer together. The average distance per minute moved differed between the groups (LMM; $P<0.001$); the cows in the virtual distance group moved on average in a much lower pace than the cows in the electric fence group (Figure 2; right panel).
Also behavioral synchronization of the group in the virtual fence condition measured during grazing ($P=0.02$) was weaker; when at least one cow in the group was grazing, more cows (on average more than 2) were grazing in the electric fence group compared to the virtual fence group (Figure 3, left panel). When at least one cow was standing more cows were standing in the virtual fence group compared to the electric fence group in the one strip area (Fence×Strip interaction, $P=0.034$; Figure 3, right panel).

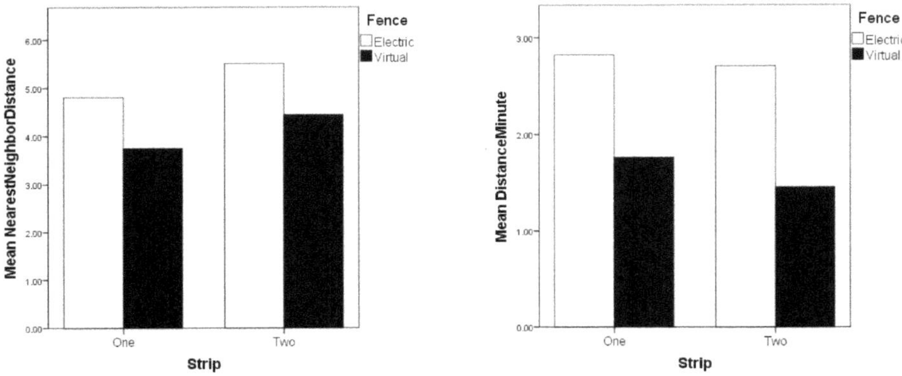

Figure 2. Mean nearest neighbour distance measured by GPS during the stay on one or two strips (left panel). Mean distance walked per minute measured by GPS during stay on one or on two strips (right panel).

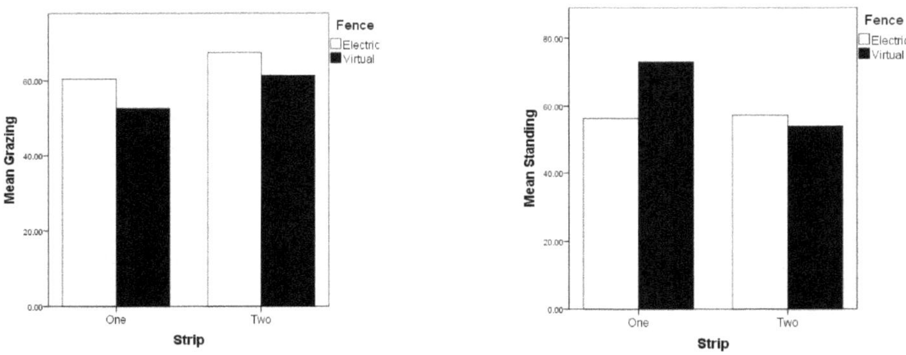

Figure 3. Mean grazing when at least one cow was grazing as indication of synchronisation (left panel). In a comparable way, the synchronisation of standing when at least one cow was standing (right panel).

4. Discussion

A virtual fence had some effects on social aspects of the behavior of cows in the pasture as compared to electric fenced cows. The nearest neighbor distance in the virtual fenced group indicates that on average cows are closer together than in the electric fence group. We have seen that sometimes during observations in the one strip condition the cows were standing – with head low – in a tight group. Coupling between the behavioral observations and the GPS-recordings will be done as the follow-up step to compare the nearest neighbor distances between the cows during the different behaviors. The synchronization of behavioral elements appeared to be different between the experimental and control group. In the follow-up analysis an overall synchronization index will be calculated and the influence of the virtual fence on this index will be calculated. Also, more detailed social network analysis will show whether the social organization of the cows is also affected.

5. Conclusions

The virtual fence may become a tool for precise allocation of grazing areas to cows. It is important to analyse social aspects of groups of dairy cows within the virtual fence. We made a start with the analysis of some social parameters. We aim to analyse the social networks of groups of the cows to reveal higher level effects on the behavior in order to adapt the virtual fence functioning and learn the cows the right tools to adapt to a virtual fence. More experiments are planned in 2016 to enhance the understanding of the cows learning process in relation to the virtual fence. Especially the effects of group size and social attraction on the effectiveness of the virtual fence and the number of escapes will be part of the experiments.

Acknowledgements

The authors wish to acknowledge the Amazing Grazing programme, Courage, the Netherlands, for funding and visiting students and their supervisors of HAS Den Bosch, the Netherlands, for support during the observations.

References

Monod, M.O., Faure, P., Moiroux, L. and Rameau, P. (2009). Stakeless fencing for mountain pastures. In: Precision Lifestock Farming '09'. Wageningen Academic Publishers, Wageningen, the Netherlands, pp. 175-181.

Noldus, L.P.J.J. (1991). The observer: a software system for collection and analysis of observational data. Behavior Research Methods, Instruments and Computers 23: 415-429.

Umstatter, C., Morgan-Davies, J. and Waterhouse, T. (2015). Cattle responses to a type of virtual fence. Rangeland Ecology and Management 68: 100-107.

Umstatter, C. (2011). The evolution of virtual fences: a review. Computers and Electronics in Agriculture 75: 10-22.

Use of mechanical rotating brushes in dairy cows – a potential proxy for performance and welfare?

L.J. Keeling[1], D. De Oliveira[1] and B.-O. Rustas[2*]
[1]*Department of Animal Environment and Health, Swedish University of Agricultural Sciences, Box 7068, Uppsala, Sweden;* [2]*Department of Animal Nutrition and Management, Swedish University of Agricultural Sciences, Box 7024, Uppsala, Sweden; bengt-ove.rustas@slu.se*

Abstract

Mechanical brushes are becoming routine on dairy farms to improve cattle welfare. Their use could potentially be automatically recorded on an individual cow level. We investigated brush use in a mixed group of 72 Holstein (32%) and Swedish Red (68%) dairy cattle over a 9-week period in a loose housing cubicle system. Milk yield and feed consumption were monitored individually and merged with data from behaviour observations of brush use by individual cows. From observations of social behaviour (giver and receiver) we determined the dominance rank (based on butting, threats, pushing and fighting) and the affiliative rank (based on social licking and sniffing) for each individual. Affiliative rank affected frequency and duration of brush use, high receivers of affiliative social behaviour visited the brush most often and middle affiliative brushed for the shortest duration. Dominance rank affected duration of a brush bout, which was shortest for middle dominance ranked cows. There was a positive relationship between the frequency of brush use, milk yield and intake of roughage. Each additional brush use was associated with a higher milk yield of 1.52 kg and a higher roughage intake of 0.67 kg dry matter per day. It can be hypothesised that the tactile stimulation from the brush is a trigger for the oxytocin release chain which drives milk yield and consequently feed consumption. If confirmed, deviations from a cow's usual brush use level may be an early indicator of future changes in performance. More research is needed to support the proposed link between changes in brush use and changes in welfare. We speculate on the steps towards automated brush monitoring as a potential management tool.

1. Introduction

In recent years several companies have started selling brushes to promote grooming activity. De Vries *et al.* (2007) reported a high proportion of cows in the group using the brush and concluded that brushes make it easier for cows to groom themselves, particularly in places that are hard to reach. Grooming is considered a potential indicator of positive welfare (Boissy *et al.*, 2007). Schukken and Young (2010) found that providing mechanical rotating brushes resulted in a significant increase in milk production of second lactation cows and a significant decrease in clinical mastitis in second and higher lactation cows.

Welfare assessment protocols are increasingly being used in commercial dairy farms. Mandel *et al.* (2013) found that brush usage decreased on days when the herd was subjected to intrusive medical procedures. It was also reduced when the location of the brushes was away from the feeder and when temperature and humidity levels were high. Given that high producing dairy cows are under time budget pressure, a measure such as brush usage that may be reflecting (even indirectly) that the cow has had time to eat, ruminate, be milked, etc. is of interest as welfare indicator.

The aim of this study was to explore the potential of mechanical brushes as a welfare indicator in dairy cattle by investigating the relationship between use of mechanical brushes, social behaviour in the group and production parameters.

2. Methods

This study was carried out on a mixed group of Holstein (32%) and Swedish Red (68%) dairy cows at the Swedish Livestock Research Centre-Lövsta (Uppsala, Sweden). The average group size was 55 cows and information was gathered on a total of 72 cows as new cows entered and others left the group during the period. The loose housing cubicle system was equipped with an individual automatic milking system (AMS; DeLaval, Tumba, Sweden) and two mechanical rotating brushes (Swinging cow brush, DeLaval, Tumba, Sweden) located in the alley opposite the roughage bins, one at each end. Each cow was equipped with a collar chip that controlled access to roughage feeding stations (Biocontrol A/S, Rakkestad, Norway) and to concentrate feeders (DeLaval, Tumba, Sweden). Milk yield, feed consumption and performance parameters were monitored individually. The number of cows lying, feeding, drinking, queuing to be milked, in the AMS, using the brush or standing/walking was noted Monday to Thursday every second week over a 9 week period to determine the overall time budget of each cow. The identity of all cows using the brush was recorded within a period of 8 hours/day on these days. The identity of the giver and receiver in all social interactions between cows were recorded every Friday in this period. Observations included agonistic interactions (threat, head butting, pushing and fighting) and affiliative interactions (licking and sniffing). These were used to classify cows as high, middle or low in a dominance rank and high, middle or low in an affiliative rank.

Data related to milk production was collected automatically in the AMS. It consisted of quantity of milk (kg/day), somatic cell count (cells/ml) and milk flow (max and mean rate, averaged over teats, in kg/min). Also available for this analysis were concentrate feed intake, frequency of visits to the concentrate feeder per day, silage eaten, total time eating silage and frequency of visits to the silage bins. Production data was averaged over the same period as the behavioural time budget observations. We used generalized mixed models to analyze the data, using breed, number of lactation, dominance rank, affiliative rank, stage of lactation (early, middle, late) and their interactions as fixed effects. Animal was included as a random effect. In order to assess the relationship between use of mechanical brushes and production variables we used frequency of brush use as a covariate. Tukey adjust was used as the post-hoc test. All the analyses were performed in SAS 9.3 (SAS Institute, Cary, NC, USA)

3. Results

The behavioural time budget for cows during our experimental days was 42% lying, 27% walking or standing, 16% feeding, 10% queuing to be milked and 1% of their time drinking, using the brush or being milked.

There was no effect of stage of lactation on brush use, but there was an effect of lactation number on both the frequency of brush use (F=2.53, P=0.03) and the duration of brush use (F=3.19, P=0.01). Third lactation cows visited the brush significantly more often (mean±se of frequency/experimental day; 1.2±0.10) than first lactation cows (0.8±0.07). Third and fourth lactation cows brushed for longer (duration in seconds/experimental day: 268±35.3; 267±85.2) than first lactation cows (140±20.0). There were no effects of dominance rank on the frequency of brush use but there was an effect of dominance on the duration of brushing bout (F=8.61, P=0.0005) with middle ranked cows using the brush for the shortest durations (seconds/experimental day; Highest ranked, 64±10.7; Middle ranked, 49±5.9; Lowest ranked, 78±11.1).

There was a significant effect of affiliative rank on both the frequency of brush use (F=7.19, P=0.001) and the duration of brush use (F=6.75, P=0.002). Cows that received most positive affiliative interactions used the brushes most often compared to cows of other affiliative ranks (frequency/experimental day: Givers, 0.8±0.13; Middle ranked, 0.8±0.08; Receivers, 1.3±0.13). Cows that received most positive interactions used the brushes for longer than middle ranked cows (duration in seconds per experimental day: Givers 165±48.7; Middle ranked 113±23.8; Receivers 218±46.3).

There was a positive relationship between the frequency of brush use and the milk yield (F=5.14; *P*=0.02). The estimate was 1.52, implying that for each additional occasion that a cow used the brush, the milk yield increased by 1.52 kg per day. There was also an increase in average milk flow associated with increased brush use (F=4.75, *P*=0.03), where the estimate was 0.05, meaning that for each additional visit to the brush the milk flow increased by 0.05 kg/min.

Relationships between feeding and the frequency of brush use were only found for eating roughage, not for eating concentrate. There was a positive relationship between frequency of brush use and intake of roughage (F=10.73, *P*=0.0012; estimate=0.67 kg DM silage per brush visit) and the duration of eating roughage (F=4.16, *P*=0.04; estimate=410 seconds per brush visit). There was no relationship between the frequency of brush use and the somatic cell count (F=0.14, *P*=0.70).

4. Discussion

This study demonstrates that brushes are an attractive resource for the cows, particularly for cows that receive most positive social interactions and for cows of higher lactation numbers, and that there is a relationship between increased brush use and increased milk production and roughage intake. Unfortunately, it is not possible from this study to identify the driver in this triad; brush use, roughage intake and milk production. It would be interesting in future studies to test the hypothesis that increased brush use triggers increased milk production via the tactile stimulation of the brush acting to increased oxytocin release and that the increase in feeding is a consequence of the increased milk production. The fact that we found a relationship between brush use and milk flow would tend to support that oxytocin is the driver.

In summary, the results of this study suggest that providing mechanical rotating brushes to cows may help them better achieve their production potential as well as provide them the opportunity to engage in a preferred activity.

5. Possibilities for automated monitoring of brush usage

Our results, in combination with previous studies, imply that cows' use of brushes has potential to serve as a proxy measure for performance and welfare. In the second part of the paper, we speculate on some of the necessary steps to be carried out before such a monitoring is implemented in practice as part of a management tool. These include: (1) developing a feasible way to validate these results in other herds; (2) developing an automated brush use monitoring system; and (3) evaluate its usefulness under commercial conditions.

1. Validating these results in other herds

   Detailed observations of brush use are time consuming. Most studies have used video cameras to record brush use (De Vries *et al.*, 2007; Mandel *et al.*, 2013; Schukken and Young, 2010; Val-Laillet *et al.*, 2008). We used direct observations of cows, which allowed us to link brush use to other characteristics at the individual cow level, however the observations were limited to only a part of the day. Some form of automated recording, that can provide information on brush use at the individual cow level over the whole 24 h period would greatly facilitate research in this area. Since most cows on modern farms, and probably all cows on research farms, are already individually identified we propose automatically detecting the identity of cows close to the brush in combination with video recording of brushing behaviour as a feasible next step to validate the results of this study. It would allow more data to be collected in a larger number of herds without a large investment in new technology at this stage. For example, with only a few sick cows during the period of our observations, we didn't find any connections between health and brush use but other investigations have found connections. The relation between brush use and illness needs to be validated on a scale that allows a sufficient incidence of the most common production diseases, e.g. mastitis, lameness and digestive disorders. With such an approach linkages between brush use patterns and the type of illness could potentially be established. The positive relation between milk production and brush use in this study supports several advantages for rotating brushes. These also need validating. Of particular interest would be to confirm if the connection to feed intake is consistent, since this has not been reported previously, and to assess the variation in the estimates linking increased brush use to increased milk yield and flow rate.

A more feasible form of data collection would also allow recording of individual variation in brush use in different experimental conditions. The lower use of mechanical brushes from middle ranked and first lactation cows in this study, together with previous findings that cows in stressful situations decrease brush use (Mandel *et al.*, 2013), strongly suggests a connection between brush use and negative emotional states. That increased brush use may be associated with positive emotional states is perhaps supported by its association with affiliative behaviour in this study, since affiliative behaviour has been suggested to be an indicator of positive welfare in cattle (Napolitano *et al.*, 2009). Both these hypotheses would need to be tested in experimental setups together with other already validated indicators of welfare. Furthermore, to fully utilize the relationship between brush use and performance on farm level an understanding of cause and effect would be desirable. The magnitude of response in milk yield to brush use could be tested by varying availability of the brushes within the barn. If brush use is driving yield, improved availability could potentially increase production. The relationship between brush use and milk flow suggests a connection between brushing, hormonal release and milk let-down. However, it could also be that brush use is a consequence of higher milk yield. If the higher milk yield from cows is a consequence of their better welfare, monitoring of brush use could be a tool for improving welfare and thereby increase production.

2. Developing an automated brush use monitoring system

   If the resulted are validated and the causal connection clarified, then there would probably be the incentive for a fully automated commercial brush use monitoring system to be developed. This is not, however, without its problems. Mechanical rotating brushes are designed to be easily accessed, which makes it a challenge to construct an identification system. The system should detect each cow using the brush while excluding cows standing close by. The fact that cows frequently brush the rear part of the body adds another difficulty as identification systems are usually built on transponders on neck collars or ear tags. A transponder around the base of the tail could perhaps be necessary. Once developed, a study comparing automatically collected data of cow identity and brushing and the golden standard of data collected by direct observations would be necessary to confirm the accuracy of the system. For example, we noted that on occasions two or even three cows may be using the brush at the same time. Good precision is needed in a research setting, but for commercial use a lower level can probably be accepted.
3. Evaluating its usefulness under commercial conditions

   Monitoring of brush use could potentially be beneficial in several ways on a dairy farm. On a herd level, brush use could be an indicator of the general welfare and used as a benchmark between herds. Absolute brush use is probably not a good measure, as several factors might influence it, e.g. brush location. The between animal variation might be a better estimate indicating the proportion of cows that would benefit from an improved situation. Our results would suggest that among these low brush users would most likely be young and sub dominant cows. Information on brush use could probably serve as a response variable in improving welfare through management changes, e.g. by adjusting the routines for milking, feeding and grouping. Probably most useful, if brush use is to be used as a proxy for welfare and performance, would be monitoring changes over time for individual cows. A decrease in brush use might give an indication of a change to suboptimal production conditions that would be difficult to identify by other measures or as early warnings for diseases. As brushing can be considered a luxury behaviour (Mandel *et al.*, 2013) it is probably one of the first behaviours to decrease if the welfare of the cow starts to be reduced. The decrease in brush use might therefore occur before a decrease in milk production or an increase in sub-clinical signs of disease (e.g. elevated somatic milk count, changes in movement or eating patterns). Cows that have a higher risk of disease (i.e. the ones in the transition period around calving, due to metabolic diseases and mastitis, for instance) could particularly benefit from such early warning. This may result in a smaller reduction in their performance than would otherwise be expected. On the other hand, an increase in brush use might give an indication of an improvement in production conditions for that individual. The fact that milk yield was higher in cows that also used the brushes more in this study implies, whatever the underlying mechanism, that brushes can be used to manage cows so that they better achieve their production potential.

References

Boissy, A., Manteuffel, G., Jensen, M.B., Moe, R.O., Spruijt, B., Keeling, L.J., Winckler, C., Forkman, B., Dimitrov, I., Langbein, J., Bakken, M., Veissier, M. and Aubert, A. (2007). Assessment of positive emotions in animals to improve their welfare. Physiology and Behaviour 92: 375-397.

De Vries, T.J., Vankova, M., Veira, D.M. and Von Keyserlingk, M.A.G. (2007). Use of mechanical brushes by lactating dairy cows. Journal of Dairy Science 90: 2241-2245.

Mandel, R., Whay, H.R., Nicol, C.J. and Klement, E. (2013).The effect of food location, heat load, and intrusive medical procedures on brushing activity in dairy cows. Journal of Dairy Science 96: 6506-6513.

Napolitano, F., Knierim, U., Grasso, F. and De Rosa, G. (2009). Positive indicators of cattle welfare and their applicability to on-farm protocols. Italian Journal of Animal Science 8: 355-365.

Schukken, Y.H. and Young, G.D. (2010). Effects of swinging cow brush on milk production and mastitis. Western Dairy News 10(5): 87-88.

Val-Laillet, D., Vieira, D.M. and Von Keyserlingk, M.A.G. (2008). Dominance in free-stall-housed dairy cattle is dependent upon resource. Journal of Dairy Science 91: 3922-3926.

Monitoring social behaviour in grazing cows

R. Gabrieli
Ministry of Agriculture and Rural Development, Extension Service, Beit Dagan, Israel; ragav@shaham.moag.gov.il

Abstract

Cows' social behaviour is crucial for welfare and production. Interactions are not easily noticeable, since hierarchy within the herd is passive. Previous studies showed irregularity in activity patterns of grazing beef cattle. Identifying uniform parameters may provide a reliable analysis tool for influential factors. We believe that grazing beef behavior can serve as a reference for dairy cows' natural behavior, allowing better adjustment of welfare management. Innovative wireless pedometric monitoring system was modified to serve as a standalone station in pasture in northern Israel. 148 cows (78 primiparous and 70 multiparous) were tagged. 46 of them were pairs of mothers and daughters. Activity was monitored hourly during 170 days. Observations were conducted daily and interactions between cows recorded. Interactions were grouped as 'positive' (grooming, proximity) or 'negative' (pushing, isolation). Pedometric data showed variability in daily activity (137±44 steps per hour) and in discrete hours' activity. Lowest variability was found in average hourly activity between 6:00-7:00 am (grazing) and 01:00-02:00 (rest) (231±56 and 41±17 steps per hour respectively). These two parameters were used for evaluating social effects. No interactions were observed between mothers and daughters, but kins' group showed higher amount of positive interactions than cows without kins (97±4.1 and 45±2.8; P=0.001) respectively. Kinship showed a tendency to affect 6:00-7:00 activity but when cows with average daily activity exceeding 400 steps per hour were excluded, no significant difference was found. Number of positive interactions was negatively correlated to average daily activity, night rest activity and noon rest activity levels but R^2 values were low suggesting additional effects. Primiparous cows showed higher average hourly activity than multiparous (178±4.6 and 109±4; P<0.001) respectively, suggesting an advantage to adult cows in a mixed age group. Past studies proved beneficial effects of positive interactions. Correlation between these and lower activity pattern found in the present study, suggests the beneficial effect of the latter.

Keywords: beef cattle, interactions, kinship, primiparous

1. Introduction

Activity patterns reflect the measure of freedom each cow has, to choose her own behavior within the herd. Assuming that establishing low energy cost routine is beneficial for each cow, identifying cows that are unable to do so can help minimize social stress and improve production. Commercial beef breeding is carried out in vast pastures. Herd management and data collection are conducted manually due to the lack of suitable technology. Lack of online information inhibits the breeders' ability to address varied, transient physiological or behavioural events and perform adequate management modifications that will result in improved production during the same production cycle. Sampling activity over fixed times and places, resulted in irregular identification and data reception, due to irregularity in behaviour patterns of grazing beef cows. This lack of regularity was reflected in fluctuated activity graphs. An innovative wireless pedometric system was modified to serve as a standalone station in a 500 ha pasture in northern Israel. Applying this system and monitoring continuously cows' activity resulted in steadier activity graphs, which allowed identification of distinct patterns expressing physiological and behavioral events. Online information minimizes the cause and consequence gap existing between initiation of an event and its detection by the breeder, for example: estrus, conception, abortion, etc. Applying this system allowed conception detection according to cessation of estrus behaviour or abortion detection by resumed estrus cycle for pregnant cows. That done long before data reception is possible in conventional management. Successful detection is based on steady activity graphs. Comparative analysis of activity graphs of cows in the study herd, revealed two different graph types. One reflecting relatively steady activity and the other reflecting fluctuated high activity. Mounaix *et al.* (2007), referred to behavior patterns which express high variability, as 'auto-centric behaviour'. Different individual adaptations, including heat resistance, feed efficiency, grazing ability and other physical, physiological and behavioural traits, some innate and some acquired, that play an important role in creating this diverse herd environment. Barroso *et al.* (2000) described the effects of variable vegetation on grazing patterns of goats in the Mediterranean area, and the influence of different ranks on selective grazing. Age was mentioned as a major factor affecting dominancy. Our previous study showed that fluctuated activity patterns characterized low ranking cows (Gabrieli and Misha, 2014). The aim of the current study was to assess effects of age, kinship and social interactions on activity patterns in order to create a management tool for the breeder.

2. Materials and methods

2.1 The Pedometric system

The system consisted of:

1. ENGS 'track a cow' system comprising a central receiving unit: PC, Eco-herd software, RS485 communication box, UHF receiver + high gain antenna, wired to electricity in office.
2. Battery powered UHF transceivers + high gain antenna units based in the pasture.
3. UHF transmitting tags (ENGS, Rosh Pina, Israel) weighing 75 g measuring acceleration (g) at a frequency of 1000 Hz.

2.2 Animals

Tags were fitted to 148 cows, 78 primiparous and 70 multiparous. 46 were pairs of mothers and daughters. The daughters were heifers, weaned and separated the previous year and reintroduced to the mother herd two weeks prior to the beginning of the study.

The common breed was Israeli Simmental, developed by repeated crossing with the local breed – the *Baladi*. The weight range was 480-530 kg, with the average being 500 kg. The average age of the multiparous cows was six years, ranging from five to twelve. The primiparous cows were two years old, maximal age difference within the group was six months. The total number of cows in the herd was 230 and the pasture area was 500 ha, allowing 2.2 ha/cow.

2.3 Observations

Observations were conducted daily. Twice a week a night observation was conducted, using a night watching device (ThermApp – more to see, Ya'ar technologies, Yokneam, Israel). Individual and group activities and physiological events (calving, estrus, insemination, etc.) were recorded. Interactions recorded included grooming, proximity, pushing and butting. Interactions were defined as 'positive' (licking, proximity) or 'negative' (pushing, butting).

3. Results

A seemingly uniform general activity pattern was observed repeatedly. Observing cows individually, revealed changing patterns. These changing patterns were expressed by high variability in the pedometrs data. Average hourly activity value (sum of steps per day divided by 24 hours) was 137±44 steps per hour. Discrete hours' activity showed even higher variability. Lowest variability was found in average hourly activity between 6:00-7:00 (grazing) and 01:00-02:00 (rest) (231±56 and 41±17 steps per hour respectively). This relative uniformity was observed during 6:00-7:00, being the only hour during the day when all the cows were engaged in the same activity simultaneously. Individual uniformity during night rest was undetectable to the observer since different cows were engaged in different activities concurrently at any given time. Few overt negative interactions were observed during the whole study period, average 1.6 per cow, therefore they weren't analyzed. Positive interactions consisted mainly of proximity since grooming was scarce as well.

Table 1 presents average values (± standard deviation) of key parameters, measured and observed, for all the cows that transmitted data throughout the whole study period (n=140).

Data was compared between age groups (adult – multiparous and young – primiparous), and between kinship groups (kins=mothers and daughters, non kins=cows with no relatives in the herd). Table 2 summarizes comparison of parameters between the groups.

In order to further test differences between the kinship groups, we refined analysis adding subdivision of each of the age groups into kinship groups. Table 3 summarizes this analysis.

Comparison between kins within the primiparous cows' group and non kins within the multiparous cows' group resulted in significant differences in the number of positive interactions, and in the activity parameters (6:00-7:00, noon rest and night rest activities). Activity parameters remained significantly higher in the primiparous cows compared to the multiparous cows even when kinship effect was considered, suggesting young kins were still inferior to adult non kins judging by higher level of activity. On the other hand, positive interactions were significantly higher for the primiparous kins, suggesting a beneficial effect of kinship between age groups.

Next we analyzed correlations. Selected correlations are summarized in Table 4.

Both noon and night rest activities were correlated significantly to 6:00-7:00 activity. Night rest activity showed a bi-modal correlation and was sub-divided into cows with activity above and below 250 steps per hour. Correlation was weaker for <250 speps/hour activity cows.

Table 1. Average values for transmitting cows.

Positive interactions	Negative interactions	Average hourly activity (steps/hour)	Average grazing activity 6:00-7:00	Average rest activity 12:00-13:00	Average rest activity 01:00-02:00
62±36	1.6±1.45	137±44	231±56	134±59	41±17

Table 2. Comparison of selected parameters between age and kinship groups.[1]

	Multiparous	Primiparous	Kins	Non kins
n	62	78	44	96
Positive interactions	80±4^{a}	47±4^{b}	97±4.1^{c}	45±2.8^{d}
Average 6:00-7:00 activity	194±29^{a}	320±13^{b}	290±17^{c2}	259±12^{c2}
Average noon rest activity (12:00-13:00)	97±6^{a}	163±6^{b}	138±9^{c}	132±6^{c}
Average night rest activity (01:00-02:00)	30±1.8^{a}	49±1.6^{b}	40±2.5^{c}	41±1.7^{c}
Average daily activity	109±6.4^{a}	178±5.9^{b}	152±6.2^{c}	147±9.2^{c}

[1] Values in the table are averages ± standard error mean. Different lettering in a row indicates a significant difference.

[2] *P*-value for the comparison between kins and non kins was 0.072.

Table 3. Comparison of selected parameters within age groups.[1]

	Multiparous kins	Multiparous non kins	Primiparous kins	Primiparous non kins
Positive interactions	110±5.8^{a}	61±4.6^{b}	82±4.5^{c}	34±2.6^{d}
Average 6:00-7:00 activity	198±6^{a}	190±4.8^{a}	401±27^{b}	292±16^{c}
Average noon rest activity	101±4^{a}	94±3.2^{a}	182±14^{b}	156±8.3^{b}
Average night rest activity	31±1.4^{a}	30±1.1^{a}	51±4^{b}	48±2.3^{b}
Average hourly activity	114±3.1^{a}	107±2.5^{a}	196±15^{b}	172±648.6^{b}

[1] Different lettering in a row indicates a significant difference. Values in the table are averages ± standard error mean.

Table 4. Selected correlations.

X	Y	Rsq	*P*-value	Remarks
Positive interactions	Average daily activity	0.03	0.039	Weak negative
Positive interactions	Noon rest activity	0.03	0.035	Weak negative
Positive interactions	Night rest activity	0.04	0.018	Weak negative
Positive interactions	6:00-7:00 activity	0.013	0.17	No effect
Noon rest activity	Night rest activity	0.61	>0.0001	
6:00-7:00 activity	Night rest activity	0.34	>0.0001	average hourly activity>250 steps/hour
6:00-7:00 activity	Night rest activity	0.17	>0.0001	average hourly activity<250 steps/hour

4. Discussion

The most surprising finding that emerged from the observations was the lack of visible interactions between mothers and their daughters. Past studies mentioned kinship as a factor in the formation of high ranking sub-groups (Keeling and Gonyou, 2001; Lazzo, 1994). Sato *et al.* (1993) found a significant effect of kinship on allogrooming in Holstein cows, but none was found in a Japanese beef herd. In our study, kinship group consisted of mothers and their daughters that were reintroduced to the herd after one year of separation. Lack of recognition is unlikely the explanation. If that was the case, we would still expect random interactions to occur in a similar frequency as with other members of the herd. One possible explanation may be that high ranking sub-groups based on age were formed in the absence of the daughters, and new kins' interactions might have disturbed their stability. Sato *et al.* (1993) stressed the importance of continuity in herd composition for maintaining kinship effect and avoiding agonistic behaviour. In our study we seldom encountered agonistic behaviour, but the lack of continuity (one year separation) might have had a negative effect on kinship behaviour, especially in the dyadic aspect.

Noon rest hours were the only time when crowding occurred voluntarily. The close physical contact during a period of time when no competition over resources occurred may be an expression of the feral instinct of prey animals, as suggested by Lazo (1994). It may also be an expression of the time needed to maintain the social organization in surroundings and at a time of day that offer no competitive challenges. Another possible explanation could be the hot climate conditions that impose crowding in a favourable place for thermoregulation.

Positive interactions were the most prominent effect found. Number of positive interactions varied significantly between age and kinship groups, expressing an advantage (higher number) to adult cows and to Kins as a group (not as dyads). This finding together with the lack of dyadic interactions between mothers and daughters may imply better developed social skills for the kinship group. Howery *et al.* (1996) suggested that sub-herds were a result of gregariousness rather than tightly knit, social interactions. Similarly, random dyadic positive interactions arising from gregariousness may result in the significant effects of positive interactions found in the current study.

Hourly activity analysis showed higher uniformity during peak activity hours (6:00-7:00; grazing). The choice of simultaneous activity, during specific times of day although inflicting stress on the low ranking cows, reflects a need that must be facilitated by the breeder by supplying adequate grazing and supplementary feed for simultaneous consumption by the entire herd.

All the activity parameters were higher for the primiparous group compared to the multiparous group. Considering the higher amount of positive interactions and the well-known beneficial effect they have (Barroso *et al.*, 2000; Bennet and Holmes, 1987; Bouissou, 1975; Kabuga *et al.*, 1991), lower activity pattern suggests an advantage in the cow herd. This assumption complies with our previous study (Gabrieli and Misha, 2014) which showed that low and steady activity graph characterized the high ranking cows.

Kinship didn't result in a pronounced beneficial effect. No significant differences were found between kin's and nonkin's groups. The tendency (P=0.072) for higher 6:00-7:00 activity of kins compared to non kins, disappeared when 13 primiparous cows with 6:00-7:00 activity exceeding 400 steps/hour were excluded from the analysis. Possible explanations may be the temperament of these cows, or the fact that they were the youngest within the primiparous group. In order to further test kinship effect, each age group was subdivided into kinship groups. The amount of positive interactions was higher in the two age groups for kins compared to nonkins suggesting an advantage to kinship. Multiparous non-kin cows showed lower activity values than primiparous kin cows, thus repeating the advantage of age over kinship. Within the primiparous group, 6:00-7:00 activity was significantly higher for the kins but when the 13 cows with activity exceeding 400 steps per hour were excluded, there was no difference in this parameter.

The correlations found between selected parameters further clarify the complexity of this social network. Significant correlations with very low R square values imply the involvement of different influencing additional factors. Thus for example the significant influence of positive interactions on rest and daily activity is low indicating that they act as a small part within other unmeasured factors. The lack of correlation between the amount of positive interactions and 6:00-7:00 activity, which is peak grazing activity, further stress that grazing activity is one of these 'intended' uniform activities which are crucial for every cow, and is unaffected by other factors. This finding stresses further the importance of the 6:00-7:00 grazing activity.

Both noon and night rest activities were correlated significantly to 6:00-7:00 activity. Night rest activity showed a bi-modal correlation and was sub-divided into cows with activity above or below 250 steps per hour. This sub division, shown in Table 4, revealed a lower correlation between peak grazing activity (6:00-7:00) and night rest (1:00-2:00) for cows with overall lower activity (<250 steps/hour). This is an example of activity pattern expressing the freedom high ranking cows have to choose their own behaviour within the herd. Their free choice imposed constraints upon the low ranking cows. The latter's high activity during peak grazing time resulted in high activity during night rest, as discovered by the relatively high correlation. Both parameters influenced by the high ranking cows' independent behaviour patterns, imposing constraints on the low ranking cows. Higher activity during night rest may also reflect the need to compensate for inferior grazing patterns during the peak grazing activity. From a breeder's management perspective, this implies the inability of these cows to rest or graze freely. Possible solution might include grouping those cows differently, or allocating excess of food.

References

Barroso, F.G., Alados, C.L. and Boza, J. (2000). Social hierarchy in the domestic goat: effect on food habits and production. Applied Animal Behaviour Science 69: 35-53.

Bennett, I.L. and Holmes, C.R. (1987). Formation of a feeding order in a group of cattle and its relationship with grazing behaviour, heat-tolerance and production. Applied Animal Behaviour Science 17: 9-18.

Bouissou, M.F. (1975). Etablissement des relations de dominance-submission chez les Bovins domestiques. III. Effect de l'experience Sociale Z. Tierpsychology 38: 419-435.

Howery, L.D., Provenza, F.D., Banner, R.E. and Scott, C.B. (1996). Differences in home range and habitat use among individuals in a cattle herd. Applied Animal Behaviour Science 49: 305-320.

Kabuga, J.D., Gari-Kwaku, J. and Annor, S.Y. (1991). Social status and its relationship to maintenance behaviour in a herd of N'dama and West African Shorthorn cattle. Applied Animal Behaviour Science 31: 169-181.

Keeling, L.J. and Gonyou, H.W. (2001). Social behaviour in farm animals. CABI publishing CAB international, Wallingford, UK, 406 pp.

Kondo, S., Sekine, J., Okubo, M. and Ashahida, Y. (1987). The effect of group size and space allowance on the agonistic and spacing behaviour of cattle. Applied Animal Behaviour Science 24: 127-135.

Lazo, A. (1994). Social segregation and the maintenance of social stability in a feral cattle population. Animal Behavior 48: 1133-1141.

Mounaix, B., Boivin, X., Brulle, A. and Schmitt, T. (2007). Cattle behavior and the human animal relationship: variation factors and consequences in breeding. Publications of the Institut de l'Elevage, Monvoisin, France.

Sato, S., Tarumizu, K. and Hatae, K. (1993). The influence of social factors on allogrooming in cows. Applied Animal Behaviour Science 38: 235-244.

Syme, L.A. (1981). Social disruptions and forced movement orders in sheep. Animal Behavior 29: 283-288.

Development of a seven-point shape model for analysis of social interactions in dairy cattle

O. Guzhva[1*], M. Nilsson[2], H. Ardö[2], A. Herlin[1], K. Åström[2] and C. Bergsten[1]
[1]*Swedish University of Agricultural Sciences, Department of Biosystems and Technology, Alnarp, Sweden;* [2]*Lund University, Center for Mathematical Sciences, Lund, Sweden; oleksiy.guzhva@slu.se*

Abstract

The efficiency of an automatic milking system (AMS) relies on a passage rate of cows through the waiting area to the AMS units. In order to assure high number of visits per cow per AMS unit it is important to understand the hierarchy and social interactions between cows that might occur during the time in the waiting area prior to milking. The competition between cows of a different rank to enter the AMS unit could result in a number of aggressive interactions, which could compromise the individual performance and endanger their health and welfare. The aim of the study was to investigate social interactions between cows in the waiting area with four AMS units and collecting data by three top-down view video cameras. A seven-point shape model for each cow was developed to manually tag cows in video sequences. Furthermore, the interactions between pair of cows were annotated into one of five states: body pushing, head butting, head pressing, body sniffing and no interaction. A classification system, investigating features from pair of cow shape models, was developed. The identification of social interactions using this system and crosscheck with ethogram containing descriptions of all the interactions was tested on a set of video sequences. The system showed a potential for further development and algorithm flexibility dependent on a number of features involved in the analysis. For the single frame analysis system showed higher accuracy with 'line border features' than without: 83.1 vs 79.2% respectively and analysis of three consecutive frames with 'line border features' showed accuracy of 85.1%. Exploration of social interactions between cows was addressed and the proposed seven-point cow shape model was investigated for this purpose. The proposed seven-point model may act as a crucial component towards building a robust system for automatic detection of social interactions in areas of interest.

Keywords: object recognition, automatic milking, shape models, social interactions, computer vision

1. Introduction

In order to fulfil a number of important parameters (e.g. animal health, animal welfare and comfort, general production traits) consistent with a sustainable dairy production, real-time continuous automatic assessment solutions within Precision Livestock Farming (PLF) research field create a basis for innovation and applied approach (Busse *et al.*, 2015). Due to the rapid development of the dairy industry and increased complexity of everyday farm work (e.g. advanced feeding and management techniques, heat detection, health control on individual and heard levels, etc.), visual assessment of individuals by farmer becomes ineffective and requires large investments of time (Busse *et al.*, 2015). There are number of products for automatic assessment of animal performance that are capable of real-time monitoring and on-the-go measurement of individual variables for every animal, both in form of stand-alone products or integrated parts of more complex farm management systems (Banhazi and Black, 2009). However, none of the products takes into consideration the complex social interactions between animals during the decision making stage, and therefore, the potential range of practical applications is limited (Nadimi *et al.*, 2012).
To achieve a better functionality and a better economic output for farms with automatic milking systems (AMS), farmers should aim to maximize the individual performance of every animal. One

of the main factors ensuring the general efficiency of an AMS and a high individual performance is the flow of animals between different sections of the dairy barn and especially surrounding the AMS (Hermans *et al.*, 2003). Wechsler and Lea (2007), as well as Herlin and Frank (2007), showed that competition for resources combined with limited space (e.g. feeding areas, areas prior to milking units) could lead to an increased number of aggressive behaviours. Von Keyserlingk *et al.* (2009) showed that aggressive interactions between animals, such as head butting and body pressing, could result in displacements from resources and if further escalated, leading to serious injuries.
A direct relationship between cows' inter-cow distance and their aggressive behavioural pattern was demonstrated by deVries *et al.* (2004). Other studies (Hemsworth, 2003; Kilgour, 2012) showed inevitable effects of inferior animal welfare connected to restrained performance of their natural behaviour. Therefore, early diagnostics of unconditional changes in animal behaviour when linked to health and welfare could not only save time and money for the farmer, but also decrease the production pressure for every animal in the barn (Polikarpus *et al.*, 2015).
A number of recent studies on animal behaviour in the dairy barn environment were based on different GPS or wireless sensor network (WSN) solutions (Nadimi *et al.*, 2012). These products for behavioural monitoring are a definite step forward, which allows us to see spatial distribution of animals and to measure different levels of activity (Nadimi *et al.*, 2012). However, the number of behaviours that could be monitored with position-based approach is limited. Therefore, new methods based on video surveillance and image analysis, which could extend the number of parameters for studying, are of great importance. By understanding underlying causes for social interactions and the outcomes of those interactions, tools for evaluation of cow-traffic could be developed in the scope of PLF.

2. Materials and methods

2.1 Setup for recordings

The experiment was carried out in a commercial free stall dairy barn. The 252 Swedish Holstein cows had free access to four AMS (VMS®, DeLaval, Sweden) with the common waiting area (6x18 meters) that had entrances to all four AMS units at any time during the day. Video setup included multi-camera system consisting of three Axis M3006-V (Axis® Communications) cameras placed at 3.6 meter height to assure top-down view over the study area and a Hewlett Packard ProLiant MicroServer (8th generation) used for data storage. All the video-segments that were recorded and used for annotations had the frame resolution 800×600 pixels and 8-bit RGB colour space.

2.2 Image segmentation approach

In order to assure correct identification of objects (cows) in every frame under different lighting conditions in the study area and to create a robust flexible algorithm, effective and reliable segmentation methods were needed. Our previous work (Nilsson *et al.*, 2015) showed that the learning based approach utilizing twelve channels of features shows higher segmentation accuracy, for extracting individuals from backgrounds with high levels of scene complexity, than compared to one proposed by Otsu (1979). In order to avoid overfitting and balance number of features needed for segmentation, elastic net regularized logistic regression was used as part of learning based framework (Nilsson, 2014). Currently, there is work ongoing on training convolutional neural networks (CNNs), which can be seen as an extension to our previous multi-part segmentation approach, but also as part of a framework for learning the features during training. However, in this study, the focus is only on the actual shape interaction, and thus, manually labelled seven landmark points are used. Further work involves investigating the method proposed here on automatically positioned landmark points, which typically involves noisier positions.

2.3 Seven-point shape model

A seven-point shape model was developed to manually tag cows in video sequences. Every cow in every frame from the training pool with video sequences was manually tagged, and seven landmark points were manually assigned to form a shape. These landmark points were consistent with following body parts: head, left shoulder, front middle (root of the neck), right shoulder, left hip, rear middle (root of the tail) and right hip (Figure 1). The model made it possible to automatically recognize different interactions and explain patterns in cow behaviour. Furthermore, the interactions between pairs of cows were annotated into one of five states: body pushing, head butting, head pressing, body sniffing and no interaction. Descriptions for all interactions used to train the detector were based on the ethogram adapted from Rousing and Wemelsfelder (2006).

A classification system, investigating features extracted from pairs of cow's shape models, was developed. Every cow shape model went through Bookstein normalization (Bookstein, 1997) and every pair of cows was investigated independently, via symmetric training, forming two training examples for input. From every training example positions and distances for every landmark point were extracted (24 positions plus 49 distances, resulting in 73 features per training example) and used for investigating possibilities of interactions (Figure 2). Furthermore, an outline of each shape

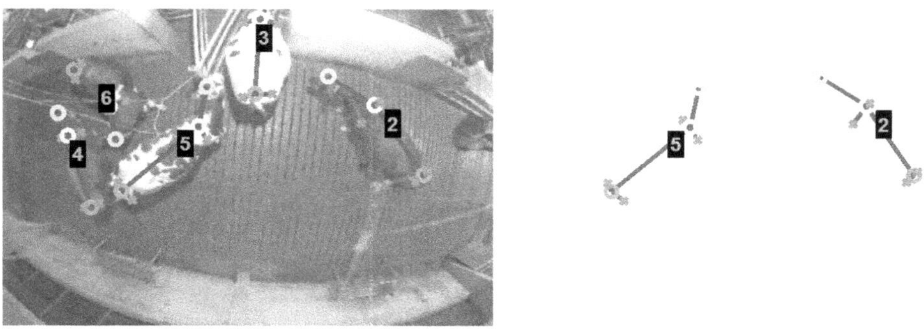

Figure 1. Left; an example of a training frame with landmark points assigned to every animal. Right; example of how cow five is related to cow two using the shapes only.

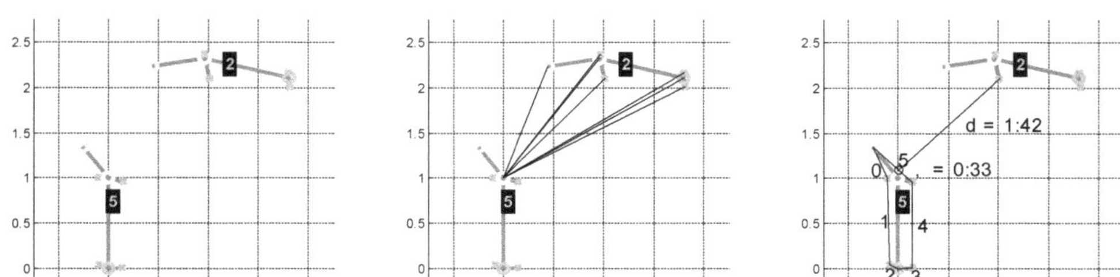

Figure 2. Example of feature calculation, see Figure 1 for cow five to cow two interaction. Left; when two cow-shapes are going through Bookstein normalization. Five x,y coordinates from cow five used (omitting 0,0 and 0,1) and seven x,y positions from cow two. Middle; all distances from every point on cow two to all other points on cow five are extracted. Right; an outline of cow five with six lines and each point (only one shown in Figure) on cow two find its closest line and stores the distance and the line index (0-5) plus the alpha value where it hits the line (between zero and one). Thus, in this example the point selected from cow two would result in two values, 5.33 and 1.42.

model formed by six line segments and the closest point to each point on the other cow is found. A single number for each such closest point is formed by unfolding the outline into a straight line and rescaling each segment to place the connecting end points of the segments at the integer positions 0, 1, ..., 5.

3. Results and discussion

The results showing performance of the classifier based on the features extracted from the proposed seven-point shape model are presented as a confusion matrix (Table 1). The results presented in the confusion matrix are from the test-set utilizing three consecutive frames with 'line boarder features'. The overall accuracy of the classifier varied depending on different test-conditions:

1. no 'line boarder features', single frame: accuracy 79.2%;
2. with 'line boarder features', single frame: accuracy 83.1%;
3. with 'line boarder features', three consecutive frames: accuracy 85.1%.

As presented in the confusion matrix, differentiation between similar object classes (e.g. body_sniff and head_press) was inconclusive as classifier blends and was based on the shape model only. However, utilization of additional features and time dimension helps in resolving aforementioned issue and shows the potential for further algorithm improvement as well as in decreasing number of false negatives. While complex social interactions involve such an important parameter as 'duration', higher classifier accuracy is difficult to achieve only from static landmark points in single frames. Our future work will concentrate on extending the pool of training data (to assure greater variability of behavioural states and reliability of classes in training sequences). Considering the performance of the classifier, similar object classes (2-D representation of behavioural state based on shape model only, single frame) could be merged to form simplified interaction descriptions (e.g. 'head_butt' and 'head_press' into general 'butting').

Table 1. Classifier performance while distinguishing between five different behavioural states and compared to GT values.

		Classification results				
		no_int	body_push	head_butt	head_press	body_sniff
Ground truth	no_int	6,719	1	0	0	2
	body_push	471	217	5	16	3
	head_butt	225	0	52	11	0
	head_press	318	18	11	96	3
	body_sniff	107	0	0	55	40

4. Conclusions

The study has shown that it is possible with the proposed approach to differentiate between different behavioural states in cows. The proposed seven-point model may act as a crucial component towards building a robust system for automatic detection of social interactions in areas of interest.

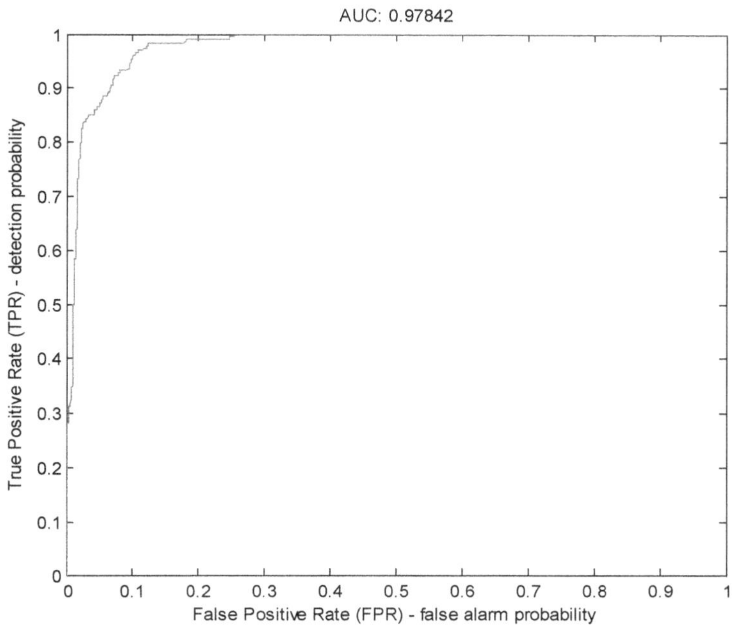

Figure 3. Receiver operation curve for body_push vs rest.

References

Banhazi, T.M. and Black, J.L. (2009). Precision livestock farming: a suite of electronic systems to ensure the application of best practice management on livestock farms. Australian Journal of Multi-disciplinary Engineering 7(1): 1-14.

Bookstein, F.L. (1997). Landmark methods for forms without landmarks: morphometrics of group differences in outline shape. Medical Image Analysis 1: 225-243.

Busse, M., Schwerdtner, W., Siebert, R., Doernberg, A., Kuntosch, A., König, B. and Bokelmann, W. (2015). Analysis of animal monitoring technologies in Germany from an innovation system perspective. Agricultural Systems 138: 55-65.

DeVries, T.J., Von Keyserlingk, M.A.G. and Weary, D.M. (2004). Effect of feeding space on the inter-cow distance, aggression, and feeding behavior of free-stall housed lactating dairy cows. Journal of Dairy Science 87: 1432-1438.

Hemsworth, P. (2003). Human-animal interactions in livestock production. Applied Animal Behavior Science 81: 185-198.

Herlin, A.H. and Frank, B. (2007). Effects of protective gates at concentrate feed stations on behaviour and production in dairy cows: a brief note. Applied Animal Behavior Science 103: 167-173.

Hermans, G.G.N., Ipema, A.H., Stefanowska, J. and Metz, J.H.M. (2003). The effect of two traffic situations on the behavior and performance of cows in an automatic milking system. Journal of Dairy Science 86: 1997-2004.

Kilgour, R.J. (2012). In pursuit of 'normal': a review of the behaviour of cattle at pasture. Applied Animal Behavior Science 138: 1-11.

Nadimi, E.S., Jørgensen, R.N., Blanes-Vidal, V. and Christensen, S. (2012). Monitoring and classifying animal behavior using ZigBee-based mobile ad hoc wireless sensor networks and artificial neural networks. Computers and Electronics in Agriculture 82: 44-54.

Nilsson, M. (2014). Elastic net regularized logistic regression using cubic majorization. IEEE, pp. 3446-3451.

Nilsson, M., Herlin, A.H., Ardö, H., Guzhva, O., Åström, K. and Bergsten, C. (2015). Development of automatic surveillance of animal behaviour and welfare using image analysis and machine learned segmentation technique. Animal 9: 1859-1865.

Otsu, N. (1979). A threshold selection method from gray-level histograms. IEEE Transactions on Systems, Man, and Cybernetics 9: 62-66.

Polikarpus, A., Kaart, T., Mootse, H., De Rosa, G. and Arney, D. (2015). Influences of various factors on cows' entrance order into the milking parlour. Applied Animal Behavior Science 166: 20-24.

Rousing, T. and Wemelsfelder, F. (2006). Qualitative assessment of social behaviour of dairy cows housed in loose housing systems. Applied Animal Behavior Science 101: 40-53.

Von Keyserlingk, M.A.G., Rushen, J., De Passillé, A.M. and Weary, D.M. (2009). The welfare of dairy cattle – key concepts and the role of science. Journal of Dairy Science 92: 4101-4111.

Wechsler, B. and Lea, S.E.G. (2007). Adaptation by learning: its significance for farm animal husbandry. Applied Animal Behavior Science 108: 197-214.

Part 5. Precision feeding

Effects of an automatic feeding system with dynamic feeding times on the behavior of dairy cows

R. Oberschätzl-Kopp[1*], B. Haidn[1], R. Peis[1], K. Reiter[1] and H. Bernhardt[2]
[1]*Institute for Agricultural Engineering and Animal Husbandry, Bavarian State Research Center for Agriculture, Prof.-Dürrwaechter-Platz 2, 85586 Poing-Grub, Germany;* [2]*Technical University of Munich, Chair of Agricultural Systems Engineering, Am Staudengarten 2, 85354 Freising-Weihenstephan, Germany; rosi.oberschaetzl@web.de*

Abstract

Automatic feeding systems (AFS) gain in importance due to rising demands on performance-related feeding of cows and animal welfare. But so far, under practical conditions, AFS are used with static settings for feeding times and frequencies without taking into account animal demands. Therefore the objective of the study was to develop a method for feeding the cows with an AFS at dynamic times concerning their behavior and to analyze the effects of this feeding strategy on the cows visits at the feed bunk and the milking behavior in an automatic milking system (AMS). 79 lactating dairy cows housed in a barn with an AFS and an AMS were subjected to each of the two treatments (each 7 static and dynamic feeding times per day). The trial was carried out in winter and summer time. Individual animal behavior was collected by a Real Time Location System whereas milking-related data were automatically collected by the AMS. Based on real-time processed positioning data regarding cows duration of stay at the feed bunk and periodically subjective evaluations of the amount of feed in the trough, decisions were made about starting time of the AFS feedings in predefined timeframes. Feedings in both treatment periods in winter and summer time resulted in an immediate increase of cows presence at the feed bunk with similar proportions of animals and with maxima of about 70-80%. Dynamization of feed delivery times might be beneficial for an even distribution of feed bunk visits throughout the day in the observed farm but only in summer time. Current results show no even spreading of milkings when AFS feedings were dynamic. The present implementation of real-time processed positioning data might be a possible approach for combining animal- and feed-related data of different sensor technologies to advance AFS for feeding cows according to their needs.

Keywords: precision dairy feeding, feed delivery times, dynamization, automatic milking system, animal behavior

1. Introduction

Feeding strategies in growing modern dairy farms are of major interest due to labor economics and aspects of animal welfare. With a share of 16% of the required total labor time in a dairy farm, feeding is the third most important work after milking and young cattle supply (Haidn and Macuhova, 2009). Because of that and due to rising demands on performance-related feeding of cows and the possibility to supply a ration with a higher frequency (Belle *et al.*, 2012) AFS gain in importance in dairy farming (DLG Committee for Technology in Animal Production *et al.*, 2014). According to information of eight different manufacturers of AFS, in 2015 more than 1,250 systems were used worldwide.

Feeding behavior of dairy cows is characterized by circadian patterns with several feeding periods spread over the day (Albright, 1993). Endogenous and exogenous factors are assumed as strongly influencing. Cows show an obvious two-phase rhythm of feed intake even when housed in a barn and fed ad-libitum (Sambraus, 1991). The delivery of feed is attributed to the greatest stimulating

effect in relation to the feed intake of dairy cows. (De Vries and Von Kyserlingk, 2005; Melin *et al.*, 2005). Several studies have focused on the effects of higher feed delivery frequencies on the behavior of dairy cows. An increased feed delivery frequency leads to longer feeding times and a higher proportion of animals feeding at the feed bunk spread over 24 h (De Vries *et al.*, 2005; Mäntysaari *et al.*, 2006; Mattachani *et al.*, 2015). The range of feeding frequencies in these studies was between once and eleven times per day. The daily cows duration of stay at the feed bunk was higher when fed six times per day whereas cows showed longer standing times in the feed alley when fed twice per day (Oberschätzl-Kopp *et al.*, 2016). Rodenburg (2002) assumed that a higher feed delivery frequency results in an increased number of voluntary visits in a milking robot. But several other studies showed that the daily milking frequency was not affected by changes in feeding frequency (Belle *et al.*, 2012; Mäntysaari *et al.*, 2015).
But so far, under practical conditions, AFS are used with static settings for feeding times and frequencies without considering animal demands. Therefore, the objective of the study was to develop a method for feeding the cows with an AFS at dynamic times concerning their behavior and to analyse the effects of this feeding strategy on the cows visits at the feed bunk and the waiting time prior to milking as well as the utilization capacity of the feed bunk and the AMS.

2. Animals, materials and methods

2.1 Animals, housing and management

The investigations were carried out in a commercial dairy farm in Bavaria (Gemany). About 79 lactating Simmental cows were milked with an AMS (DeLaval VMS, DeLaval International AB, Tumba, Sweden) and were fed by an AFS (DeLaval Optimat Master, DeLaval International AB, Tumba, Sweden). The cows, 19 primiparous and 60 multiparous (parity=3.3±1.75, milk yield=22.0±4.7 kg/d), were housed in a barn with a loose-housing layout with a total of 80 straw bedded cubicles and 72 feeding places (1.1 cow/feeding place ratio) separated by a feeding fence. The milking area was composed of one AMS and a closed-off waiting area in front of the entrance of the AMS. The entrance and exit to the waiting area were controlled by one-way gates. Cows had access to the AMS for 24 h/d, except at times for system cleaning twice per day (10:00 and 21:00 h). The lactating dairy cows were fed a partial mixed ration (PMR) (average of 43.4 kg/d per cow on fresh matter (FM) basis) seven times per day and with an average share of feed delivery quantity of 14%, respectively 15%. Concentrate was supplied in the AMS and by a concentrate feeder. The AFS mixed a PMR by a stationary mixer (8 m^3 filling volume) and the ration was delivered by a rail-mounted feed wagon. Furthermore, grass silage, corn silage, hay and straw were stored in feed bunkers, which were filled by a telescopic handler once every 2 days in winter time or once per day in summer time. Minerals and concentrate were also stored in silos. The PMR consisted of 53.6% grass silage, 38.9% corn silage, 1.5% hay, 0.9% straw, 2.7% whole grain, 2.3% rape and soy shred, 0.2% mineral feed. Every day in the morning between 06:15 and 07:30 h and in the evening between 17:15 and 18:30 h the farmer was in the barn and amongst others he fetched cows not attending for milking.

2.2 Experimental design

The cows were subjected to each of two treatment periods, period A and period B. During period A, the cows were fed statically seven times per day. This was the feeding routine practiced since the installation of the AFS in March 2014. During period B, the cows were fed dynamically seven times per day. These two treatment periods were carried out in winter 2014/2015 and in summer 2015 (Table 1). Feed was only pushed up during the feeding times. Trial days reached from 8:15 to 8:15 h the following day because every day at 7:45 h the trough was cleaned and the first feeding was at 8:15 h in the morning.
Initially, timeframes for shifting feed delivery times were defined in accordance with the feeding schedule of the AFS in the farm. Based on preliminary studies (Oberschätzl-Kopp *et al.*, 2016) the

Table 1. Trial parameters of the treatment period A and B in winter and summer time.

Season	Winter		Summer	
Period	A	B	A	B
Feeding times	7 static	7 dynamic	7 static	7 dynamic
Start date	15.12.2014	20.02.2015	7.06.2015	6.07.2015
End date	30.12.2014	25.02.2015	28.06.2015	12.07.2015
Number of selected days	3	3	3	3
Ø number of animals	79	79	78	79
Feed delivery times	8:15, 11:15, 14:00, 16:00, 19:45, 23:15, 3:45	d1 8:15, 10:51, 13:36, 16:21, 19:45, 23:27, 3:05	8:15, 11:15, 14:00, 16:00, 19:45, 23:15, 3:45	d1 8:15, 10:56, 13:41, 16:00, 19:45, 23:45, 2:40
		d2 8:15, 11:36, 13:35, 15:41, 19:45, 22:36, 2:26		d2 8:15, 10:28, 14:13, 16:34, 19:58, 22:46, 2:30
		d3 8:15, 10:51, 13:47, 16:22, 19:45, 23:39, 3:35		d3 8:15, 10:25, 13:56, 16:33, 19:56, 23:30, 3:25
Ø time of sunrise	07:59	07:03	05:09	05:20
Ø time of sunset	16:26	17:46	21:55	21:53
Ø ambient temperature	2.5	0.8	15.4	19.5

d1, d2, d3; abbreviations for the three selected days of the treatment periods.

cows duration of stay at the feed bunk and accordingly an adequate amount of fresh feed in the trough were found as crucial parameters for influencing behavior of dairy cows. As a result the cows duration of stay at the feed bunk was defined as a control parameter for finding decisions about starting times of the AFS feedings. Based on real-time processed positioning data the cumulative cows duration of stay at the feed bunk was put out in intervals of 15 min. The averaged maxima of the cumulative duration of stay at the feed bunk were calculated as boundary. Its value was between 1,800 and 2,100 min. If this value was exceeded the AFS feeding was started. Additionally, during trial periods B, percentage variation of amount of feed in the trough was evaluated periodically every five meters of the trough according to a predefined scheme considering the height and width of the feed swath.

2.3 Data collection, data editing and statistical analysis

Individual animal positioning data were collected by a Real Time Location System (RTLS) (Ubisense Series 7000, Ubisense GmbH, Düsseldorf, Germany). Ten sensors were installed in the barn and neck collars with identification units (Industrial Tags) were attached to each cow dorsally. Radio signals were sent between the sensors and the identification units with a frequency between six and eight GHz. The cows positions in the barn were detected by the angle of arrival and the timing differences of the sensors signals. Raw data (x- and y-coordinates) were imported in a PostgreSQL-database for further processing. The barn layout was compiled in vector format and areas in the barn (e.g. feed bunk, feed allay, waiting area, lying cubicles) were defined for structured data processing. The positioning data were summarized by a programmed algorithm using the median with the boundaries *time* (60 sec) and *way* (1 m). These data were used to determine feeding behavior. A cow was recorded as feeding when its head was behind the feeding fence (according to De Vries *et al.*, 2005). Proportion of animals at the feed bunk was recorded once every 5 min. Based on that, the median of relative proportions of animals at the feed bunk was calculated over three days in treatment period A. Furthermore total daily time at the feed bunk (h/cow), frequency of visits (n/d), the average visit duration (h/n) and the daily duration of stay in the feed alley (h/cow) were calculated. Daily waiting time was defined as the time a cow spent in the closed-off waiting area in front of the AMS and was also determined by positioning data. Milking-related data for all cows were saved periodically as log files from the AMS software and were also imported into the PostgreSQL-database for further processing. The mean milk yield (kg/d), milking frequency (n/d) and refusal frequency (n/d) were calculated on per cow basis.

Activities in the barn and daily working routines were recorded by daily protocols. Based on these information three days were selected for the analyzis of the criteria named above.

The statistical analyzes were performed with the open-source software 'R'. Normal distribution of the data was disproved by the Shapiro-Wilk test. To compare the parameters between the treatment periods A and B in winter and summer time respectively the Wilcoxon signed rank test was used (Hedderich and Sachs 2012). Therefore, only cows were selected that have been in the herd during period A and B in winter (n=60) and summer time (n=63) respectively. For investigating the differences of the parameters between the seasons only cows were selected, that have been in the herd during winter and summer time (n=30). In the statistical analyzes, significance was declared when $P<0.05$ and tendency was declared when $P<0.10$.

3. Results

Feedings in both treatment periods in winter and summer time led to similar proportions of animals at the feed bunk with maxima of about 70-80% (Figure 1) while cows left the feed bunk significantly ($P<0.05$) earlier in the first 60 min after feeding if the feeding times were dynamically in winter. In summer period B cows stayed longer at the feed bunk in the first 60 min after dynamic feeding times. The presence of cows at the feed bunk increased immediately following delivery of fresh feed, both in winter and summer periods A and B. This increase was just shifted according to the changed feeding intervals during period B. The proportion of cows at the feed bunk tended to be on a higher level during the daytime compared to the night. But even the feeding at 3:45 h led to a higher proportion of animals feeding up to 40-50% (Figure 1). In this context, the high correlation of 0.95 between the estimated amount of feed in the trough and the proportion of animals at the feed bunk should be mentioned.

Results for the behavioral parameters of the cows observed in the treatment periods are presented in Table 2. In winter time the total time per cow and day at the feed bunk was significantly shorter in period B (5.06 h/d) than in period A (5.59 h/d). In summer time a higher total time per cow and day at the feed bunk was observed when the AFS feeding times were performed dynamically (5.46 vs 4.93 h/d). But overall the daily total time a cow spent at the feed bunk was significantly higher in winter than in summer time. The proportion of utilized feeding places over 24 h was significantly lower in summer period B than in A. Dynamic feed delivery times increased the cows daily frequency of visits at the feed bunk in winter as well as in summer time. This parameter showed a wide range from 6 to 65 single visits per cow and day at the feed bunk. Cows fed dynamically decreased the average duration per visit at the feed bunk only in winter time. The animals tended to stand longer in the feed alley in period B (winter: $P<0.1$; summer: $P<0.05$).

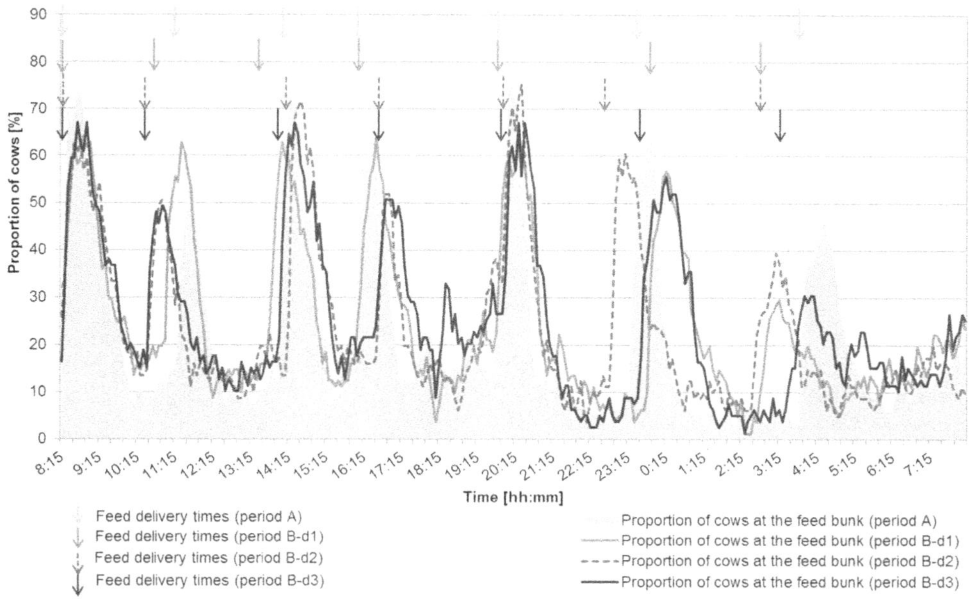

Figure. 1 Proportion of cows at the feed bunk over a 24 h period during summer time treatment period A (averaged for n=3 d and n=78 cows) and during period B showing every day (n=79 cows)

Dynamic feeding times by an AFS affected the milking frequency of the cows significantly only in winter time, whereas no difference was observed in the refusal frequency per cow and day neither in winter nor in summer time. The daily milk yield was significantly lower in both treatment periods B. But total time per cow and day in the waiting area prior to milking did not differ significantly neither between treatment periods A and B nor between the seasons.

4. Discussion

The proportion of animals at the feed bunk over 24 h shows a circadian rhythm with a lower feeding activity at night according to the results of Albright (1993). Corresponding with the studies of De Vries and Von Keyserlingk (2005) and Melin *et al.* (2005) the delivery of fresh feed can be described as an important exogenous influencing factor in terms of stimulating dairy cows to feed. The feeding time late in the morning (usually at 11:15 h) should be started earlier during treatment periods B (83% of this feeding times). It can be derived that an additional feeding time or a higher amount of delivered feed in the morning might encourage the biological rhythm of a dairy cow housed in the observed farm. Dynamization of AFS feed delivery times resulted in longer daily total times and a higher frequency of single visits at the feed bunk in summer period with high ambient temperatures despite a lower degree of utilization of the feeding places in the course of the day. This might indicate that cows distribute their visits at the feed bunk more evenly throughout the day with reduced competition during feeding when delivery times are controlled dynamically. Present results show no more evenly spreading of milkings during the day and night when AFS feedings were dynamic. But corresponding with the results of Belle *et al.* (2012), milking frequency might be influenced amongst others by the number of animals milked with an AMS and in the present studies a rather high number of cows were milked with one milking unit.

Table 2. Specific parameters for feeding and milking behavior in treatment periods A and B and in winter (n=60 cows) and summer time (n=63 cows) (n=3 d per treatment period).

Season	Winter						Summer					
Period	A			B			A			B		
Parameter	Median	Min	Max	Median	Min	Max	Median	Min	Max	Median	Min	Max
Feeding behavior												
Total time at the feed bunk/cow/d [h]	5.59[ae]	2.07	8.71	5.06[be]	2.12	8.63	4.93[cf]	1.83	9.02	5.46[df]	2.17	9.55
Frequency of visits at the feed bunk/cow/d [n]	26.50[ae]	6.00	63.00	29.00[be]	8.00	60.00	26.00[ce]	8.00	64.00	33.00[de]	8.00	65.00
Duration of visit at the feed bunk/cow/n [h]	0.20[ae]	0.08	0.57	0.17[be]	0.08	0.40	0.17[cf]	0.07	0.63	0.17[cf]	0.06	0.54
Total time in the feed alley/cow/d [h]	0.94[ae]	0.02	4.28	1.12[ae]	0.15	4.00	1.05[cf]	0.17	3.04	1.12[df]	0.21	4.53
Milking behavior												
Milking frequency/cow/d [n]	2.00[ae]	1.00	5.00	2.00[be]	1.00	4.00	2.00[cf]	1.00	4.00	2.00[cf]	1.00	4.00
Refusal frequency/cow/d [n]	4.00[ae]	0.00	13.00	4.00[ae]	0.00	15.00	3.00[cf]	0.00	18.00	3.00[cf]	0.00	15.00
Milk yield/cow/d [kg]	23.23[ae]	1.82	44.98	19.54[be]	0.71	54.49	20.70[cf]	1.27	43.16	19.85[df]	1.81	39.01
Total time in the waiting area/cow/d [h]	0.76[ae]	0.04	5.62	0.90[ae]	0.08	5.11	0.67[ce]	0.03	4.90	0.66[ce]	0.06	5.24

a, b; different letters show significant differences within winter time ($P<0.05$).
c, d; different letters show significant differences within summer time ($P<0.05$).
e, f; different letters show significant differences between the seasons ($P<0.05$, n=30 cows).

5. Conclusions

Based on present results, dynamization of feed delivery times according to cows duration of stay at the feed bunk can be beneficial for even distribution of feed bunk visits throughout the day in the observed farm and in summer time. However, to achieve this effect a sufficient amount of feed in the trough is required during the day. The present implementation of real-time processed positioning data might be a possible approach for combining animal- and feed-related data of different sensor technologies to advance AFS for feeding cows according to their needs in terms of precision feeding. Based on this information, recommendations for farmers regarding the AFS settings of feeding times and amounts of feed can be derived. In further investigations, precise measuring of amount and quality of feed in the trough as well as a dynamic adjustment of feeding frequency and the delivered amount of feed should be additionally technically feasible and tested.

References

Albright, J. (1993). Nutrition, feeding, and calves – feeding behavior of dairy cattle. Journal of Dairy Science 76: 485-498.

Belle, Z., André, G. and Pompe, J. (2012). Effect of automatic feeding of total mixed rations on the diurnal visiting pattern of dairy cows to an automatic milking system. Biosystems Engineering 1: 33-39.

DeVries, T. and Von Keyserlingk, M.A.G. (2005). Time of feed delivery affects the feeding and lying patterns of dairy cows. Journal of Dairy Science 88: 625-631.

DeVries, T., Von Keyserlingk, M.A.G. and Beauchemin, K.A. (2005). Frequency of feed delivery affects the behaviour of lactating dairy cows. Journal of Dairy Science 88: 3552-3562.

DLG Committee for Technology in Animal Production, Oberschätzl, R. and Haidn, B. (2014). DLG expert knowledge series 398 automatic feeding systems for cattle-technology,performance, notes on planning. Frankfurt am Main, DLG, Germany.

Haidn, B. and Macuhova, J. (2009). Arbeitsentwicklung in bayerischen milchviehbetrieben-analyse und entwicklung. In: Wendl, G. (ed.) Strategien für zukunftsorientierte milchviehbetriebe in Bayern. Proceedings of Landtechnisch bauliche Jahrestagung. LfL, November 25, 2000. Triesdorf, Germany, pp. 37-53.

Hedderich, J. and Sachs, L. (2012). Angewandte statistik – methodensammlung mit R. Springer Verlag, Berlin/ Heidelberg, Germany, 479 pp.

Mattachani, G., Riva, E., Pompe, J. and Provolo, G. (2015). Automatic monitoring of cow behaviour to assess the effects of variations in feeding delivery frequency. In: Precision Livestock Farming, Milano, Italy, pp. 473-481.

Mäntysaari, P., Khalili, H. and Sariola, J. (2006). Effect of feeding frequency of a total mixed ration on the performance of high-yielding dairy cows. Journal of Dairy Science 89: 4312-4320.

Oberschätzl-Kopp, R., Haidn, B., Peis, R., Reiter, K. and Bernhardt, H. (in press). Untersuchungen zum verhalten von milchkühen bei automatischer fütterung in einem AMS-Betrieb. Landtechnik.

Rodenburg, J. (2002). Robotic milkers: what, where, ... and how much!!??. In: Proceedings of the Ohio Dairy Managament Conference. The Ohio State University, Columbus, USA, pp. 1-18.

Sambraus, H. (1991). Nutztierkunde. Biologie, Verhalten, Leistung und Tierschutz. Verlag Eugen Ulmer, Stuttgart, Germany, 181 pp.

Dynamic concentrate feeding and the benefits in different parity groups on dairy farms

J.W. Van Riel* and C. Lokhorst
Livestock Research, Wageningen UR, P.O. Box 338, 6700 AH Wageningen, the Netherlands; johan.vanriel@wur.nl

Abstract

The daily estimation of the curvilinear response on concentrate intake and milking interval, using adaptive dynamic models, provides information for optimal settings for these traits for each cow. This type of precision feeding and milking has been implemented in available software from 2008 and economic benefits by using this software has been proven in farm comparison studies. However, it seems that the potential economic benefits of the software are often not reached in practice. This may be caused by adjustments of the optimum settings by the farmer. Therefore it is interesting whether these adjustments by the farmer are one of the reasons for lower economic benefits in practice. In a case control study with 24 farms, the effect of adjusting advised concentrate levels was investigated by comparing farms with minor adjustments for concentrate settings, therefore using optimal settings (OS), with farms using major adjusted settings (AS). Productivity and gross margins were compared for the first 100 days post-calving. The 12 randomly selected OS farms and AS farms have comparable productivity in the first 3 weeks after calving and comparable milk production of the multiparous cows at day 100 after calving. The results show that AS farms have lower gross margins (milk returns minus concentrate costs). The AS farmers provide a higher average concentrate level in the period from calving until day 40 after calving. After day 40, the difference (compared to OS farms) in average concentrate level is reverse. For heifers this difference in slope of the average concentrate level is more extreme, and significantly different compared to multiparous cows. Average milk production of heifers is significantly lower in the AS farms. This study provides evidence for the economic benefit on the short term of using the optimum settings, estimated by the adaptive dynamic models. Also it is clear now, that there is a tendency in practice to feed much more concentrate in the first weeks after calving, especially for heifers.

Keywords: concentrate allocation, dynamic linear models, Bayesian forecasting

1. Introduction

Economic profit of dairy farms largely depends on milk returns and feeding costs, therefore optimization of feeding and milking for an individual cow has substantial influence on economic profit. Hogeveen *et al.* (2001) presented ideas to maximize economic profit by continuously optimizing individual settings of concentrate allocation and milking frequency for all cows in the herd given the available robot capacity. For this optimization André *et al.* (2007) introduced the use of dynamic linear models (DLM), developed by West and Harrison (1997), for online estimation of individual parameters that describe curve-linear response to concentrate intake and milking interval. Therefore, computer controlled milking and concentrate feeding as parts of precision livestock farming (Wathes *et al.*, 2005) or precision dairy farming (Doluschitz, 2003) was established. A schematic overview of the components of precision dairy farming is given in Figure 1, which is after Aerts *et al.* (2003) and André *et al.* (2007).
Current systems of concentrate allocation, using adaptive (DLM) models of André *et al.* (2011) can estimate the parameters of the quadratic response curve to concentrate intake per cow per day (*C*). Input of the model are real time process data and prior information of the individual cow. Based on

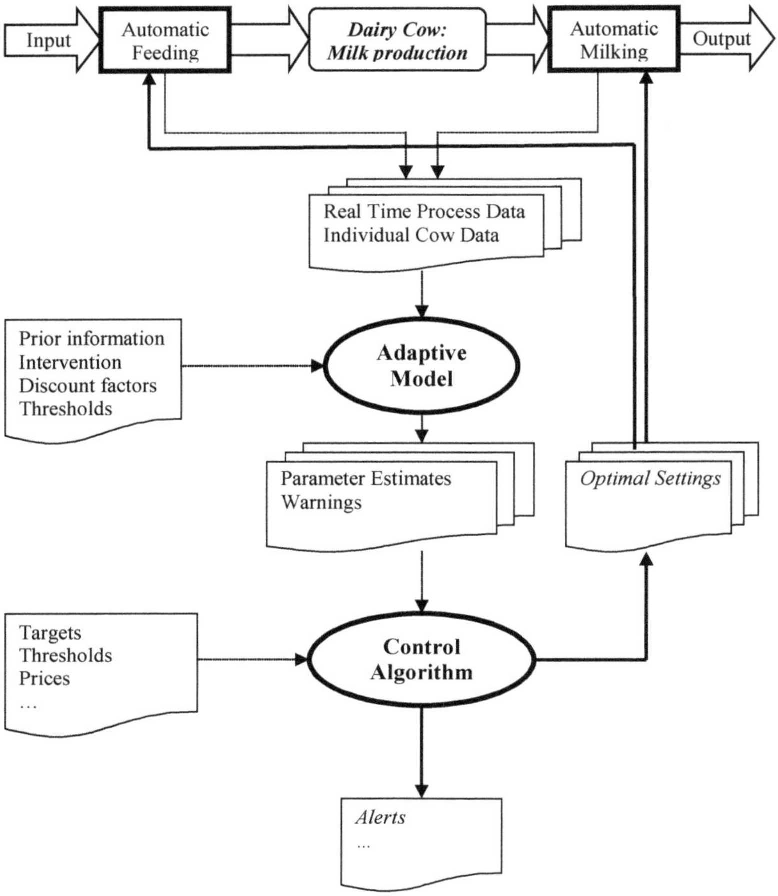

Figure 1. Schematic overview of the components of precision dairy farming.

daily parameter estimates and actual feed and milk prices the optimal individual settings of milking frequency and concentrate allocation (C_{Opt}) can be determined automatically and daily. The number of Dutch dairy farmers using these systems remains constant after a few years of growth after the introduction in 2008. Saving labour and using variability among cows are important reasons for farmers to use this tool. Although economic benefits have been proven in farm comparison studies, it seems that potential economic benefits of the tool are often not reached in practice. This may be caused by adjustments of the optimum settings by the farmer. Some farmers try to challenge the cows by feeding more concentrate in the first weeks after calving. Therefore it is interesting to have more information about adjustments and its results in practice.

2. Materials and methods

In a case-control study 12 dairy farms using the optimum settings (OS) of concentrate level were selected, where each paired ad random with comparable farms from a group of farmers that make adjustments in the tool to generate adjusted settings (AS) for concentrate level. Pairing of farms was

based on milk production level in the first 3 weeks of lactation and milk production of multiparous cows at day 100 after calving.
Milking data and concentrate intake of 2,925 cows were available from November 2013 until March 2015. As major adjustments were expected in the first part of lactation, only cow data until day 100 in lactation were used.
Although data were available on a daily level, a subset of the cow data with time interval of 5 days was analysed. Data of parity 2 and 3 were excluded pre-analysis. Milk production, concentrate intake and estimated optimum value of concentrate intake were analysed in a restricted maximum likelihood (REML) analysis, using the statistical software package GenStat for Windows (VSN International Ltd., Hemel Hempstead, United Kingdom). Random effects of herd and cow were included in the model, as well as an auto-regression term to account for correlation of data within cow.

3. Results and discussion

The average adjustments, $(C\text{-}C_{opt\_recalculated})$ of farms are given in Figure 1. The value of C_{opt}, developed by André *et al.* (2007), which is normally calculated in the practical tool from day 21 after calving using actual prices according to the farmer, was recalculated with standard prices of milk (0.38 €/kg) and concentrate (0.28 €/kg). The peak of adjustments for AS farms was around day 30 in lactation, for both parity groups. Primiparous AS cows receive much higher adjustments, up to 3 kg per cow per day. Also OS farms have differences in actual concentrate intake, with a negative value for multiparous cows at day 25. However, differences are mainly comparable after day 35. On average, adjustments for both treatment groups were negligible after day 55. This means that the difference in concentrate allocation between AS en OS farms is mainly a difference in concentrate level of primiparous cows in the first month after calving.
The higher level of concentrate, respectively +1.4 (primiparous) and +1.6 kg (parity >3) of AS farms was already present at day 20 (Table 1). This confirms that AS farms have a higher speed of increasing concentrate after calving, leaving OS farms with a fairly low concentrate level in the first 3 weeks. Milk production and are comparable for AS en OS, as a result of the design. In the following period from day 25 to day 100 the difference in average concentrate intake becomes reverse due to a lower

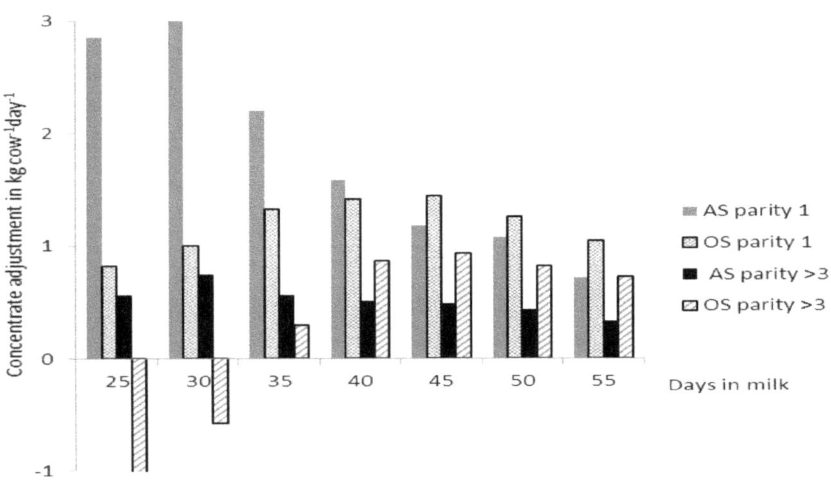

Figure 2. Average early lactation adjustments of concentrate settings by farmers.

C_{Opt} of the AS farm at the end of this period. The reverse effect is more extreme for primiparous cows resulting in a significant interaction effect for treatment by parity group.
Average milk yield in this period is significant lower (-1.8 kg) for AS primiparous cows, but not for AS cows (parity >3), meaning that young cows have a relative lower productivity in AS farms compared with old cows in the same herd. Using average prices of milk (0.38 €/kg) and concentrate (0.28 €/kg) in this period results in an advantage of respectively €0.53 (primiparous) and €0.47 (parity >3) per cow per day for OS farms. When the assumption of comparable roughage quality is made, this indicates that adjustments of the optimal settings by the farmer are not needed for short-term economic reasons. Figure 2 presents the slope of concentrate intake of primiparous cows. Although long-term effects are not included in this study so far, it is clear that AS primiparous cows end with a lower concentrate level at day 100. It seems that this group has developed a status with lower response to concentrate. Zom *et al.* (2002) describe the importance of early development of roughage intake. AS primiparous cows may have problematic development of sufficient roughage intake in the first weeks of lactation with a chance for rumen acidosis. This difference between OS en AS farms does not clearly occur for multiparous cows (Figure 3). Although AS farms also have a higher concentrate level for multiparous cows at an early stage, it seems that these cows end up with the same response to concentrate at day 100. As these cows also have a comparable milk production at day 100, it is not very likely that average roughage quality of both treatment groups is very different.

Table 1. Predicted results from farms using adjusted settings (AS) of concentrate allocation compared with predicted results from farms using the optimal settings (OS).

	Adjusted settings	Optimal settings	Adjusted settings	Optimal settings
	parity 1	parity 1	parity >3	parity >3
Concentrate intake at day 20 (kg/cow/day)	5.8[a]	4.4[b]	6.4[a]	4.8[b]
Milk yield day 20 (kg/cow/day)	25.9	26.4	37.0	37.9
Optimum concentrate level day 25	4.6	4.6	7.2	7.2
Concentrate intake day 25-100	4.9	5.4	7.3	7.5
Milk yield day 25-100	27.7[b]	29.5[a]	38.3	39.7
Balance day 25-100 (€/cow/day)	9.15	9.68	12.54	13.01

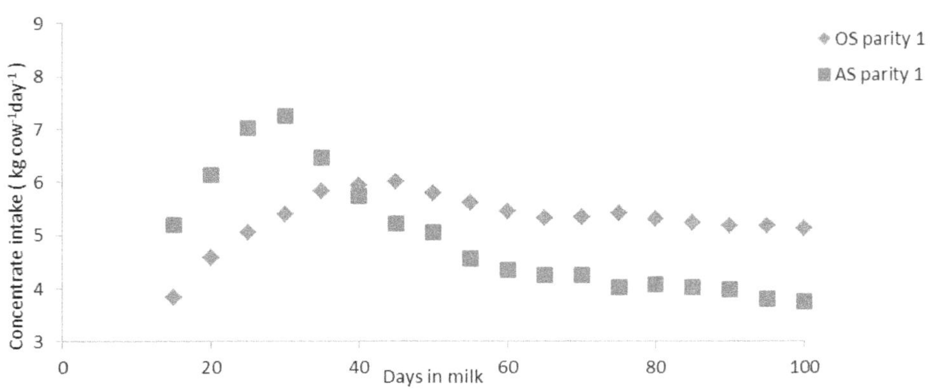

Figure 3. Concentrate intake (Y-axis) vs days in milk (X-axis) of cows in parity 1.

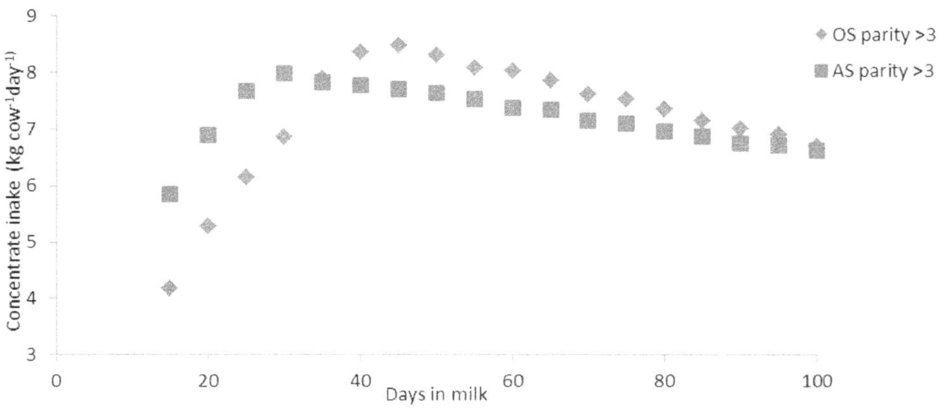

Figure 4. Concentrate intake (Y-axis) vs days in milk (X-axis) of cows in parity >3.

4. Conclusions

On average, adjustments of individual dynamic optimal settings, in order to provide a higher concentrate intake, do not give higher economic profit until day 100. These adjustments by the farmers in practive are relatively higher for primiparous cows.

Acknowledgements

This research was a cooperation between Agrovision B.V. and Wageningen UR Livestock Research. Therefore we thank Gerrit Braakman, Fokko de Vries and Remco de Veer for their valuable contributions.

References

André, G., Ouweltjes, W., Zom, R.L.G. and Bleumer, E.J.B. (2007). Increasing economic profit of dairy production utilizing individual real time process data. In: Cox, S. (ed.) Precision livestock farming '07. Wageningen Academic Publishers, Wageningen, the Netherlands, pp. 179-186.

André, G., Engel, B., Berentsen, P.B.M., Van Duinkerken, G. and Oude Lansink, A.G.J.M. (2011). Adaptive models for on-line estimation of individual milk yield response to concentrate intake and milking interval length of dairy cows. Journal of Agricultural Science, published on-line.

Aerts, J.-M., Wathes, C.M. and Berckmans, D. (2003). Dynamic data-based modelling of heat production and growth of broiler chickens: development of an integrated management system. Biosystems Engineering 84: 257-266.

Doluschitz, R. (2003). Precision agriculture – applications, economic considerations, experiences and perspectives. EFITA Conference paper, pp. 541-546.

Hogeveen, H., Ouweltjes, W., De Koning, C.A.J.M., Stelwagen, K. (2001). Milking interval, milk production and milk-flow rate in an automatic milking system. Livestock Production Science 72: 157-167.

Wathes, C.M., Kristensen, H.H., Aerts, J.-M. and Berckmans, D. (2005). Is precision livestock farming an engineer's daydream or nightmare, an animal's friend or foe, and a farmer's panacea or pitfall? Precision livestock farming '05, Wageningen Academic Publishers, Wageningen, the Netherlands, pp. 33-47.

West, M. and Harrison, J. (1997). Bayesian forecasting and dynamic models. Springer-Verlag, New York, NY, USA.

Zom, R.L.G., Van Riel, J.W., André, G. and Van Duinkerken, G. (2002). Voorspelling voeropname met Koemodel 2002. Praktijkrapport 11, Praktijkonderzoek Veehouderij, Lelystad, the Netherlands.

Cow feed intake prediction with machine learning

B.D. Van der Waaij[1*], R.L. Feldbrugge[1] and R.F. Veerkamp[2]
[1]Monitoring and Control Services, TNO, Eemsgolaan 3, 9727 DW Groningen, the Netherlands; [2]Animal Breeding and Genomics Centre, Wageningen UR Livestock Research, Droevendaalsesteeg 1, Wageningen, the Netherlands; bram.vanderwaaij@tno.nl

Abstract

In a world where data availability is growing, the analysis of data moves from traditional model based methods to 'big data' methods; like data mining and machine learning. These techniques enable researchers to model the complex processes, like the relations between the cow, her milk production, context and behavior. This paper presents a machine learning technique to predict the actual feed intake of individual cows. This prediction can be used as input for farmers or advisors where individual feed intake is of interest, but too costly to measure. The data driven model was trained using a unique dataset, which contains the specific roughage and concentrate intake from cows on the Dairy Campus in the Netherlands. This data is combined with other routinely sensor data and weather data coming from the KNMI. The data was cleaned, and divided into a training and test set. The training data was used to learn an artificial neural network. Artificial Neural Networks (ANN) are computational models that find their inspiration in biological neural networks. These networks can be used as universal function approximators. They are good at inter- and extrapolating highly dimensional nonlinear relationships, but require large amounts of data to learn their internal settings. A test dataset was used to calculate the performance of our model. An effective machine learning model was constructed to predict the feed intake of cows with a precision of 7.7% (RMS error of 1,7 kg), based on cow number, concentrate, milk yield, parity, weight, rumination, lactation day, fat%, protein%, outdoor temperature and outdoor humidity. This research forms a good basis to develop proxies for total daily feed intake of individual dairy cows, based on routinely available data on the farm. These proxies can be used to optimize the feed management of individual or groups of dairy cows or be aggregate to predict full lactation intake, when no or partial feed intake records are available.

Keywords: feed intake prediction, cow feed management, neural network, big data

1. Problem statement

This paper focusses on prediction of the actual feed intake of cow using multiple sensor data and big data machine learning techniques. The feed intake of cows consists out of a combination of concentrate and roughage. The amount of concentrate a certain cow is eating is sometimes known because it is delivered via a feeding machine. The roughage is almost never know for individual cows because it is delivered in the stable to all the cows at once.

The research question of this paper is 'Can the feed intake of cow be predicted, based on multiple sensor data and machine learning techniques?' This question will be answered including and excluding the concentrate intake.

In a world where data is growing, the analysis of data moves from traditional methods to 'big data' methods, like data mining and machine learning. Although the fields of statistics and machine learning are converging, there remain important differences. The largest of them being that the core of traditional statistical techniques is making assumptions about the distribution of the data and the underlying model (and verify these hypothesis), where in machine learning no prior assumption about the model and the distribution are being made.

Machine learning focusses on modeling the data, learning it's patterns, the algorithms use large amounts of data to train algorithms that are able to predict the data. In general they require much more data than standard statistical techniques (since no pre-assumptions about the model are being made). The advantages of machine learning techniques are that the amount of domain knowledge required is minimal and that the resulting models (when set up and validated correctly) can model very complex (nonlinear) systems. Because it is possible to model non-linear systems, the models are good at generalizing new data points (inter and extrapolating).

2. Data

The data used in this research paper comes from two sources. Data about the behavior of cows, especially the feed intake, comes from the Dairy Campus. Outdoor weather information comes from a weather station of KNMI. Both located in the Netherlands. Both datasets cover the period May 21st 2012 until April 5th 2015.

The Dairy Campus data set contains 209 unique cows that cover 7 parities, containing 7,281 measurement-sets. Each measurement-set consists out of: 10 parameters, 8 on a daily basis, 2 on a weekly basis. To extend the size of the data set, necessary for big data analyses, the weekly data is linear interpolated up to a daily basis. After removing too small and incomplete measurement-sets, this results in 5,461 measurement-sets (75% of the original data) with at least 10 days of data containing for each day values for all parameters.

In order to learn about the sensor data of the cow during a single lactation, the measurement-sets are grouped into lactations periods. Because all the measurements are now grouped into lactations, the lactation day can be used as an unified time index for all the parameter measurements. Unfortunately for most cows the measurement-sets are not covering entire lactations, which brings the total down to 194 partial lactations.

The KNMI data set contains only data for the Lelystad station, the station closest to the Dairy Campus in Lelystad. It is limited to the same period as the Dairy Campus data set (21-5-2012 till 15-3-2015) and contains two parameters: maximum daily outdoor temperature, and Air humidity daily average. The maximum daily outdoor temperature is used with the idea that the maximum temperature would be to most influential daily outdoor temperature reading. The daily average of the air humidity is chosen as an indicator of the type of day.

Lastly the cow number is added to keep the relation between the measurements and the cow. An additional reason to add the cow number is to start investigating the influence of genotype on the roughage intake prediction. This results in a dataset containing:

- 7 sensors from the Dairy Campus dataset (roughage intake, concentrate intake, milk yield, weight, rumination, fat%, protein%);
- 2 sensors from the KNMI dataset (temperature, humidity);
- the cow number;
- the lactation number and the lactation day.

3. Methods

The general approach is to train a neural network on predicting feed intake based on sensor data about the physical state of the cow (parity, weight), her milk production (milk yield, fat, protein), the environment (temperature, humidity) and her behavior (rumination). Two different neural networks are trained, one with concentrate and one without (Figure 1).

A distinction must be made between training the neural network, and using the trained neural network into production. In the training phase the previously described data set is used to train and validate the neural network. The data set is spilt into an input and output set. The output is the feed (=concentrate+roughage) intake, which the neural network should train to. The other parameters are

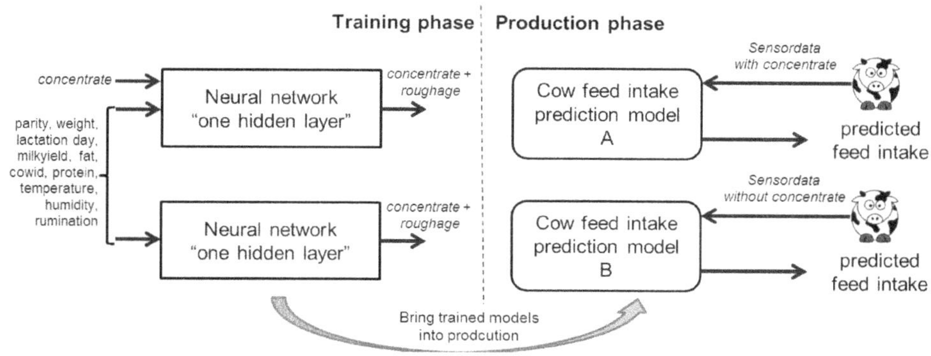

Figure 1. Training and production phase of the cow feed intake prediction model.

used to feed the inputs of the neural network. During the training phase all the data (both input and output set) are offered to the network in order to learn the offered feed intake pattern as close as possible. As the schematic shows, after training, both models can be applied on new cows to predict their feed intake. For these cows the same parameters must be measured (by sensors) as being used in the training phase. The feed intake can then be predicted by one of models. If the concentrate intake is available, model A should be used, otherwise model B must be used.

3.1 Initial setup

Since the number of inputs is small a feed forward neural network with one hidden layer, and a single output is used. The activation function is chosen to be a hyperbolic tangent (tanh) for the hidden layers, and linear for the input and output layers. This setup is standard for a Neural Network used for curve fitting (Russell, 2003).

The number of hidden neurons is determined using an experiment in which all 11 input parameters are selected. The number of hidden neurons are incrementally increased and the resulting RMS error (Root Mean Squared) is measured on a separate test set. This resulted in 35 being the optimum number of hidden neurons. Choosing the number of hidden neurons too high will result in a high probability of overfitting the model on the data (Russell, 2003).

The weight of the network were learned using the Levenberg Marquardt optimizer (Hagan, 1994; Levenberg, 1944; Marquardt, 1963). The advantage of the Levenberg Marquardt training method is that it can learn very quickly, compared to traditional backpropagation. Scalability of this method (memory consumption) will be an issue when moving towards larger datasets or distributed systems, in that case a move to incremental learning algorithms like backpropagation is possible.

3.2 Validation

The dataset is divided into a train (90%) and a validation/test (10%) set. Since the amount of data was marginal a k-fold cross validation (McCormick, 2013) was used on the dataset to select the train and validation set. K-fold cross validation means subdivide the data into k subsets, and then for each set use the kth set as our validation/test set and the remainder for the training set. This method provides better insight into how the model performs on an independent (unknown) dataset when the amount of data is sparse.

3.3 Performance of the models

The performance of both models were calculated using a brute force approach, calculating all possible combinations of parameters and then analyzed the results for the best model with and without concentrate, see Table 1.

Table 1. The best performances of the model with and without concentrate. The black fields are fields intentionally left out. An 'x' means the input is used, an 'o' means it has no positive influence on that scenario.

Model	Cow number	Milk yield	Parity	Weight	Concentrate intake	Rumination minutes	Lact day	Fat%	Protein%	T	Hum	RMS error (kg)	% RMS error
A	x	x	x	x	x	x	x	x	x	x	x	1.72	7.67
B	x	x	x	x		x	x	x	x	x	o	2.92	13.04

The best performance was found in model A, the model with concentrate. This resulted in an RMS error of 1.72 kg (approximately 7.67% with respect to the average feed intake of 22.4 kg in this dataset).

4. Results

Both models were evaluated with 5-fold-crossvalidation. This resulted for model A in an average RMS error of 1.72 kg, or 7.67% based on an average feed intake of 22.4 kg per cow. Some farms do not have concentrate intake measured. For this model B is designed, working without concentrate. Model B's best achieved RMS error is 2.92 kg, or 13.04% on average. Concentrated food is a dominant parameter, with an effect on the performance of 5,37% (=13.04-7.67).

For display purposes two random cows were intentionally left out from both the training and the test/validation set to see how our model A would perform on them (Figure 2A and B). Note that cow 2 has some gaps in the predicted values. Due to the scarcity in the training dataset, the model is not always able to predict a feed intake for each moment in a reliable manner. On those moments no predictions are being made, which shows as gaps in the Figure.

When applying these two cows on the model B, excluding concentrate intake as an input, the model has more trouble following the peaks. Resulting in a smoother curve (but it still follows the general trend (Figure 3A and B).

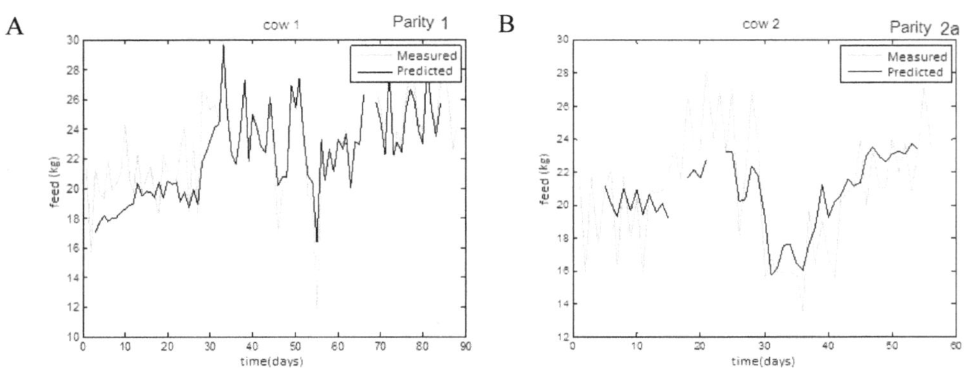

Figure 2. (A) Results model A, cow 1. (B) Results model A, cow 2.

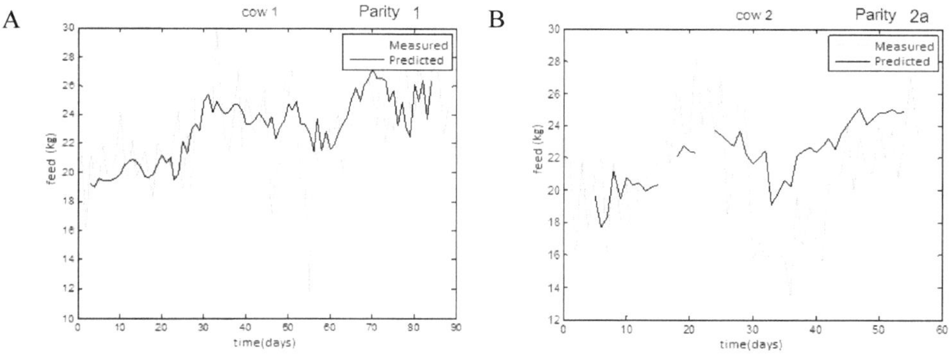

Figure 4. (A) Results Model B without concentrated feed intake, cow 1. (B) Results model B without concentrated feed intake, cow 2.

5. Conclusions

We were able to construct an effective machine learning model to predict the feed intake of cows in the described Dairy Campus dataset. We are able to predict this with a precision of approximately 7.67% when concentrate intake data is available. If concentrate intake is not available, the precision drops to 13.04%. This research is based on a small dataset, to upscale the results to be applicable also for other farms, a larger dataset is needed to (re)train and validate the model.
This means we are able to accurately predict the feed intake of cows that do not have labor expensive systems to monitor the roughage intake. This model forms a good basis to develop proxies for daily feed or roughage intake of individual dairy cows, based on routinely available data on the dairy farm. These proxies for daily feed or roughage intake can be used in management modules to optimize the feed management of individual or groups of dairy cows or be aggregate to predict full lactation intake, when no or partial feed intake records are available. The value of this research for genetic improvement needs further investigation

Acknowledgements

Contributions, ideas and comments from all members of this project are acknowledged. In particular R. de Mol and R. Goselink. Special thanks goes to the Dairy Campus for letting us use their feed intake dataset.

References

Hagan, M.T. and Menhaj, M.B. (1994). Training feedforward networks with the marquardt algorithm. IEEE Trans. on Neural Networks 5(6): 989-993.

Levenberg, K. (1944). A method for the solution of certain nonlinear problems in least squares. Quarterly Applied Mathematics 2: 164-168.

Marquardt, D.W. (1963). An algorithm for least squares estimation of nonlinear parameters. SIAM Journal 11: 431-441.

McCormick, C. (2013). K-fold cross-validation, with MATLAB code. Available at: http://tinyurl.com/j25gebq.

Russell, S.J. and Norvig, P. (2009). Artificial intelligence: a modern approach. Pearson Education, New York, NY, USA.

Creating value from data: milking order and its role in future precision dairy feeding systems

K.M. Dias[1], S.G. Garcia[2] and C.E.F. Clark[2*]
[1]*Animal Science Department, Santa Catarina State University (UDESC), Lages, SC, Brazil;* [2]*Dairy Science Group, Faculty of Veterinary Science, University of Sydney, Camden, NSW 2570, Australia; cameron.clark@sydney.edu.au*

Abstract

Pasture varies in its chemical composition from its leaf to stem and cattle prefer to eat the leaf fraction. Dairy cattle in pasture-based systems when milked typically walk to pasture with the first cattle arriving to pasture hours before the last. Here, we discover the impact of milking order on milk yield for dairy cattle offered grazed pasture. For the 6 farms evaluated over 8 months, cows milked first produced, on average, 4.5 l/cow/day more than those cows milked last in the order, equating to 20% more milk volume. Differences in milk production from the first to last cattle between farms ranged from 14 to 29%, presumably due to differences in management practice and associated pasture allowance and nutritive value. Our work will now determine the ability to significantly increase nutrient use efficiency for improved milk production and environmental outcomes through strategic pasture allowance and supplementation strategies.

Keywords: pasture, milk yield, nutrients

1. Introduction

The depletion of a pasture sward canopy by cattle typically occurs in successive layers (Wade and Carvalho, 2000) which varies in chemical composition, with the top of the fraction typically containing more crude protein (CP) and less neutral detergent fibre (NDF) than lower fractions (Scott *et al.*, 2014). Dairy cattle also have a defined milking order (Scott *et al.*, 2014) meaning that for a defined management routine, the state of pasture accessed by individual cows will be consistent. With the trend towards larger herd sizes (Dairy Australia, 2014) and greater milking times (Dias and Clark, unpub data) the differences in nutritive value accessed between cattle and associated differences in milk yield and milk composition may be increasing. The objective of this research was to evaluate the impact of milking order on milk yield and milk composition for dairy cattle offered grazed pasture.

2. Materials and methods

We collected daily milking order and milk yield data of predominantly Holstein-Friesian cattle from six commercial large herds (Table 1) located in Tasmania, Australia, from January to August 2015. Animals were classified in milking order groups of 50 animals for a.m. and p.m. milking (e.g. group number 1: 1st to 50th cow; group number 2: 51st to 100th, etc.). Daily milking order was the average of the a.m. and p.m. milking order. Factors in the model were grouped accordingly: Period (week of the year), parity, milking order (cows were grouped 1-50, 51-100, …) and concentrate intake (groups 0-3.0, 3.1-6.0…). Cows were included as a random effect and DIM as a covariate, using the statistical program SAS, MIXED procedure. Individual cows were considered as an experimental unit and period as a repeated measurement. To determine the effects of milking order (MO) (11 to 24 groups), period (P) (31 groups), parity (PA), DIM (D) and concentrate intake (C) (4 groups) on milk yield (MY) (Model 1), a mixed linear model was fitted to the data as follows:

$$MY_{ijklm} = \mu + MO_i + P_j + PA_k + C_m + b1(D)_{ijklm} + \varepsilon_{ijklm}$$

Where MY is the dependent variable, μ is the overall mean, $b1(D)_{ijklm}$ is days in milk as a covariate and ε_{ijklm} is the random experimental error.

3. Results

There was a strong association between a.m. and p.m. milking order ($y=0.51+0.92x$; $R^2=0.87$, $P<0.001$). Milking order had a significant impact on milk yield ($P<0.01$) for all farms. The mean milk yield difference between the first 50 and last 50 cows for all farms was 4.5 l/cow/day, equating to 20% less milk (Table 2).

Table 1. Number of cows, days in milk (DIM), milk yield (l/cow/day), body weight (kg/cow) and grain-based concentrate offered (kg/cow/day) for experiment 1.

Farm	Number of cows		DIM[1]		Milk yield[2] l/cow/day		Body weight[2] kg/cow		Grain-based concentrate[2] kg/cow/day	
	Average	SD[3]	Average	SD	Average	SD	Average	SD	Average	SD
1	451	40	192	97	27.1	8	555	88	9.8	3
2	519	50	227	119	21.2	6	-	-	5.2	1
3	704	33	178	127	24.6	8	-	-	6.9	2
4	770	41	187	121	26.4	8	-	-	6.6	2
5	497	5	253	35	13.0	3	-	-	1.5	1
6	774	36	278	186	15.5	5	475	77	3.6	3

[1] DIM=days in milk.
[2] Average=7 days average.
[3] SD=standard deviation.

Table 2. Milk yield difference (MYdif) between first 50 and last 50 cows and milking information of each farm.

Farm	MYdif (%)	MYdif (l/cow/d)	Cattle		Milking throughput (cows/h)		Milking duration (h/d)	
			Average	SD[1]	Average	SD	Average	SD
1	17	-5.0	476	115	210	38	2.3	0.9
2	19	-3.9	568	172	212	40	2.7	0.9
3	29	-7.5	700	111	231	42	3.2	0.9
4	15	-4.1	776	118	178	29	4.5	0.9
5	14	-2.3	618	82	272	47	2.3	0.6
6	16	-3.4	763	210	232	55	3.4	1.2

[1] SD=standard deviation.

4. Discussion

This work is the first to show the impact of milking order on milk yield for dairy cattle in pasture-based systems. The first cows to be milked in our study produced, on average, 20% more milk than those milked last with a range for individual farms of between 14 and 29% between farms. Previous work from our group (Scott *et al.*, 2014) with an intermediate herd size (350 cows) showed crude protein levels to decrease by 21%, and acid detergent fibre to increase by 15%, between the first cow accessing pasture and the last. Our findings highlight the requirement for detailed studies to elucidate opportunities to improve the efficiency of milk production from pasture through capitalising on milking order data. Foremost of these opportunities is to determine the direct link between pasture nutritive and the milk production across dairy herds. Our work will now determine how to optimize pasture nutrient use efficiency for milk production to enable a step change in farm productivity and profitability. The impact of varying management practices associated with pasture allowance, along supplementation strategies will now be determined.

Acknowledgements

The authors acknowledge the financial support of CAPES (Process number: 99999.004580/2014-01), Dairy Australia and the collaboration of the farmers and team from the University of Sydney Dairy Science Group.

References

Dairy Australia (2014). 'Dairy 2014: situation and outlook'. Report by Dairy Australia, Melbourne, Australia.

Scott, B.A., Clark, C.E.F., Camacho, A., Golder, H., Molfino, J., Kerrisk, K.L., Lean, I., García, S.C., Chaves, A.V. and Hall, E. (2014). The nutritive value of pasture ingested by dairy cows varies within a herd. 6th Australiasian Dairy Science Symposium, Hamilton, New Zealand.

Wade, M.H. and Carvalho, P.F. (2000). Defoliation patterns and herbage intake in grazed pastures. Ecophysiology of Grasslands and the Ecology of Grazing. CAB International, Wallingford, UK, pp. 233-248.

The relation between eating time and feed intake of dairy cows

R.M De Mol[1*], R.M.A. Goselink[1], J.W. Van Riel[1], H.M. Knijn[2] and A.T.M. Van Knegsel[3]
[1]*Wageningen UR Livestock Research, P.O. Box 338, 6700 AH, Wageningen, the Netherlands;* [2]*CRV, Wassenaarweg 20, 6843 NW, Arnhem, the Netherlands;* [3]*Wageningen University, Adaptation Physiology Group, P.O. Box 338, 6700 AH, Wageningen, the Netherlands; rudi.demol@wur.nl*

Abstract

The feed intake of individual dairy cows is important information for dairy farmers to monitor health and to evaluate feed efficiency. It is difficult to measure roughage intake in practical circumstances without dedicated equipment. An alternative for measuring intake in practice may be measuring the duration of feeding visits with sensor technology. The relation between feeding duration and feed intake is studied by using data from dairy research farm Dairy Campus, where feed intake and visit duration are recorded individually by transponder-controlled feeding bins. Data from trials performed in the years 2012 through 2015, were available. Some cows were fed concentrates and roughage separately (partially mixed ration, PMR), others were fed a total mixed ration (TMR). In all trials recordings of visits to feeding bins were available with starting time, ending time and intake, besides information on individual cows like parity, milk yield, body weight and rumination time. For the analysis all visits, with or without effective feed intake, were included. The total data set included 37,233 cow days with 1.251,156 visits. Visits were aggregated to meals by combining visits with an interval lower than a threshold (29 minutes), resulting in 286,576 meals with on average 7.7 meals per cow per day. The correlation between eating time (defined as duration of visits) and total feed intake per cow per day depended on the type of ration (PMR or TMR) and differed between cows. The average correlation between eating time and total feed intake was 0.59 (median 0.65) for PMR and 0.53 (median 0.56) in case of TMR feeding. Median values of feed intake and other cow characteristics, like parity, milk yield, body weight, rumination time, eating time, and number of meals were used in a REML model to predict feed intake. Parity, milk yield, body weight were relevant to estimate feed intake in both a PMR and a TMR system. Eating time was a significant term, but only in case of a PMR system. Rumination time and number of meals were not relevant in the current study to predict feed intake. These results show that data on feeding visits can be used to estimate eating time and number of meals per cow per day. Eating time can be used to improve the estimation of feed intake in a PMR system.

Keywords: feed intake, REML model, meal determination, rumination

1. Introduction

Efficiency is important in dairy husbandry due to limited availability of natural resources and statutory regulations, like formerly milk quota and nowadays constraints in land use or fertilizer use. It is in practice difficult to quantify individual efficiency, because available data on individual feed intake is limited. Feed intake per cow can be measured in experimental conditions using dedicated feeding equipment: the Roughage Intake Control (RIC) system (Hokofarm, Marknesse, the Netherlands) recognizes cows when they enter a feeding place and the difference in bin weight between start and end of a visit is recorded.

Sensors are available to support dairy management. Automatic milk yield measurements and computerized concentrates feeders are being used in practice. Sensors for measuring feeding and rumination behavior are available. Results on the value of data on rumination behaviour to estimate feed intake are ambiguous. On the one hand, Schirmann *et al.* (2012) and Clément *et al.* (2014) found that rumination time is a bad indicator for dry matter intake of dairy cows. On the other hand, Byskov *et al.* (2015) concluded that the daily intakes of forage NDF and starch were positively related to rumination time, whereas intakes of sugar and the remaining fraction were negatively related to rumination time. Differences among studies could be related to the method to determine rumination behavior or differences in feed composition.

Rumination or eating sensor data, in combination with data on body weight, milk yield, concentrate intake and parity might be useful to estimate the individual feed intake of dairy cows. The objective of this research is to explore the possibilities of estimating feed intake by using sensor data of feeding and rumination behavior in addition to cow characteristics like parity, milk yield and body weight.

2. Materials and methods

2.1 Experimental data

Data from seven feeding experiments on the experimental farm in Lelystad of Wageningen UR Livestock Research were available. In all experiments data on visits to the RIC system were available. Recordings of these visits included time of arrival and time of departure, bin weight at arrival and bin weight at departure (kg) and feed intake (kg). The duration of a visit was based on arrival and departure time. Total duration per day was the sum of all visit durations per day (but overlapping visits were counted only once); this was considered as the eating time per day. Eating time was not used for the first and last day in a sequence of days for which RIC data were available as these days might be incomplete. Visits could be combined into meals if the interval length between successive visits was below a threshold. For the determination of this threshold the method of Tolkamp (1998) was used. RIC data were not available on some days due to failures. The intervals between successive visits of a cow were calculated to determine which visits belong to the same meal.

Other data available in all experiments included milk yield, body weight, parity and rumination time. Milk yield (kg) was recorded per milking (twice a day) and totalized to day level. Body weight (kg) was recorded once a day. Rumination was measured by rumination sensors (SCR HR tags, SCR Engineers Ltd., Netanya, Israel) attached to neck collars (Schirmann *et al.*, 2009). Rumination time was recorded in minutes per 2 hour interval and totalized to daily rumination time if at least 10 recordings per day were available. Daily totals based on 10 or 11 intervals were corrected proportionally to estimate daily rumination time.

The experiments differed in lactation stage of the cows and feeding method (Table 1). The feeding method was partially mixed ration (PMR) or totally mixed ration (TMR). In a PMR system a roughage mixture is fed at the feeding fence, concentrates are supplied in a computerized concentrates feeder or in the milking parlor according to an individual daily maximum which is generally predetermined by the farmer. In a TMR system all ration components are supplied at the feeding fence in a predetermined mixture of roughage and concentrates.

2.2 Statistical analysis

All data were available on a daily level; this made it possible to calculate to correlations between variables.

Variables like milk yield, feed intake, mostly show a more or less constant level per cow when restricted to a project and a treatment. The median values per cow, project and treatment were calculated to represent this level. The median was used instead of the average to compensate for possible outliers. Medians were calculated for milk yield, total feed intake, roughage intake, concentrate intake, body weight, rumination time, eating time, number of meals per day in all experiments. The medians were explored in a restricted maximum likelihood (REML) analysis using the statistical software package GenStat for Windows (VSN International Ltd., Hemel Hempstead, UK).

Table 1. Characteristics of seven feeding experiments on the experimental farm in Lelystad of Wageningen UR Livestock Research; ration type is partially mixed ration (PMR), totally mixed ration (TMR) or a combination.

Experiment	Number of cows	Number of cow days	Lactation stage	Feeding method
1	29	1,236	beginning	PMR
2	54	5,224	middle	PMR/TMR
3	72	4,104	beginning	PMR
4	49	4,030	end	PMR/TMR
5	182	7,728	middle	TMR
6	60	3,420	beginning	PMR
7	69	11,491	complete	PMR

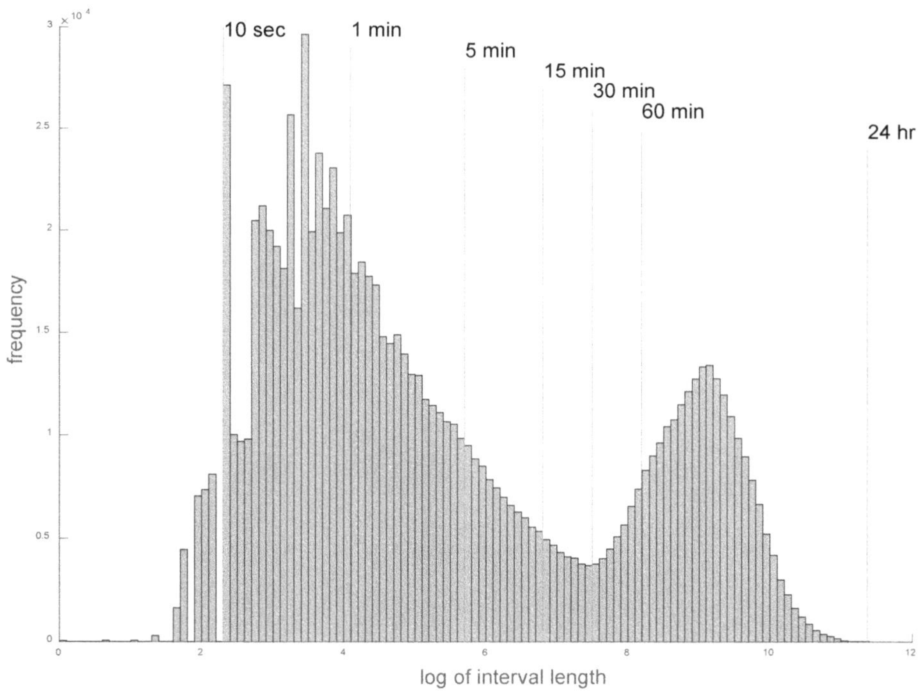

Figure 1. Histogram of log-transformed interval between visits of all cows.

3. Results

3.1 Visits and meals

The total number of cow days was 37,233 (Table 1), the corresponding number of visits was 1.251,156. A histogram of the logarithms of the interval between visits is depicted in Figure 1. There are two peaks in the histogram; the minimum between these two peaks was around 7.45, corresponding with an interval length of 1,720 seconds (28.7 minutes). This minimum value in interval length, the 'meal criterion' was used to distinguish meals (Tolkamp *et al.*, 1998): a successive visit with an interval less than 1,720 seconds belonged to the same meal; otherwise it was the beginning of a new meal. This resulted in 286,576 meals, with on average 7.7 meals per cow per day.

3.2 Correlation between rumination behavior, eating time and feed intake

Correlations between feed intake and rumination or eating time differed between cows and depended on parity. Histograms of the correlation per parity are included in Figure 2. The average correlation between feed intake and rumination per day was 0.06 (0.00, 0.07, 0.08 and 0.09 for parity 1, 2, 3 and 4 or more respectively). The average correlation between feed intake and eating time per day was 0.59 in a PMR system (0.56, 0.58, 0.60 and 0.61 for parity 1, 2, 3 and 4 or more respectively). The corresponding median was 0.65 (0.65, 0.59, 0.66, 0.67 per parity). The average correlation between feed intake and eating time per day was 0.53 in a TMR system (0.46, 0.57, 0.59 and 0.54 for parity 1, 2, 3 and 4 or more respectively). The corresponding median was 0.56 (0.50, 0.57, 0.57, 0.59 per parity).

3.3 REML analysis

Median values were calculated per cow, project and treatment for all variables. These medians were interpreted as a measure for the level of a variable for that cow in that project under that treatment. In total there were 680 sets of medians, each representing a combination of cow, project and treatment. Medians from treatment with a mixture of roughage and PMR were excluded, as well as medians from experiment 2 with an extreme low roughage share in the diet. 519 sets of medians have been

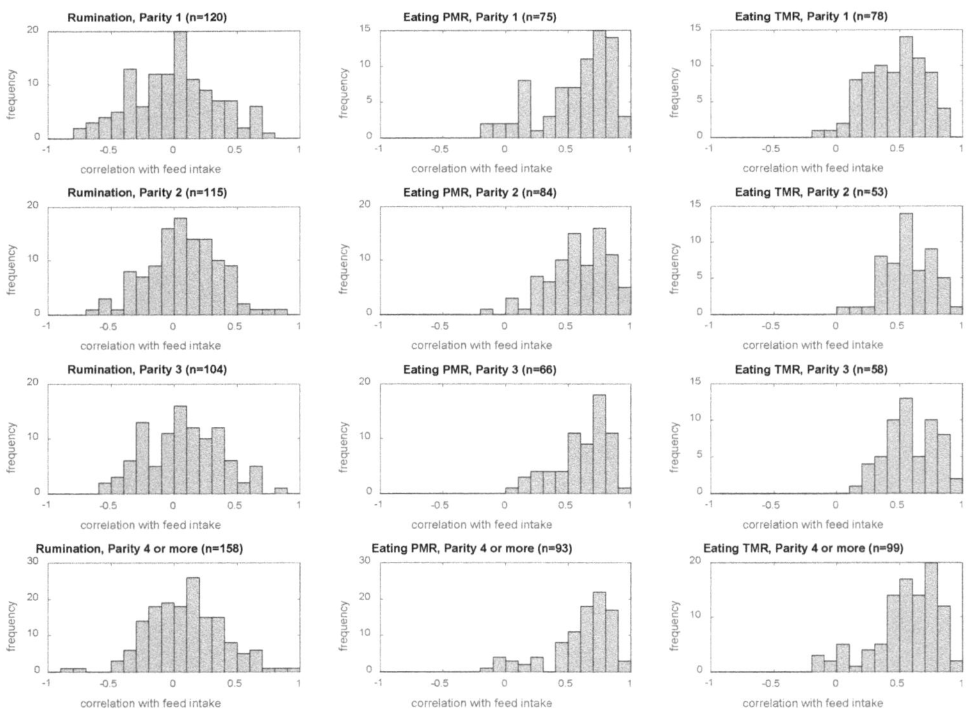

Figure 2. Histograms of the correlations between daily feed intake and rumination (left column), eating time in PMR system (middle) and eating time in TMR system (right) for parity 1, 2, 3 and 4 or more (downwards).

used (278 with PMR and 241 with TMR). The global results of the REML analysis are presented in Table 2. In a PMR system parity, concentrates intake, milk yield, body weight and eating time are significant terms in predicting the feed intake, while number of meals and rumination time are not significant. In a TMR system parity, milk yield, body weight are significant, while eating time, number of meals and rumination time are not.

Table 2. Results of the REML model for feed intake, terms in the model are successively added (with significance) and the explained part of the variance is given; both for partial mixed ration (PMR) and total mixed ration (TMR) systems.

Terms	PMR system (n=278)		TMR system (n=241)	
Constant	0.0%		0.0%	
+Parity	23.6%	***	47.7%	**
+Concentrates intake	34.3%	***	n/a	
+Milk yield	62.6%	***	79.0%	***
+Body weight	70.2%	***	80.7%	***
+Eating time	73.2%	***	80.6%	
+Number of meals	73.1%		80.6%	
+Rumination time	73.9%		80.7%	

* $P \leq 0.05$; ** $P \leq 0.01$; *** $P \leq 0.001$.

4. Discussion

Data on visits from the RIC system could be used to calculate the eating time per cow per day. The histogram of the log-transformed interval length between visits gave a meal criterion. The histogram and criterion are comparable with results from literature (e.g. Maselyne *et al.*, 2015). The histograms per cow resulted in comparable meal criterion per cow (data not shown). The number of meals was not a significant term in predicting the feed intake.

Eating time and rumination time were correlated with feed intake, but only eating time in PMR system was a significant term in the REML model for feed intake. The difference concerning the added value of eating time to explain feed intake between the TMR and PMR system might be due to the high proportion of explained variation in the basic model for the TMR system (>80%) and the fact the PMR system mostly concerned cows in early lactation, with possibly more variation in feed intake between cows.

The significance of terms appeared to be depending on parity (data not shown). Body weight was not significant for cows with parity 4 or higher. Rumination time was significant (*) in TMR system for parity 4 or higher.

Connor (2015) states in her review that because data on dairy cattle are limited, additional research is needed to determine how much feeding behavior and physical activity contribute to variation in residual feed intake within dairy populations. She concluded that more efficient cows (heifers and dairy cows) spent less time eating. The results in the REML model were similar when body weight was replaced by metabolic weight, and also when feed intake was replaced by residual feed intake (data not shown).

These results showed that it can be useful to estimate individual feed intake based on the eating time. In practice, this may be measured by applying a location sensor (is the cow near the feeding gate) or an eating sensor estimating eating time by head movements. These alternatives are easier to implement than measuring individual feed intake with RIC bins, while it may help to determine individual feed efficiency and thereby improve cow and farm performance.

5. Conclusions

Eating time can be estimated by using visit data from the RIC system, the number of meals can be derived from these visits. The current study shows that eating time, combined with parity, milk yield and body weight, is relevant for estimating feed intake in lactating dairy cows fed PMR. Rumination time and number of meals did not have an added value to predict feed intake.

Acknowledgements

This research was a cooperation between Wageningen UR Livestock Research, Dairy Campus and CRV with financial support from SNN (The Northern Netherlands Provinces Alliance).

References

Byskov, M.V., Nadeau, E., Johansson, B.E.O. and Nørgaard, P. (2015). Variations in automatically recorded rumination time as explained by variations in intake of dietary fractions and milk production, and between-cow variation. Journal of Dairy Science 98(6): 3926-3937.

Clément, P., Guatteo, R., Delaby, L., Rouillé, B., Chanvallon, A., Philipot, J.M. and Bareille, N. (2014). Added value of rumination time for the prediction of dry matter intake in lactating dairy cows. Journal of Dairy Science 97(10): 6531-6535.

Connor, E.E. (2015). Improving feed efficiency in dairy production: challenges and possibilities. Animal 9(3): 395-408.

Maselyne, J., Saeys, W. and Van Nuffel, A. (2015). Quantifying animal feeding behaviour with a focus on pigs. Physiology and Behavior 138: 37-51.

Schirmann, K., Chapinal, N., Weary, D.M., Heuwieser, W. and Von Keyserlingk, M.A.G. (2012). Rumination and its relationship to feeding and lying behavior in Holstein dairy cows. Journal of Dairy Science 95(6): 3212-3217.

Schirmann, K., Von Keyserlingk, M.A.G., Weary, D.M., Veira, D.M. and Heuwieser, W. (2009). Validation of a system for monitoring rumination in dairy cows. Journal of Dairy Science 92(12): 6052-6055.

Tolkamp, B.J., Allcroft, D.J., Austin, E.J., Nielsen, B.L. and Kyriazakis, I. (1998). Satiety splits feeding behaviour into bouts. Journal of Theoretical Biology 194(2): 235-250.

Part 6. Grazing management

Using behaviour in dairy cows as a predictor of grass intake

C.G. Van Reenen*, J.T.N. Van der Werf, B. Timmer, D.L. Hoeksma and R.L.G. Zom
Wageningen UR Livestock Research, P.O. Box 338, 6700 AH, Wageningen, the Netherlands; kees.vanreenen@wur.nl

Abstract

At present, there are no practically feasible methods to reliably estimate the grass intake of individual cows kept on pasture. The aim of the present study was to examine whether cow behaviours continuously recorded with the use of sensors, in combination with other potential predictors related to the pasture and to production characteristics of the cow, could be used to estimate grass intake by grazing dairy cows. Behavioural and grass intake data were collected during a one-week experimental period in 60 cows involved in a grazing trial. All cows were equipped with two commercially available 3D accelerometers: an 'IceQube' (IceRobotics, UK) attached to the leg, and a 'Smarttag Neck' (NEDAP, the Netherlands) attached to the neck, recording standing/lying/number of steps and grazing behaviour, respectively. Grass intake (kg DM/day) was determined in individual cows with the use of the n-alkane method. Grass intake was considered as response variable (y), and stepwise multiple regression analysis was used to identify the best fitted models based on percentage of adjusted variance explained (adj R Square). The best model to predict individual grass intake (adj R Square=82%) had the following predictors: grazing area/cow, days in milk (DIM), milk protein (%), milk urea (mg/kg), milk yield (kg) and time spent grazing (min/hour spent on pasture). When milk protein and urea were left out of the analysis, the best model to predict grass intake (adj R Square=71%) had the following predictors: milk yield, DIM, grazing area/cow, time spent grazing, and the total number of steps while on pasture. These findings demonstrate that grass intake of individual cows on pasture can be reliably estimated from a number of behavioural and other predictors. These models may represent potential management tools in grazing systems.

Keywords: grazing, behaviour, estimating grass intake

1. Introduction

Currently, approximately 70% of the total number of dairy cows in the Netherlands are allowed to graze on pasture (CBS, 2015). From an economic perspective, grazing systems are generally profitable because fresh grass represents a relatively cheap feed for dairy cows (Finneran *et al.*, 2010; Van den Pol-Van Dasselaar *et al.*, 2010). Moreover, in comparison with zero grazing, providing access to pasture is associated with cow health and welfare benefits, including reduced prevalences of hoof disorders, lameness, and hock, knee and skin lesions and swellings (De Vries *et al.*, 2015; Olmos *et al.*, 2009).

However, the percentage of Dutch dairy cows allowed to graze is steadily declining over the last decades, in particular on farms with large herds. This is likely due to a more difficult grazing management on farms with large herds. In the Netherlands, the availability of sufficient pasture is a limiting factor on many farms. Therefore, maintaining a large herd in a pasture-based system would frequently result in high stocking densities during grazing as well as a high need for supplementary feeding with forage and concentrate in order to maintain sufficient dry matter intake. At the same time, there are no practically feasible methods available to reliably estimate the grass intake of individual cows kept on pasture. Under these circumstances, it would be difficult to monitor energy intake during grazing, and to optimize the amounts and composition of supplementary forage and concentrates in terms of milk production and feed efficiency.

The objective of this research was to examine whether cow behaviours continuously recorded with the use of sensors, in combination with other potential predictors related to the pasture and to production characteristics of the cow, could be used to estimate grass intake by grazing dairy cows in intensive systems.

2. Materials and methods

Behavioural and grass intake data were collected during a one-week experimental period in 60 cows involved in a grazing trial. Cows were blocked according to parity, days in milk (DIM), body weight, and fat and protein corrected milk production, and randomly divided within blocks across three grazing systems: (1) rotational grazing; (2) strip grazing; and (3) rotational continuous grazing. Grazing areas per cow in these grazing systems were 0.0069, 0.0046, and 0.0275 ha per cow, respectively. Cows were machine-milked twice a day, and had daily access to pasture during day time between morning and afternoon milking. Cows in each grazing system were kept separate both outdoors and indoors, and were supplemented in the barn with silage when herbage allowance (estimated from sward characteristics at paddock level) in the pasture restricted herbage intake. In addition to daily milk yield (kg), milk composition in terms of protein (%), fat (%), urea (mg/kg) and SCC (cells/ml) was determined in the experimental week during 4 successive milkings.
All cows were equipped with two commercially available 3D accelerometers: an 'IceQube' (IceRobotics, UK) attached to the leg, and a Smarttag Neck (NEDAP, The Netherlands) attached to the neck. The 'IceQube' recorded standing/lying behaviour and number of steps (Munksgaard *et al.*, 2006). The 'Smarttag Neck' reliably recorded grazing behaviour (Van Reenen *et al.*, 2016). The time spent grazing was expressed as sec/hour spent on pasture. Grass intake (kg DM/day) was determined in individual cows with the use of the n-alkane method (Mayes *et al.*, 1986).
In the statistical analysis, grass intake determined with the n-alkane method was considered as response variable (y), and a range of behavioural and other cow- and pasture-related variables were considered as potential predictor or covariables (x). Stepwise multiple regression analysis was used to indentify the best fitted models based on the percentage of adjusted variance explained (adj R square). Two cows were excluded from the analysis because of missing data.

3. Results and discussion

The best model (Model 1) to predict individual grass intake (adj R Square=82%) had the following predictors: grazing area/cow (ha), DIM, milk protein, milk urea, milk yield and time spent grazing while on pasture. Estimates of parameters of Model 1 with corresponding standard error (SE) and statistical significance are given in Table 1.

Table 1. Multivariate regression model (Model 1) for grass intake. Estimates of intercept and regression coefficients of predictors, with standard error (SE) and statistical significance.

Parameter	Estimate	SE	*P*-value
Constant (intercept)	-23.72	3.68	<0.001
Grazing area/cow	-162.6	21.2	<0.001
Milk yield	0.4677	0.0438	<0.001
DIM	0.01433	0.00444	0.002
Milk protein	3.79	1.12	0.001
Milk urea	0.2388	0.0522	<0.001
Time spent grazing	0.003288	0.000741	<0.001

When milk protein and urea were left out of the analysis, the best model (Model 2) to predict grass intake (adj R Square=71%) had the following predictors: milk yield, DIM, grazing area/cow, time spent grazing while on pasture, and the total number of steps while on pasture. Estimates of parameters of Model 2 with corresponding SE and statistical significance are given in Table 2. The estimated grass intake for each cow was calculated with the use of the regression equations obtained in Model 1 and Model 2. Relationships between grass intake determined with the n-alkane method and the estimated grass intake either based on Model 1 or Model 2 are presented in Figure 1 and Figure 2, respectively. The Pearson correlation between grass intake determined with the n-alkane method and the estimated grass intake according to Model 1 was 0.92 (n=58, $P<0.001$). The Pearson correlation between grass intake determined with the n-alkane method and the estimated grass intake according to Model 2 was 0.87 (n=58, $P<0.001$).
The overall average (n=58) of grass intake determined with the n-alkane method was 13.172 kg DM per cow per day. Overall averages (n=58) of estimated grass intakes based on Model 1 and Model 2 were 13.171 and 13.167 kg DM per cow per day, respectively. This suggests that these prediction models provide highly accurate estimates of grass intake at herd level.

Table 2. Multivariate regression model (Model 2) for grass intake. Estimates of intercept and regression coefficients of predictors, with standard error (SE) and statistical significance.

Parameter	Estimate	SE	*P*-value
Constant (intercept)	-13.35	2.54	<0.001
Grazing area/cow	-116.2	25.9	<0.001
Milk yield	0.4287	0.0528	<0.001
DIM	0.03033	0.00419	<0.001
Number of steps	0.001591	0.000382	<0.001
Time spent grazing	0.003032	0.000891	0.001

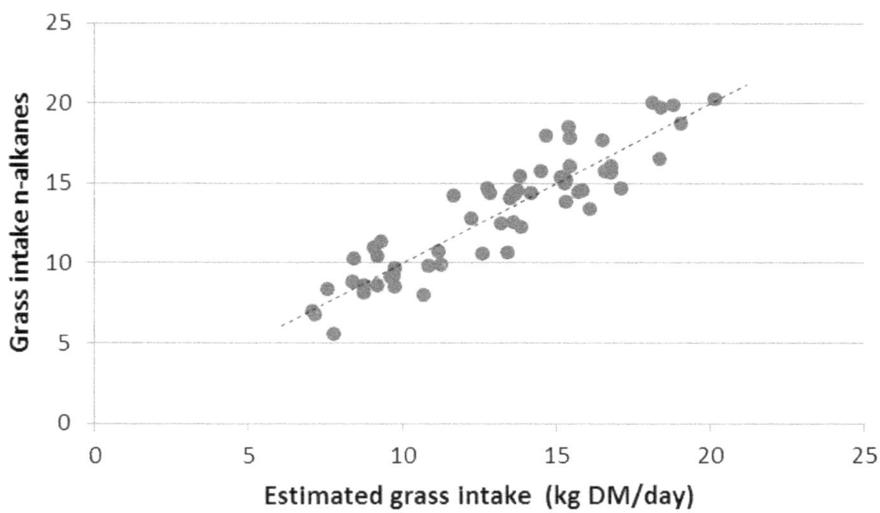

Figure 1. Relationship between grass intake determined with the n-alkane method and grass intake estimated with the use of regression Model 1.

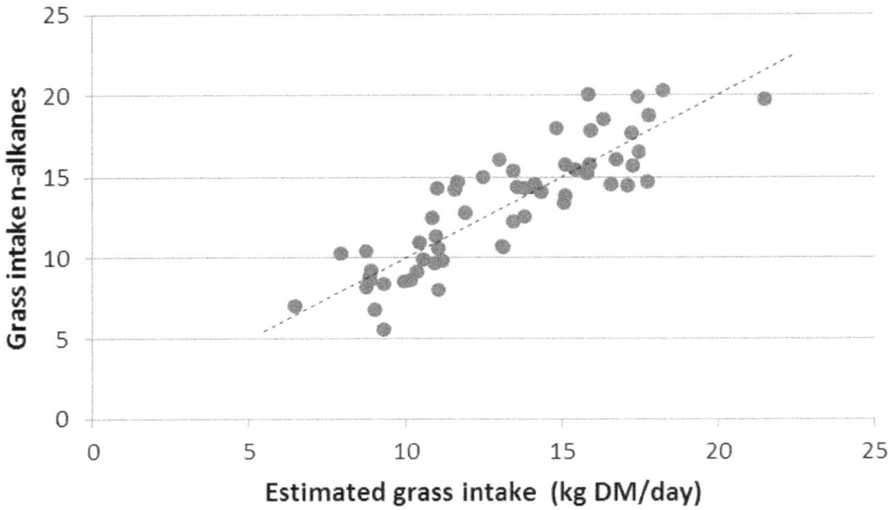

Figure 2. Relationship between grass intake determined with the n-alkane method and grass intake estimated with the use of regression Model 2.

Model 1 had the highest adj R square, and required milk constituents as input. The adj R square of Model 2 was slightly lower, but might be useful in situations when no information about the composition of the milk is available.

4. Conclusions

These findings demonstrate that grass intake of individual cows on pasture can be reliably estimated from a number of behavioural and other predictors. Different prediction models may be selected in practice based on the availability of predictor data. These models may represent potential management tools in intensive grazing systems for dairy cows, and may help to promote grazing in practice.

Acknowledgement

This research is part of the project 'Amazing Grazing', which is commissioned by 'Duurzame Zuivelketen' (Sustainable Dairy Chain), and funded by ZuivelNL and the Dutch Ministry of Economic Affairs. NEDAP is gratefully acknowledged for providing 'Smarttag Neck' sensors.

References

Centraal Bureau voor de Statistiek (CBS) (2015). Weidegang van melkvee. Available at: http://statline.cbs.nl/Statweb/.

De Vries, M., Bokkers, E.A.M., Van Reenen, C.G., Engel, B., Van Schaik, G., Dijkstra, T. and De Boer, I.J.M. (2015). Housing and management factors associated with indicators of dairy cattle welfare. Preventive Veterinary Medicine 118: 80-92.

Finneran, E., Crosson, P., O'Kiely, P., Shalloo, L., Foristal, D. and Wallace, M. (2010). Simulation modelling of the cost of producing and utilising feeds for ruminants on Irish farms. Journal of Farm Management 14: 95-116.

Mayes, R.W., Lamb, C.S. and Colgrove, P.M. (1986). The use of dosed and herbage N-Alkanes as markers for the determination of herbage intake. Journal of Agricultural Science 107: 161-170.
Munksgaard, L., Van Reenen, C.G. and Boyce, R. (2006). Automatic monitoring of lying, standing and walking behavior in dairy cattle. Journal of Animal Science 84: 304.
Olmos, G., Boyle, L., Hanlon, A., Patton, J., Murphy, J.J. and Mee, J.F. (2009). Hoof disorders, locomotion ability and lying times of cubicle-housed compared to pasture-based dairy cows. Livestock Science 125: 199-207.
Van den Pol-van Dasselaar, A., De Haan, M., Evers, A. and Philipsen, A.P. (2010). Simulation of the effect of grass intake on the farmer's income. Grassland Science in Europe 15: 100-102.
Van Reenen, C.G., Fiechter, T. and Van der Werf, J.T.N. (2016). Validation of the NEDAP Smarttag Neck for measuring eating behaviour in dairy cows. In: Proceedings of the International Precision Dairy Farming Conference. June 21-23, 2016. Leeuwarden, the Netherlands.

Recording grazing time of dairy cows in AMS farms with the Lifecorder+® sensor

C. Allain[1*], S. Danilo[2], J. Raynal[1], C. Beck[3], R. Delagarde[4] and V. Brocard[1]
[1]*Institut de l'Elevage, Le Rheu, France;* [2]*Institut de l'Elevage, Angers, France;* [3]*Department of Farm Systems, Wageningen UR Livestock Research, Lelystad, the Netherlands;* [4]*Institut National de la Recherche Agronomique, UMR1348 Pegase, Saint-Gilles, France; clement.allain@idele.fr*

Abstract

The Lifecorder+® is a uniaxial neck mounted activity-meter. It was first tested to assess grazing time in two French experimental automatic milking system (AMS) farms (20 cows equipped in Derval farm, 14 cows equipped in Trévarez farm) in 2014. The Lifecorder+ raw signal (from 0 to 9) was converted into a grazing yes/no information over a certain threshold. The results show a high correlation of grazing time between the visual observations of activity and the information from the sensor (R2=0.93 in Derval and 0.82 in Trévarez) with a mean prediction error of 18 min (9%) in Derval and of 29 min (20%) in Trévarez. Lifecorder+® appears to be a possible cheap, easy and precise tool to record grazing time at pasture. Then, experiments were conducted during spring 2015 on those two experimental AMS farms to assess grazing behaviours in two different grazing systems. In a system with feed supplementation (Derval), daily intake time decreases together with the decline of grass quantity and quality and an increase in the amount of supplementation. Grazing time is about 43% of the access time to pasture (319/775 minutes respectively). In a 100% grazing system (Trévarez), where daily pasture allowance is constant, daily grazing time is about 9 hours. It is longer when grass quality improves. The cows' hierarchy of access to the AMS is a predominant feeding behaviour factor. 'Dominant' animals have longer grazing time and higher animal performances (milk production, milking frequency) compared to 'dominated' animals. The use of grazing time data could be of interest in addition to other decision support tools already used by farmers (pasture probe, grazing calendars) to better combine AMS and grazing.
Keywords: grazing time, accelerometer, Lifecorder+

1. Introduction

The increasing number of automatic milking systems (AMS) in France and in Europe has resulted in a reduction of grazed grass in the diet of dairy cows. To improve the competitiveness of the European dairy farming by combining grazing and AMS, the European project Autograssmilk (www.autograssmilk.eu) aimed to enhance the use of new technologies to succeed in this association. Recently, Ueda *et al.* (2011) and Delagarde and Lamberton (2015) showed that a human activity-meter named Lifecorder+® could be used to assess cows grazing time.
Therefore, it was decided to check the possibility to use this sensor to record grazing behaviour in two systems combining AMS and grazing, one with supplementation and the other one in full grazing.

2. Materials and methods

2.1 Sensor evaluation
The Kenz Lifecorder+® (LC+; Suzuken Co. Ltd., Nagoya, Japan), a device for monitoring uniaxial acceleration, has recently been developed as a commercially available tool for the management of and research on human health.

The LC+ not only provides a step count per minute and estimates energy expenditure, but also records the intensity of physical activity at 4-s intervals. The raw data are summarized into 2-min average activity levels ranging from 0 to 9.

To assess grazing time on dairy cows, the sensors were mounted on neck collars on cows from 2 AMS experimental farms (20 cows equipped in Derval farm, 14 cows equipped in Trévarez farm) at spring 2014. The data from the sensors were then converted into grazing time by an Excel tool when the activity level was exceeding a certain threshold (configurable – different threshold were tested in Derval farm). Intra-meals intervals (≤4 min) are included in grazing time and inter-meals parasite activities (≤4 min) are excluded from grazing time (Figure 1; Rook and Huckle, 1995). The sensor data when the cows are in the barns are also excluded. After treatment, the data from the sensors were compared with visual observations as reference: trained observers recorded activities with a scanning every 10 min in the pastures. The recorded activities were the following: grazing / ruminating and standing / lying / walking. 1 observation session was performed in Derval and 12 observation sessions were performed in Trévarez on 7 days. Finally, 20 recordings were available for Derval farm (121 h of cumulated observation time in pasture) and 91 for Trévarez farm (336 h of cumulated observation time in pasture).

The accuracy of the LC+ device was studied by calculating the coefficient of determination of the regression between observed and predicted (LC+) grazing time and the mean prediction error (MPE) which is the square root of the mean squared prediction error (MSPE).

2.2 Grazing behaviour assessment

To assess grazing behaviour on dairy cows, the sensors were mounted during 2 months on neck collars on cows from 2 AMS experimental farms in France (23 out of 67 cows equipped in Derval farm, 14 out of 46 cows equipped in Trévarez farm).

In Derval farm, the grazing system is simplified (3 paddocks of 8, 10 and 10 ha) with buffer feed supplementation. In Trévarez farm, a rotational 100% grazing system with 3 paddocks per day

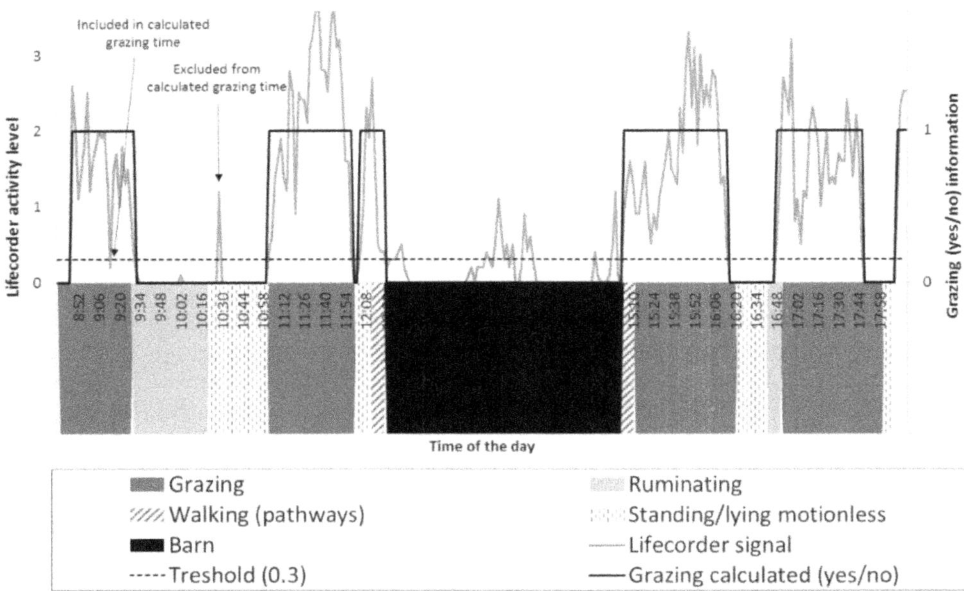

Figure 1. Example of conversion of the LC+ signal (grey line) into grazing information (black line) and comparison to observed behaviours (colored stripes) for cow 3439 in Derval farm.

is applied. The grazing organisations on the 2 farms are presented in Figures 2 and 3. The cows equipped with the LC+ were chosen to be representative of the herd in terms of production level, parity and lactation stage.

The data from the sensors were converted into grazing time by the previously mentioned Excel tool when the activity level was exceeding a certain threshold. In Derval farm, an electronic identification system at the entrance and the exit of the barn was used to calculate the time spent by cows in the pastures.

Finally, 1,323 recordings were available for Derval and 518 for Trévarez. A typology of animals according to their grazing behavior was achieved through the SPAD 8.2.10® software. A principal components analysis (PCA) was used to calculate the correlations between quantitative variables and similarities between individuals. Hierarchical ascending classifications (HAC) followed by descriptions of the classes were achieved to make groups of individuals with common features. Animal characteristics of the different classes of cows were tested by an ANOVA (proc GLM) with the SAS 9.4® software. The values presented in the Table 1 are adjusted averages.

3. Results and discussion

3.1 Sensor evaluation

The results of the comparison between measured and observed grazing time are presented in Table 1. The average grazing times observed were 196 min in Derval farm (65 h of cumulated time) and 147 min in Trévarez farm (222 h of cumulated time).

The results show a high correlation of grazing time between the visual observations of grazing activity and the information from the sensor. In Derval farm, the best correlation was found when the activity level of 0.3 was used as threshold (R^2=0.93). In this case, the average bias was 3 min (1.5% of the observed grazing time) and the mean prediction error (MPE) was 18 min (9%). In Trévarez

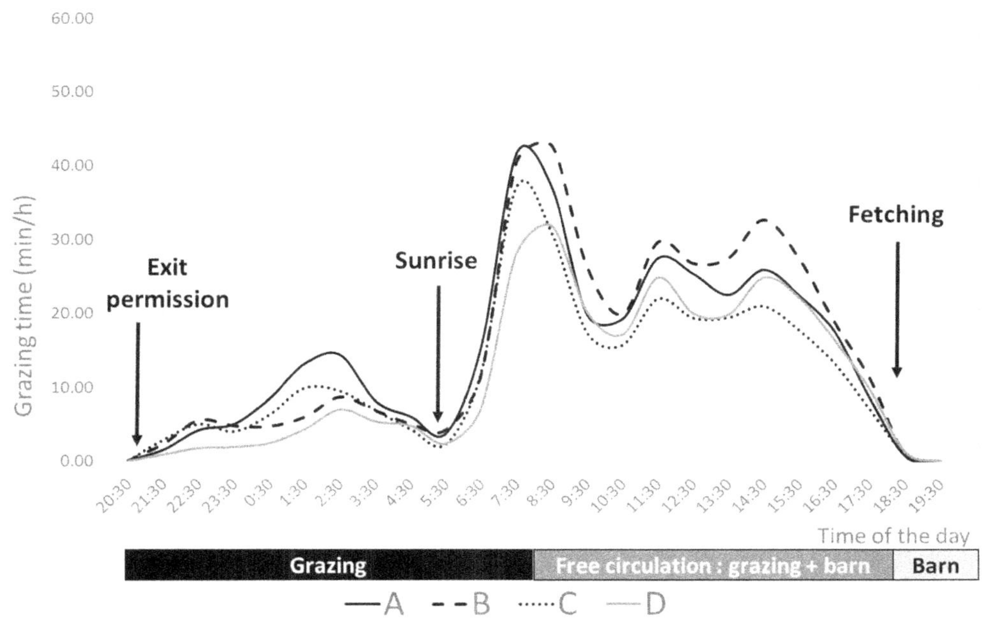

Figure 2. Grazing organisation and grazing behaviours of the cow groups in Derval.

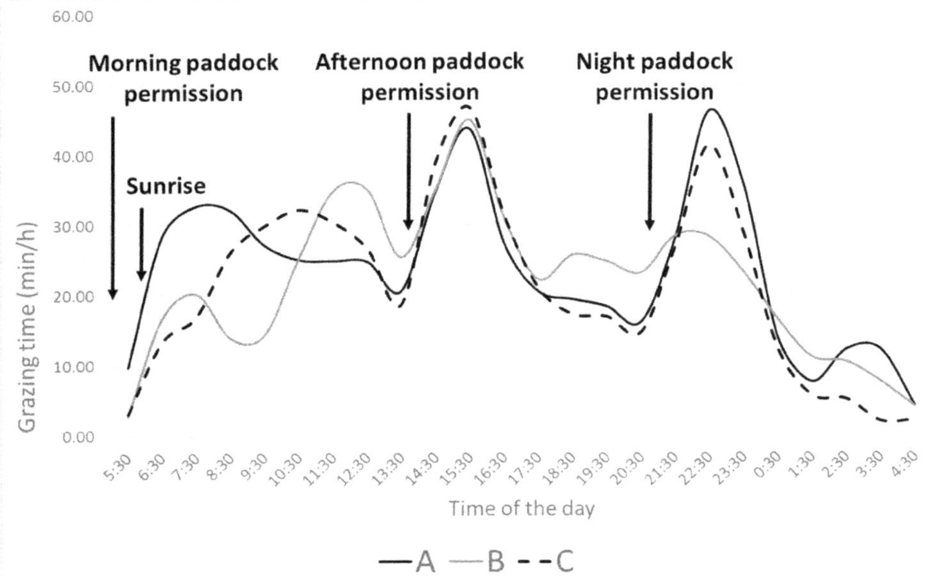

Figure 3. Grazing organisation and grazing behaviours of the cow groups in Trévarez.

Table 1. Results of the comparison between observed and measured grazing time.[1]

Farm	n	Threshold	Obs. grazing time (min)	LC+ grazing time (min)	Ave. bias (min)	R^2	MPE	
							min	% obs
Derval	20	0.3	196	199	3	0.93	18	9
		0.5		188	-8	0.84	27	14
		0.7		177	-19	0.75	37	19
		1		162	-34	0.69	49	25
Trévarez	91	0.3	147	153	6	0.82	29	20
All data	111	0.3	155	161	5	0.84	27	17

[1] Average bias = observed – LC+ grazing time; R^2 = coefficient of determination of the regression; MPE = mean prediction error in min and % of the observed grazing time.

farm, with a threshold of 0.3, the R^2 of the correlation was 0.82, the average bias was 6 min (4.1%) and the MPE was 29 min (20%). The positive biases were mostly related to walking in the pathways that sometimes generated a signal on the LC+ sensors (Figure 1). In Derval farm, the correlations were lower and the MPE higher with higher thresholds (0.5, 0.7 or 1).

These results confirm, with a lower accuracy, the good results obtained by Delagarde and Lamberton (2015) with the same sensor but with a threshold of 0.5. In a previous study, Ueda *et al.* (2011) observed that the best results were obtained with a threshold of 1. The differences between these studies concerning the best threshold can probably be explained by environmental effects and especially the position of the neck collar. This involves that the use of the LC+ sensor to record grazing time needs an adjustment to each farm situation.

3.2 Grazing behaviour assessment

Table 2 shows the descriptive statistics in the 2 experimental farms. In a full grazing system (Trévarez farm), cows spent 533 minutes grazing per day in average. In a system where they are supplemented with maize silage (Derval Farm), the pasture access is limited and they spent only 319 minutes grazing. The difference between the 2 systems was also observed for the number of meals (6.8 vs 5.3) and the meals duration (84 vs 64 min). The ratio between the grazing time and the access time to pasture was quite similar in the two systems (37% vs 43%) and close from the one observed by Kaufman *et al.* (2009). In both farms, the results showed a great individual and temporal variability among the data. In a simplified rotational grazing system (Derval), grazing time decreased when there was a decline of grass quantity and quality and an increase in the amount of maize silage delivered. In a rotational grazing system (Trévarez), where pasture allowance remains at the same level every day, grazing time is longer when grass quality is better.

Table 2. Descriptive statistics and groups' comparison in Derval and Trévarez farms.

Derval farm (n=1,323)[1]						
Groups	A	B	C	D	Herd ave. (SD)	*P*[2]
# of animals	6	4	8	5	23	
Grazing time (min)	360 [a]	365 [a]	294 [ab]	274[b]	320 (102)	<0.001
# of meals	5.6 [a]	5.7 [a]	5.2 [b]	4.8 [b]	5.3 (1.8)	0.0064
Meals duration (min)	69 [a]	69 [a]	60 [b]	59 [b]	64 (23)	0.0003
Access time (min)	817 [a]	776 [a]	801 [a]	686 [b]	775 (237)	0.001
Grazing time/access time (%)	44 [a]	47 [a]	37 [b]	40 [b]	43 (14)	<0.001
Exit time[3]	0 h 45 [a]	2 h 00 [b]	0 h 50 [a]	3 h 00 [b]	1 h 20	<0.001
Yield (kg/cow/day)	30.0 [a]	21.2 [b]	29.4 [ab]	24.8 [ab]	27.1	0.047
Milking frequency (milk./day/cow)	2.03	1.84	1.96	1.77	1.92	0.16

Trévarez farm (n=518)					
Groups	A	B	C	Herd ave. (SD)	*P*
# of animals	4	5	5	14	
Grazing time (min)	565 [a]	525 [b]	508 [b]	530 (79)	<0.001
# of meals	7.2	6.9	6.3	6.8 (1.8)	0.16
Meals duration (min)	82	81	85	84 (22)	0.64
Grazing time/access time (%)[4]	39 [a]	36 [b]	35 [b]	37 (6)	0.0004
Access time 1st paddock[5]	8 h 00 [a]	9 h 45 [b]	8 h 20 [a]	8 h 45	0.018
Yield (kg/cow/day)	21.1 [a]	13.9 [b]	18.9 [a]	17.7	0.045
Milking frequency (milk./day/cow)	2.01 [a]	1.57 [b]	1.77 [ab]	1.77	0.023

[1] a,b,c = Tukey test, significantly different at threshold of 5%.

[2] Fischer test.

[3] In Derval, cows are allowed to graze at 21:00 h, after they have been milked by the AMS. This parameter gives an idea of the cow dominance.

[4] In Trévarez, animals are in full grazing so AT=1,440 min.

[5] There are 3 paddock changes per day. The access time to the 1st one gives an idea of the cow dominance.

Results of the CPA and the HAC are also presented in Table 2 with the comparison between the different groups of animals. In both farms, we observed that feeding behaviours enabled to bring out the place of animals in the herd social order. 'Dominant' animals, which have the first access to the AMS and therefore to the pastures, have longer grazing time and higher performances (groups A and C in Derval and group A in Trévarez) compared to 'dominated' animals (groups D in Derval and B in Trévarez). The other groups (B and D in Derval and C in Trévarez) are intermediary. Figure 2 and 3 show the grazing kinetics of the different groups. In Derval, we can see that the 'dominant' animals have a significant meal at night and a consequent one when the sun is rising up. The 'dominated' cows are penalized by their late access to pasture and are not able to compensate at the sunrise and in the afternoon. In Trévarez, the graph in Figure 3 shows that the 'dominated' animals have a late access to the first paddock and often do not go to the 3rd one. Consequently they have to graze the left-overs of the 2nd paddock.

4. Conclusions

The use of grazing time data could be of interest in addition to other decision support tools already used by farmers (pasture probe, grazing calendars) to better combine AMS and grazing. In order to maximize grass ingestion, several recommendations can be provided. First of all, forage supplementation must be calibrated to the grazing allowance and not the opposite. Then, the access to AMS, therefore to pasture, could be regulated according to the animals' social hierarchy in order to give an advantage to 'dominated' cows.

References

Delagarde, R. and Lamberton, P. (2015). Daily grazing time of dairy cows is recorded accurately using the Lifecorder Plus device. Applied Animal Behaviour Science 165: 25-32.

Kaufman, L.D., Muenger, A., Rerat, M., Junghans, P., Goers, S., Metges, C.C. and Dohme, F. (2009). Dépense énergétique, activité physique et comportement alimentaire des vaches laitières au pâturage ou consommant la même herbe à la crèche. Rencontres Recherche Ruminants 16: 33-36.

Rook, A.J. and Huckle, C.A. (1995). Criteria for identifying meals from continuous recordings of the behaviour of grazing dairy cows. Annales de Zootechnie 44: 105-105.

Ueda, Y., Akiyama, F., Asakuma, S. and Watanabe, N. (2011). The use of a physical activity monitor to estimate the eating time of cows in pasture. Journal of Dairy Science 94: 3498-3503.

Grazing management and welfare: effects of a virtual fence on dairy cow behavior

P. Koene*, P. Hogewerf and B. Ipema
Department of Animal Welfare, Wageningen Livestock Research, De Elst 1, 6708 WD, Wageningen, the Netherlands; paul.koene@wur.nl

Abstract

Dairy cows may graze more efficiently when allowed grazing in predefined areas with predictable sward height and grass quality. Moving the animals daily or preferably more frequent is needed. Using traditional fencing so-called strip grazing causes high labor requirements. Also other unwanted effects of electric fences might be prevented by the use of virtual fencing. We used a non-GPS system with underground wires producing a boundary signal that could be partly and differentially activated by remote control. The cows had a collar in which the boundary signal is received and from which a warning signal was given to the cow when approaching the boundary and a correction signal within 0.5 meter from the boundary. Questions are whether the virtual fence works as expected and what the impact of the signals is on the behavior and welfare of cows. Dairy cows – grazing in a group of four – in virtual fenced area were compared with cows in the electric fenced area. This was replicated once with two other groups (n=16). During 12 days the groups were compared in several aspects. One of the aspects was grazing on a single strip (60×10 m) compared to a double strip (60×20 meter). The behaviour, position and activity were monitored during 5 hours per day. Cows learned the virtual fence system in the same pace as the electric fence. During the 12 day period all cows stayed in the strips with only few temporary boundary crossings. The cows in the electric fence system came closer to the visible fence while the virtual fence cows kept more distance to the virtual fence. The activity of the cows was significantly influenced by the strip size. In a 1-strip area the virtual fence cows showed more standing and less lying. In a 2-strip area no difference was found for standing and lying. Our preliminary conclusion is that a virtual fence is effective and has some effects on the behavior different from an electric fence. More detailed analysis will show whether the behavior differences imply welfare differences between the groups.

1. Introduction

In this project, the question is examined whether it is possible to train cattle using pre-defined signals that indicate boundaries of – for example, pastures – instead of physical and visible fences. Such a 'virtual fence' could make welfare-friendly and grassland-use-efficient grazing possible and are still in development (Umstatter, 2011). There are systems developed and made available in the market to keep cattle with signals within a fixed predefined area (Monod *et al.*, 2009; Umstatter *et al.*, 2015). In this pilot the focus is on strip-grazing. Each cow gets a collar with a box and sensor to determine the position of the animal with respect to an underground wire (a 'virtual fence') in the pasture and can, moreover, produce warning or correction signals when approaching the ring wire. The warning signal (sound) is intended to alert the cows on the approaching a virtual boundary; this signal will have little impact on welfare. The correction signal (an electric signal) is only given if the animal persists in further approaching the virtual fence. This may have a greater effect on well-being of the cow. By learning the animals to associate the warning signal with the correction signal it is expected that the corrective signal is not given often. The feasibility of the development of a method which is based on an area that can be adjusted over time (instead of a fixed area) is estimated to be high and will facilitate the development of efficient strip-grazing methods. This pilot experiment will be reviewed as being successful when the main part of the animals learn the association between the warning signal and location or behavior and when the experimental -virtual fence - group in terms of well-being scored not significantly lower than the control - electric fence - group.

2. Materials and methods

2.1 Experimental setup

Animals were heifers without grazing experience and had always been stabled. They had to learn to graze and learned to deal with a fence (barbed wire) in a separate grazing area. The experiment is done with small groups of animals (Noldus, 1991) that are allowed to graze on strips of grass which during the test are shifted every day. We used the Boviguard system, described in detail elsewhere (Monod *et al.*, 2009). In short, when approaching a virtual boundary, the animals receive a first warning signal. When they do not stop or return the warning signal is followed by a correcting signal. The research focuses primarily on whether and how fast a cow learns to avoid the correcting signal in small areas of pasture. The animals learn in trials and get in the course of the trials more freedom of movement. This experiment will be conducted in 2 replications with an experimental group and a control group (a total of 4 groups of 4). Groups differ in the manner in which the border of their grazing area is indicated, either by virtual fence (experimental group with warning and correction signal via the collar), or by the traditional electric fence (control group with exposed wires and correction in contact with the wire). Both groups are treated in the same way and the same observations took place in both groups. The experiment is carried out in four phases, which are identical in setup. Every day 2 strips are made available (each 10×60 meter) - an old (grazed the day before) and a new (with fresh grass were available) one. In the first phase of 3 days (three repeats to allow trial learning) an electric wire was placed above the underground hidden virtual fence. In the second phase of three days the electric fence was placed 2 meter past the virtual fence. After these 2 training phases the final test was done, i.e. the electric fence was removed and in fact placed at the actual boundary of the field. In the control group the conventional electric fence, of course, is always on the borders of the area, i.e. the border of the two strips. In the phases 1-3 the animals are from 9:00 am until 15:00 in the test strips. In the remainder of the time the animals are as a group (experimental and control group) kept in separate pastures with fixed fence (not virtual or electric fence) under conventional husbandry conditions. After each trial day after 15:00 the conditions for the next day are prepared. Before 9:00 on test days the animals are prepared to enter the test area. GPS sensors are placed in the collar and observers are preparing the behavioural observations. Observers are present during the test days in all phases during the period of 9:00-15:00. The observers taske care of the welfare of the animals and will intervene when an animal is showing obvious signs of stress. Per animal responses will be recorded and correction signals - if possible - are registered. An animal that continuously ignores signals, and wherein no learning effect can be observed will be removed from the test, if more than light distress is detected. In this paper we focus mainly on the following 4th phase. This phase was somewhat different from the first 3 in that only one strip was available on day 1, two strips on day 2 and one strip on day 3 between 9:00 hrs. and 15:00 hrs. Furthermore, the animals stayed on the strip for 24 hours. However, we report the observations and GPS recordings of the animals between 15:00 and 9:00 hrs.

2.2 Measurements and statistics

Animals are monitored individually and as a group. Daily in each group each animal is observed with continuous protocols (5 protocols with a duration of 10 min per animal) and the group is scanned behaviorally (25 times per day). The ethogram included the behaviors walking, lying, standing, grazing, social positive, social negative behaviors and the reactions of the animal on the signals recorded at the borders of the grazing area using the Observer (Noldus, 1991). The experimental area is divided in several subareas (Figure 1) and also the locations are recorded by the observers. The locations and movements of the cows (x, y co-ordinates) are recorded by GPS (Extreme 66

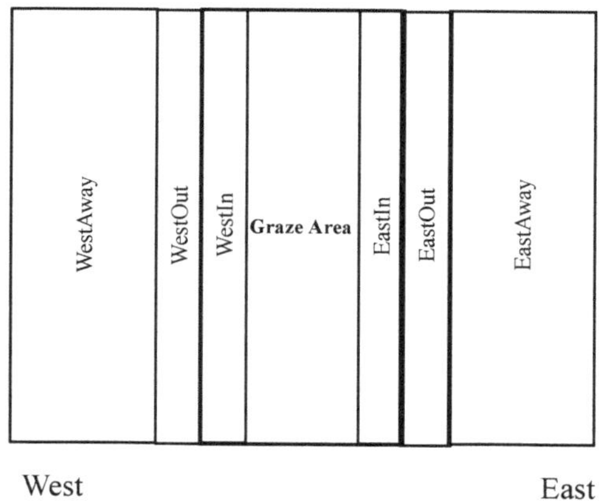

Figure 1. The black line indicates the fenced area (20×60 meter). The presence near the fence (WestIn, EastIn), over the fence (WestOut, EastOut) or escaped (WestAway, EastAway) in measured in the observations and by GPS.

GPS travel recorder attached to the collar of the animal). This GPS system gives no inconvenience for the animal.

Primary outcome parameters from the observations (visual observations and automated) are derived per cow per day. The behavioral observations provide insight into the nature, intensity and duration of the responses to the signals. Performance of the experimental group (virtual fence) is compared with that of the control group (electric fence) using Linear Mixed Models (LMM) and repeated measures GLM of SPSS (IBM, 2016). In this paper special attention is given to individual aspects of the cows behavior, while in a twin paper the focus is on social aspects of behavior. Analyzed are aspects of the use of the strips and several individual behaviors, i.e. grazing, ruminating, standing and lying.

3. Results

The cows in the control group (electric fence) did not escape, as expected. In total, five escapes between 9:00 and 15:00 were recorded in the experimental group during the 12 days study. The escapes were on day 1 (2×), day 10 and day 12 (2×). After the escapes the cows were returned to the graze area and stayed there. The escapes on day 1 and day 12 were both the same cows, suggesting individual differences and social influences in escape behavior.

The grazing activity was significantly influenced by the condition (P=0.034); the electric fence group spent more time grazing than the virtual fence group (62% vs 53%). The grazing activity of the cows was significantly influenced by the strip size (P=0.013); in the two strip area 63% of the time grazing was found vs 52% in the one strip area. Ruminating was not influenced by both factors. Based on a uniform distribution in the field the expectation is that the cows are on average 20% of the time in the border area (WestIn + EastIn of the total area; 2×60+2×60/20×60), i.e. 10% in WestIn and 10% in EastIn. In fact electric, fenced cows are less than expected in the border area (Figure 2). Moreover, the cows in the virtual fenced area are significant less in the border area (LMM; P<0.001). Probably the cows in the electric fence system came closer to the visible fence while the virtual fence cows kept more distance to the virtual fence (LMM; P<0.01).

In the one strip area the virtual fence cows showed more standing idle (rmGLM; P<0.001) and less lying (rmGLM; P<0.01). In the 2-strip area no difference was found for standing and lying (Figure 3).

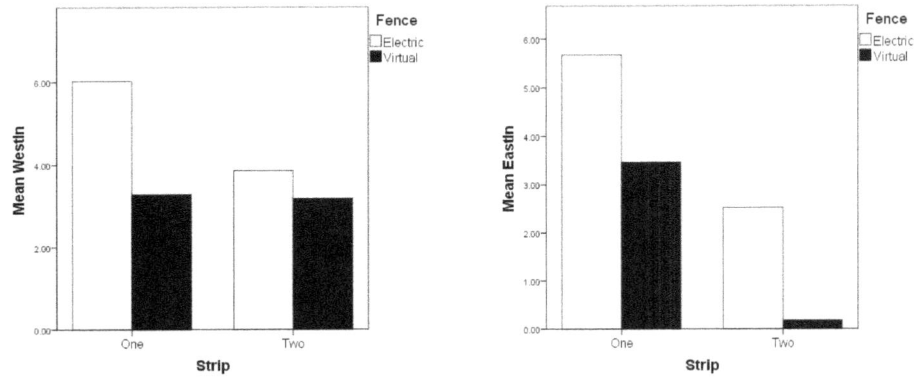

Figure 2. In the left panel the average percentage time spent in the WestIn area (Figure 1); in the right panel the time in the EastIn area.

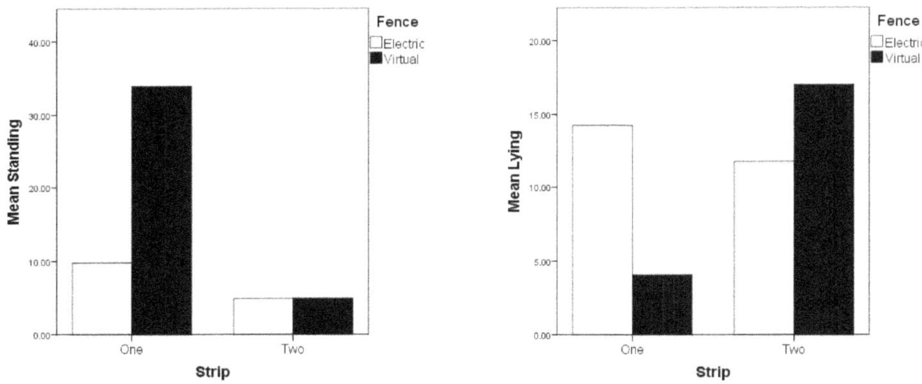

Figure 3. Standing idle (as ruminating, grazing or being alert) sometimes with head low (left panel). Lying idle is shown in the right panel. More standing and less lying behavior may indicate restlessness in cows.

4. Discussion

The escapes on day 1 were in the learning phase of the virtual fence; the other escapes were during the days that only one strip was available meaning a limited area was available and a relatively small graze area (60%) and large border area (40%) was present in the virtual fence group. The cows remained mostly within the one strip (with virtual fence), although a total of three escapes were recorded. Despite the pressure of grazing on one strip the virtual fence group spend less time grazing in the border area and avoided the border area more than the electric fence groups (Figure 2). This suggested that the cows avoided the border are, the location more than the audio signal. Although precise recording of the signals given were missing. The time spend grazing was affected in the virtual fence groups on one strip, because the safe graze area became quite small. The cows showed more standing idle and less lying idle and both behaviors may be indicative of restlessness in the virtual fence group. This result implies that there is minimum strip size needed for the virtual fence groups that is larger than the minimum strip size for the electric fence groups. Especially the behavior standing idle may be a welfare indicator of grazing cows.

5. Conclusions

The virtual fence may become a tool for precise allocation of grazing to cows. It is important that the minimum strip size is corrected for the virtual border making this minimum size larger than needed for grazing in an electric fenced area. More experiments are planned in 2016 to enhance the understanding of the cows learning process in relation to the virtual fence. Especially the effects of group size and social attraction on the effectiveness of the virtual fence and the number of escapes will be part of the experiments.

Acknowledgements

The authors wish to acknowledge the Amazing Grazing programme, Courage, the Netherlands, for funding and visiting students and their supervisors of HAS Den Bosch, the Netherlands for support during the observations.

References

Monod, M.O., Faure, P., Moiroux, L. and Rameau, P. (2009). Stakeless fencing for mountain pastures. In: Precision lifestock farming '09'. Wageningen Academic Publisers, Wageningen, the Netherlands, pp. 175-181.

Noldus, L.P.J.J. (1991). The observer: a software system for collection and analysis of observational data. Behavior Research Methods, Instruments and Computers 23: 415-429.

Umstatter, C. (2011). The evolution of virtual fences: a review. Computers and Electronics in Agriculture 75: 10-22.

Umstatter, C., Morgan-Davies, J. and Waterhouse, T. (2015). Cattle responses to a type of virtual fence. Rangeland Ecology and Management 68: 100-107.

Part 7. Socio-economic impact

Farmers' preference towards heat detection sensor technology

S. Van Weyenberg[1*], L.M. Pluym[1], K. Piccart[1], A. Van Nuffel[1], B. Aernouts[2], T. Huybrechts[2], E. De Busser[2], K. Mondelaers[3] and W. Saeys[2]
[1]*Institute for Agricultural and Fisheries Research – ILVO, Technology and Food Sciences Unit, Burgemeester van Gansberghelaan 115, 9820 Merelbeke, Belgium;* [2]*KU Leuven Department of Biosystems, MeBioS, Kasteelpark Arenberg 30 Box 2456, 3001 Leuven, Belgium;* [3]*Institute for Agricultural and Fisheries Research – ILVO, Social Sciences Unit, Burgemeester van Gansberghelaan 115, 9820 Merelbeke, Belgium; stephanie.vanweyenberg@ilvo.vlaanderen.be*

Abstract

A wide range of sensors is available for dairy cattle. Yet, only a limited amount are efficiently used in practice. Not including farmers' preference towards sensor is one of the factors that impede the use at dairy farms. This study aims to reveal Flemish farmers' preferences for heat detection sensors and to estimate how sensor performance and farmer characteristics can influence this preference. First, general information about the participating farmers and their farms was collected (age, herd size, milking system, milk production, reproduction parameters and if they already use a detections system on their farm). Further, a Discrete Choice Experiment was added to assess farmers' preference, presented by the utility. Each farmer had to choose between 2 types of heat detection sensors (activity sensor or milk-progesterone-sensor, with varying performance characteristics) or to choose visual heat detection (no sensor). During statistical analysis, these choices are related to the price, the detection rate and the amount of false positive alarms, and to farmer characteristics. Most farmers were between 40 and 49 year old. The average herd size was 95 cows, with a mean milk production of 8,878 kg/cow/year. A detection system was present in 40% of the farms. The utility of the sensors relative to the utility of visual detection is 1.05 and 0.42 for activity and progesterone sensors respectively. For both sensors, utility increases with increasing % detection, decreasing % false alarm and decreasing price. Younger farmers, who have bigger farms, with an interval calving-first insemination <100 days, who already had a detection system, allocate higher utility to the proposed sensors. These results can be used when developing new heat detection sensors. Also targeted communication campaigns can be set up.

Keywords: technology adoption, activity sensor, progesterone, farmer involvement

1. Introduction

A wide range of sensors and technology is available for dairy cattle. Yet, only a limited amount of developments become real Precision Livestock Farming (PLF) applications in practice. A Flemish information platform (www.koesensor.be) focusses on the spread of information concerning the use of sensors and technology to dairy farmers. It purport to close the gap between technology developers and dairy farmers.

One of the most widespread PLF technology at dairy farms is the activity sensor. Changes in individual cow movement are recorded as indirect indication that the cow is in heat. Also more direct automatic heat detection sensors that measure the progesterone level in milk, are commercially available (HerdNavigator©, Delaval) or being developed in research (Mariën *et al.*, 2015). Nevertheless, to date most farmers perform heat detection by visual cow monitoring.

Excluding farmers' preference towards sensor at the development phase is mentioned as one of the factors that impede the use of PLF systems in dairy farms (Van Weyenberg *et al.*, 2014). Conjoint analysis reveals that Dutch farmers want, at least for a mastitis sensor, a high level of specificity (low amounts of false positive cases) and a small time window (no more than 24 h between alarm

and the onset of the disease, so no early detection). No information of farmers' preference for heat detection sensors is available.
The aim of this study is to assess farmers' preference for sensor technology to perform heat detection and to estimate how this preference is influenced by sensor performance characteristics and farmer characteristics.

2. Materials and methods

A Web-based questionnaire was send out by newsletters of MCC Vlaanderen (Milk Control Center Flanders) and of the project 'Koesensor'. First, some general information about the participating farmers and their farms was collected through a standard questionnaire. Farmers were asked for their age, herd size, milking system, milk production (kg/cow/year), reproduction parameters and if they already use a detections system on their farm.
The second part of the survey aimed to assess the farmers' preference for performance characteristics, presented by the utility, by Discrete Choice Analyses (DCA). DCA is a survey technique used for marketing research that aims to quantify people's preferences. First, 20 different choice sets were generated and randomly divided over four blocks using Ngene (ChoiceMetrix, Australia). Further, each farmer was randomly provided with one block of five different choice sets (Limesurvey, Schmitz, 2015). Each set consists of two sensors (one activity sensor and one milk-progesterone-sensor), with varying performance characteristics (called attributes in DCA). Sensor performance characteristics can take different levels: 70, 80, 90 or 99% for detection rate, 1, 5, 10 or 15% for amount of false positive alarms and 15, 55, 95 or €135 for price. Besides the two types of sensors, each choice set also provide farmers an opt-out (the option to perform heat detection by visual detection instead of using one of the two proposed sensors) (Table 1).
During statistical analysis, the choice made by the farmer within each choice set is related to the presented level of price, detection rate and amount of false positive alarms and to farmer characteristics using Nlogit (Econometric software, Inc., USA). For each sensor type within each choice set, so with its specific levels of sensor characteristics, a utility value relative to the utility of visual heat detection was determined (called marginal utility). The mean marginal utility of each type of sensor was calculated as the average of all sensor specific marginal utility in all choice sets of all farmers.

3. Results

The survey was completed by 135 farmers. Farmers' characteristics are given in Table 2, 3 and Figure 1. The attribute model (containing the two attribute specific constants and the three attributes) shows that utility increases with increasing % detection, decreasing % false alarm and decreasing price (Table 4).

Table 1. Example of one choice set.

	Activity sensor	Milk-progesterone-sensor
Cows in heat detected	90%	95%
False alarms	5%	1%
Price	€135	€135
You prefer:	○ Activity sensor ○ Milk-progesterone-sensor ○ Visual heat detection, without one of these sensors	

Table 2. Farmers' characteristics in terms of herd size and production level.

	Mean	Median	Minimum	Maximum
Amount of cows	93	75	15	560
Kg milk/cow/year	8,878	8,900	6,100	12,000

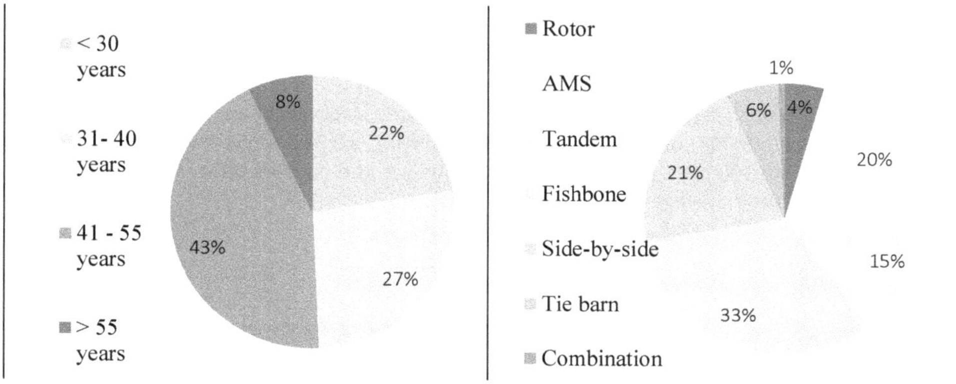

Figure 1. Age (left) and milking system (right) of the 135 responders.

Table 3. The number of responders within each category of the reproduction parameters.

Reproduction parameters		n
Calving interval	<400 days	58
	Between 400 and 420 days	60
	>420 days	16
Interval calving-1st insemination	<100 days	119
	>100 days	8
	I don't know	7
Number of insemination/conception	1.8 to 2	59
	2.2 to 2.2	44
	>2.2	31

Table 4. Regression results from the discrete choice experiment: the attribute model (model with the two alternative specific constants and the three attributes as continuous variables).

	Coefficients	Standard error	*P*-value
Constant Activity sensor	-2.95	0.56	/
Constant Milk-Progesterone-sensor	-3.54	0.57	/
Attributes			
Detected (%)	0.072	0.008	<0.001
False alarms (%)	-0.068	0.013	<0.001
Price	-0.021	0.002	<0.001

Mean marginal utility was higher for the activity sensor than for the milk-progesterone-sensor (Figure 2).
Marginal utility was also significantly influenced by farmers' characteristics (determined in the univariable model). Farmers who are 55 years or older have a lower marginal utility (β=-0.445±0.169, P=0.008), compared to younger farmers (β=0.445±0.169, P=0.008). Utility was not significantly influenced by herd size. Farms with an average production higher than 8,500 kg/cow/year have an additional utility of 0.191±0.107 (P=0.075) compared to farms with an average milk production lower than 8,500 kg/cow/year (β=-0.191±0.107, P=0.075). If the farmer is familiar with a heat detection system on his farm, the utility of both sensors increases with 0.689±0.124 (P<0.001), compared to farmers who don't have an automatic heat detection system (β=-0.689±0.124, P<0.001). Utility of the heat detection sensors compared to visual heat detection was not significantly influenced by milking system. From all reproduction parameters, only the interval calving-1st insemination had a significant effect (P=0.008) on utility (β=1.24±0.19, β=-0.17±0.34 and β=-1.07±0.28, for average intervals shorter than 100 days, average intervals >100 days and farmers indicating that they don't know the length of their average interval, respectively).

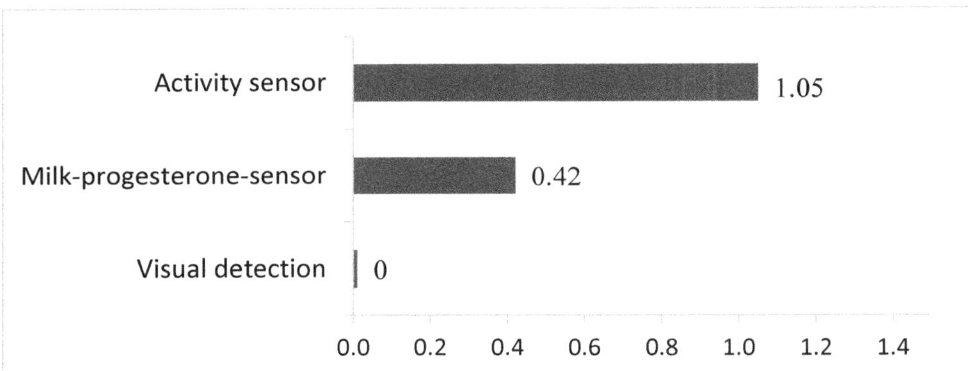

Figure 2. Marginal utility of the two heat detection sensors relative to 'visual heat detection without sensor' as reference, determined in the attribute model (model with the two alternative specific constants and the tree attributes as continuous variables).

4. Discussion

A response of 135 farmers is in line with other amount of responders in other questionnaires, but still might be seen as rather low, compared to the total amount of dairy farmers in Flanders (4,965 dairy farms). Therefore, the representation of the study might be questioned. However, most farm and farmer characteristics approximate Flemish averages. Only the average herd size and the percent of farmers with an automatic milking system (AMS) are higher than the average in Flanders (65 cows/farm on average and 8.8% AMS). No information about the use of heat detection systems in Flanders are available.
The influence of the attributes on the marginal utility was as expected: higher utility with higher performance (higher detection and lower amount of false alerts) and with lower price. Other questionnaires also revealed that farmer based their sensor investment decision on the benefit-to-cost ratio and the total investment cost (Borchers and Bewley, 2015). Although for mastitis detection, Mollenhorst *et al.* (2012) also found that farmers prefer systems with low amount of false alerts.

Technology preference and adoption increases with decreasing age both in our study and in literature (Russell and Bewley, 2013). Farmers with a larger amount of cows are more likely to adopt new technology (Jackson-Smith and Barham, 2000). No significant effect of herd size on utility was found in our study. But average herd size was much lower is our study. Russel and Bewley (2013) found that not being familiar with technologies was one of the most important reasons for not buying sensor technology. This was confirmed in our study: lower utility for both sensor types in case farmers don't have a heat detection system already.

5. Conclusions

DCA was proved to be a valuable technique to get insight into farmer preference towards heat detection systems in particular and PLF applications in general. Farmers allocate positive utility to heat detection sensors, compared to visual detection without sensor technology. The utility significantly increases with higher detection rates, lower false alerts and lower prices, but is also influenced by farmer characteristics. These results can be used when new heat detection sensors are being developed. Also targeted communication campaigns can be set up. More in depth analyses can give information on the relationship between the attributes in terms of willingness to pay and how changes in attributes levels can influence the market shares of commercial sensors and sensors under development.

References

Borchers, M.R. and Bewley, J.M. (2015). An assessment of producer precision dairy farming technology use, prepurchase considerations, and usefulness. Journal of Dairy Science 98: 4198-4205.

Jackson-Smith, D.B. and Barham, B. (2000). The changing face of Wisconsin dairy farms: a survey of PATS' research on structural change in the 1990s. Program on Agricultural Technology Studies, College of Agricultural and Life Sciences, University of Wisconsin, Madison, WI, USA.

Mariën, N., Daems, D., Delport, F., Spasic, D., Adriaens, I., Saeys, W. and Lammertyn, J. (2015). Fiber Optic SPR biosensor for progesterone detection in raw milk as a tool to improve reproduction. 2nd Dairycare Conference. March 3-4, 2015. Cordoba, Spain.

Mollenhorst, H., Rijkaart, L.J. and Hogeveen, H. (2012). Mastitis alert preferences of farmers milking with automatic milking systems. Journal of Dairy Science 95: 2523-2530.

Russell, R.A. and Bewley, J.M. (2013). Characterization of Kentucky dairy producer decision-making behavior. Journal of Dairy Science 96: 4751-4758.

Van Weyenberg, S., Pluym, L., Mertens, K., Van de Gucht, T., Vangeyte, J. and Van Nuffel, A. (2014). Research towards successful application of PLF in dairy farms. Proceedings of the 1st Dairycare Conference. August 22-23, 2014. Copenhagen, Denmark, pp. 28.

Understanding the roles of farm advisors in precision dairy farming

C.R. Eastwood[1*], S. Chaplin[2], B.T. De la Rue[1], N.A. Lyons[3] and D. Gray[4]
[1]*DairyNZ, Hamilton 3240, New Zealand;* [2]*Department of Economic Development, Jobs, Transport and Resources, 255 Ferguson Rd, Tatura, Victoria 3616, Australia;* [3]*Intensive Livestock Industries, NSW Department of Primary Industries, Elizabeth Macarthur Agricultural Institute, Menangle NSW 2568, Australia;* [4]*Massey University, Palmerston North, New Zealand; callum.eastwood@dairynz.co.nz*

Abstract

Farmer networks have been highlighted as important for the implementation of precision dairy systems however little is known of the roles farm advisors perform for precision dairy farmers. The aims of this research were to: understand how farm advisors interact with their clients' on-farm data and identify opportunities for developing farm advisor capability to improve adoption and effective use of precision technology. In-depth semi-structured interviews focused on grazing management technology, and an online survey related to animal monitoring technologies, were used to explore the perceptions of farm advisors to the use of precision farming data in New Zealand (interviews) and Australia (online survey). Farm advisors contacted in this study included farm consultants, nutritionists, and veterinarians. Key themes explored with participant advisors were the current use of precision dairy data, knowledge of different animal and grazing precision technologies, and preferred learning approaches. We found a variety of levels of interaction between advisors and such tools, and advisors indicated they had different levels of skill in accessing and using data. Resources are required to expand awareness of different data gathering technology and software, specifically targeted at farm advisors. More information is needed to highlight to farmers and advisors where the main areas of value/opportunity exist in use of these tools. Further research is needed to identify the appropriate processes for engaging with advisors to improve their knowledge and capability with using precision dairy data.

Keywords: farm advisors, capability, data, decision making, networks

1. Introduction

Precision dairy technology development and use on farm has recently been examined (Borchers and Bewley, 2015; Jago *et al.*, 2013). Farmer networks have been highlighted as important for the implementation of precision farming (Eastwood *et al.*, 2012) but little is known of the roles farm advisors perform for precision dairy farmers. In research that examined New Zealand precision dairy farmers, Eastwood and Yule (2015) identified a lack of capability among the dairy service sector in regard to use of data derived from precision technologies.
The aims of this research were to: understand how farm advisors interact with their clients' on-farm data and identify opportunities for developing farm advisor capability to improve adoption and effective use of precision technology. The research focused on the use of data and software associated with animal and pasture monitoring and management technologies.

2. Methods

The interaction of farm advisors with precision dairy technologies and related data was investigated via two case studies. The first case study involved interviews of New Zealand farm management consultants to understand their interaction with farmer clients in respect to precision grazing management approaches and use of decision support software (DSS). An in-depth semi-structured interview methodology was used, with interviews lasting up to two hours. Six consultants representing the major New Zealand farm consultancy firms were interviewed in January and February 2016. Interviews were recorded, transcribed and then analyzed using NVivo™ software.

The second case study focused on the interaction of farm service providers such as farm consultants, nutritionists, and veterinarians with precision dairy data and herd management systems (HMS). An online survey was designed in consultation with industry experts, and piloted with five service providers. It focused on animal monitoring technologies, and the questions explored the current use of precision dairy data, knowledge of different animal and grazing precision technologies, and preferred learning approaches. The survey was undertaken in late 2015 using SurveyMonkey™ via a link distributed through industry contacts and email lists. Data were collated in Excel and analyzed using means, sums, and percentages.

3. Results

3.1 New Zealand interviews

The New Zealand farm advisors interviewed ranged in experience from 15 to 34 years as consultants and interacted with up to 50 farmer clients each. They reported a range of farming philosophies from experiential management to more objective and data-driven approaches. Three of the advisors used sophisticated optimization modelling tools such as Farmax™ or Udder™ to determine strategic options for farmer clients. For this they had developed approaches to encourage their clients to capture regular per paddock pasture data (often through visual assessment or tools such as rising plate meters or a CDax™ pasture meter).

Grazing management in New Zealand's seasonal, pasture-grazed dairy systems has variable demands throughout the year as pasture growth and animal demand fluctuate. Advisors identified key periods, such as spring, where regular data was most needed, and where rapid decision making was required. Increased use of data and decision making tools would help many farmers be more proactive with their grazing decisions, according to advisors interviewed. One advisor noted that 'the difference between a good farmer and an average farmer is three days', identifying the value of timely decision making. Another advisor said that the better farmers were thinking 'three months ahead' and making decisions accordingly. For this level of management they needed regular data. However, it was also noted by an interviewee that 70% of good grazing management could be achieved with numbers and decision rules, while the final 30% still needed extra skills based on experience, intuition, and acceptance of risk.

Among their clients there was a wide variety of pasture data capture and use of DSS. According to those interviewed, larger farm businesses where farmers owned more than one property tended to use more grazing management tools, and corporate farms were the most engaged with pasture data capture and use of DSS tools. These corporate businesses were demanding more integrated grazing management systems that translated data capture at the farm level through to strategic management decision making at a company board level. This therefore requires capture of good quality data at the per paddock level which in turn requires motivated operational staff.

There was a segment of the advisors' clients who collected regular pasture data, but did not know how to get value from the data. Advisors said that it was still a challenge to show farmers the value from DSS analyses of regular capture of per paddock pasture data, given the significant effort required to collect the data. One advisor commented 'I struggle to get people to record or report stuff if they don't get value out of it'.

Advisors were asked about opportunities for future use of grazing DSS. One noted that there is an increasing amount of farm staff with no historic connection to dairy that have little learned experience of dairy systems. Increased use of decision support tools could aid these staff in more rapidly upskilling in grazing management. Larger farms were also identified as having more specialized roles for staff, who therefore don't see system-wide implications of changes. Multi-farm businesses were also becoming smarter with use of data, one client had installed large television screens in each parlor where key performance indicators for each farm were displayed. This resulted in more staff engagement with data, and more discussion about farm performance. The advisor noted that there was a risk of flooding staff with too much information. Some of the grazing management key performance indicators identified by advisors were pasture harvested per hectare, average farm pasture cover, grazing rotation speed, current post-grazing residuals, up-to-date monthly growth rates and tonnes of supplementary feed used.

In terms of advisor use of DSS tools, some of those interviewed were using commercially available tools, while others had made their own specialized Excel™-based spreadsheets and benchmarking services. The need for more 'pain-free' information sharing was identified, with seamless data transfer between software systems. In general, advisors felt it saved time for them to be able to access pasture data before they had meetings with their clients, in order to know what would need to be discussed and to be more efficient with their time. They also noted that younger consultants might be more in need of systems-based DSS to guide their advice, until they became more experienced. In the future, farm owners will employ 'hands-on skilled people' above data collectors and demand better data recording from them, according to one advisor. This will therefore require more training in how to turn the data into better grazing management decisions. Currently, some training organizations make trainees aware of basic tools and fundamentals, but are not adapting their training to include more detailed decision making processes. Opportunities exist for improved tools related to farmer practice such as allocating grazing areas on a daily basis, and improved tools for measuring pasture cover.

3.2 Australian online survey

Almost of quarter of the 83 survey respondents (24%) worked in a veterinary practice, 13% represented dairy processors, 11% worked in extension, and the remaining respondents were from feed companies, herd improvement organizations, finance/banking, farm equipment suppliers and education/training organizations. There was a mix of respondents working for businesses with more than 5 advisors (41%), 2-5 advisors (31%), and sole operators (28%). Most respondents came from the state of Victoria (73%), with northeastern Australia and Tasmania represented by 11% and 9% of respondents respectively.

Through interaction with their clients, the advisors were exposed to a variety of HMS, with almost one-fifth (19%) having experience with seven or more different HMS. In total 16 different types of software were used among advisors' clients. The most common was EasyDairy™, an Australian-designed HMS linked to a herd improvement organization. Some users expressed a 'high degree' of comfort for each of the main HMS, and 14% of respondents were at least 'comfortable' with all the types of software that they used.

Advisors were asked about how they used data generated by different animal monitoring technologies on their client's farms. Data from in-line milk meters was seen as the most useful for providing advice on production, analyzing milking times, and providing advice on nutrition, drying off and culling. Data from other devices such as heat detection and mastitis detection were used by 18 and 15% of respondents, respectively. Data from automated weighing and calf feeders was used the least. The main challenges identified by the respondents for accessing and using data included incompatibility between different technology platforms, and differences in data formats. Learning to use the software was also a key issue, with limited access to training, difficulty justifying the time to learn many different software platforms, needing to learn how to use the data effectively and to evaluate the usefulness of different data.

Around 40% 'frequently' accessed data while on a farm visit, and 55% of respondents said they never access data remotely. In general the number of respondents accessing data remotely was low – only 45% had accessed data remotely in any form. A quarter (25%) of respondents frequently get their clients to send them data. Around half of the respondents were confident of their ability to analyze and interpret precision dairy data, use data to make daily management decisions and to guide seasonal and annual planning. Advisors had limited confidence in their ability to advise on precision dairy technologies (38% were confident) but feedback from the pilot testing suggested that these types of advisors did not see this as part of their role.

Half of respondents felt there would be 'a lot' of value to their consultancy services in accessing and using data more effectively. However, 15% felt there would be only 'a little' value. Examples of the value perceived for their consultancy business were better quality of advice (15%); time saving, efficiency and value for money (15%); and ability to anticipate problems (9%). In the words of some respondents, some of the benefits of accessing data are 'Being able to get more intimate with a business means you are able to make more informed decisions' and 'Give advice before there is a major problem rather than only being alerted once the problem has occurred'

Most respondents (69%) were interested in improving their skills around precision dairy and farm data use. Improved guidance around client management decisions (75%), strategic planning (66%), analyzing and interpreting data (60%) were the main areas where advisors would like to improve their skills. Personal networks (experienced farmers, colleagues and industry contacts) were an important source of advice when advisors want to know more about precision dairy technologies or HMS. Advisors were asked to indicate their preferred learning methods. Short training courses/workshops (74%), online learning (70%), and hands-on learning (67%) were the most popular methods selected. Current sources of information on precision technologies were technology developers (46%), internet (35%), colleagues/other service providers (22%), industry sources (19%).

4. Discussion

The studies outlined in this paper identify that, in the two precision dairy contexts studied, farm advisors are interacting with pasture and animal data of their clients, but that there is opportunity for greater use of this data. The New Zealand interviews highlighted that the level of interaction with a client's data depends on the farm management philosophy of the advisor. Those who were more engaged with accessing and using pasture data were also those who used more structured methods in their consultancy. Advisors in this study were faced with a wide range of HMS and grazing DSS among their clients, meaning there can be a significant learning demand to become familiar with each system. Facilitation of standardized outputs from the data and software may enable advisors to more readily engage with different systems, for example pasture feed wedges or animal health reports. The research also showed that there wasn't a grazing DSS that supported all of the advisors needs, therefore they often transferred data between systems, or created their own spreadsheets. This highlights the ongoing need to improve data exchangeability, and design of systems which not only match farmer needs, but the needs of their advisors.

Data quality was identified by advisors as a factor influencing their engagement. Uncertainty related to walk over weighing data was one example, while some pasture growth data was not seen as accurate by advisors. Some of the data quality issues are related to technological design, and these may be solved over time by new approaches. A limitation can be the time required to collect good quality data, and there are opportunities for automation or simplification of the process. However, other data quality is related to the value that farmers perceive from taking the time to ensure accurate measurement and recording (eg pasture data and recording of animal treatments). In these cases improvement is not just a technological issue, but is also linked to making the role of collected data in better decision making clearer for users such as farmers and advisors. Building capability of farmers and their networks to interpret data is also vital.

Opportunities for greater use of data from precision dairying were identified through this study. Remote access of data was not widespread among these advisors and presents a large opportunity. New Zealand farm advisors who were using this approach noted how it helped them prepare for on-farm meetings with clients, and to keep up to date between interactions with the client. In cases where advisors also had a role in reporting to a board in large farming operations, this access to data simplified and sped up benchmarking and reporting. With dairy farm sizes increasing in both countries, and corporate style farming becoming more prevalent, this access and use of data will prove more important.

The collection of data on-farm, and the interaction of advisors with this data, also presents a potential disruptive influence, leading to adaptations in the advisor-client relationship. For example if an advisor accesses data remotely, this may mean less face-to-face visits overall and evolution of advisory services to engagement primarily via the Internet. Face-to-face interactions may become more focused on strategic planning, rather than operational issues which can be monitored and discussed from distance. Such adaptation of advisory processes could give farmers access to international experts in aspects such as data analysis, nutrition and animal health. This may also aid farmers in more remote location where local service support is reducing. Additional advisory models in some sectors are already changing with the help of precision dairy technologies, for examples veterinarians evolving more into animal health and reproduction advisors, as technologies provide more basic per cow health, reproduction and production information.

There appeared to be learning and training opportunities for farm advisors in the decision support tool space. Advisors appeared engaged and willing to enhance their knowledge around the use HMS and DSS. Based on the responses of the advisors, peer-to-peer learning would be a key approach. Another approach would be to provide case studies of how data management and analysis has been successfully used by others in the industry. Understanding which data are most valuable for improved farmer decision making, and how to use this data to make better decisions, would help both farmers and advisors identify tools and approaches valuable in in their own context.

5. Conclusions

The research in this paper provided information about farm advisor use of data and decision tools for pasture and animal management. We found a variety of levels of interaction between advisors and such tools, and advisors indicated they had different levels of skill in accessing and using data. The design of future data management tools such as HMS and grazing DSS needs to take into account more of the inherent variability in farming systems. For example the design of grazing DSS tends to be based on a consistent approach and toolbox across the year. This does not take into account changes in management in different seasons such as spring and autumn. Resources are required to expand awareness of different data gathering technology and software, specifically targeted at farm advisors. More information is needed to highlight to farmers and advisors where the main areas of value/opportunity exist in use of these tools. Additionally the importance of good data management for farmers and advisors should be promoted in industry programs.

Acknowledgements

The New Zealand study was funded by New Zealand dairy farmers through DairyNZ Inc. in partnership with Ministry of Primary Industries Primary Growth Partnership funding. The Australian study was supported by Dairy Australia, and The State of Victoria Department of Economic Development, Jobs, Transport and Resources. We would like to thank the farm advisors in both studies for their time and enthusiasm.

References

Borchers, M.R. and Bewley, J.M. (2015). An assessment of producer precision dairy farming technology use, prepurchase considerations, and usefulness. Journal of Dairy Science 98(6): 4198-4205.

Eastwood, C.R. and Yule, I. (2015). Challenges and opportunities for precision dairy farming in New Zealand. Farm Policy Journal 12(1): 33-41.

Eastwood, C.R., Chapman, D.F. and Paine, M.S. (2012). Networks of practice for co-construction of agricultural decision support systems: case studies of precision dairy farms in Australia. Agricultural Systems 108: 10-18.
Jago, J., Eastwood, C.R, Kerrisk, K. and Yule, I. (2013). Precision dairy farming in Australasia: adoption, risks and opportunities. Animal Production Science 53(9): 907-916.

Smart dairy farming in practice: design requirements for user-friendly data based services

C.A. Van der Weerdt[1*], J. Kort[1], J. De Boer[2] and G.L. Paradies[2]
[1]*Department of Human Behaviour and Organisational Innovation, Earth, Life and Social Sciences, TNO, Eemsgolaan 3, 9727 DW, Groningen, the Netherlands;* [2]*Department of Human Behaviour and Organisational Innovation, Earth, Life and Social Sciences, TNO, Kampweg 5, 3769 DE, Soesterberg, the Netherlands; caroline.vanderweerdt@tno.nl*

Abstract

Key for precision dairy farming is the management of different kinds of (sensor) data from multiple sources and its disclosure to users. The data can be used to implement services that support farmers to care for individual animals and work more effectively. Naturally, these services need to provide added value to be successful and adopted. Our results show that added value can be found in services that are fine-tuned to a farmer's daily business/business processes and that are easily applicable and usable in daily business. The specific design of the front-end and the functionality of these precision dairy farming services is most important in creating the above addressed added value. This paper addresses research in understanding what added value means to farmers in the context of precision dairy farming and how this insight can best be translated into concrete functionalities and front-end designs of precision dairy farming services. Through contextual inquiry insights in a farmer's values in his/her work and preferences therein are explored based on the User eXperience (UX) framework. This framework focusses on three levels of experiences: (1) The values important to a user that should be supported; (2) the functionality, interactions and behaviors that support and belong to these values in practice; and (3) the aesthetic experience (e.g. form) of the front-end. The results of these contextual inquiries combined with earlier research results were translated into different storyboards expressing different values (meaning) and functionalities (interactics) important to different kinds of farmers. The next step is the validation of these storyboards in workshops and the translation of the results into specific requirements for the development of precision dairy farming services that provide added value for different kinds of farmers.

Keywords: adoption, user experience, value, scenarios, storyboards

1. Introduction and background

Successful services are characterized by their adoption rate by the target users. Services are only adopted when they provide added value for these users, and if the values are in balance with the costs. Another crucial factor of creating added value is designing a desirable product with the right functionalities and aesthetics. Actively involving users in research, innovation and product and service development ensures a close fit of development and design with users' needs and make adoption of new products and services more viable (De Boer and Kuiper, 2008; Steen *et al.*, 2011). There are many design methodologies available that either directly or indirectly include users in the design process itself. These methodologies are generally referred to as User Centered Design (UCD) methodologies (Steen *et al.*, 2011) and help identifying added value for users, translating this value into concepts and final designs and evaluating and fine-tuning these concepts and designs towards a viable service. Several researches in precision livestock or dairy farming are calling for active involvement of farmers in the development of these services; The European Innovation Partnership 'Agricultural Productivity and Sustainability' (EIP-AGRI) for instance states that 'the innovation process in Precision Farming should concentrate on developing farm management solutions that are focused on today's farming reality and on perceive challenges for the majority of farmers'. It is therefore advised that farmers actively participate in the research and innovation process (Anonymous, 2015).

This paper addresses UCD activities applied within the context of a precision dairy farming project in the Netherlands called Smart Dairy Farming (SDF). In general, UCD methodologies consist of different phases which are iteratively revisited a number of times throughout projects. Most commonly recognized phases are: problem analysis; concept development; and testing or evaluation. During each of these phases (potential) users can and should be included to ensure the necessary feedback and insights needed to further progress in development. We have currently conducted activities in problem analysis and concept development in the Smart Dairy Farming project and will share our methodology and results from those activities.

2. Problem analysis

Precision dairy farming is based on the management of different kinds of (sensor) data from multiple sources. Precision farming products and services turn these data into information for decision-making to enhance care and health of livestock and support work and business processes and insights more effectively and efficiently. The potential of precision farming as driver for a thriving dairy sector is significant, and was thus the basis for the Smart Dairy Farming (SDF) project which started in the Netherlands four years ago. The aim of the SDF project was to develop precision dairy farming services that facilitate farmers to make decisions that support a longer productive livestock stay at the farm and an increased lifetime milk production due to improvement of individual health of livestock (Knijn *et al.*, 2013). Within this project different concepts for data gathering, management and flows were developed resulting in a platform called the InfoBroker. This platform enables different parties to use several building blocks based on these data for development of SDF services. In addition, value networks were studied to obtain the necessary insights in potential business models for further provisioning and commercialization of these SDF services (Van der Weerdt and de Boer, 2015). Due to the promising results of the first SDF project, a follow-up project started in 2015, referred to as SDF2. In SDF2 the goal is to also focus on application development and include more farmers.

During the problem analysis phase in a UCD methodology the focus is on (potential) users and how they relate to what is being designed; the things they value, and their accompanying aspirations, motivations, beliefs, but also their current behaviors and practices, challenges, and capabilities, etc. Extracting these factors during interactions with (potential) users is obviously not a straightforward matter. Guiding this process of problem analysis in SDF2 therefore was the UX framework, which supports with addressing the relevant factors in UCD that constitute a good overall experience a user has with given technology (Kort in Bulterman, 2012). The UX framework is composed of three dimensions: (1) the dimension of meaning and value; (2) the dimension of interactics; and (3) the dimension of aesthetics.

The UX dimension of meaning and value addresses current users' aspirations, motivations, and beliefs such as: security and safety (e.g. financial security); competence and effectiveness (e.g. feeling that one is capable and effective in one's own actions, affecting results); stimulation (e.g. having pleasure, joy and feeling stimulated in events, to explore, advance, etc); respect and (social) connectedness (e.g. feeling appreciated by others, one's contributions valued and feeling connected to and respected by others such as other farmers and family). These aspects directly relate to universal psychological needs that determine well-being and are sometimes called 'be-goals' (Hassenzahl *et al.*, 2010). Users' current behaviors and practices, challenges, capabilities, etc. are in the UX framework addressed as *Interactics*, the functional elements and flow in processes, behaviors, a user's capabilities therein and the challenges encountered. Interactics refer to do-goals (Hassenzahl *et al.*, 2010); to getting the job done in an effective, efficient and emotionally satisfying way. These factors may come forth from analyzing a daily routine in light of the new service, such as: when will you use it? And how? The third dimension in the UX framework is *Aesthetics*, which addresses the physical design of products and interfaces in the context in which they are foreseen to be used. Examples of Aesthetic aspects are form, size (of devices and interface elements) and the use of sound and visuals.

The meaning and interactics levels of the UX framework was used in SDF2 to create an open, unstructured interview protocol addressing possible relevant aspects for exploring added value in SDF services. This protocol was consequently used for a contextual inquiry study with two farmers and a farmer advisory body (i.e. visiting farmers and observing them doing their work, while they explain what they are doing (show and tell)). Furthermore, desk research and results from the first SDF project (Van der Weerdt and de Boer, 2015; Van der Weerdt and Vonder, 2013) were applied for thorough insight into the topic of added value for the farmer. These activities together lead to detailed insights in farmers' current values, experiences, as well as in their current behaviors, capabilities, and challenges they encounter and possible improvements therein.

First results of this phase show that value can be added in several ways, ranging from financial benefits, more efficient ways of working and managing one's business, improved animal welfare and early identification of problems, including better monitoring of problems. Furthermore, 'softer' aspects were recognized related to a farmer's work or life such as being able to handle and act within work and time or place more flexibly, to have the feeling your expertise as a farmer is continuously respected and complemented by the use of technology. To realize added value through service development it is also important to address several 'hygiene factors' that can prevent adoption, such as needed investments in technology (e.g. in time, effort, finances, learning and understanding); undesirable changes in business processes and ways of working; and trust in the technology and services (e.g. transparency regarding the source of data and data-driven advice (Van Hyfte *et al.*, 2015; Van der Weerdt and de Boer, 2015; Van der Weerdt and Vonder, 2013), etc.

Insights obtained during this phase also offer the opportunity to distinguish different types of user groups which have different preferences. Archetypical descriptions of a user representing a specific user group are called Personas (Anonymous, 2015) and are very valuable during concept development to create concepts that live up to users' expectations for different user groups. In turn however, Personas representing different preferences between groups also help to identify the common denominator and steer design and development towards an innovation that suits all or most identified Personas. In SDF1 Personas have already proven their worth to 'secure a design of the platform that would suit the farmers, but also helped with the adoption of the concept of the SDF platform as such' (Van der Weerdt and de Boer, 2015). In SDF1, one Persona depicted the SDF farmer. However, based on our SDF2 research, more farmer archetypes can be formulated. First, the progressive farmer, often running a larger business (e.g. 400 cows), sometimes with additional employees besides involved family members and always looking and closely monitoring new precision farming developments which are easily bought and tried out when added value in terms of business management and increased efficiency are foreseen. Second, the follower. This farmer is often running a smaller business (e.g. 150 cows) but is steadily growing over time. Business is mostly done by the farmer him/herself. External personnel are only hired when absolutely needed and on a temporary base. He/she is interested in new developments but less quick to try them out. Often technology is adopted after having seen it successfully working for other farmers. In that respect it fulfills basic requirements and no additional features are needed. Third, the reactive farmer, often running a smaller business (e.g. 100 cows), content with the way things are working now, without foreseen problems in the future in terms of scaling up one's business, etc. This farmer will only be interested in precision farming technology when it becomes an absolute necessity and he/she could not continue business without it. The Persona description are used in the next phase Concept Development to create stories and scenarios in which functionalities are illustrated.

3. Concept development

During the concept development phase, the insights obtained in the previous phase need to be turned into fitting requirements for SDF product and service concepts. In other words, the added value identified in the previous phase needs to be transformed into concrete functionalities and interactions for service concepts. We have used scenarios illustrated with storyboards to show the possibilities of SDF for each Persona mentioned earlier.
Scenarios and storyboards are simplified representations or stories about how users will interact with service concepts. They are fine-tuned and designed for specific Personas and describe SDF concept encounters and interactions in specific contexts. They go beyond mere service interactions and can include the complete journey a user makes from first hearing about a product or service, seeing other use a product/service, acquiring information about a product/service to actually signing a contract, having the product/service installed, learn to use it and adopt it in daily practice, etc. Scenarios are textual descriptions of these stories describing the possible added value and experience of the whole journey from first connecting with a product or service to the end of a product's or service's life cycle. Storyboards are often built based on scenario descriptions and often represented as drawing or sketches with additional explanatory texts. Both scenarios and storyboards have two main advantages: first, they can be easily discussed and evaluated with potential users gathering feedback on ideas and input for redesign or design refinements; second, they are a great way to communicate requirements to support the (ICT) engineering process because the added value and experience is described as a process which provides more detail for actual implementation than does an abstract list of requirements. In this phase in SDF2, 6 different scenarios were developed for each identified Persona describing the most important needs and foreseen added values of SDF services

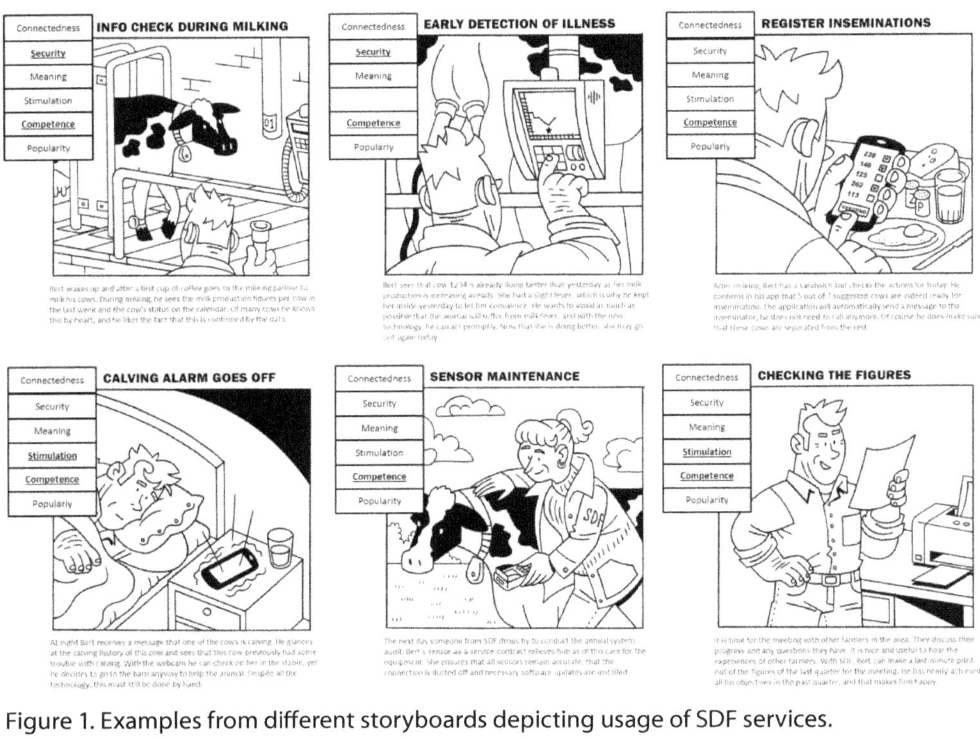

Figure 1. Examples from different storyboards depicting usage of SDF services.

addressing all steps in adopting such services, from 'getting interested in', to 'actual daily use' and 'service maintenance by the provider'. In Figure 1 an example is given of one storyboards developed, indicating specific usage of SDF services and the values that are addressed by these services.
The next phase in concept development is to discuss the storyboards with a larger group of farmers. As the storyboards are illustrative they will induce responses as to what extent they recognize themselves or can imagine themselves acting the way as illustrated. This creates a lower threshold in making explicit what they want, thus eliciting values and behaviors, but also feelings and beliefs even better.

4. Results and discussion

A preview on the applied methodology in SDF2 and brief description of some preliminary results indicate that actively involving the user perspective in the design process of smart farming services firstly provides insights in what can be understood by added value in terms of these services. Secondly, the insights created in the problem analysis phase provides a deeper understanding of valid and valuable service concepts and speeds up the process of concept generation, without forgetting about the user needs. Thirdly, by structurally applying UCD methods throughout the design process feedback is gathered to fine-tune concepts towards final designs with a better understanding of adoption and added value. Looking forward, one could also view this as a form of risk management for eventually successful product/service launchings. And finally, we believe communicating service design requirements is much easier through storyboards describing the desired interactions and functionalities in a flow or process to engineering. This will be applied in practice and in 2017 SDF2 will end with an evaluation of the UCD process for SDF service design. The methodology and results in this paper and our future work in terms of requirements and guidelines for SDF service design will be made available to other parties developing precision farming services to enable them to not only gain insight in the added value their services could provide but to realize the added value itself in their service design as well through direct application of already identified requirements and guidelines.

Acknowledgements

This research was supported and financed by the Dutch Top Sector High Tech Systems and Materials and Friesland Campina (Amersfoort, the Netherlands), CRV (Arnhem, the Netherlands), and Agrifirm (Apeldoorn, the Netherlands). The researchers would also like to thank the farmers that have participated in the research and Accon AVM (Arnhem, the Netherlands) for their review of the storyboards.

References

De Boer, J. and Kuiper, L. (2008). Challenges and merits of co-design of mobile concepts. In: Proceedings of MobileHCI, Amsterdam, the Netherlands.

Hassenzahl, M., Diefenbach, S. and Göritz, A. (2010). Needs, affect, and interactive products – facets of user experience. Journal of Interacting with Computers 22: 353-362.

Holtzblatt, K., Wendell, J.B. and Wood, S. (2005). Rapid contextual design: a how-to guide to key techniques for user-centered design. Morgan Kaufmann, San Francisco, CA, USA.

Knijn, H., Taweel, H., Van Wichen, H., Wulfse, B.J. and Vonder, M. (2013). Smart Dairy Farming program in the Netherlands. In: Proceedings of the precision dairy conference and expo Precision Dairy 2013, Paris, France, pp. 141-142.

Bulterman, D.C.A. (2012). Together anywhere, together anytime – technologies for intimate interactions. Centrum Wiskunde en Informatica, Amsterdam, the Netherlands.

Steen, M., Manschot, M. and De Koning, N. (2011). Benefits of co-design in service design projects. International Journal of Design 5(2): 53-60.
The Organisation for Economic Co-operation and Development (OECD) (2014). Data-driven innovation for growth and well-being. Paris, France.
The European Innovation Partnership 'Agricultural Productivity and Sustainability' (EIP-AGRI) (2015). Focus group precision farming – final report.
Van der Weerdt, C. and De Boer, J. (2015). Focusing on behaviour to ensure adoption of big data information services in precision livestock farming. In: Proceedings of the 7th European Conference on Precision Livestock Farming. September 15-18, 2015. Milan, Italy.
Van der Weerdt, C. and Vonder, M. (2013). Smart dairy farming, requirements-analyse (eisen en wensen). TNO, Groningen, the Netherlands.
Van Hyfte, H., Bruggeman, G., Benmeridja, N., Mahdavi, H., Kamphuis, C., Lehr, H., Hogeveen, H., Lokhorst, L. and Vermeij, I. (2015). Report on the impact of animal key indicators on the value chain. EU-PLF project co-funded by the European Commission within the Seventh Framework program.

Relation between sensor performance and willingness of farmers to pay for lameness detection systems

T. Van de Gucht[1,2], A. Van Nuffel[1*], L.M. Pluym[1], K. Piccart[1], L. Lauwers[3,4], W. Saeys[2] and S. Van Weyenberg[1]

[1]*Institute for Agricultural and Fisheries Research – ILVO, Technology and Food Sciences Unit, Burgemeester van Gansberghelaan 115, 9820 Merelbeke, Belgium;* [2]*KU Leuven Department of Biosystems, MeBioS, Kasteelpark Arenberg 30 Box 2456, 3001 Leuven, Belgium;* [3]*Institute for Agricultural and Fisheries Research – ILVO, Social Sciences Unit, Burgemeester van Gansberghelaan 115, 9820 Merelbeke, Belgium;* [4]*Ghent University, Faculty of Bio-Engineering, Department of Agricultural Economics, Coupure Links 653, 9000 Gent, Belgium; annelies.vannuffel@ilvo.vlaanderen.be*

Abstract

Several automatic lameness detection systems are described in literature, but only one system is commercially available. At development stage, research first focuses on excellent sensor performance. As a result, proof of concepts and prototypes mostly consist of state-of-the-art sensor technology able to provide very detailed raw data. The resulting high costs of these prototypes impede them to become real applications in dairy practice. Downscaling of sensors as an option to reduce high cost might be possible, but could result in decreased sensor performance. Information about the relation between the willingness of farmers to pay extra for more precise sensor technology and hence higher detection performance might therefore be useful during the downscaling process. A Discrete Choice Experiment was set up to assess farmers' willingness to pay for different sensor performance in terms of % missed, % false alarms and indication of the lame leg. In an online survey, farmers were asked to choose between three types of lameness detection systems (sensors attached to the cows, camera systems and walk-over devices – all with varying performance characteristics) or to opt for their own visual lameness detection without sensors. During statistical analysis, these choices are related to price, detection rate, amount of false positive alarms and lame leg indication. In general, i.e. for the three types of lameness detection systems, farmers are willing to pay €2.3 extra per % less missed lame cows, €1.8 per % less false alarms and €11.6 for lame leg indication. These results can be helpful during further development of lameness detection prototypes in order to increase their adoption in dairy practice.

Keywords: technology adoption, investment, lameness, farmer involvement

1. Introduction

Automatic lameness detection systems are currently being developed to provide a solution for the problems incurred when using visual lameness detection (Van Nuffel *et al.*, 2015a,b). Such systems could possibly decrease the workload of the farmer, reduce the losses caused by lameness and improve the animal welfare (Leach *et al.*, 2012). However, except for one, these systems are prototypes and not yet commercially available. Since development is still going on, these sensor systems still provide very detailed information, but do so at a high cost.

The future adoption in practice of these systems partially depends on their cost. Up till now, it is not known how much farmers would want to pay for automatic lameness detection. Possibly, their willingness to pay does not equal current system costs. Therefore, researchers could adapt their prototypes in such way that it is more likely that they get adopted in practice. Adapting the systems could be done by downscaling the technology, which can have a negative effect on the performance of the lameness detection system. Therefore, it is important to know which system features are important, and how much a farmer wants to pay more for a system with a higher performance.

The goal of this study was to find out the willingness to pay of farmers for different system features. This could be explored by using a discrete choice experiment to reveal farmers' preferences for automatic lameness detection systems.

2. Materials and methods

A discrete choice experiment was executed using an online survey. In such experiments, respondents are asked to choose their preferred option between possible alternatives within a choice set. Each of the alternatives is described by attributes (features) and attribute levels (quantification of the features). In this study, three alternatives (lameness detection systems) were proposed: a walkway or pressure mat on the floor, a system attached to the cow and a camera system. Respondents were also given the possibility to opt for not using a sensor system and to detect lame cows visually without the help of a sensor. The attributes were the percentage false alarms, the percentage missed lame cows, the cost and the ability to indicate which leg is lame. An example of a choice set can be found in Table 1. In total, 32 choice sets divided over eight blocks of four choice sets each were generated using Ngene (ChoiceMetrix, Australia). These blocks were incorporated in an online survey using limesurvey (Schmitz, 2015). Each respondent was presented one block of four choice sets. Farmers were informed of the survey by mail and through announcements on websites. Responses were collected during two months and were gathered in a dataset.
For each choice set in the dataset, the utility value of each alternative with its specific attributes and attribute levels was determined using Nlogit (Econometric software, Inc, USA), based on the choice made by the respondent. A regression model describing the utility of each proposed detection system was determined. With this model, the willingness to pay was determined for each % point improvement of the percentage false alarms and the percentage missed lame cows, and for the ability to indicate the lame leg. The model was also used to perform simulations used to indicate how the market share of each system changes in case the detection performance and cost are reduced due to downscaling of the technology. Two simulations were performed by increasing both the percentage false alarms and missed lame cows by 5 and 10%. This scenario was repeated, but in this case also the price was reduced by €10 and €20. In all cases, the system cost was equal for all systems.

Table 1. Example of a choice set as presented to the farmer.

	System attached to the cow	Walkway or pressure mat on the floor	Camera system
Missed lame cows (%)	10	5	20
False alarms (%)	5	10	15
Cost/cow/year (€)	75	15	35
Indication lame leg	yes	no	no
Opt out: I detect lame cows myself visually without the help of a lameness detection sensor.			

3. Results

In total, 135 respondents filled out the survey completely. Table 2 contains the regression coefficients after analysis of the responses. All four attributes had a significant contribution to the utility. A specific constant was defined for each alternative, and coefficients for each attribute were determined as well. The willingness to pay was higher for improvements in the percentage missed lame cows compared to the percentage false alarms. This indicates that farmers want to pay €2.33 extra per cow per year if the percentage missed lame cows decreases by one percent. Farmers were also willing to pay €11.58 extra per cow per year if the automatic lameness detection system is able to indicate the lame leg. In Table 3, the results of the simulations are presented. These results show that in general, farmers opted more to keep detecting lameness visually in case the % missed lame and % false alarms are increased by 5%. Similarly, when the performance was reduced with 10%, the market share changed more or less proportionally to the first situation. Systems attached to the cow lost the largest market share, while camera systems lost the smallest market share. When these changes in performance were combined with a lower system cost, the market shares changed less, but there was still a general trend towards more visual lameness detection. In case the systems become quite cheap, but with a low performance, farmers still tended to choose more for visual lameness detection.

Table 2. Regression coefficients of the model after analysis of the choices made by the farmers. The coefficients, standard error and *P*-value are given. For each of the attributes, the willingness to pay (€ per cow per year per % improvement) was calculated as well.

	Coefficients	Standard error	*P*-value	Willingness to pay
Constant activity meter	7.75	0.21	/	
Constant Gaitwise	1.42	0.21	/	
Constant camera system	1.16	0.21	/	
Attributes				
Missed lame cows (%)	-0.056	0.006	<0.001	€2.33
False alarms (%)	-0.044	0.011	<0.001	€1.83
Price	-0.024	0.003	<0.001	/
Indication of lame leg	0.278	0.072	<0.001	€11.58

Table 3. Simulation results of a decreasing performance and a decreasing performance combined with a lower cost on the market share of each alternative.

Attribute	Current level	Simulated level	% farmers changed towards			
			System attached to cow	Pressure mat or walkway	Camera system	Visual detection
Influence of decreasing performance on market share						
Missed (%)	10	15	-4.7	-3.7	-2.8	+11.2
False alarms (%)	5	10				
Missed (%)	10	20	-9.7	-7.7	-5.9	+23.3
False alarms (%)	5	15				
Influence of decreasing performance and price on market share						
Missed (%)	10	15				
False alarms (%)	5	10	-3.3	-1.7	-1.3	+5.4
Price (€)	35	25				
Missed (%)	10	20				
False alarms (%)	5	15	-4.8	-3.6	-2.8	+11.2
Price (€)	35	15				

4. Discussion

The results obtained in this study showed that farmers find it more important not to miss lame cows than to have false alarms and are thus willing to pay more for a system that misses less lame cows. This could be because farmers might think that it is more important not to miss a lame cow due to high financial losses that are caused by not treating it timely. False alarms require time of the farmer to check up on the claws and are hence also unwanted, but are less costly than missing a lame cow and not treating it timely. Farmers were found to prefer solutions with a high return on investment (Borchers and Bewley, 2015), hence it is likely that they go for systems with low rates of missed lame cows.

Farmers also want to pay maximally €11.58 more per cow per year for a system that can indicate on which leg the cow is lame compared to a system that can only detect the cow as being lame, regardless of the location of the injury. They may find this important because lame cows are difficult to spot, and hence assigning the lameness to a specific leg is difficult as well, even if a detection system indicates that the cow is lame.

The market shares of the different systems changed with relatively large percentages when the performance is reduced. This indicated that farmers find the performance of the system important. Even when the price is reduced greatly, bad system performance prompts farmers to choose visual lameness detection over automatic lameness detection. Probably farmers are not aware of the fact that they miss a large number of the lame cows themselves when performing visual lameness detection. This is supported by the finding that farmers tend to underestimate the lameness prevalence in their herd (Leach *et al.*, 2010; Wells *et al.*, 1993; Whay *et al.*, 2003). It would therefore be interesting to investigate whether the preferences of farmers change after being educated more about lameness and its consequences on animal and farm level.

5. Conclusions

The conclusion of this study was that the detection performance of automatic lameness detection systems is important to farmers, and that they are willing to pay €2.3 extra per % less missed lame cows, €1.8 per % less false alarms and €11.6 for lame leg indication. It was concluded that the performance remains important, even if the system's cost is decreased. These results are important during downscaling of lameness detection technologies in order to improve the adoption in practice.

References

Borchers, M.R. and Bewley, J.M. (2015). An assessment of producer precision dairy farming technology use, prepurchase considerations, and usefulness. Journal of Dairy Science 98: 4198-4205.

Leach, K.A., Tisdall, D.A., Bell, N.J., Main, D.C.J. and Green, L.E. (2012). The effects of early treatment for hindlimb lameness in dairy cows on four commercial UK farms. Veterinary Journal 193(3): 626-632.

Leach, K.A., Whay, H.R., Maggs, C.M., Barker, Z.E., Paul, E.S., Bell, A.K. and Main, D.C.J. (2010). Working towards a reduction in cattle lameness: 1. Understanding barriers to lameness control on dairy farms. Research in Vetereniary Science 89(2): 311-317.

Schmitz, C. (2015). LimeSurvey: an open source survey tool. LimeSurvey Project Team, Hamburg, Germany.

Van Nuffel, A., Zwertvaegher, I., Pluym, L., Van Weyenberg, S., Thorup, V.M., Pastell, M., Sonck, B. and Saeys, W. (2015a). Lameness detection in dairy cows: part 1. How to distinguish between non-lame and lame cows based on differences in locomotion or behavior. Animals 5(3): 838-860.

Van Nuffel, A., Zwertvaegher, I., Van Weyenberg, S., Pastell, M., Thorup, V.M., Bahr, C., Sonck, B. and Saeys. W. (2015b). Lameness detection in dairy cows: part 2. Use of sensors to automatically register changes in locomotion or behavior. Animals 5(3): 861-885.

Wells, S.J., Trent, A.M., Marsh, W.E. and Robinson, R.A. (1993). Prevalence and severity of lameness in lactating dairy-cows in a sample of Minnesota and Wisconsin herds. Journal of the American Veterinary Medical Association 202(1): 78-82.

Whay, H.R., Main, D.C.J., Green, L.E. and Webster, A.J.F. (2003). Assessment of the welfare of dairy cattle using animal-based measurements: direct observations and investigation of farm records. Veterinary Record 153(7): 197-202.

Development of an investment model for automated mastitis detection systems

C.R. Eastwood[1*], J. DeVeer[2], M. Neal[1], B.T. De la Rue[1] and C. Kamphuis[2]
[1]*DairyNZ, Hamilton 3240, New Zealand;* [2]*Chair Group Business Economics, Wageningen University, Wageningen, the Netherlands; callum.eastwood@dairynz.co.nz*

Abstract

Analyses that combine the costs of mastitis with performance characteristics of automated mastitis detection (AMD) systems would aid farmer investment decisions. A partial budget modelling approach was developed to compare an AMD system with visual detection of clinical mastitis (CM) in pasture-based dairy systems. The analysis indicated that while some AMD systems perform better than visual inspection at CM detection due to higher sensitivity and frequency of the measurement, the value of investment in AMD is difficult to justify on a financial basis. The lack of value was primarily influenced by initial cost of investment and/or the performance of the technology (related to labor costs associated with checking (false) alerts and treating cows). Under the models default settings, investing in a high performing AMD system using quarter-level electrical conductivity (ECq) resulted in an internal rate of return (IRR) of -1.11% and a 15 year net present value (NPV) of €-5,443. In this case, the 'benefit' relates to a decrease in labor requirement for detecting CM episodes, while maintaining or improving the percentage of cows detected and treated for CM. In well performing AMD systems, the cost of treating CM cases had a negative financial impact, raising questions on the profitability of treating CM episodes (other than for animal welfare reasons). The investment analysis model was reliant on expert opinions when reliable data were absent or insufficient and indicating a need for future research.

Keywords: in-line sensors, investment analysis, precision dairy

1. Introduction

The management of clinical mastitis (CM) continues to be an animal welfare issue and significant animal health cost for dairy farmers. Automated mastitis detection (AMD) systems have been developed as management aids for farmers and previous studies have investigated the performance and on-farm use of these devices (Kamphuis *et al.*, 2016). There has been a range of adoption rates reported in surveys, including use on 6% of New Zealand (NZ) farms (Edwards *et al.* 2015), on 35 and 93% of conventional and automatic milking systems farms, respectively, in the Netherlands (Steeneveld and Hogeveen, 2015), and ~26% of USA farmers surveyed (Borchers and Bewley, 2015). However, farmers have also indicated uncertainty regarding the value of precision dairy technology investment (Borchers and Bewley, 2015; Steeneveld and Hogeveen, 2015; Eastwood *et al.*, 2015). Generalized total costs associated with mastitis have been calculated through several studies and range from €61 to €97 per cow/year (Hogeveen *et al.*, 2011). Underlying these cost estimates, however, are complicated factors determining the actual cost faced by a farmer such as individual costs, the relative influence of mastitis on strategic decisions, such as culling, and when it is economically feasible to treat a cow with mastitis (Hogeveen *et al.*, 2011). Analyses that combine the costs of mastitis with performance characteristics of automated mastitis detection systems would aid farmer investment decisions. In this study we used a partial budget modeling approach to compare AMD with visual detection in the context of a New Zealand pasture-based dairy system.

2. Methods

A stochastic modelling approach developed in MS Excel (Microsoft Corp., Redmond, WA) was used to compare measuring electrical conductivity (EC) at whole udder and quarter level with farmer visual detection performance. Sensitivity analysis and Monte Carlo simulation were performed using @Risk for Microsoft Excel (Palisade Corporation, 2014). The farm modeled was an average pasture-grazed NZ dairy farm; 402 spring calving cows, 42 week lactation, producing 346 kg of milksolids (MS) /cow/year. Key inputs to the model were: (1) dynamics of mastitis states (clinical mastitis, subclinical mastitis, healthy); (2) effect of mastitis states on milk production losses, number of treatments, discarded milk and total use of antibiotics; (3) CM detection performance (detection rate and the number of false alerts per 1000 cow milkings (FAR1000) for AMD and visual detection); (4) the delay from start of a CM episode until treatment and the effect of the delay on use of antibiotics per treatment; (5) labor requirement (checking cows, treating cows, discarding milk); and (6) investment costs.

In this study, CM and the duration of CM episodes were defined according to Kamphuis *et al.* (2016) where CM occurs when there is observation of clots at two out of three, or all three, consecutive milkings and the CM episode continues until clots are followed by 4 consecutive cow milkings without clot observation. A CM incidence rate of 14.8% of the entire herd per year (Petrovski *et al.*, 2009) was used, distributed throughout the lactation according to McDougall and Compton (2005). The model used the seasonal distribution of *Streptococcus uberis*, the most common cause of CM in NZ, particularly prevalent during early lactation and the dry period. The equation calculated the daily CM incidence which determined the start of a CM episode by:

$$\text{CM incidence} = 0.264 * \text{day}^{-0.426}$$

It was assumed that a proportion of the CM episodes would be detected and treated, and of the undetected CM cases, 20% would remain CM, 48% become subclinical mastitis (SCM), and 32% self-cure (Lacy-Hulbert, 2014; McDougall, 2014 – personal communications). The SCM and undetected CM cows propose a risk of pathogen transmission to healthy cows, introducing new CM episodes. A transmission model based on the Susceptible-Infected-Removed (SIR) model of Kermack and McKendrick (1927) was developed to calculate the rate of mastitis spreading. Of the total number of cows (N), cows are either susceptible (S), infected (I) or removed (R) by treatment or self-cure. β is the transmission rate for the combination of pathogens at that time window of ten milkings.

$$\frac{dI}{dt} = \frac{\beta * S * I}{N} - R$$

Treatment cure rates vary and will depend on a range of variables including the type of bacteria present, the antibiotic used and the duration of treatment, the number of recurring infections that a cow has been exposed to, and the delay in treating infection. For simplicity, the current model assumes that all CM treatments resulted in cure, subclinical mastitis was not treated during lactation, and all CM and SCM infections were assumed to be cured during dry cow therapy at the end of lactation. New CM episodes caused by transmission incurred costs such as increased likelihood of culling and the default total costs per new CM episode was set at NZD$306 (~€187) (Malcolm, 2006). When initially identified (either by visual detection or AMD) the existence of CM needs to be confirmed by teat-stripping the cow at subsequent milkings. Clinical mastitis episodes were only treated after the second observation of clots in the CM episode. A standard CM treatment typically consists of 3 intramammary treatments at 12-24 hour intervals (3 consecutive milkings). An extended treatment may be required which is typically a further 3 intramammary treatments. Early treatment of CM has been shown to significantly reduce the use of antibiotics and degree of milk production losses (Milner *et al.*, 1997). However, in the current model only early treatment effects on antibiotic use were included. Using an adaption of the model described by Milner *et al.* (1997), the number of doses used per CM episode ranged between 3.0 and 3.8 depending on when detection occurred in relation to the start of the CM episode. Treatment is followed by a milk withholding period of 5-8 milkings, which is drug dependent. We used a default withholding period of 13.5 milkings including milkings where CM is diagnosed in this period milk was discarded. The annual use of antibiotics was calculated at the herd level by multiplying the number of treated animals by antibiotic use per cow (default cost NZD$7.08 per dosage (~€4.32)).

A milksolids production curve, based on Wood (1967) was modeled to estimate costs associated with discarded milk at any time over the lactation. The curve was fitted to available data on milk production of cross-bred herds, milked twice a day, lactation length of 42 weeks, peak milk production at 55 days in milk (DIM), default production of 346 kg MS/cow/year. The cost of discarded milk was set at NZD$2.37/kg MS (~€1.45), calculated by subtracting average production costs (NZD$4.13/kg MS, ~€2.52) from the average milk price (NZD$6.50/kg MS, ~€4.0). Production losses associated with elevated SCC for cows with undetected CM were estimated at 2.1% for each doubling of cow SCC for the remainder of the lactation (Malcolm, 2006). Cows that developed SCM were assumed to have an average SCC of 800,000 cells per ml (from a base of 150,000 cells per ml).

In New Zealand, standard practice for visual mastitis detection is based on periodic teat stripping during milking, augmented by up to four test milk recording events per lactation. In the model, the default teat-stripping frequency was set at once every 30 milkings, although this can be modified in the model. Average farmer visual detection rates using clots observed during individual cow teat-stripping has been estimated at 80% (Claycomb *et al.*, 2009) however a lower detection rate of 70% was used in the model to account for higher errors when teat-stripping all cows in a herd during milking. For consistency with CM detection using AMD, it was assumed that cows with clots observed at a milking were marked and teat-stripped at the next 1 or 2 consecutive milkings (average 1.5 milkings) to determine whether two out of three consecutive milkings exhibited clots as required for CM confirmation. If a CM cow was not detected at a teat-strip event, then the cow could remain CM or change to SCM or self-cure. Cows continuing to have CM therefore could be detected at the next teat-strip event, or at herd testing.

AMD systems are expected to generate at least one alert within a time window for overall detection (OD) starting from 2 milkings before the first clot observation to the end of the CM episode, with a maximum of 10 milkings in total (Kamphuis *et al.*, 2016). The overall detection rate (ODR) is a percentage of the CM episodes with an alert with the OD 10-milking time window and represents the sensitivity of the device. There are a number of false alerts (specificity) associated with the detection rate. A trade-off exists between detection rate and false alert rate (FAR1000), where changing the settings of an AMD system to increase detection rate will result in an increased FAR1000, which at some point will become impractical. The early detection and late detection rates were also considered. Early detection (ED) is defined as an alert generated by the AMD system within the first 5 milkings and late detection (LD) an alert within milkings six to ten of the OD time window. The late detection rate (LDR) is the difference between the ODR and the EDR. An AMD system using EC measurements at quarter level (ECq) was compared with visual detection. Data on potential performance of this device was derived from a field study by Claycomb *et al.* (2009). The ODR default value was 88% with a FAR1000 of 6. The EDR and LDR both had a default value of 44%. The investment modeling indicated that this was the optimal ODR setting for ECq as increasing detection rate resulted in large increases in FAR1000 and hence the labor costs of checking cows.

Initial investment cost was estimated based on the 402 cows in the farm model, a 25 milking point milking platform, with an EC sensor at every milking point. The total investment cost was estimated at NZD$28,000 (~€17,080) for ECq, with a 15 year lifetime and declining depreciation with a diminishing value of 30%. Saved tax over depreciation was added (at a marginal tax rate of 17.5%) and tax paid over the residual value was subtracted. The labor requirements for CM detection using AMD and visual inspection were based on time required for teat-strip checks, treating of

CM cows and discarding milk withheld from supply. The default time was set at twenty seconds per cow for teat-stripping of either the entire herd for visual detection at the designated interval (default=every 30 milkings), or the alerted cows at each milking for AMD systems (plus follow-up checks at consecutive milkings). The number of teat-stripping events for a false alert was set to three consecutive milkings. Time required for antibiotic treatment of CM cows was set at 5 min per dose, and the time taken to discard milk was set at 2 min per treated cow for milkings during the withholding period at a rate of NZD$18.66 (~€11.40) per hour.

The marginal profitability of using the AMD system compared with visual detection was investigated in terms of the total annual costs of CM management (cash flow), net present value (NPV) and the internal rate of return (IRR). The annual cost of CM management was categorized into 4 areas and included costs associated with: (1) CM treatment (e.g. antibiotics, discarded milk); (2) labor (e.g. checking cows, treating cows, discarding milk); (3) untreated CM cows (e.g. production losses, transmission of CM); and (4) maintenance (e.g. operational and maintenance cost of AMD).The NPV considered the total annual operating cost and the investment cost, calculated at a discount rate of 5%.

3. Sensitivity analysis and Monte Carlo simulation

A sensitivity analysis and Monte Carlo simulation were undertaken to determine the impact on annual CM management costs of: (1) AMD overall detection rate (ODR) when compared with visual detection methods; (2) visual detection rate, frequency of manual detection. Visual detection rates of 40 to 100% were modeled. The impact of teat-stripping frequency was also investigated by simulating checks every 10, 20, 30, 40 or 50 milkings. All other inputs were left as the default values. A Monte Carlo simulation was performed with the CM incidence and milk price as risky variables, with all other inputs set at the default values. For the incidence of CM, a normal distribution was used (mean=0.148, sd=0.03) (Petrovski *et al.* 2009). A uniform distribution was used for milk price (interval NZD$4.35-$8.70 (~€2.65-~€5.31) per kg MS) (LIC, 2013). The input milk price was maintained for each iteration across the 15 year lifespan, no year-to-year variation was modeled. Simulations (n=1000) were performed for ECq.

4. Results

Based on assumptions made in the model, the analysis indicated that the installation of ECq sensors provided a positive mean annual cash flow of NZD$1,736 (~€1,059) when compared with visual detection (after a Monte Carlo simulation). Including initial investment costs (around NZD$1,000 (~€610) per milking point) indicated a mean net present value of NZD$-8,923 (~€-5443) over the 15 year period. Therefore, investment in an AMD system was not profitable. However, key factors influencing this outcome were the initial cost of investment and the performance of the technology (related to associated labor costs of checking alerts and treating cows).

The impact of changes to the visual detection rate had little impact on the annual cash flow when compared with using ECq systems. At a visual detection rate of 40%, the difference in annual cash flow was NZD$1,500 (~€915) and the NPV was NZD$-11,000 (~€-6,710). At 70% and 100% detection rates annual cash flows were NZD$1,700 (~€1,037) and NZD$1,600 (~€976) and the NPV values were NZD$-8,900 (~€-5,429) and NZD$-9,600 (~€-5,856), respectively.

5. Discussion

We simulated the impact of CM on a typical herd, and the comparative impact on costs between visual and sensor-based mastitis detection. The research highlighted a lack of certainty around specific costs, such as the cost of sub-clinical mastitis, the impact of early treatment on milk production losses and reduced use of antibiotics, and the costing of farm staff time allocated to mastitis detection and treatment. The current method of costing the inspection and treatment of CM indicated that the cost is greater than the benefit (excluding non-financial benefits); therefore, AMD systems that perform poorly at detecting cows with CM will show a greater return on investment. This is based on the cost of an untreated mastitis event and needs further exploration and refinement around which animals to treat. Also, definition of the pathways of IMIs and the proportion of cows that develop CM, change state to subclinical or self-cure and transmit infections is critical to calculation of the economic value of mastitis detection systems. There are few published studies relevant to New Zealand farm systems that provide this information and much of data used in this current study relied on unpublished data and expert opinions.

The effect of improved detection performance of an AMD system relative to visual detection was more labor to confirm CM cases, more cows treated and therefore more treatment costs, and more discarded milk. This lead to overall increased costs with higher detection rates. Also, as false alerts (depending on the AMD system) increased there was more labor required to confirm CM. Improved detection also led to a higher proportion of CM cows cured, therefore less impact of reduced milk production through clinical or subclinical cows in the herd and fewer new CM cases via the infection transmission model.

It should be noted that the AMD performance values related to detection of clots, but it is unlikely that all cows with true positive alerts will be treated by the farmer using antibiotics. Also presence of clots is not always linked to a CM case. No information is available on the relationship between true positive alerts and cows considered to require antibiotic treatment. Therefore, the cost of treatment of alerted cows is likely to be overestimated compared with visual detection rates. Through the model, we aimed to more clearly define value for farmers investing in AMD. The inability to put costs on the impact of SCM on the bulk milk SCC, and the role of AMD in end of lactation decision making, may lead to underestimating the value of AMD to farmers.

6. Conclusions

The analysis indicated that AMD investment was not profitable, primarily because initial cost of investment combined with the performance of the technology (related to associated labor costs of checking (false) alerts and treating cows) was greater than the estimated return. Investing in AMD may be most favorable for farmers to decrease the labor requirement of detecting clinical mastitis (CM) episodes while maintaining or improving the percentage of detected and treated CM episodes. There may be additional benefits for farmers with high staff turnover or staff without sufficient skills to identify mastitis.

Acknowledgements

The authors would like to thank Scott McDougall and Jane Lacy-Hulbert for their input in this study. This study was funded by New Zealand dairy farmers through DairyNZ Inc. in partnership with Ministry of Primary Industries Primary Growth Partnership funding.

References

Borchers, M.R. and Bewley, J.M. (2015). An assessment of producer precision dairy farming technology use, prepurchase considerations, and usefulness. Journal of Dairy Science 98(6): 4198-4205.

Claycomb, R.W., Johnstone, P.T., Mein, G.A. and Sherlock, R.A. (2009). An automated in-line clinical mastitis detection system using measurement of conductivity from foremilk of individual udder quarters. New Zealand Veterinary Journal 57(4): 208-214.

Eastwood, C.R., Jago, J.G., Edwards, J.P. and Burke, J.K. (2015). Getting the most out of advanced farm management technologies: roles of technology suppliers and dairy industry organisations in supporting precision dairy farmers. Animal Production Science.

Edwards, J.P., Dela Rue, B.T. and Jago, J.G. (2015). Evaluating rates of technology adoption and milking practices on New Zealand dairy farms. Animal Production Science 55(6): 702-709.

Hogeveen, H., Huijps, K. and Lam, T.J.G.M. (2011). Economic aspects of mastitis: new developments. New Zealand Veterinary Journal 59(1): 16-23.

Kamphuis, C., Dela Rue, B.T. and Eastwood, C.R. (2016). Field validation of protocols developed to evaluate in-line mastitis detection systems. Journal of Dairy Science 99(2): 1619-1631.

Kermack, W.O. and McKendrick, A.G. (1927). A contribution to the mathematical theory of epidemics. Proceedings of the Royal Society of London A: Mathematical, Physical and Engineering Sciences 115(772): 700-721.

LIC (2013). New Zealand Dairy Statistics 2012-13.

Malcolm, D. (2006). The cost of mastitis; Dairy Insight Research 2005/2006; final report. National Mastitis Advisory Committee.

McDougall, S. and Compton, C. (2005). Bacteriology, cure rate and herd level risk factors for clinical mastitis in dairy herds. In: Proceedings of the Society of Dairy Cattle Veterinarians of the NZVA Annual Conference. VetLearn Foundation.

Milner, P., Page, K.L. and Hillerton, J.E. (1997). The effects of early antibiotic treatment following diagnosis of mastitis detected by a change in the electrical conductivity of milk. Journal of Dairy Science 80(5): 859-863.

Petrovski, K.R., Heuer, C., Parkinson, T.J. and Williamson, N.B. (2009). The incidence and aetiology of clinical bovine mastitis on 14 farms in Northland, New Zealand. New Zealand Veterinary Journal 57(2): 109-115.

Steeneveld, W. and Hogeveen, H. (2015). Characterization of Dutch dairy farms using sensor systems for cow management. Journal of Dairy Science 98(1): 709-717.

Wood, P.D.P. (1967). Algebraic model of the lactation curve in cattle. Nature 216(5111): 164-165.

Current and future adoption of milking related technologies on Australian and New Zealand dairy farms

N.A. Lyons[1*], C.R. Eastwood[2], B.T. De la Rue[2] and K.L. Kerrisk[3]
[1]*Intensive Livestock Industries, NSW Department of Primary Industries, Elizabeth Macarthur Agricultural Institute, Menangle NSW 2568, Australia;* [2]*DairyNZ, Hamilton 3240, New Zealand;* [3]*Dairy Science Group, Faculty of Veterinary Science, School of Life and Environmental Sciences, University of Sydney, Camden 2570, New South Wales, Australia; nicolas.lyons@dpi.nsw.gov.au*

Abstract

Despite use of advanced milking technologies in Australia (AU) and New Zealand (NZ), there is limited information regarding farmers' choices. Appropriate investment in RD&E needs more information on the current state of milking related technology on farms, future intentions of farmers, and the potential knowledge gaps. Surveys were conducted in both countries to ascertain milking technology practices. The AU 2015 online survey and the NZ 2013 phone survey gathered 301 and 500 responses respectively. In AU the top 5 currently installed milking-related technologies were automatic cup removers (ACRs) (71%), herd management software / computers at the dairy (60%), automatic milk plant wash system (43%), automatic in-parlor feeding (37%) and electronic identification (35%). In the next decade farmers expected greater use of automatic heat detection systems (79%), automatic sorting gates (77%), mastitis detection tools (66%), automatic (robotic) milking systems (60%) and in-line milk meters (58%). In NZ the top 5 currently installed technologies were, electronic identification (37%), automatic in-parlor feeding (33%), ACRs (29%), automatic teat spraying (27%), and automatic sorting gates (15%). Desired technologies included: automatic sorting gates (42%), ACRs (21%), automatic teat spraying (19%), mastitis detection tools (19%) and in-line milk meters (11%). Overall, more automation technologies that saved labor were installed compared to technologies that collect data to inform decision-making on-farm (e.g. liveweight monitoring). There is an association between technology adoption and larger herd size, so continued uptake of technologies can be expected if herd sizes increase further. Greater use of data collecting devices will depend on the value perceived by farmers and the fit with their operational management. Future RD&E investment should focus on integrating data and making data interpretation easier for users with guided management recommendations.

Keywords: automation, data, labor

1. Introduction

Since 1990 the number of dairy farms in Australia (AU) and New Zealand (NZ) has reduced by 18 and 60% respectively and the average herd size has increased by around 160% (Dairy Australia, 2015a; DairyNZ and LIC, 2015). Most of these operations are family owned and operated with variable levels of external labor. The availability of skilled labor and the demands of animal management under increasing herd sizes have been identified as two key variables that might limit future increases in production and profitability in Australasian dairy farm systems (Dairy Australia, 2015b).

In recent decades, milking technology has evolved with increasing adoption of automation/sensor systems on-farm. Such technology presents an opportunity to address labor issues and increase production efficiency. Despite limited surveys on milking and technology use in AU (Dharma *et al.*, 2012; Watson, 2009) and NZ (Edwards *et al.*, 2015), little is known about farmers' technology choices and their opinions about future adoption of technology. To make appropriate investment

decisions (in research, development and extension) at an industry level, more information is needed about the current state of milking technology on farms, and future intentions of farmers.
This paper describes data from recent surveys on adoption of milking related technologies in Australasia and discusses implications for future research, development and extension investment.

2. Materials and methods

Dairy farmer surveys were undertaken in AU and NZ, in 2015 and 2013 respectively. Both surveys were focused on the adoption of milking related technologies but the individual questions differed. The surveys were designed by multidisciplinary teams including researchers, farmers and extension personnel, and piloted with a small farmer group to refine methodology and logic flow. The surveys covered a broad range of questions including but not limited to farmer and farm demographics, current levels of technology adoption and future intentions. In this paper we focus on responses from farm owners or managers as primary decision makers on technology investments.
The AU survey was performed between May and August 2015. It was an Internet-based survey using Google Forms and included up to 32 questions depending on each respondent's profile. The survey was distributed online with an email and a link. Participants were recruited using existing industry networks and a snowball method. Eight responses (<3%) were from farmers operating automatic (robotic) milking systems, and are not included in this paper as they represent less than 1% of industry farms. The NZ survey was carried out in March – April 2013, using a telephone survey method. On average, interviews lasted 16 minutes. Criteria of a minimum farm size of 100 cows and operating either a rotary or herringbone parlors were used and participants were sourced from an industry list. Selected farmers were spread across the nine major dairy regions of NZ. A randomized approach was used to gather responses from dairy farmers in order to represent industry demographics. The survey built on a previous industry survey conducted in 2008. Mean values for all farms were adjusted to account for the proportion of rotary and herringbone parlors on farms nationally.
For the purpose of this paper the results of these two surveys are presented as summary means and percentages. Potential survey limitations include the voluntary nature of the AU survey and the farm size cut-off of the NZ survey.

3. Results

3.1 Australian survey: respondent demographics
A total of 169 farm managers or owner operators, representing each dairy region, responded to the survey. Milking facilities used by respondents were: herringbone (73%) and rotary (27%). Across the respondents, 74% of dairy facilities were commissioned more than 10 years ago (51% more than 15 years ago) with 60% of them not having had an upgrade within the past 5 years. Responses were across all farm sizes: 11% less than 150 cows, 35% between 151 and 300 cows, 37% between 301 and 500 cows, 7% between 501 and 700 cows and 10% more than 701 cows.

3.2 Australian survey: technologies installed and currently used by respondents
Overall, the five technologies most commonly installed on respondents' farms were automatic cup removers (ACRs) (71%), herd management software / computers at the dairy (60%), automatic milk plant wash systems (43%), automatic in-parlor feeding (37%) and electronic identification (EID) (35%) (Table 1). Rotary parlors had higher levels of automation than herringbone parlors.

3.3 Australian survey: technologies that respondents believed will have an increased adoption rate in the next 10 years
Overall the top five technologies respondents believed will have an increased adoption rate in the next 10 years were automatic heat detection systems (79%), automatic sorting gates (77%), mastitis

Table 1. The top five technologies installed and currently used, as well as those that respondents believe will have an increased adoption rate in the next 10 years, listed in proportion across all farms with proportion for herringbone and rotary farms itemized.

Technologies currently installed (%) Proportion (herringbone, rotary)	Technologies with expected increased adoption (%) Proportion (herringbone, rotary)
Automatic cup removers (ACRs): 71 (71, 71)	Automatic heat detection systems: 79 (75, 89)
Herd management software / computers: 60 (50, 87)	Automatic sorting gates: 77 (76, 80)
Automatic milk plant wash systems: 43 (34, 64)	Mastitis detection tools: 66 (63, 73)
Automatic in-parlor feeding: 37 (20, 82)	Automatic (robotic) milking systems: 60 (55, 73)
Electronic identification: 35 (20, 73)	In-line milk meters: 58 (52, 73)

detection tools (66%), automatic (robotic) milking systems (60%) and in-line milk meters (58%) (Table 1). A larger proportion of farmers operating with rotary parlors believed that technologies would have an increased adoption rate in the next 10 years.

3.4 New Zealand survey: respondent demographics

In total 500 farmers were surveyed including 265 rotary and 235 herringbone parlor operators. Responses were across all farm sizes: 6% less than 150 cows, 38% between 151 and 300 cows, 29% between 301 and 500 cows, 15% between 501 and 700 cows and 12% more than 701 cows. The average farm size was 131 ha and herd size was 363 cows (302 and 598 cows on farms with herringbones and rotaries respectively). Approximately half of parlors were built in the last 18 years, with 73% of rotary parlors and 43% of herringbone parlors either being built or substantially upgraded in the 10 years prior to the survey.

3.5 New Zealand survey: technologies installed and currently used by respondents

Electronic identification was a core technology for farmers, with 37% overall, and 51% of rotary parlors, having EID. Farmers had primarily installed automation technologies aimed at labor saving, such as ACRs and automatic teat spraying (Table 2). In general, farmers with rotary parlors had a higher technology adoption compared to farmers with herringbones. Thirty-three percent of all parlors had automatic in-parlor feeding. Most only had the ability to flat-rate feed, but 7% of all parlors could feed at the individual animal level.

In comparison to automation technologies, the use of data capture devices was lower among those interviewed but most popular among users of rotary parlors. These devices included mastitis detection tools (6%, 14%), in-line milk meters (4%, 14%), automatic animal weighing (3%, 8%), and automatic heat detection systems (3%, 4%) (numbers in italics represent rotaries).

3.6 New Zealand survey: technologies that respondents would like to have in the future

Farmers were asked about the technologies on their 'wish list' and overall the top five desired technologies were automatic sorting gates (42%), ACRs (21%), automatic teat spraying (19%), mastitis detection tools (19%) and in-line milk meters (11%).

The most beneficial technologies as selected by respondents who had installed them (herringbone and rotary respectively) were automatic sorting gates (100%, 80%), ACRs (91%, 80%), automatic teat spraying (78%, 46%), while automatic heat detection systems were seen to be somehow beneficial particularly by rotary users (12%, 70%) although the number of farms was low.

Table 2. The proportion of farmers who 'have' or 'want' automation and information technologies, with proportion for herringbone and rotary farms itemized.

	Have technology (%) Average (herringbone, rotary)	Want technology (%) Average (herringbone, rotary)
Automation technologies		
Automatic in-parlor feeding	33 (22, 62)	10 (10, 10)
Automatic teat spraying	27 (13, 66)	19 (21, 13)
Automatic cup removers (ACRs)	29 (13, 70)	21 (22, 19)
Automatic sorting gates	15 (6, 39)	42 (43, 42)
Information technologies		
Electronic identification (EID)	37 (32, 51)	3 (3, 4)
Mastitis detection tools	6 (3, 14)	19 (17, 22)
In-line milk meters	4 (1, 14)	11 (8, 20)
Automatic animal weighing	3 (1, 8)	11 (9, 19)
Automatic heat detection systems	3 (3, 4)	7 (5, 15)
None of these technologies	16 (20, 4)	7 (8, 6)

4. Discussion

The aim of this paper was to describe adoption rates of milking related technologies in Australasia to inform future RD&E investment. The two surveys outlined here captured between three and four percent of industry farms, surveying a greater proportion of farms operating larger dairies (larger herd sizes, dairy parlors or higher proportion of rotary sheds) in AU (Dairy Australia, 2015a; Dharma *et al.*, 2012; Watson, 2009), and on average smaller farms than the national average in NZ (DairyNZ and LIC, 2015). The age of dairy parlors was similar between AU and NZ. Larger farms and herds, as well as old dairy parlors which have not had a significant recent upgrade create a great challenge for farm owners and managers. Under these operating conditions, it is not surprising for farmers to mention the difficulty to attract and retain talented staff to work on farms and that labor will be one of the main constraints for growth in the future (Dairy Australia, 2012; Nettle and Oliver, 2009). Some dairy technologies such as automatic in-parlor feeding systems had similar adoption rates in AU and NZ. Conversely, farms in AU had lower adoption of other technologies such as EID. On AU dairy farms, most of the technology that has been adopted addresses either labor (ACRs, automatic milk plant wash systems or automatic in-parlor feeding) or data and management challenges (herd management software / computers at the dairy and EID). Both countries reported higher levels of automation and technology adoption in rotary parlors in comparison to herringbone parlors. This might be related to the scale of the operation, the ease of automation in these types of parlors or the age of each type of parlor.

Farmers in both countries identified technologies they expected would be desirable in the future. Most of the technologies mentioned would be focused particularly around managing large herds (through the adoption of automatic sorting gates), improving reproductive performance (through the adoption of automatic heat detection systems) and the challenges associated with producing a high quality product whilst improving animal welfare and longevity (through the adoption of mastitis detection tools). Farmers with rotary facilities were more inclined to future technology investment. It is surprising that despite the slow adoption of automatic milking systems in AU (currently less than 40 farms) almost 60% of farmers surveyed believed that this technology will have an increased adoption rate in the future. This is probably seen as a technology that will address some of the current challenges related to labor quality and quantity. Automatic cow weighing scales did not appear high in the list of priorities for farmers in the future. This is possibly related to a lack of understanding at a farm level on how to use such information (Eastwood *et al.*, 2015), perceived lack of accuracy and value, and the familiarity of farmers to operate with measurements such as body condition scoring (Dairy Australia, 2013). There have recently been some commercial developments in technologies to measure body condition score which may offer significant opportunities.

Increased use of herd management software was not identified by farmers but is a key component of data compilation and analysis (Eastwood *et al.*, 2015). Future progress on the usefulness of sensor technologies is based around their integration in decision support software that bring together data produced or captured by different devices and sensors and are coupled with a powerful data analytics tool. This will allow farmers to make better use of captured data to improve on-farm decisions, leading to enhanced productivity and profitability. Anecdotally, farmers highlighted the need for greater awareness of the potential and challenges associated with these technologies. Usually farmers learnt from technology suppliers, other farmers, and the internet when making decisions about new technology.

5. Conclusions

Australasian farmers have been active in adopting advanced technologies related to milking and animal monitoring. However, significant potential exists for increased uptake and utilization of these technologies. To date, automation technologies that are labor-saving are more popular with farmers than those designed to collect data for decision making. Further adoption can be expected if herd sizes continue to increase, and with continued difficulty to recruit capable farm staff.

Future RD&E investment should focus on integrating data via decision support software, or integrated dashboards, aligning data more closely to farmer practice such as seasonal decisions making, and making data interpretation easier for users. Providing the user with guided management recommendations is likely to be more appealing that simple data reports that require significant interpretation.

Acknowledgements

The authors would like to thank the farmers who took part in both surveys, and industry representatives involved with designing, piloting and conducting the surveys. The studies were funded by New Zealand dairy farmers through DairyNZ, and the NSW Department of Primary Industries in Australia.

References

Dairy Australia (2012). Dairy 2012: situation and outlook. Victoria, Australia, pp. 67.

Dairy Australia (2013). Cow body condition scoring handbook. Victoria, Australia.

Dairy Australia (2015a). Australian dairy industry in focus 2015. Victoria, Australia.

Dairy Australia (2015b). Sustainable farm profitability 2015. Victoria, Australia, pp. 48.

Dairy NZ and LIC (2015). New Zealand Dairy Statistics 2014-15.

Dharma, S., Shafron, W. and Oliver, M. (2012). Australian dairy: farm technology and management practices 2010-11. Available at: http://tinyurl.com/ztg38h2.

Eastwood, C.R., Jago, J.G., Edwards, J.P. and Burke, J.K. (2015). Getting the most out of advanced farm management technologies: roles of technology suppliers and dairy industry organisations in supporting precision dairy farmers. Animal Production Science. DOI: http://dx.doi.org/10.1071/AN141015.

Edwards, J.P., Rue, B.T. and Jago, J.G. (2015). Evaluating rates of technology adoption and milking practices on New Zealand dairy farms. Animal Production Science 55(6): 702-709.

Nettle, R. and Oliver, D. (2009). Workforce planning and action for the Australian dairy industry. Parkville, VIC, Australia, pp. 32.
Watson, P. (2009). Cowtime tracking survey 2009. Available at: http://tinyurl.com/jlnzgp3.

Technical, economic and sociological impacts of an automated estrus detection system for dairy cows

C. Allain[1*], A. Chanvallon[2], R. Courties[1,2,3], D. Billon[3] and N. Bareille[3]
[1]*Institut de l'Elevage, BP 85225, 35652 Le Rheu, France;* [2]*Institut de l'Elevage, Oniris, CS 40706, 44307 Nantes, France;* [3]*BIOEPAR, INRA, Oniris, La Chantrerie, 44307, Nantes, France; clement.allain@idele.fr*

Abstract

The use of a sensor system for detecting estrus of dairy cows is increasing rapidly in Europe. Many studies have evaluated these sensors ability to detect estrus, but there is a lack of information on their economic and sociological impacts on dairy farms. The objective of this study was first to assess the technical and economic consequences of an investment in an automated estrus detection (AED) system on a farm. A dynamic stochastic simulation model was used to simulate all the physiological and management processes in a dairy farm. For a herd of 125 cows, the increase of the annual profit (excluding labour time) was €2007 when estrus detection rate increased from 50 to 80% (leading to a decrease of the calving interval from 433 to 412 days). This amount was equivalent to the annual cost of the equipment (€2,100). In smaller herds (35 or 65 cows), simulations showed negative profit investments (from -€800 to -€1,600 per year). The consequences on the farmer's work and practices were also evaluated. Qualitative interviews were conducted on 20 farms equipped with AED sensors. Two-thirds of the farmers invested mostly to improve reproductive performances and 1/3 to improve labour time and quality of life. Eighteen out of 20 farmers maintained visual estrus detection (occasionally or frequently) in supplement of the alerts. Concerning labour, positive impacts were mentioned by 17 farmers on the reduction in stress caused by AI decision making, by 9 of them on the reduced labour time and the improved relationships with colleagues (because of decision making facilitation), and by 6 of them on increased schedule flexibility. This first analysis in a French context points out that the economic relevance of AED systems is obtained only in large herds with low estrus detection rates and provides information on the farmers' motivations for adoption.

Keywords: estrus detection, work comfort, economics, dairy cows

1. Introduction

The reproductive management in dairy cattle herds is based on the detection of estrus behaviour in order to implement the AI. This detection is made difficult by the discrete behavioural changes associated with estrus in high producing Holstein dairy cows (Roelofs *et al.*, 2010). In addition, the time that the farmers devote to estrus detection tends to decrease as the herd size increases (Chanvallon *et al.*, 2014). In this context, the use of a sensor system (pedometers, activity-meters) for detecting estrus is increasing rapidly, with usually better performances than farmers' ones (Rutten *et al.*, 2013).
Many studies have evaluated these sensors ability to detect estrus, but there is still a need for information on their economic relevance. This lack of demonstration of a positive cost/benefit ratio is one of the main explanations for slow adoption rates of precision dairy farming technologies (Bewley and Russel, 2010). Indeed, the economic consequences of poor reproductive performances are high in dairy farms (Fourichon *et al.*, 2001; Inchaisri *et al.*, 2010). The improvement in estrus detection contributes to shorten the calving intervals and consequently to decrease the economic impact (Seegers *et al.*, 2006). Some studies have investigated whether the gains due to improved reproductive performances exceed the cost of the sensor system (Dolecheck *et al.*, 2013; Rutten *et al.*, 2014). The response is mainly positive but depends on the technology employed, on the prices and on the farms

characteristics (herd size and, estrus detection rate by the farmer). However, none of these estimations were done in herds with less than 100 cows while, in France, the average herd size is 50 cows. Furthermore, most of the studies neglect the sociological impacts of precision farming technologies. These impacts are related to labour time, labour organisation and quality of life. However, sociological factors are essential for investment decision making in precision dairy farming technologies (Bewley and Russel, 2010).

The objectives of this study were to assess in the French context (1) the technical and economic consequences of an investment in an AED system at the farm level; (2) the implications on the farmer's work.

2. Materials and methods

2.1 Technical and economic assessment

The individual-based mechanistic model was used to simulate all the physiological events occurring on each female (calf, heifer and cow) and all the management processes occurring in a dairy farm. The model is stochastic; consequently, different runs of the model will result in variable outcomes; the results shown are the mean of 250 runs. Each animal is defined by a set of state variables (age, milk yield, reproduction, health, etc.). Herd-level state variables are also defined by cumulating the individual simulated events and performances.

Postpartum ovarian activity of cows starts after a minimum interval after calving. For each cow, the estrus occurs according to the distribution of interval between calving and first estrus (a truncated gamma distribution) or of interval between estrus (a normal distribution). Estrus is detected according to a detection sensitivity rate. Two levels of estrus detection rate were compared: 50% (frequent values for farmers' detection (Roelofs *et al.*, 2010)) and 80% (maximum values for the accelerometers (Rutten *et al.*, 2013; Chanvallon *et al.*, 2014)). The positive predictive value of estrus detection was kept at an average level of 90%. Then, decision to inseminate is based on reproductive strategy variables, the main was 'no AI before 45 days postpartum'. The conception rate after AI of cow in estrus depends on the production level of the cow at this time.

Scenarios varied in terms of herd size and milk production level, but the herds were all characterized by all year round calving with a 5 months grazing period. Consequently, the number of individual automated sensors needed was estimated to 40% of the number of lactating cows (Table1).

The purchase cost of the equipment was based on experts' information: €100 /sensor, €3,000 for a system for data transmission and software, €500 for annual maintenance. We considered that the technology was paid off in 5 years. Pricing and costs parameters were set at year 2013 values for western France. The main utility criterion was the annual profit, that is to say incomes minus expenses (excluding the labour costs).

Table 1. Characteristics of the herds used for the implementation of the simulation model and needs for individual sensor equipment.

Herd code	35H	65H	65L	125H
Herd size (no. of cows)	35	65	65	125
Production level	High[1]	High	Low[2]	High
No. of individual equipment	14	26	26	50

[1] Annual milk production per cow near 10,700 kg.

[2] Annual milk production per cow near 8,200 kg.

2.2 Sociological assessment

Qualitative interviews were conducted on 20 farms from western France equipped with AED sensors for 2 to 4 years. Interviews were based on semi directed surveys with the farmers. The objective was to assess the their motivation to invest in an automated estrus detection system, their estrus detection practices before and after investing in the sensors and the impact of the sensors on labour time, organization and quality of life. The farms were chosen regarding their type of equipment and the herd size (Table 2) to be representative of the diversity of the French context.

3. Results and discussion

3.1 Technical and economic assessment

For a herd of 125 cows, the improvement of the estrus detection rate from 50 to 80% allowed by the AED system shortened both the interval calving – first AI (104 to 90 days) and the calving interval (433 to 412 days) even if the conception rate at first AI decreased slightly (33.7 to 31.8%). The average increase of the annual profit (excluding labour time) was €2,007 (Table 3). This amount was comparable to the annual cost of the equipment (€2,100).

In small herds, the technical effects of the estrus detection sensors are similar to those of large herds, but the cost of the heat detection system is relatively higher per cow.

In smaller herds (35 or 65 cows), simulations showed slightly negative profit investments (from -€800 to -€1,600 per year). In the herd with lower production level, the investment was less profitable.

The comparison of our results with previous studies is made difficult by the variability of the assumptions made and of the costs included. For example, the effect of the shorter interval calving – first AI on fertility was neglected by Dolecheck *et al.* (2013) and Rutten *et al.* (2014). Furthermore, Rutten *et al.* (2014) considered that labour time was saved with AED system but our results (see below) show that it was not clearly the case. Our results are less optimistic than previous estimations in large herds: the improvement of estrus detection rate needed to have a positive annual profit was at least 30% whereas it was 12 to 21% (Dolecheck *et al.*, 2013) or 15% (Rutten *et al.*, 2014). One possible explanation is that we considered that the AED system was paid off in 5 years whereas Rutten *et al.* (2014) calculated that the investment was paid back in 8 years.

3.2 Sociological assessment

Before investing in AED sensors, 19 farmers out of 20 used only visual detection and one was equipped with the Kamar® system. The farmers could not assess the daily observation duration because it was always included in another activity (feeding, milking, etc.) but the number of observation periods was evaluated. Twelve farmers had defined systematic daily observation periods (1 to 3), at the same time than another activity. The other eight farmers had no systematic periods and did it randomly.

Table 2. Farm sample used for the interviews.

		Standalone system (ex: Heatime®, HeatPhone®)	Integrated system[1] (ex: Lely, DeLaval)
Herd size (number of cows)	n≤50	2	0
	51≤n<70	7	4
	n≥70	4	3

[1] Sensors included in a global offer, often with an automatic milking system.

Table 3. Technical and economic impact of an investment in an AED system (considering that estrus detection increases from 50 to 80%). The results are expressed relatively to the situation before investment.

Herd code	35H	65H	65L	125H
Technical consequences				
Interval calving-first AI (days)	-13.8	-14.4	-13.9	-13.9
Conception rate at 1st AI (%)	-2.2	-2.0	-2.4	-1.9
Calving interval (days)	-19.4	-22.4	-17.4	-21.1
Economic consequences				
Variation of annual farm income (€)				
Milk	242	1,714	1,097	3,291
Culled cows	401	879	414	1,784
Calves	37	400	63	335
Variation of annual farm expenses (€)				
Automated sensor system	1,380	1,620	1,620	2,100
AI	198	315	248	647
Treatments for reproductive disorders	-34	-40	-117	-146
Concentrates (cow /heifer)	670	1,730	943	2,707
Other costs for cows	5	32	46	24
Other costs for calves & heifers	54	93	84	171
Variation of annual profit (€)	-1,593	-757	-1,250	-93

The main reasons to invest in AED sensors were: to improve estrus detection or reproductive performances (13), to save time (5) or to improve the quality of life/labour organisation (2). These results are consistent with surveys made in recent studies (Neves and Leblanc, 2015; Steeneveld and Hoogeveen, 2015).

After investing, 2 farmers totally delegated estrus detection to the sensors and 18 of them maintained visual estrus detection (occasionally or frequently) in supplement of the alerts. Amongst these 18, twelve of them checked the sensors alerts by visual observations and 6 of them never checked the alerts. The effects of AED systems on labour and mental workload are presented in Figure 1. Almost half of the farmers considered that they saved labour time. The effects on labour organisation were moderate with a majority of insignificant impacts. Almost half of the farmers noticed improvements in the relationships with their colleagues, especially because they had an objective information in case of disagreements. The main impact was clearly on decision making. Indeed, 17 of the farmers mentioned that the AED sensors help them to reinforce their decision to inseminate or not and therefore reduce the stress caused by the decision making process.

Figure 2 shows the satisfaction of the farmers regarding their equipment for estrus detection. A high majority of them was satisfied of the easiness of use and the dealer support. They were also satisfied with the performances and the robustness. Regarding the impacts on reproduction performances and the economic impact, the farmers where moderately satisfied while it was the two main reasons for investing. The farmers' motivations to invest in AED sensors and the impacts on labour time and quality of life were closely related to their estrus detection practices before and after investment and from the farm features. Firstly, regarding investment motivations, the 13 farmers who invested to improve reproduction performances used more frequently sexed semen (6/13 vs 1/7 for farmers who invested to improve labour or quality of life). Considering the price of sexed semen, they were looking for a tool to reduce false positive alerts. They also had less specific observation periods (8/13 had only

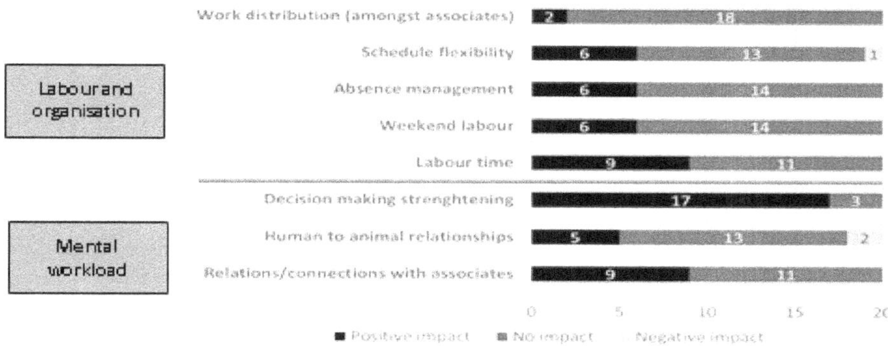

Figure 1. Impacts of an AED system on labour and mental workload.

Figure 2. Farmers' satisfaction regarding their equipment.

1 observation period or less vs 2/7 for other farmers) so probably worse detection performances before investing. Secondly, regarding the impacts of AED systems on labour time, the 9 farmers who estimated to save time spent generally more time in estrus detection before investing: they had less workforce (62 cows/farmer vs 41/farmer for other farmers) and more observation periods (6/9 had 2 or 3 daily observation periods vs only 4/11 for other farmers). They also had less time consuming practices: the equipment rate was higher (8/9 had more than 40% of the cows equipped with sensors, limiting the time necessary to change the collars vs 6/11 for other farmers) and they were less frequently checking the sensors' alerts by visual observations (5/9 never checked vs 3/11 for other farmers). Finally, the main impact of automated estrus detection systems mentioned by the farmers (17/20) was the possibility to be reinsured by the alerts for the AI decision making. This was completely consistent with their practices as 18 farmers continued visual detection to complement the alerts.

4. Conclusions

This first analysis in a French context points out that the economic relevance of AED systems is obtained only in large herds with low estrus detection rates and provides the first information on sociological impacts of an AED system. In our case, we observed that the farmers' motivations to invest, the impacts on labour time and on quality of life were closely dependent from their estrus detection practices before and after investment.

References

Bewley, J.M. and Russel, R.A. (2010). Reasons for slow adoption rates of precision dairy farming technologies: evidence from a producer survey. In: Proceedings of 1st North American Conference on Precision Dairy Management, Toronto, Canada.

Chanvallon, A., Coyral-Castel, S., Gatien, J., Lamy, J.M., Ribaud, D., Allain, C., Clément, P. and Salvetti, P. (2014). Comparison of three devices for the automated detection of estrus in dairy cows. Theriogenology 82(5): 734-741.

Dolecheck, K.A., Heersche, G. and Bewley, J.M. (2013). Investment analysis of automated estrus detection technologies. In: Proceedings of Precision Dairy Conference Minnesota, USA, pp. 103-104.

Fourichon, C., Beaudeau, F., Bareille, N. and Seegers, H. (2001). Incidence of health disorders in dairy farming systems in western France. Livestock Production Science 68: 157-170.

Inchaisri, C., Jorritsma, R., Vosa, P.L.A.M., Van der Weijden, G.C. and Hogeveen, H. (2010). Economic consequences of reproductive performance in dairy cattle. Theriogenology 74: 835-846.

Neves, R.C. and LeBlanc, S.J. (2015). Reproductive management practices and performance of Canadian dairy herds using automated activity-monitoring systems. Journal of Dairy Science 98(4): 2801-2811.

Roelofs, J., Lopez-Gatius, F., Hunter, R.H.F., Van Eeerdenburg, F.J.C.M. and Hanzen, C. (2010). When is a cow in estrus? Clinical and practical aspects. Theriogenology 74: 327-344.

Rutten, C.J., Velthuis, A.G.J., Steeneveld, W. and Hogeveen, H. (2013). Sensors to support health management on dairy farms. Journal of Dairy Science 96: 1928-1952.

Rutten, C.J., Steeneveld, W., Inchaisri, C. and Hogeveen, H. (2014). An ex ante analysis on the use of activity meters for automated estrus detection: to invest or not to invest? Journal of Dairy Science 97: 1-19.

Seegers, H., Grimard, B., Billon, D., Beaudeau, F., Ponsart, P. and Humblot, P. (2006). Economic losses and profitability of control for reproductive problems in dairy herds. In: Proceedings of 10th Conference of the European Society for Domestic Animal Reproduction. Portoroz-Lipica, Slovenia.

Steeneveld, W. and Hogeveen, H. (2015). Characterization of Dutch dairy farms using sensor systems for cow management. Journal of Dairy Science 98: 709-717.

Authors index

www.ingramcontent.com/pod-product-compliance
Lightning Source LLC
Chambersburg PA
CBHW081124180926
51611CB00004B/43
* 9 7 8 9 0 8 6 8 6 2 8 3 2 *